QUARKS AND LEPTONS:
An Introductory Course in Modern Particle Physics

QUARKS AND LEPTONS: An Introductory Course in Modern Particle Physics

Francis Halzen

University of Wisconsin
Madison, Wisconsin

Alan D. Martin

University of Durham
Durham, England

JOHN WILEY & SONS,

New York • Chichester • Brisbane • Toronto • Singapore

Library of Congress Cataloging in Publication Data:

Halzen, Francis.
Quarks and leptons.

Includes index.
1. Quarks. 2. Leptons (Nuclear physics)
I. Martin, Alan D. (Alan Douglas) II. Title.
QC793.5.Q2522H34 1984 539.7′21 83-14649
ISBN 0-471-88741-2

Printed in the United States of America

10 9 8 7 6 5 4 3 2 1

To
Nelly and Penny
Rebecca, Robert, Rachel, and David

Preface

Dramatic progress has been made in particle physics during the past two decades. A series of important experimental discoveries has firmly established the existence of a subnuclear world of quarks and leptons. The protons and neutrons ("nucleons"), which form nuclei, are no longer regarded as elementary particles but are found to be made of quarks. That is, in the sequence molecules → atoms → nuclei → nucleons, there is now known to be another "layer in the structure of matter." However, the present euphoria in particle physics transcends this remarkable discovery. The excitement is due to the realization that the dynamics of quarks and leptons can be described by an extension of the sort of quantum field theory that proved successful in describing the electromagnetic interactions of charged particles. To be more precise, the fundamental interactions are widely believed to be described by quantum field theories possessing local gauge symmetry. One of the aims of this book is to transmit a glimpse of the amazing beauty and power of these gauge theories. We discuss quarks and leptons, and explain how they interact through the exchange of gauge field quanta (photons, gluons, and weak bosons).

We are very conscious that this book has been written at a crucial time when pertinent questions regarding the existence of the weak bosons and the stability of the proton may soon be decided experimentally. Some sections of the book should therefore be approached with a degree of caution, bearing in mind that the promising theory of today may only be the effective phenomenology of the theory of tomorrow. But no further apology will be made for our enthusiasm for gauge theories.

We have endeavored to provide the reader with sufficient background to understand the relevance of the present experimental assault upon the nature of matter and to appreciate contemporary theoretical speculations. The required core of knowledge is the standard *electroweak model*, which describes the electromagnetic and weak interactions of leptons and quarks; and *quantum chromodynamics* (QCD), which describes the strong interactions of quarks and gluons. The primary purpose of this book is to introduce these ideas in the simplest possible way. We assume only a basic knowledge of nonrelativistic quantum mechanics and the theory of special relativity. We spend considerable time introducing quantum electrodynamics (QED) and try to establish a working familiarity with the Feynman rules. These techniques are subsequently generalized and applied to quantum chromodynamics and to the theory of weak interactions.

The emphasis of the book is pedagogical. This has several implications. No attempt is made to cover each subject completely. Examples are chosen solely on pedagogical merit and not because of their historic importance. The book does not contain the references to the original scientific papers. However, we do refer to books and appropriate review articles whenever possible, and of course no credit for original discovery is implied by our choice. A supplementary reading list can be found at the end of the book, and we also encourage the students to read the original papers mentioned in these articles. A deliberate effort is made to present material which will be of immediate interest to the student, irrespective of his experimental or theoretical bias. It is possible that aspiring theorists may feel that an injustice has been done to the subtle beauty of the formalism, while experimentalists may justifiably argue that the role of experimental discoveries is insufficiently emphasized. Fortunately, the field is rich in excellent books and review articles covering such material, and we hope that our guidance toward alternative presentations will remedy these defects.

Although the book is primarily written as an introductory course in particle physics, we list several other *teaching options*. The accompanying flow diagram gives some idea of the material covered in the various chapters.

A One teaching option is based on the belief that because of its repeated phenomenological successes, modern particle physics, or at least some aspects of it, is suitable material for an advanced quantum mechanics course alongside the more traditional subjects such as atomic physics. For this purpose, we suggest Chapters 3 through 6, with further examples from Chapter 12, together perhaps with parts of Chapter 14.

B An undergraduate course on the introduction of the Feynman rules for QED could be based on Chapters 3 through 6.

C The sequence of Chapters 3 through 11 could serve as an introduction to quantum chromodynamics.

D A course on weak and electromagnetic interactions could cover Chapters 3 through 6 and 12 and 13, perhaps supplemented with parts of Chapters 14 and 15.

E For a standard introductory particle physics course, it may not be possible to cover the full text in depth, and Chapters 7, 10, 11, 14, and 15 can be partially or completely omitted.

Exercises are provided throughout the text, and several of the problems are an integral part of the discussion. Outline solutions to selected problems are given at the end of the book, particularly when the exercise provides a crucial link in the text.

This book was developed and written with the encouragement of students and friends at the Universities of Durham and Wisconsin. Many colleagues have given

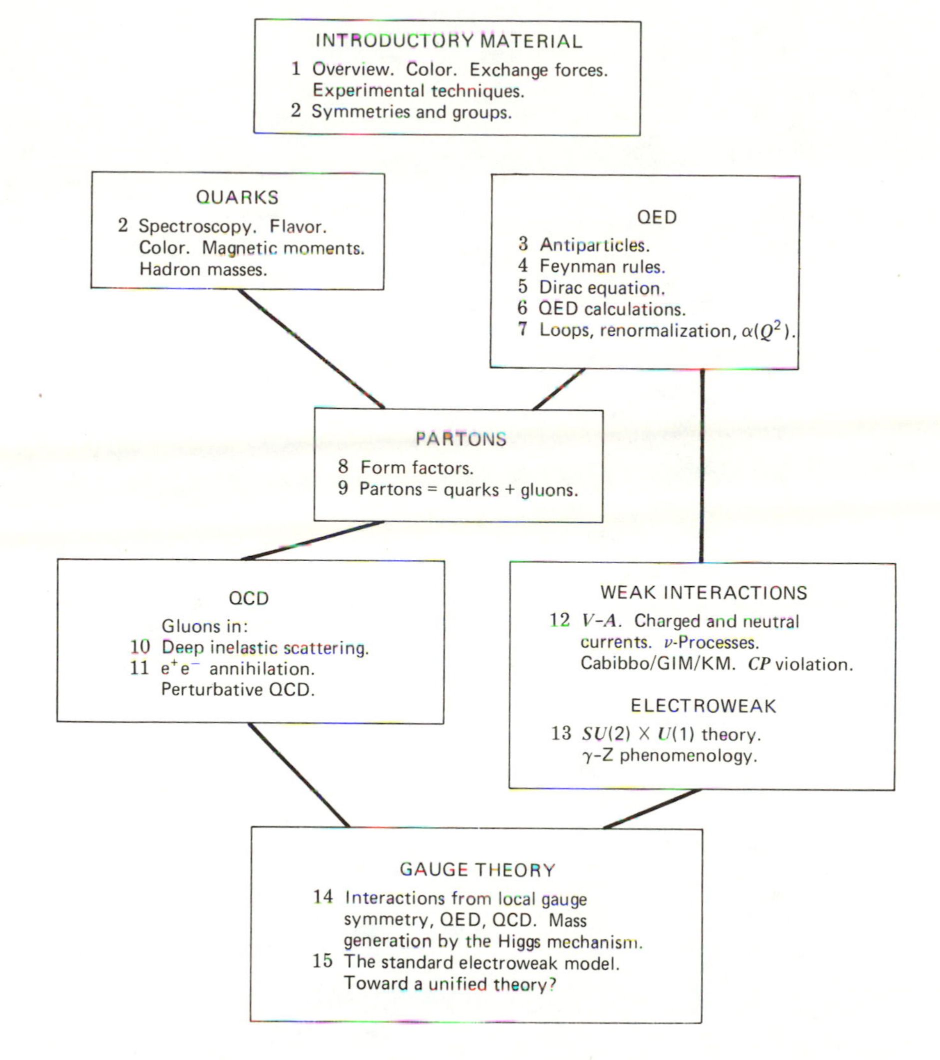

us valuable assistance. In particular, we acknowledge our special debt to Peter Collins and Paul Stevenson. They read through the entire manuscript and suggested countless improvements. We also thank our other colleagues for their valuable comments on parts of the manuscript, especially D. Bailin, V. Barger, U. Camerini, C. Goebel, K. Hagiwara, G. Karl, R. March, C. Michael, M. Pennington, D. Reeder, G. Ross, D. Scott, T. Shimada, T. D. Spearman, and B. Webber.

We thank Vicky Kerr, and also Linda Dolan, for excellent typing of a difficult manuscript. By so doing, they made our task that much easier.

Francis Halzen and Alan D. Martin

Durham, England
January 12th, 1983

Contents

QUARKS AND LEPTONS:
An Introductory Course in Modern Particle Physics

1

A Preview of Particle Physics

1.1 What is the World Made of?

Present-day particle physics research represents man's most ambitious and most organized effort to answer this question. Earlier answers to this riddle included the solution proposed by Anaximenes of Miletus, shown in Fig. 1.1. Everyone is familiar with the answer Mendeleev came up with 25 centuries later: the periodic table, a sort of extended version of Fig. 1.1, which now contains well over 100 chemical elements. Anaximenes's model of the fundamental structure of matter is clearly conceptually superior because of its simplicity and economy in number of building blocks. It has one fatal problem: it is wrong! Mendeleev's answer is right, but it is too complicated to represent the "ultimate" or fundamental solution. The proliferation of elements and the apparent systematics in the organization of the table strongly suggests a substructure. We know now that the elements in Mendeleev's table are indeed built up of the more fundamental electrons and nuclei.

Our current answer to the question what the world is made of is displayed in Table 1.1. It shares the conceptual simplicity of Anaximenes's solution; it is, however, just like Mendeleev's proposal, truly quantitative and in agreement with experimental facts. The answer of Table 1.1 was actually extracted step by step from a series of experiments embracing the fields of atomic, nuclear, cosmic-ray, and high-energy physics. This experimental effort originated around the turn of this century, but it was a sequence of very important discoveries in the last decade that directly guided us to a world of quarks, leptons, and gauge bosons. Previously, this picture had just been one of many competing suggestions for solving the puzzle of the basic structure of matter.

The regularities in Mendeleev's table were a stepping-stone to nuclei and to particles called protons and neutrons (collectively labeled nucleons), which are "glued" together with a strong or nuclear force to form the nuclei. These subsequently bind with electrons through the electromagnetic force to produce the atoms of the chemical elements. Conversion of neutrons into protons by so-called weak interactions is responsible for the radioactive β-decay of nuclei, including the slow decay of the neutron into a proton accompanied by an electron

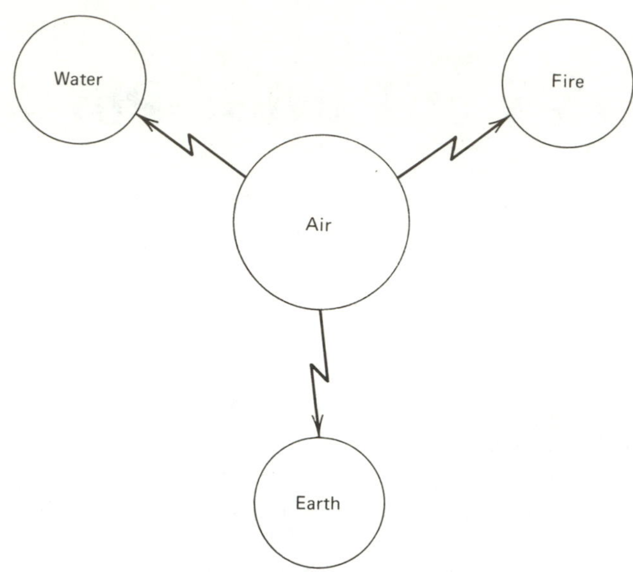

Fig. 1.1 In the original version of the theory, all forms of matter are obtained by condensing or rarefying air. Later, a "chemistry" was constructed using the four elements shown in the figure.

and an antineutrino. At this point, the world looked very much like Table 1.1 but with the nucleons p and n playing the role of the quarks u and d.

But the neutron and proton were not alone. They turned out to be just the lightest particles in a spectrum of strongly interacting fermion states, called *baryons*, numbering near 100 at the latest count. An equally numerous sequence of strongly interacting bosons, called *mesons*, has also been discovered, the pion being the lightest. Fermions (bosons) refer to particle states with spin $J = n(\hbar/2)$, where n is an odd (even) integer. All the particles which undergo strong interactions, baryons and mesons, are collectively called "*hadrons*."

This proliferation of so-called "elementary" particles pointed the way to the substructure of the nucleons (the quarks) in a rather straightforward replay of the arguments for composite atoms based on Mendeleev's table. Also, the π-meson and all other hadrons are made of quarks. The electron and neutrino do not experience strong interactions and so are not hadrons. They form a separate group of particles known as *leptons*. The neutrino participates exclusively in the weak interactions, but the charged electron can of course also experience electromagnetic interactions. Leptons have not proliferated like hadrons and so are entered directly into Table 1.1 as elementary point-like particles along with the quarks. The pion, neutron, proton... are not part of the ultimate pieces of the puzzle; they join nuclei and atoms as one more manifestation of bound-state structures that exist in a world made of quarks and leptons.

TABLE 1.1
Building Blocks of Elementary Particles and Some of Their Quantum Numbers[a]

Name	Spin	Baryon Number B	Lepton Number L	Charge Q
Quarks				
u (up)	$\frac{1}{2}$	$\frac{1}{3}$	0	$+\frac{2}{3}$
d (down)	$\frac{1}{2}$	$\frac{1}{3}$	0	$-\frac{1}{3}$
Leptons				
e (electron)	$\frac{1}{2}$	0	1	-1
ν (neutrino)	$\frac{1}{2}$	0	1	0
Gauge bosons				
γ (photon)	1	0	0	0
$W^{\pm}, Z$ (weak bosons)	1	0	0	$\pm 1, 0$
g_i ($i = 1, \ldots, 8$ gluons)	1	0	0	0

[a] The spin is given in units of $\hbar$. The charge units are defined in such a way that the electron charge is -1. Not listed in the table are the antiparticles of the quarks and leptons [$\bar{u}$, $\bar{d}$ (antiquarks), e^+ (positron), $\bar{\nu}$ (antineutrino)]. Although they will not be carefully defined until Chapter 3, they are for present purposes identical to the corresponding particles except for the reversal of the sign of their B, L, and Q quantum numbers. For example, the $\bar{u}$ has $B = -\frac{1}{3}$, $L = 0$, $Q = -\frac{2}{3}$, and the e^+ has $B = 0$, $L = -1$, and $Q = +1$.

A theoretical framework was needed that could translate these conceptual developments into a quantitative calculational scheme. Clearly, Schrödinger's equation could not handle the creation and annihilation of particles as observed in neutron decay and was furthermore unable to describe highly relativistic particles as encountered in routine cosmic ray experiments. In the early 1930s, a theory emerged describing the electromagnetic interactions of electrons and photons (quantum electrodynamics) that encompassed these desired features: it was quantized and relativistically invariant. Even though it has become essential to include quarks, as well as leptons, and to consider other interactions besides electromagnetism, relativistic quantum field theory, of which quantum electrodynamics is the prototype, stands unchanged as the calculational framework of particle physics. The most recent developments in particle physics, however, have revealed the relevance of a special class of such theories, called "gauge" theories; quantum electrodynamics itself is the simplest example of such a theory. The weak and strong interactions of quarks and leptons are both believed to be described by gauge theories: the unified electroweak model and quantum chromodynamics.

The interplay of models and ideas, formulated in the general framework of gauge theories, with new experimental information has been the breeding ground for repeated success. The purpose of this book is to introduce the reader to this form of research which is entering new energy domains with the completion and imminent construction of a new generation of particle accelerators. These accelerators, together with very sophisticated particle detectors, will probe matter to previously unexplored submicroscopic distances.

1.2 Quarks and Color

The overwhelming experimental evidence that the nucleons of nuclear physics are made of particles called *quarks* is reviewed in Chapter 2. The *baryons* are bound states of three quarks; the *mesons* are composed of a quark and an antiquark. The proton is a uud bound state; the additive quantum numbers of the two kinds of quarks, u and d, listed in Table 1.1, correctly match the fact that the proton is a baryon ($B = 1 = \frac{1}{3} + \frac{1}{3} + \frac{1}{3}$) and not a lepton ($L = 0$) and that it has total charge 1 ($Q = \frac{2}{3} + \frac{2}{3} - \frac{1}{3}$). By analogous arguments, the neutron is obtained as a udd bound state. The π^+-meson is a $u\bar{d}$ state; it is a meson, in the sense that $B = L = 0$, and its charge is indeed 1 [$Q = \frac{2}{3} - (-\frac{1}{3})$]. In passing, we should note that the charge Q, the baryon number B, and the lepton number L are more than simply labels; they are *conserved* additive quantum numbers. That is, a particle reaction (e.g., $\pi^- p \to \pi^0 n$ or $n \to pe^-\bar{\nu}$) can only occur if the sum of the B values in the initial state equals that in the final state; similarly for L and Q.

By the conventional rules of addition of angular momenta, the total spin $J = \frac{1}{2}$ for the nucleon and $J = 0$ for the π-meson can be constructed from their $J = \frac{1}{2}$ constituents. The quark scheme naturally accommodates the observed separation of *hadrons* into baryons (three quark fermion states) and mesons (quark–antiquark boson states).

An immediate success of the quark model is theoretical in nature. Protons and neutrons are relatively complicated objects, with a size and a rich internal quark structure. Quantum field theory, on the other hand, deals with point-like elementary particles, that is, with structureless objects like the electron, for example. The structureless quarks, rather than the nucleons, are the fundamental entities described by quantum field theory. Their introduction enables us to explore the other interactions with the same powerful theoretical techniques that were so successful in describing the properties and electromagnetic interactions of electrons (*quantum electrodynamics*).

When implementing the quark scheme, however, one runs into trouble at the next logical step:

$$\begin{aligned} p &= uud \\ n &= udd \\ \Delta^{++} &= uuu. \end{aligned} \tag{1.1}$$

The uuu configuration correctly matches the properties of the doubly charged

Δ^{++}-baryon (the π^+p resonance originally discovered by Fermi and collaborators in 1951). Its spin, $J = \frac{3}{2}$, is obtained by combining three identical $J = \frac{1}{2}$ u quarks in their ground state. That is, the quark scheme forces us to combine three identical fermions u in a completely symmetric ground state uuu in order to accommodate the known properties of the Δ-particle. Such a state is of course forbidden by Fermi statistics.

Even ignoring the statistics fiasco, this naive quark model is clearly unsatisfactory: it is true that the qqq, $\overline{\text{q}}\overline{\text{q}}\overline{\text{q}}$, and q$\overline{\text{q}}$ states reproduce the observed sequence of baryon, antibaryon, and meson states, as will be demonstrated in the next chapter, but what about all the other possibilities such as qq, $\overline{\text{q}}\overline{\text{q}}$,..., or single quarks themselves? No uu particles with charge $\frac{4}{3}$ have ever been observed. Both problems can be resolved by introducing a new property or quantum number for quarks (not for leptons!): "*color.*" We suppose that quarks come in three primary colors: red, green, and blue, denoted symbolically by R, G, and B, respectively. "Color" has of course no relation to the real colors of everyday life; the terminology is just based on the analogy with the way all real colors are made up of three primary colors. If we then rewrite the quark wavefunction for the Δ-state in (1.1) as $u_R u_G u_B$, we have clearly overcome the statistics problem by disposing of the identical quarks. The three quarks that make up the Δ-state are now distinguishable by their color quantum number. This solution might appear very contrived; we request, however, that you reserve judgement until, well, maybe Chapters 10 and 14, where the color degree of freedom will acquire a "physical" meaning in the context of a gauge theory. There remains a more immediate problem: if $u_R u_G u_B$ is Fermi's Δ^{++}, then we appear to have many candidate states for the proton: $u_R u_G d_B$, $u_R u_G d_G$, $u_B u_R d_R$, and so on. Yet only one proton state exists; we have to introduce our color quantum number without proliferating the number of states, since this would lead us to a direct conflict with observation. The way this is done is to assert that all particle states observed in nature are "colorless" or "white" (or, to be more precise, unchanged by rotations in R, G, B space). It is easy to visualize the color quantum number by associating the three possible colors of a quark with the three spots of primary red, green, and blue light focussed on a screen, as shown in Fig. 1.2. The antiquarks are assigned the complementary colors: cyan ($\overline{\text{R}}$), magenta ($\overline{\text{G}}$), and yellow ($\overline{\text{B}}$). If you do not know color theory, it may be more helpful to just think of the complementary colors as antired, antigreen, and antiblue. The colors assigned to the antiquarks appear in Fig. 1.2 in those parts of the screen where two and only two primary beams overlap. There is now a unique set of ways to obtain colorless (white) combinations by mixing colors (quarks) and complementary colors (antiquarks):

- Equal mixture of red, green, and blue (RGB).
- Equal mixture of cyan, magenta, and yellow ($\overline{\text{R}}\overline{\text{G}}\overline{\text{B}}$).
- Equal mixtures of color and complementary color (R$\overline{\text{R}}$, G$\overline{\text{G}}$, B$\overline{\text{B}}$).

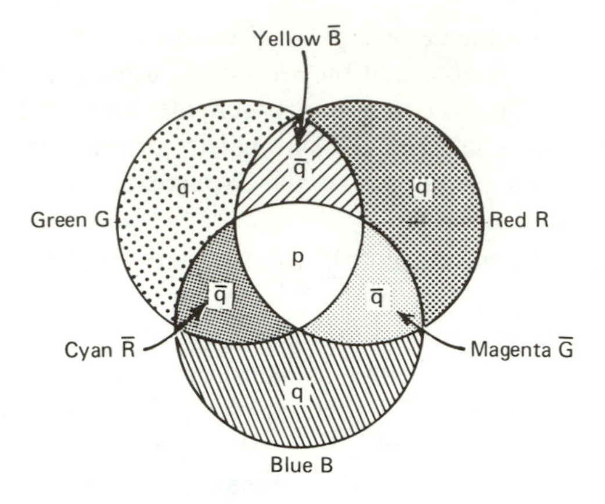

p = "RGB" $\bar{p}$ = "$\bar{R}\bar{G}\bar{B}$"

π = "$R\bar{R} + B\bar{B} + G\bar{G}$"

Fig. 1.2 Color composition of hadrons.

These possibilities correspond respectively to the particle states observed in nature: baryons, antibaryons, and mesons. For example,

$$\begin{aligned} \text{p} &= \text{``RGB''} \\ \bar{\text{p}} &= \text{``}\overline{\text{R}}\overline{\text{G}}\overline{\text{B}}\text{''} \\ \pi &= \text{``R}\overline{\text{R}} + \text{G}\overline{\text{G}} + \text{B}\overline{\text{B}}\text{''}. \end{aligned} \tag{1.2}$$

In other words, the proton is still a uud state as in (1.1), but with the specific color assignments to the quarks exhibited in (1.2). The quotation marks remind us that these wavefunctions eventually have to be properly symmetrized and normalized. We will do so in Chapter 2, and indeed show that the combination "RGB" is antisymmetric under interchange of a pair of color labels as required by the Fermi statistics of the quarks.

The analogy we have developed between the color quantum number and color is not perfect. The three $q\bar{q}$ states $\overline{\text{R}}$R, $\overline{\text{G}}$G, and $\overline{\text{B}}$B are colorless, but it is only the combination $\overline{\text{R}}\text{R} + \overline{\text{G}}\text{G} + \overline{\text{B}}\text{B}$, unchanged by rotations in R, G, B color space, which can form an observed meson. In other words, we use "colorless" to mean a singlet representation of the color group.

Let us briefly recapitulate. We have assigned a "hidden" color quantum number (no connection with hidden variables implied!) to quarks; it is hidden from the world in the sense that all the particles or quark bound states that hit the experimentalists' detectors are colorless (color singlets). It solves the embarrassment that our successful (see Chapter 2) quark model appears to violate Fermi statistics, but it does much more. Notice indeed that, for example, qq states of the type RG, GB... are necessarily colored and therefore cannot occur according to our dogma that only colorless bound states exist. The color scheme explains the exclusive role played in nature by qqq, $\bar{q}\bar{q}\bar{q}$, and $q\bar{q}$ quark combinations. The quarks themselves are colored and therefore hidden from our sight. But, as we discuss later on, there are nevertheless a multitude of ways to infer experimentally their existence inside hadrons. We are now ready to introduce, however, the most profound implication of the color concept.

1.3 Color: The Charge of Nuclear Interactions

In Maxwell's theory of electromagnetism, charged particles such as electrons interact through their electromagnetic fields. However, for many years it was difficult to conceive how such action-at-a-distance between charges came about. That is, how can charged particles interact without some tangible connection? In quantum field theory, we have such a tangible connection: all the forces of nature are a result of particle exchange. Consider first the event taking place at point A in Fig. 1.3. An electron emits a photon (the quantum of the electromagnetic field) and as a result recoils in order to conserve momentum. It is clearly impossible to conserve energy as well, so the emitted photon is definitely not a real photon. It is a photon with "not quite the right energy"; we call it a "*virtual*" photon (Chapters 4 and 6). An electron can nevertheless emit such a photon as long as it is sufficiently quickly reabsorbed. Because of the uncertainty inherent in quantum mechanics, the photon can in fact live for a time $\Delta t \lesssim (\hbar/\Delta E)$, where ΔE is the "borrowed" or missing energy. However, suppose that instead of being reabsorbed by the same electron, the photon is absorbed by another electron, as in Fig. 1.3. The latter electron will recoil in the act of absorbing the virtual photon at point B. The net result is a repulsive force between the two electrons. In quantum field theory, such exchanges are responsible for the Coulomb repulsion of two like charges!

The dramatized picture of the force sketched in Fig. 1.3 might lead one to believe that only repulsive forces can be described by the exchange of particles. This is not so; the impulse of a virtual particle can have either sign in quantum field theory because its momentum vector does not necessarily have the orienta-

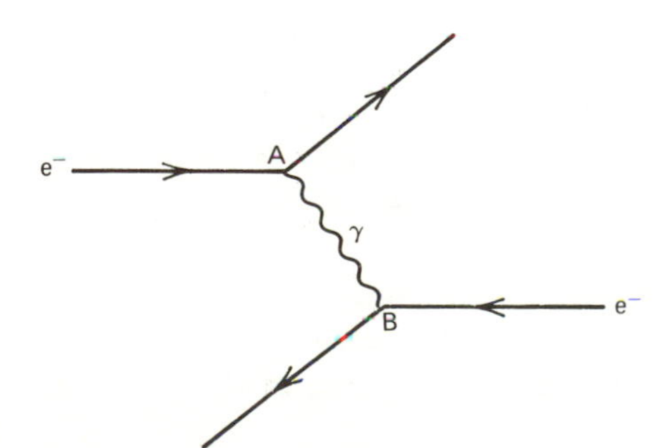

Fig. 1.3 Electrons repel by exchanging a photon.

tion prescribed by classical physics. This should not be too surprising, as clearly the whole discussion of the exchange force of Fig. 1.3 is classically impossible. The exchanged "virtual" photons are in other respects different from freely propagating real photons encountered in, for example, radio transmission, where the energy of the oscillating current in the transmitter is carried away by the radio waves. Virtual photons cannot live an existence independent of the charges that emit or absorb them. They can only travel a distance allowed by the uncertainty principle, $c\,\Delta t$, where c is the velocity of light.

From now on, electromagnetic interactions of charged particles are represented by pictures (*Feynman diagrams*, see Chapter 3 onward) like the one shown in Fig. 1.4a. The charged quarks will also interact by photon exchange. Although we are familiar with the fact that electromagnetic interactions bind positronium (e^-e^+), it is clear that the electromagnetic interaction cannot bind quarks into hadrons. A "*strong*" force, overruling the electromagnetic repulsion of the three (same-charge) u quarks in the Δ^{++}-particle, must be invoked to bind quarks into hadrons. In fact, the color "charge" endows quarks with a new color field making this strong binding possible. The interaction of two quarks by the exchange of a virtual "gluon" is shown in Fig. 1.4b. Gluons are the quanta of the color field that bind quarks in nucleons and also nucleons into nuclei. It is useful to follow the flow of color in the diagram of Fig. 1.4b. This is shown in Fig. 1.4c. The red quark moving in from the left of the page switches color with the blue quark coming from the right. Quarks interact strongly by exchanging color. Now match Figs. 1.4b and 1.4c. The gluon, shown as a curly line in Fig. 1.4b, must itself be colored: it is in fact a bicolored object, labeled ($B\bar{R}$) in Fig. 1.4c. Based on the analogy pictured in Fig. 1.4, one can construct a theory of the color, strong, or nuclear force, whatever you prefer to call it, by copying the quantized version of Maxwell's theory (quantum electrodynamics, or QED). The theory thus obtained is called *quantum chromodynamics*, or QCD (Chapters 10 and 11).

This theory is fortunately sufficiently like QED to share its special property of being a *renormalizable* (i.e., calculable) gauge theory. Although we postpone explanation of this statement to much later (Chapter 14), it should preempt any other statement in a presentation of QCD. We are equally fortunate that it is sufficiently different from QED to play the role of the strong force. First, nine bicolored states of the type depicted in Fig. 1.4c exist: $R\bar{R}$, $R\bar{G}$, $R\bar{B}$, $G\bar{R}$, $G\bar{G}$, $G\bar{B}$, $B\bar{R}$, $B\bar{G}$, and $B\bar{B}$. Notice that the gluon depicted in Fig. 1.4c is labeled $B\bar{R}$, not BR. Indeed, when in Chapter 3 an operational definition is given to these diagrams, it will be clear that the different directions of the arrows on the color lines of the exchanged quantum in Fig. 1.4c imply that they represent a color–anticolor combination. One of the nine combinations, $R\bar{R} + G\bar{G} + B\bar{B}$, is a color singlet [see (1.2)] which lacks any net color charge and therefore cannot play the role of a gluon carrying color from one quark to another. Chromodynamics is therefore a theory like electromagnetism, but with eight gluons instead of a single photon. Since the gluons themselves carry a color charge, they can directly interact with other gluons, as depicted in Fig. 1.4d. This possibility is not

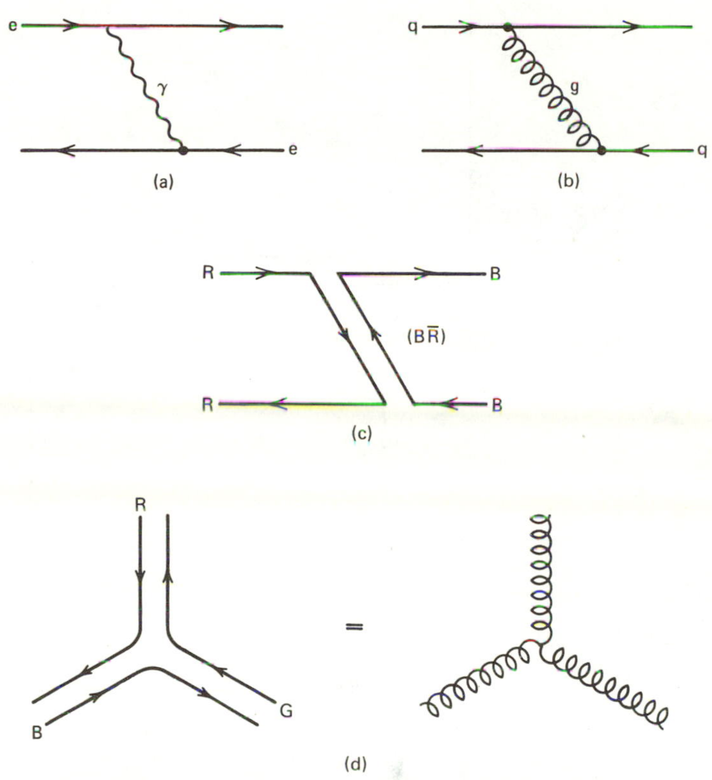

Fig. 1.4 (a) Electromagnetic interaction by photon exchange. (b) Strong interaction by gluon exchange. (c) Flow of color in (b). (d) Self-coupling of gluons.

available in electrodynamics, as photons do not have electric charge. Theories in which field quanta may interact directly are called "*non-Abelian*" (Chapter 14).

The existence of this direct coupling of gluons has dramatic implications that become evident if one contrasts the effects of *charge screening* in both QED and QCD. Screening of the electric charge in electrodynamics is illustrated in Fig. 1.5. In quantum field theory, an electron is not just an electron—it can suddenly emit a photon, or it can emit a photon that subsequently annihilates into an electron–positron pair, and so on. In other words, an electron in quantum field theory exhibits itself in many disguises, one of which we show in Fig. 1.5a. Note that the original electron is surrounded by e^-e^+ pairs and, because opposite charges attract, the positrons will be preferentially closer to the electron. Therefore, the electron is surrounded by a cloud of charges which is polarized in such a way that the positive charges are closer to the electron; the negative charge of the electron is thus screened, as shown in more detail in Fig. 1.6. Suppose that we want to determine the charge of the electron in Fig. 1.6 by measuring the Coulomb force experienced by a test charge. The result will depend on where we

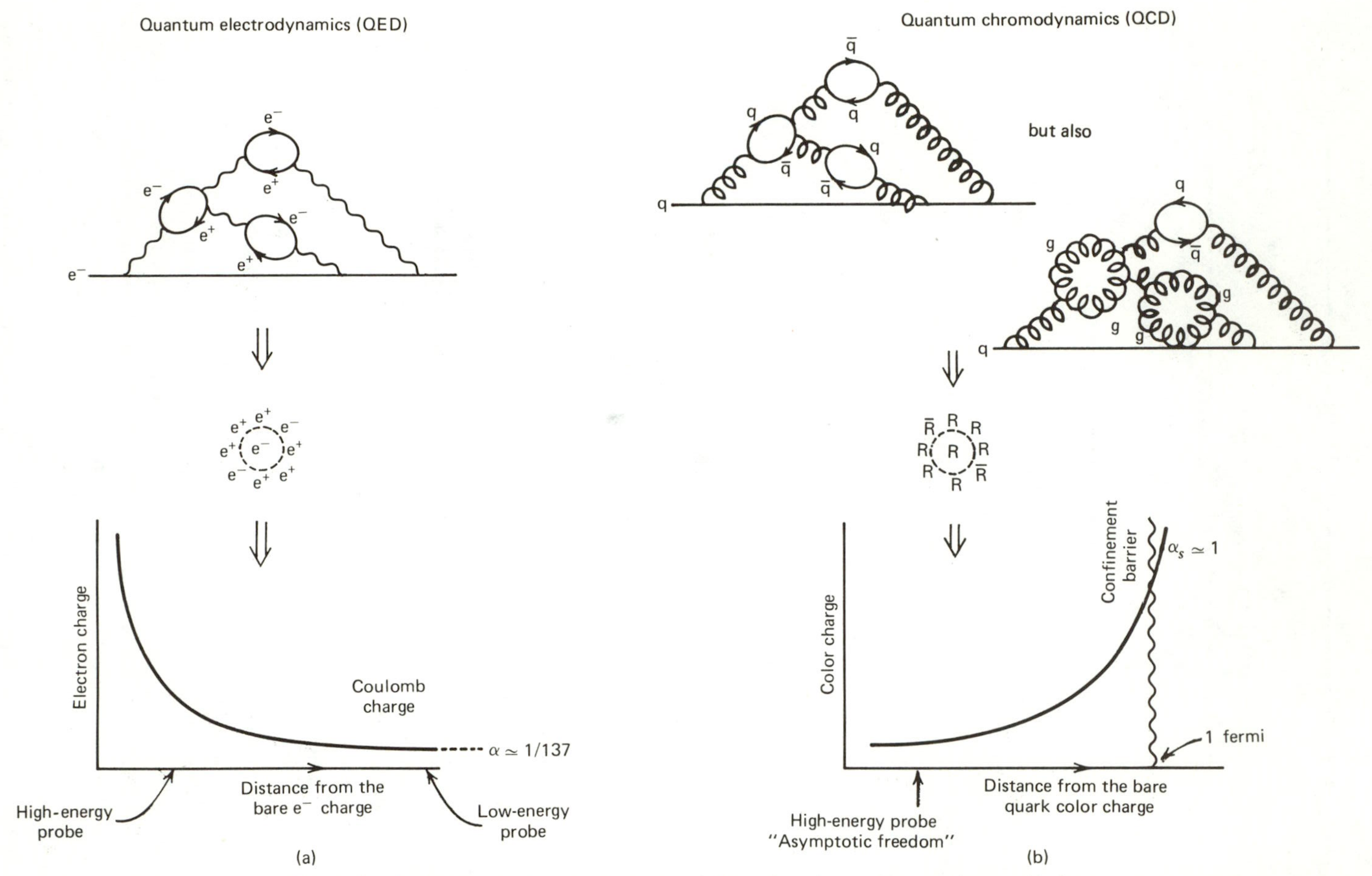

Fig. 1.5 Screening of the (a) electric and (b) color charge in quantum field theory.

place the test charge; when moving the test charge closer to the electron, we penetrate the cloud of positrons that screens the electron's charge. Therefore, the closer one approaches the electron, the larger is the charge one measures. In quantum field theory, the vacuum surrounding the electron has become a polarizable medium. The situation is analogous to that of a negative charge in a dielectric medium: the electron–positron pairs in Fig. 1.6 respond to the presence of the electron like the polarized molecules do in the dielectric. This effect is known as charge screening; as a result, the "measured charge" depends on the distance one is probing the electron; the result is shown pictorially in Fig. 1.5a. In QED, this variation of the charge is calculable by considering all possible configurations of the electron's charge cloud, only one of which is shown in Fig. 1.5a (see Chapter 7).

One can carry through the same calculation for the color charge of a quark. Color screening would be a carbon copy of charge screening if it were not for the new configurations involving gluons turning into pairs of gluons, as shown in Fig. 1.5b. The gluons, themselves carriers of color, also spread out the effective color charge of the quark. It turns out that the additional diagrams reverse the familiar result of quantum electrodynamics: a red charge is preferentially surrounded by other red charges, as shown in Fig. 1.5b. We now repeat the experiment of Fig. 1.6 for color charges. By moving our test probe closer to the original red quark, the probe penetrates a sphere of predominantly red charge and the amount of red

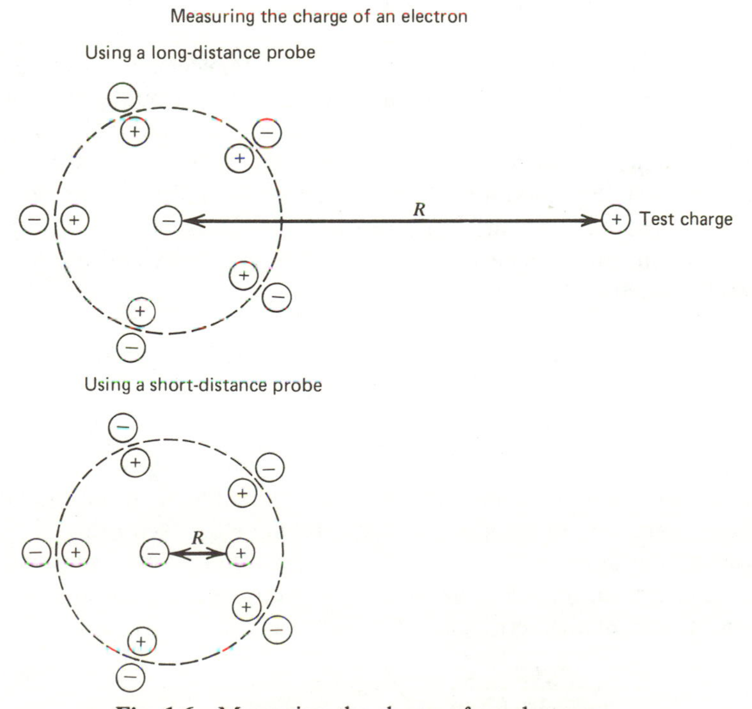

Fig. 1.6 Measuring the charge of an electron.

charge measured decreases. The resulting "antiscreening" of the red color is pictured in Fig. 1.5b and is referred to as "*asymptotic freedom*" (Chapter 7). Asymptotically, two red quarks interact (i.e., for very small separations) through color fields of reduced strength and approach a state where they behave as essentially free, noninteracting particles.

You might wonder why we have immediately emphasized this rather exotic property of color theory. As we see further on, it is asymptotic freedom that turns QCD into a quantitative calculational scheme (Chapters 10 and 11).

1.4 Natural Units

At this stage, it is necessary to break the flow of the physics discussion and to introduce units appropriate to particle physics. The two fundamental constants of relativistic quantum mechanics are Planck's constant, h, and the velocity of light *in vacuo*, c:

$$\hbar \equiv \frac{h}{2\pi} = 1.055 \times 10^{-34}\ \text{J sec}$$

$$c = 2.998 \times 10^{8}\ \text{msec}^{-1}.$$

It is convenient to use a system of units in which $\hbar$ is one unit of action (ML^2/T) and c is one unit of velocity (L/T). Our system of units will be completely defined if we now specify, say, our unit of energy (ML^2/T^2). In particle physics, it is common to measure quantities in units of GeV (1 GeV $\equiv 10^9$ electron volts), a choice motivated by the fact that the rest energy of the proton is roughly 1 GeV.

By choosing units with $\hbar = c = 1$, it becomes unnecessary to write $\hbar$ and c explicitly in the formulas, thus saving a lot of time and trouble. We can always use dimensional analysis to work out unambiguously where the $\hbar$'s and c's enter any formula. Hence, with a slight but permissible laziness, it is customary to speak of mass (m), momentum (mc), and energy (mc^2) all in terms of GeV, and to measure length ($\hbar/mc$) and time ($\hbar/mc^2$) in units of GeV^{-1}. Table 1.2a displays the connection between GeV units and mks units and Table 1.2b lists some useful conversion formulae.

EXERCISE 1.1 Cross sections are often expressed in millibarns, where 1 mb $= 10^{-3}$ b $= 10^{-27}$ cm^2. Using GeV units, show that

$$1\ \text{GeV}^{-2} = 0.389\ \text{mb}.$$

So far, we have not considered the elementary charge e, which measures how strongly electrons, say, interact electromagnetically with each other. To obtain a dimensionless measure of the strength of this interaction, we compare the electrostatic energy of repulsion between two electrons one natural unit of length apart with the rest mass energy of an electron:

$$\alpha = \frac{1}{4\pi}\frac{e^2}{(\hbar/mc)}\Big/ mc^2 = \frac{e^2}{4\pi\hbar c} \approx \frac{1}{137}. \tag{1.3}$$

TABLE 1.2a
Conventional Mass, Length, Time Units, and Positron Charge in Terms of $\hbar = c = 1$ Energy Units

Conversion Factor	$\hbar = c = 1$ Units	Actual Dimension
$1\ \text{kg} = 5.61 \times 10^{26}$ GeV	GeV	$\dfrac{\text{GeV}}{c^2}$
$1\ \text{m} = 5.07 \times 10^{15}\ \text{GeV}^{-1}$	GeV^{-1}	$\dfrac{\hbar c}{\text{GeV}}$
$1\ \text{sec} = 1.52 \times 10^{24}\ \text{GeV}^{-1}$	GeV^{-1}	$\dfrac{\hbar}{\text{GeV}}$
$e = \sqrt{4\pi\alpha}$	—	$(\hbar c)^{1/2}$

TABLE 1.2b
Some Useful Conversion Factors

$1\ \text{TeV} = 10^3\ \text{GeV} = 10^6\ \text{MeV} = 10^9\ \text{KeV} = 10^{12}\ \text{eV}$
$1\ \text{fermi} \equiv 1\ \text{F} = 10^{-13}\ \text{cm} = 5.07\ \text{GeV}^{-1}$
$(1\ \text{F})^2 = 10\ \text{mb} = 10^4\ \mu\text{b} = 10^7\ \text{nb} = 10^{10}\ \text{pb}$
$(1\ \text{GeV})^{-2} = 0.389\ \text{mb}$

In (1.3), we have adopted the rationalized Heaviside–Lorentz system of electromagnetic units. That is, the 4π factors appear in the force equations rather than in the Maxwell equations, and ε_0 is set equal to unity. This choice, which is conventional in particle physics, reduces Maxwell's equations to their simplest possible form. The value of α is, of course, the same in all systems of units, but the numerical value of e is different.

For historical reasons, α is known as the fine structure constant. Unfortunately, this name conveys a false impression. We have seen that the charge of an electron is not strictly constant but varies with distance because of quantum effects; hence α must be regarded as a variable, too. The value $\frac{1}{137}$ is the asymptotic value of α shown in Fig. 1.5a.

EXERCISE 1.2 Show that, in $\hbar = c = 1$ units, the Compton wavelength of an electron is m^{-1}, the Bohr radius of the hydrogen atom is $(\alpha m)^{-1}$, and the velocity of an electron in its lowest Bohr orbit is simply α; m is the mass of the electron or, more precisely, the reduced mass $m_e m_p/(m_e + m_p)$.

EXERCISE 1.3 Justify the statements, that due to the weakness of the electromagnetic interaction ($\alpha \approx \frac{1}{137}$), the hydrogen atom is a loosely bound extended structure and that the nonrelativistic Schrödinger equation is adequate to describe the gross structure of atomic energy levels.

1.5 Alpha (α) is not the Only Charge Associated with Particle Interactions

In Fig. 1.3, we pictorially associated the electromagnetic interaction of two charges with the emission and reabsorption of a field quantum γ. We can somewhat "quantify" this picture by interpreting some familiar examples of classical electromagnetic phenomena in terms of this language. If you feel uneasy about the vagueness of the arguments below, we would like to point out that an exact operational definition of these so-called Feynman diagrams is forthcoming in Chapters 3 through 7. There they will represent the probability amplitudes for the process pictured and will be calculated using relativistic quantum mechanics.

As a first example, consider the diagram of Fig. 1.7a, representing the Thomson scattering of photons off electrons. For long-wavelength photons, the cross section is given by

$$\sigma_{TH} = \frac{8\pi}{3}\left(\frac{\alpha}{m_e}\right)^2 = \frac{2}{3}\alpha^2\left(4\pi R_e^2\right), \tag{1.4}$$

where R_e is the Compton wavelength of the electron,

$$R_e = \frac{\hbar}{m_e c} = \frac{1}{m_e} \tag{1.5}$$

in natural units. Not surprisingly, the Thomson cross section can be obtained classically as well as from quantum electrodynamics. In fact, it is this long-wavelength limit that is used to define the charge $-e$ of the electron. It is the value of the charge (or α) when probed with a low-energy probe from a large distance, see Fig. 1.5a.

Another example of an electromagnetic interaction is the Rutherford scattering of an electron of energy E on a nucleus of charge Ze, for which the differential cross section is

$$\frac{d\sigma_R}{d\Omega} = \frac{Z^2\alpha^2}{4E^2}\frac{1}{\sin^4(\theta/2)}. \tag{1.6}$$

This process is pictured by the Feynman diagram of Fig. 1.7b. We now want to use the two processes pictured in Fig. 1.7 to illustrate the fact that α is a measure

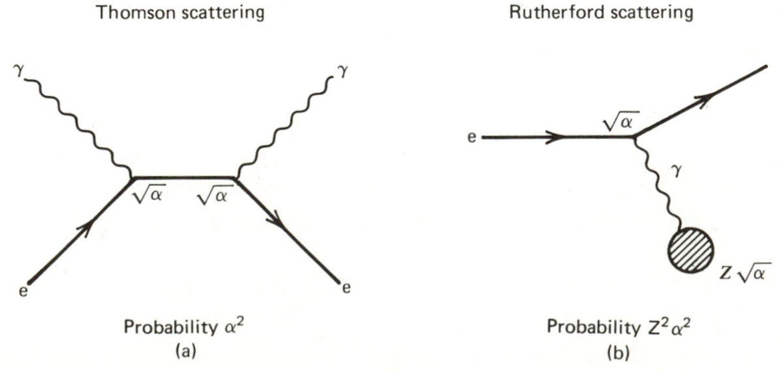

Fig. 1.7 (a) Thomson scattering. (b) Rutherford scattering.

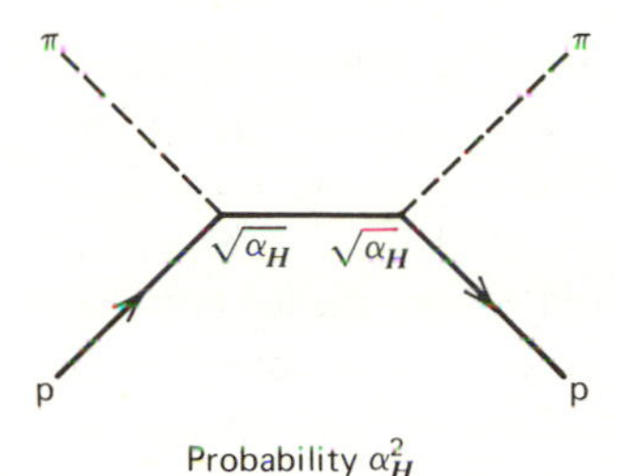

Fig. 1.8 πp scattering.

of the strength of the electromagnetic interaction or, in particle exchange language, the probability for emitting or absorbing a photon. As we eventually want the Feynman diagrams to represent probability *amplitudes*, a factor e or $\sqrt{\alpha}$ is associated with each absorption or emission of a photon by a charge e. Inspection of the diagrams in Figs. 1.7a and 1.7b makes it clear that two such factors $(\sqrt{\alpha}\,\sqrt{\alpha}\,)^2$ should appear in the cross section (1.4) and a factor $(\sqrt{\alpha}\,Z\sqrt{\alpha}\,)^2$ in (1.6).

Armed with this concept, we now return to Thomson scattering but replace the photon beam by π-mesons. First, we note that, in analogy to (1.4), the Thomson cross section for the scattering of long-wavelength photons off a proton target is

$$\sigma_{TH} = \tfrac{2}{3}\alpha^2\left(4\pi R_p^2\right), \tag{1.7}$$

where $R_p = 1/m_p$ in natural units. Our analyzer photon beam sees an effective area $(4\pi R_p^2)$ of the target proton, and α^2 is, as before, the probability of absorbing and emitting a photon. The factor $\frac{2}{3}$ is immaterial for present considerations; it actually reflects the fact that for a photon, one of the three polarization states expected for a particle of spin 1 is missing.

If we now repeat the measurement of the proton's radius using a π-meson instead of a photon beam (Fig. 1.8), we find that the coupling (emission or absorption) of π-mesons to nucleons cannot possibly be electromagnetic. The cross section for the process of Fig. 1.8 is readily found to be

$$\sigma_T(\pi \mathrm{p}) = \alpha_H^2\left(4\pi R_p^2\right), \tag{1.8}$$

where, as before, the factor $4\pi R_p^2$ is the effective area of the target; $\sigma_T(\pi\mathrm{p})$ is measured to be well over 1 mb. From a comparison of this result with the much smaller measured values of σ_{TH}, given by (1.7), we are forced to conclude that α_H, the probability for absorbing and emitting π-mesons, exceeds α by two to three orders of magnitude. That is,

$$\alpha_H \sim 1 \text{ to } 10. \tag{1.9}$$

The conclusion that a new "charge" and a new field has to be invoked to explain πN interaction cross sections is therefore inevitable. A substantial effort was made to construct an appropriate quantum field theory. More quantitative analyses of the type presented above actually show that the nucleon is not a simple object with classical radius m_N^{-1} but a complicated structure with a size

$$\langle r^2 \rangle \simeq \frac{1}{m_\pi^2}.$$

This was for a long time taken as a hint that π-mesons are in fact the field quanta associated with this new charge. These historic attempts failed for several reasons. First, the π-meson itself possesses just as rich a substructure as the nucleon. It therefore does not fit well into its role as the "photon" of the nuclear interactions. The other problem is that actually $\alpha_H \simeq 15$. This immediately precludes any hope of repeating the successes of quantum electrodynamics, which are anchored in the use of perturbation theory exploiting the numerical smallness of α to ensure that the perturbation series converges rapidly.

Nowadays, we associate the structure of the proton with quarks; the quark color charge is the "true" charge of strong interactions. Gluons are the "photons" of strong interactions. Just as the probability for photon emission by a charged particle is given by $\alpha = e^2/4\pi$, so the probability for gluon emission by a colored quark is characterized by α_s, where α_s is essentially the square of the color charge divided by 4π (see Fig. 1.9). We no longer regard α_H as a fundamental number but as an "epiphenomenon" that reflects the structure of pions and nucleons as well as α_s.

Scattering of low-energy π-mesons probes in some way the color charge at the surface of the nucleon, that is, over a typical distance $1/m_\pi \simeq 1.4$ F. Here, even $\alpha_s \gtrsim 1$, as can be seen from Fig. 1.5b; α_s is too large to permit the use of perturbation theory. Using the modern technology of very high-energy particle beams, we can however probe the color charge of individual quarks deep inside the nucleon where α_s is smaller (see again Fig. 1.5b). A large variety of experimental probes has been invented to reach energies and distances where $\alpha_s \simeq 0.2$; perturbation theory then is a workable approximation. Asymptotic freedom of QCD is the key to the use of perturbation theory (see Chapters 10 and 11).

Examples of experimental probes are shown in Figs. 1.10 and 1.11. Fig. 1.10a depicts the familiar technique of mapping the structure of an atom by scattering a beam of electrons off it. In elastic collisions, where the exchanged photon in Fig. 1.10a carries very little momentum and has therefore a relatively long wavelength,

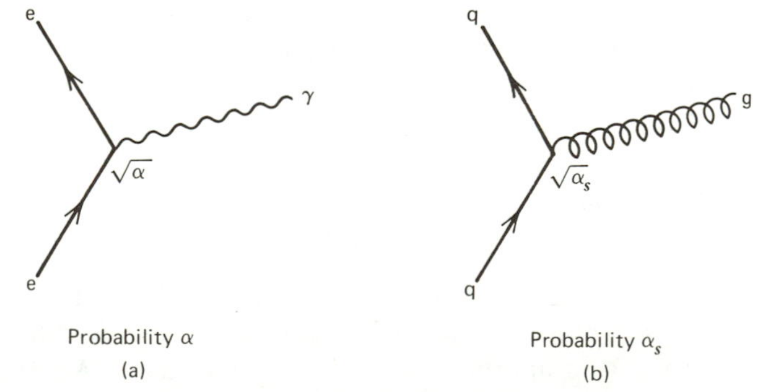

Fig. 1.9 (a) The probability amplitude for emitting a photon is proportional to e or $\sqrt{\alpha}$. (b) The probability amplitude for emitting a gluon is proportional to $\sqrt{\alpha_s}$.

the "optics" is such that we are illuminating the complete atom. In such collisions, one is therefore measuring the size $\langle r^2 \rangle$ of the electron cloud. But for very inelastic interactions of the electron beam, the exchanged photon now acquires a large momentum. The photon in Fig. 1.10a has turned into a short-wavelength probe illuminating the atom with high resolution. The cross section takes the form of (1.6). Figure 1.10c shows the result of such an experiment scattering α-particles on a gold target. The large cross section observed at large angles is due to α-particles rebounding off nuclei deep inside the gold atoms. More than 60 years after these pioneering atomic structure experiments, a very similar phenomenon was observed in high-energy proton–proton scattering. This time, the enhancement (Fig. 1.10c) is indicative of a hard-core substructure inside

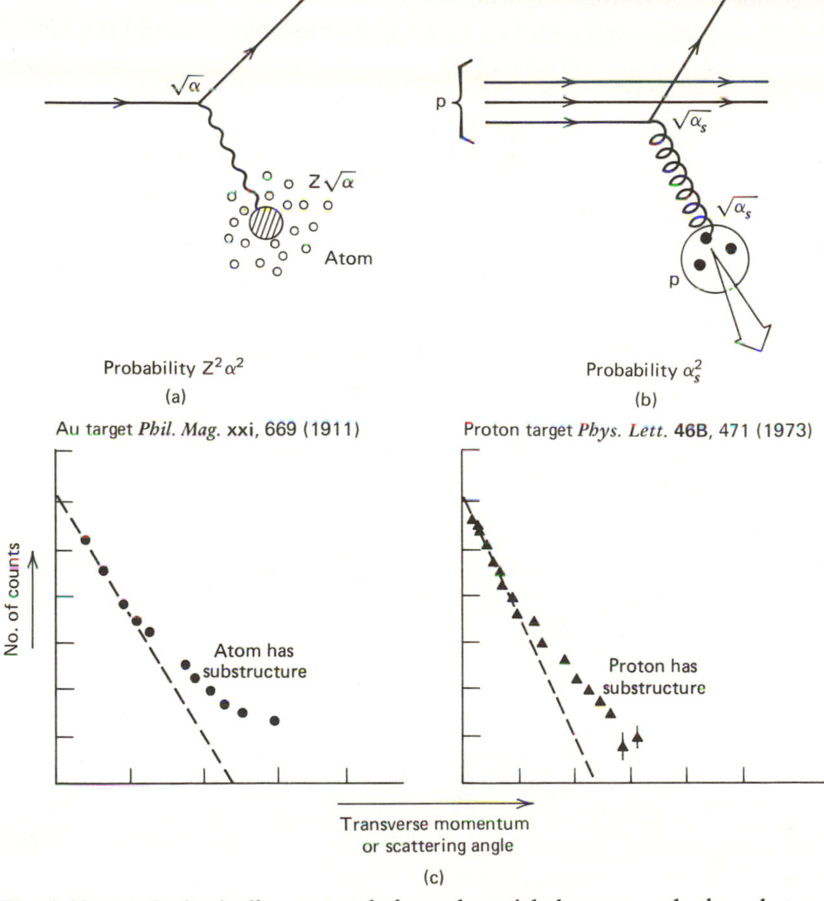

Fig. 1.10 (a) Inelastically scattered charged particle beam reveals the substructure of the atom. (b) Inelastically scattered proton beam reveals the quark structure of the proton target. (c) Experimental results.

the proton. A direct analogy between the atomic and hadronic situation is made in Fig. 1.10b. In inelastic collisions of very high-energy protons, individual quarks in the beam and the target "Rutherford scatter" off one another. As the interactions involve color charges over distances smaller than the size of a nucleon, α_s is small, and quantitative predictions can be made. An analysis of data based on the diagram shown in Fig. 1.10b shows indeed that quarks follow the Rutherford formula of (1.6):

$$\frac{d\sigma_R}{d\Omega} \sim \frac{\alpha_s^2}{4E^2} \frac{1}{\sin^4(\theta/2)}, \tag{1.10}$$

with $\alpha_s \simeq 0.2$. The presence of gluons inside the nucleon have to be taken into account, and this complicates the analysis; but in principle, these experiments are exact analogues of Rutherford scattering.

This experimental approach for resolving quarks inside nucleons is not unique. There are other techniques that allow us to illuminate hadrons with photons of very large momentum Q and therefore small Compton wavelength $\lambda = 1/Q$. Imitating the atomic physics experiments, such photons are often "prepared" by the inelastic scattering of high-energy electron and muon beams off nuclear targets (see Fig. 1.11a). Colliding very high-energy electron and positron beams also provides us with an extremely "clean" technique to probe quarks (see Fig. 1.11b). The photon, produced at rest by e^+ and e^- beams colliding head on, can decay into a $q\bar{q}$ pair as shown in the picture (Fig. 1.11b). When the pair separates by distances of order 1 F, α_s becomes large; that is, the color interactions between the quark and antiquark become truly strong, and these violent forces decelerate the quarks. The decelerated quarks radiate hadrons (mostly light π-mesons) just

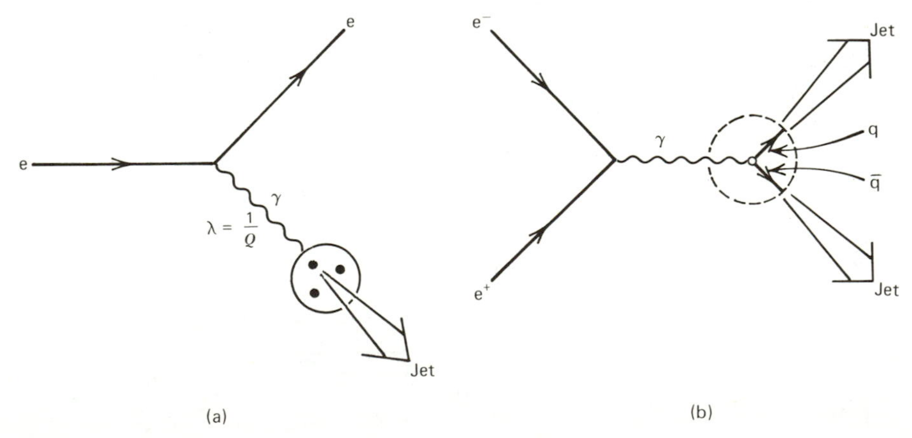

Fig. 1.11 (a) Virtual photons with a short wavelength resolve quarks in the proton target. (b) Virtual photons, obtained by annihilating e^- and e^+ beams, decay into quark–antiquark pairs.

like a decelerated charge emits photons by bremsstrahlung. The original quark is never seen in its "free" state; only these π-mesons and other (colorless) hadrons hit the experimentalist's detector. The widening arrows in Figs. 1.10b and 1.11 indicate the showers or jets of hadrons radiated from, and traveling more or less in the direction of, the struck quark. The quark thus "escapes," but only as a constituent of one of the radiated hadrons.

The central role played by the property of color screening (Fig. 1.5b) becomes more and more crucial. But is the argument that quarks (or color) are confined within hadrons any more than an excuse for our failure to observe free quarks? The answer is that this picture makes unique and dramatic predictions; as the virtual photon is produced at rest in Fig. 1.11b, the quark and antiquark should emerge in opposite directions to conserve momentum. Therefore, the two sprays of hadrons (jets) should be observed on opposite sides of the annihilation point where the photon is produced. The verification of this prediction was a real boost for the quark theory; any pre-quark theory of elementary particles had led us to expect a uniform, isotropic distribution of the emerging hadrons. The experimental verdict is unambiguous; Fig. 1.12 shows an example of a two-jet event seen by the struck wires of a detector. Notice the back-to-back orientation of the two jets.

The most striking (and certainly far from obvious) aspect of two-jet events is that the bremsstrahlung products, and never the original quark, reach the detector. Even with our very qualitative insight of color theory, it is possible to "rationalize" this experimental fact. It is once more crucial to recall the difference between QED and QCD recorded in Fig. 1.5. When a quark and antiquark separate, their color interaction becomes stronger. Through the interaction of gluons with one another, the color field lines of force between the quark and the antiquark are squeezed into a tube-like region as shown in Fig. 1.13a. This has to be contrasted with the Coulomb field where nothing prevents the lines of force from spreading out. There is no self-coupling of the photons to contain them. This is yet another way to visualize the different screening properties of QED and QCD. If the color tube has a constant energy density per unit length, the potential energy between the quark and the antiquark will increase with separation, $V(r) \sim \lambda r$, and so the quarks (and gluons) can never escape. This so-called "*infrared slavery*" is believed to be the origin of the total *confinement* of quarks to colorless hadrons. But how do they materialize as hadron jets? The answer is shown in Fig. 1.14. The separating $q\bar{q}$ pair stretches the color lines of force until the increasing potential energy is sufficient to create another $q\bar{q}$ pair. These act as the end points for the lines of force, which thus break into two shorter tubes with lower net energy despite the penalty of providing the extra $q\bar{q}$ mass. The outgoing quark and antiquark continue on their way (remember that they originally carried the momentum of the colliding e^- and e^+), further stretching the color lines. More $q\bar{q}$ pairs are produced until eventually their kinetic energy has degraded into clusters of quarks and gluons, each of which has zero net color and low internal momentum and therefore very strong color coupling. This coupling turns them

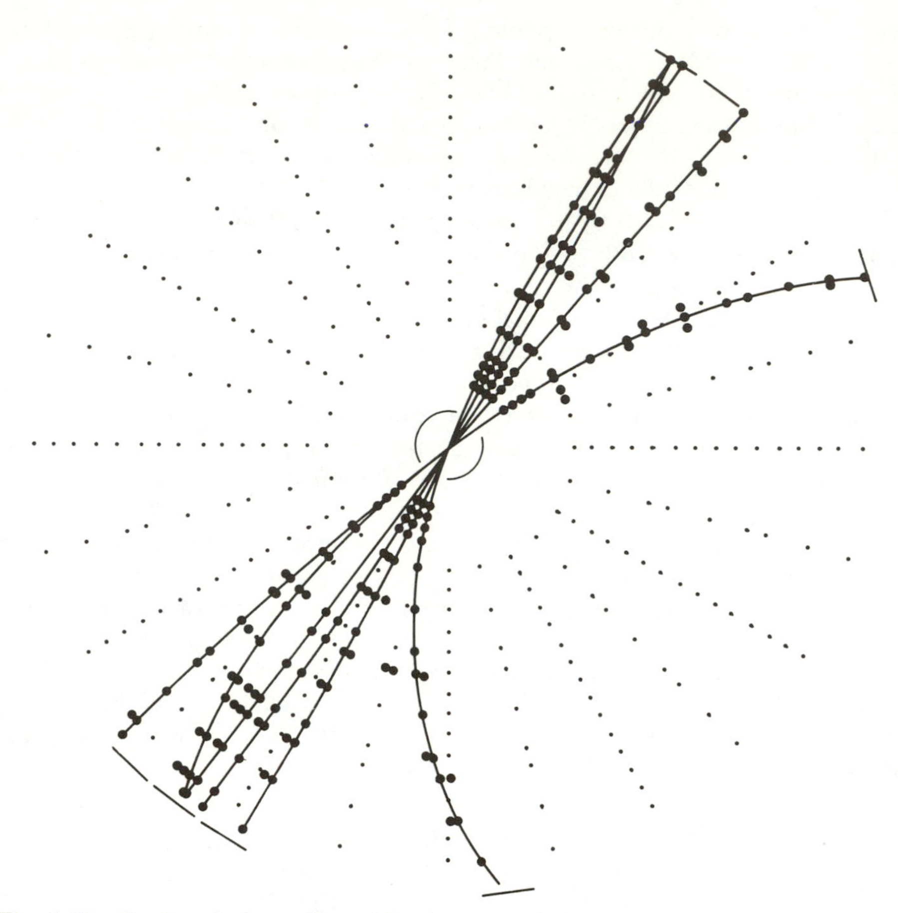

Fig. 1.12 Tracks of charged particles in a quark and antiquark jet. The TASSO detector at PETRA observes the products of a very high-energy e^- and e^+ head-on collision in the center of the picture.

into the hadrons forming two jets of particles traveling more or less in the direction of the original quark and antiquark (see Fig. 1.14).

In summary, we have argued that quarks have color as well as electric charge. The experimental evidence is compelling. Color is the same property previously introduced in an *ad hoc* way for solving a completely unrelated Fermi-statistics problem. We have argued that the theory of color (QCD) and of electrodynamic (QED) interactions are much alike in that the massless gluons exchanged between colored quarks are very similar to the massless photon exchanged between charged electrons. The crucial difference is the magnitude and screening properties of α_s and α (the strengths of the respective interactions) as indicated by the sketches in Figs. 1.5 and 1.13.

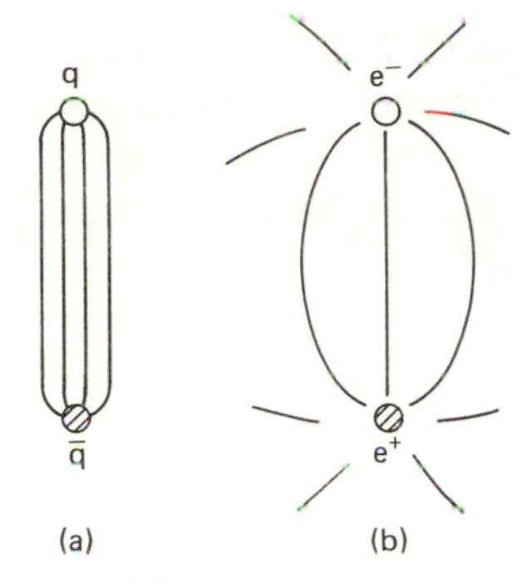

Fig. 1.13 The $q\bar{q}$ color field with $V(r) \sim r$, and the e^+e^- Coulomb field with $V(r) \sim 1/r$.

Where does all this leave the short-range nuclear force that binds neutrons and protons to form nuclei, which was in fact the original motivation for introducing the strong interaction? The answer is best seen in analogy with chemical binding due to the electromagnetic forces between two neutral atoms as they approach each other. As the atoms are electrically neutral, very little force is experienced until their electron clouds start to overlap. The force, known as the van der Waals force, increases gradually at first, and then very rapidly as the interpenetration increases. It is responsible for molecular binding and involves the exchange of electrons between the atoms. In itself it is not a fundamental interaction, but is instead a complicated manifestation of the basic electromagnetic interactions between two extended charged systems. In the same manner, the nucleon–nucleon force can be viewed as a complicated manifestation of the basic interaction between colored quarks. The nucleons, which are color neutral, only experience strong interactions at short distances when the quarks in one nucleon can "sense" the presence of the quarks in the other.

1.6 There are Weak Interactions, too

The Δ^{++}-particle, which gave us the crucial hint to introduce the color quantum number, was discovered by Fermi and collaborators by bombarding π^+-mesons

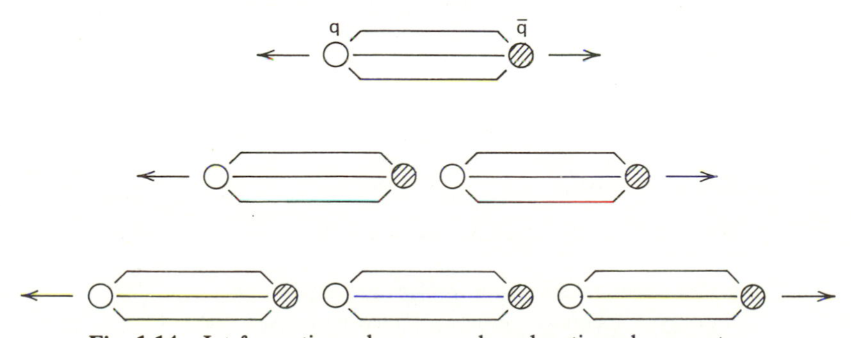

Fig. 1.14 Jet formation when a quark and antiquark separate.

on protons. The doubly charged Δ^{++} only lives for about 10^{-23} sec, then decays back into a π^+ and a proton. In the quark picture, one could imagine a decay mechanism of the type shown in Fig. 1.15a. Strong interactions, with range about 1 F, are responsible for the decay of the Δ^{++}; therefore, the typical decay time is the time it takes the π^+ and the proton to separate by a distance $R \simeq 1$ F:

$$\tau = \frac{R}{c} \simeq 10^{-23} \text{ sec.} \tag{1.11}$$

Other "resonant" baryon states formed in the scattering of π-mesons and protons also have lifetimes of order 10^{-23} sec. In contrast, the proton itself is known to

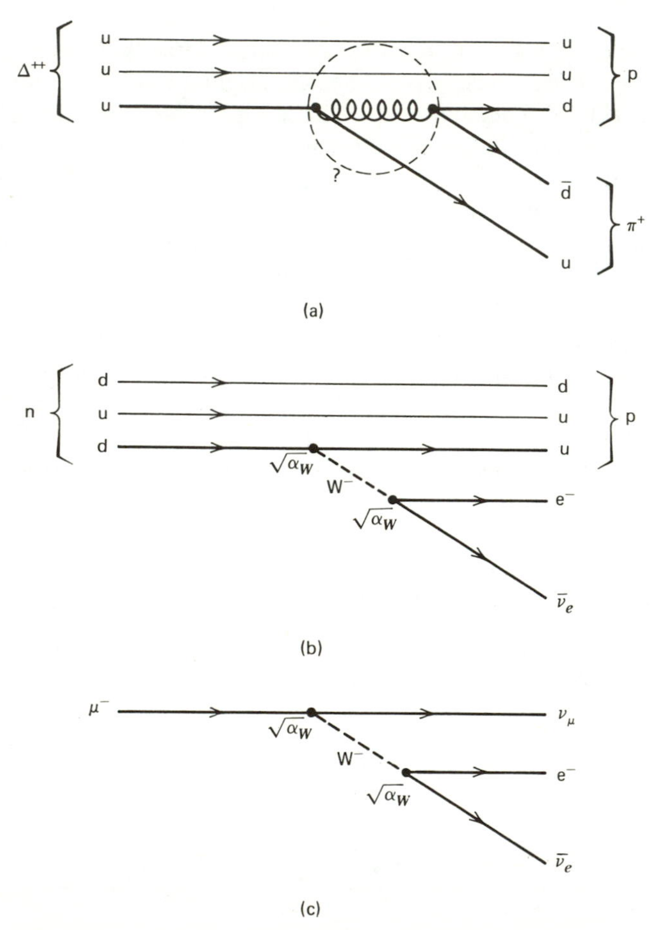

Fig. 1.15 (a) Decay of $\Delta^{++} \to \pi^+ p$. The decay mechanism is only symbolic; it reminds us that the Δ^{++} decays by strong interactions (with range about 1 F). (b) Neutron β-decay is mediated by weak interactions. A massive weak boson W is emitted and absorbed with probability α_W. (c) Muon decay.

have a lifetime in excess of 10^{30} years. To cover this special case, we have to rely on baryon number conservation. Without it, the proton could be unstable and decay; for example,

$$p \to e^+ \pi^0.$$

This decay is forbidden (but see Chapter 15). Indeed, using Table 1.1, we notice that the total baryon number before decay is $3 \times \frac{1}{3}$ for the uud quarks in the proton and zero after decay because of the cancellation of the q and $\bar{q}$ baryon number making up the π^0. The electron, the lightest charged particle, is of course stable by charge conservation.

The neutron's decay is familiar from radioactivity:

$$n \to p + e^- + \bar{\nu}.$$

This transition is energetically allowed and is not excluded by baryon number conservation. But the neutron lives 15 minutes! This is a puzzle. We can imagine that some particles have longer lifetimes than the 10^{-23} sec obtained in (1.11) because they decay exclusively through the weaker electromagnetic interactions and not through the color interaction sketched in Fig. 1.15a. The lifetime of the π^0, which decays electromagnetically via $\pi^0 \to \gamma\gamma$, is 10^{-16} sec. This is what we expect, as

$$\frac{\tau_{\text{electromagnetic}}}{\tau_{\text{strong}}} \simeq \left(\frac{\alpha_s}{\alpha}\right)^2 = 10^4 \sim 10^6, \tag{1.12}$$

and therefore electromagnetic lifetimes should be about $10^4 \sim 10^6$ times longer than the typical 10^{-23} sec lifetime of particles decaying via the strong interaction. To obtain (1.12), we took the hint from Fig. 1.15 that the hadronic decay probability is proportional to $(\sqrt{\alpha_s}\sqrt{\alpha_s})^2$. However, in our world with two scales, α and α_s, nothing can account for lifetimes substantially longer than 10^{-16} sec. The proton and electron are "protected" by conservation laws, but what about the 15-minute lifetime of the neutron?

The problem is not limited to the neutron. The charged π^- decays into $\pi^- \to e^- + \bar{\nu}$ in 10^{-12} sec, and a series of "*strange*" particles exists with lifetimes of order 10^{-10} sec, for example, $\Sigma^+ \to n + \pi^+$. The $\Sigma \to n + \pi$ decay demonstrates the puzzle in a very striking way. Energetically, the decays $\Sigma \to n + \pi$ and $\Delta \to n + \pi$ are almost identical. The available phase space for $\Sigma \to n + \pi$ is 0.12 GeV kinetic energy, approximately the same as that for $\Delta \to n + \pi$. Nevertheless, their lifetimes differ by about 13 orders of magnitude. We are forced to introduce a new scale α_W and to invent a new "*weak*" *interaction*, arguing in analogy with (1.12) that

$$\frac{\tau(\Delta \to n + \pi)}{\tau(\Sigma \to n + \pi)} \simeq \frac{10^{-23}\,\text{sec}}{10^{-10}\,\text{sec}} \simeq \left(\frac{\alpha_W}{\alpha_s}\right)^2 \tag{1.13}$$

where α_W is the probability for emitting or absorbing some "*weak quantum*" W. Neutron β-decay is then represented by the Feynman diagram shown in Fig.

1.15b. Equation (1.13) implies that

$$\alpha_W \simeq 10^{-6} \tag{1.14}$$

compared to $\alpha_s \simeq 1$ and $\alpha \simeq 10^{-2}$.

Implicit in the above discussion is that the weak field quanta (unlike gluons) couple with equal strength, α_W, to leptons (e, ν) and quarks (u, d) (see Fig. 1.15b). There is direct experimental support for this from the comparison of β-decay with purely leptonic weak processes, such as $\mu^- \to e^- \bar{\nu}_e \nu_\mu$ (compare Figs. 1.15b and 15c). The weak interaction has another novel feature: it changes a d quark into a u quark (Fig. 1.15b) and a muon into a neutrino. We say that weak interactions change quark and lepton "*flavor*." The weak field quanta W in Fig. 1.15 have electric charge, unlike the photon and gluons. In fact, they exist with positive and negative, as well as neutral, electric charge (see Chapters 12 and 13).

We can now summarize the exchanged bosons and the associated charges by which quarks and leptons interact. This is done in Tables 1.3 and 1.4.

Although we had no choice but to introduce a new interaction to explain radioactive β-decay and other "weak" phenomena, this need not imply the existence of a new field with its own *ad hoc* coupling strength α_W. There is a more pleasing and more appropriate (Chapter 15) interpretation for (1.14). One can insist that the coupling of the W in the diagrams of Figs. 1.15b and 1.15c is in fact electromagnetic in strength. The probability of emitting a W is then essentially the same as that for emitting a photon, namely, α. The slow rate of weak decays is achieved instead by giving the W a large mass. The probability of exchanging a W is small compared to that for exchanging a photon, not because it is less likely to be emitted, but because it is massive. Equation (1.14) is then reinterpreted to read

$$\alpha_W = \frac{\alpha}{\left(M_W/m_p\right)^2} \simeq 10^{-6}. \tag{1.15}$$

The qualitative statement expressed by (1.15) is already clear. The large mass M_W suppresses the coupling strength α of the weak bosons to a given "effective" coupling strength α_W. For dimensional reasons, M_W has to be expressed in terms of some reference mass; our choice is the proton mass. That the suppression is quadratic is not obvious at all but depends on the details of the "electroweak" theory we construct. An equation looking like (1.15) is derived in Chapter 15. Since $\alpha \simeq 10^{-2}$, we have the prediction that

$$M_W \simeq 10^2 m_p. \tag{1.16}$$

This result also implies that the weak interactions have a short range. A minimum energy $M_W c^2$ is necessary to emit a virtual W. It can therefore only live a time $\Delta t \lesssim \hbar/M_W c^2$, after which it has to be reabsorbed (remember the discussion of Fig. 1.3). During that time, it can travel at most a distance $c\,\Delta t = \hbar/M_W c \simeq 10^{-3}$ F, using (1.16). This range is much smaller than the 1-F range of a strong interaction. The important observation is, however, that no new charge need

TABLE 1.3
Diagrams Showing Typical Interactions

Interaction	Charge	Quarks	Leptons
Strong	Color	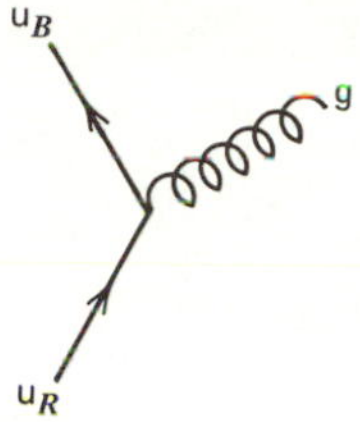 u_B, g, u_R	–
Electromagnctic	Electric charge (e)	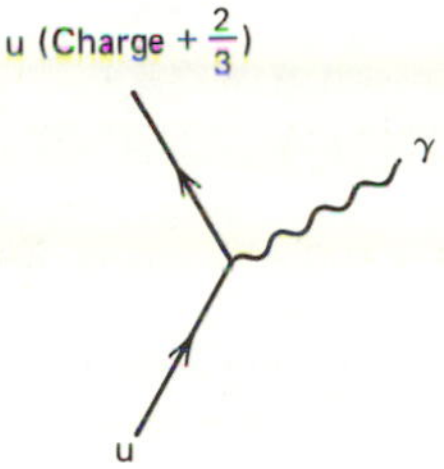 u (Charge $+\frac{2}{3}$), γ, u	e^-, γ, e^-
Weak	Weak charge (g), giving u → d or $\nu \to e^-$ flavor-changing transitions	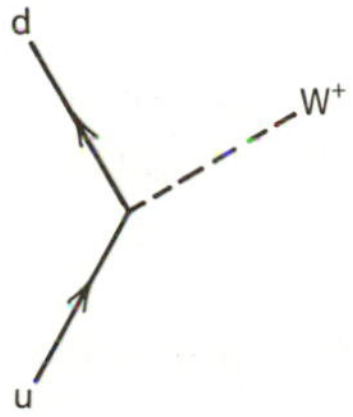 d, W^+, u	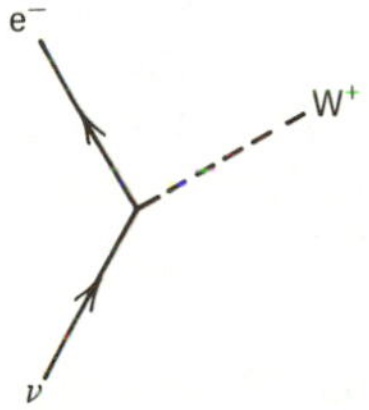 e^-, W^+, ν

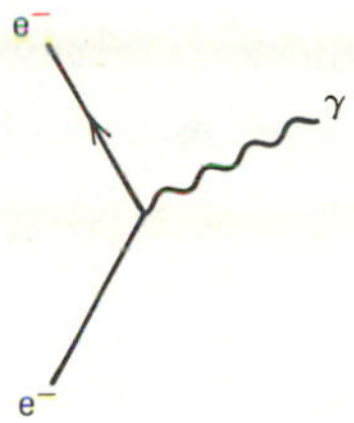

necessarily be introduced to accommodate weak interactions. Only the electric charge appears in (1.15). One might question the conceptual superiority of introducing a new mass scale instead of a new charge. A discussion of *spontaneous symmetry breaking* in Chapters 14 and 15 addresses this question. Of course, no matter how aesthetically satisfying the idea, experiment will be the final arbiter.

The previous discussion should be somewhat familiar. Maxwell's theory of electromagnetism also unifies two interactions: electricity and magnetism. The force on a charged particle moving with velocity v is

$$\mathbf{F} = e\mathbf{E} + e_M \mathbf{v} \times \mathbf{B}.$$

In Maxwell's theory one does not introduce a new charge to accommodate magnetic interactions. It *unifies* the two by stating that $e = e_M$. At low velocities,

TABLE 1.4
Summary of Chapter 1

Interaction	Range	Typical Lifetime (sec)	Typical Cross Section (mb)	Typical Coupling α_i
Strong	$1\ \mathrm{F} \simeq \dfrac{1}{m_\pi}$	10^{-23}	10	1
	Color confinement range[a]	e.g., $\Delta \to p\pi$	e.g., $\pi p \to \pi p$	
Electromagnetic	∞	$10^{-20} \sim 10^{-16}$ e.g., $\pi^0 \to \gamma\gamma$ $\Sigma \to \Lambda\gamma$	10^{-3} e.g., $\gamma p \to p\pi^0$	10^{-2}
Weak	$\dfrac{1}{M_W}$ with	10^{-12} or longer	10^{-11}	10^{-6}
	$M_W \simeq 100 m_p$	e.g., $\Sigma^- \to n\pi^-$ $\pi^- \to \mu^-\bar{\nu}$	e.g., $\nu p \to \nu p$ $\nu p \to \mu^- p \pi^+$	

[a]"van der Waals" manifestation of massless gluon exchange (see end of Section 1.5).

the magnetic forces are very weak; but for high velocities, the electric and magnetic forces play a comparable role. Unification of the two forces introduces a scale in the theory: the velocity of light. The velocity of light, c, is the scale which governs the relative strengths of the two forces. We also introduced a scale when unifying weak and electromagnetic interactions. It is an energy scale, M_W.

1.7 Down Mendeleev's Path: More Quarks and Leptons

The discovery of so-called "*strange*" particles such as the Σ, providing us with a striking example of a weak decay, has a more significant implication. They do not fit into the scheme of colorless qqq and $q\bar{q}$ excited states of u and d quarks. The

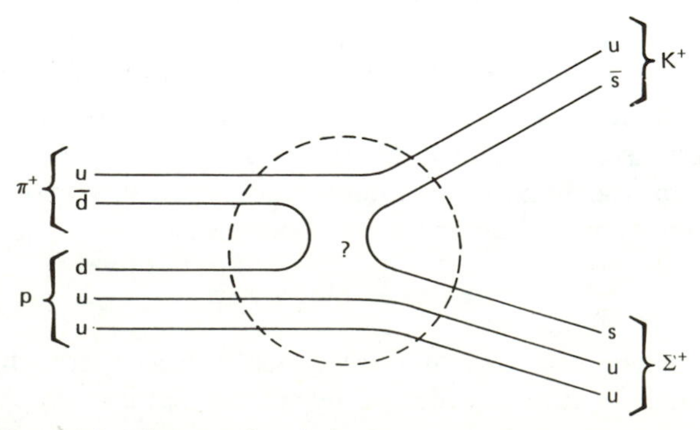

Fig. 1.16 The associated production of strange particles.

TABLE 1.5
Present Proliferation of Quarks and Leptons[a]

Quarks		
u (up)	c (charm)	t (top)?
d (down)	s (strange)	b (beauty)
Leptons		
e (electron)	μ (muon)	τ (tau)
ν_e (e neutrino)	ν_μ (μ-neutrino)	ν_τ (τ-neutrino)

[a] The masses increase from left to right, and no direct evidence for the heaviest entry (the t quark) exists at present.

experimental observation that strange particles are produced in pairs in the strong interactions of nonstrange hadrons is a clear hint that they contain a new quark. Figure 1.16 illustrates this point using the interaction

$$\pi^+ + \mathrm{p} \rightarrow \mathrm{K}^+ + \Sigma^+$$

as an example. The blob in the center represents the complicated (1-F range) color interactions before the pair of strange particles emerges. Turning our attention to Table 1.1, one concludes from inspection of such data that the new strange quark has not only the electric charge (see Fig. 1.16), but also the other quantum numbers of the d quark. The measured masses of the strange particles imply that the s quark is heavier: the $\mathrm{K}(\mathrm{u\bar{s}})$ meson is heavier than the $\pi(\mathrm{u\bar{d}})$, the Σ(uus) is heavier than the p(uud). Known look-alikes of the d quark now also include the even heavier b quark. The u quark has its heavier partner: the c quark. The history of heavier look-alikes to the quarks and leptons of Table 1.1 goes back to the muon, which appears in all respects to be an electron except for its mass, which is 200 times larger. The present evidence is that Table 1.1 should be expanded to Table 1.5.

Nuclei, atoms, and molecules, that is, "our world," is built up out of the first column of Table 1.5. The particles in adjacent columns are identical in all properties. They differ only in mass. Why our world is doubled, tripled,... is one of the major unanswered questions. A sentence from the beginning of this chapter, referring to Mendeleev's table, comes to mind: "Proliferation of elements and the apparent systematics in the organization of the table strongly suggests a substructure of more fundamental building blocks." But with five (six?) quarks and leptons, is it too soon to worry?

1.8 Gravity

It may be surprising that the force most evident in everyday life, the gravitational force, has not been mentioned. The reason is simply that it is by far the weakest force known. As a result, it has no measurable effects on a subatomic scale and no manifestations that can guide us to a quantum field theory. Then why is it so evident? That is because the gravitational force is cumulative, unlike the other long-range interaction, the electromagnetic force, and we live adjacent to an astronomical body: the Earth! Otherwise, gravity would only be apparent to astronomers. Large bodies are usually electrically neutral so that their electromagnetic forces cancel, whereas the gravitational pull of two bodies is the cumulative sum of the attractions between their constituent masses. The reader can however imagine that constructing a gauge theory for gravity has become one of the prime goals of particle physics. The problem is not easy. "Gravitons," unlike the spin 1 quanta in Table 1.1, have spin 2. The discussions are beyond the scope of this book.

1.9 Particles: The Experimentalist's Point of View

If electrons emitted from a vacuum tube, for example, are subjected to electric or magnetic fields, they follow trajectories as given by classical electrodynamics applied to a particle of charge $-e$ and mass m. This is also true for nucleons or α-particles emitted by a radioactive source, even though from the high-energy physicist's point of view they are complicated structures with a spatial extension. The experimental physicist's working definition of a particle is therefore an object to which he can assign a well-defined charge and mass and which behaves, for all practical purposes, like a point particle in the macroscopic electromagnetic fields of his accelerators and detectors.

Suppose we accelerate an electron through a potential difference of 1 volt between a cathode and an anode. If extracted from the vacuum tube "accelerator," it will emerge with an energy of 1 eV. In high-energy physics, we require, however, beams of energy $E \simeq 10^9$ eV, or 1 GeV, so that our "electron microscope" can achieve a spatial resolution of order

$$\begin{aligned}\Delta x &\sim \frac{\hbar c}{E} \\ &\sim \frac{(6.6 \times 10^{-16}\,\text{eV sec}) \times (3 \times 10^{8}\,\text{m/sec})}{10^{9}\,\text{eV}} \\ &\sim 10^{-15}\,\text{m, or 1 fermi.}\end{aligned}$$

Obviously, a vacuum tube will not do! We cannot create sufficiently great potential differences. There are two solutions to this problem: either put a (very) large number of them in series, one perfectly lined up behind the other (linear accelerators), or place a number of them in a circle so that a circulating electron can be repeatedly passed through them and subjected to their acceleration (synchrocyclotrons).

The first solution was adopted for the construction of the 2-mile linear accelerator at the Stanford Linear Accelerator Laboratory where electrons reach energies of up to 20 GeV. Electrons emerging from an ion source are accelerated by a series of radio-frequency (rf) cavities. These contain an oscillating electric field which is parallel to the electron's velocity when it enters each cavity. The electron therefore "rides on the crest of a wave" along the 2-mile accelerator. In order to confine it to a straight line, further electric or magnetic fields are needed to focus the beam, in the same way that lenses guide or focus a beam of light.

The circular accelerator solution was chosen for the construction of synchrotrons at Fermilab near Chicago (U.S.A.) and at CERN in Geneva (Switzerland). These are 2 km in diameter, and protons are accelerated up to 500 GeV. The physical principles involved are familiar from the old-fashioned cyclotron (see Fig. 1.17). In a cyclotron, a proton in a magnetic field B describes an orbit with radius

$$R = \frac{mv}{eB} \tag{1.17}$$

which increases as the velocity v increases. The proton thus spirals (see, e.g., Halliday and Resnick; see also Omnès). At relativistic speeds,

$$m = \gamma m_0 = \frac{m_0}{\sqrt{1 - (v^2/c^2)}}, \tag{1.18}$$

where m_0 is the proton rest mass, and the rotation frequency ω diminishes:

$$\omega = \frac{eB}{m}. \tag{1.19}$$

Remember that ω is a constant at nonrelativistic velocities. To achieve perfect acceleration, it is therefore sufficient to tune the alternating E field in Fig. 1.17 to the fixed frequency ω.

A synchrocyclotron, using a series of accelerating rf cavities placed in a circle, differs from a cyclotron in two ways: (1) the B field, supplied by dipole bending magnets interspaced between the rf cavities, is not constant but is continuously adjusted to force the particles into a fixed orbit; (2) the frequency of the oscillating electric field in the rf cavities is "synchronized" with the changing

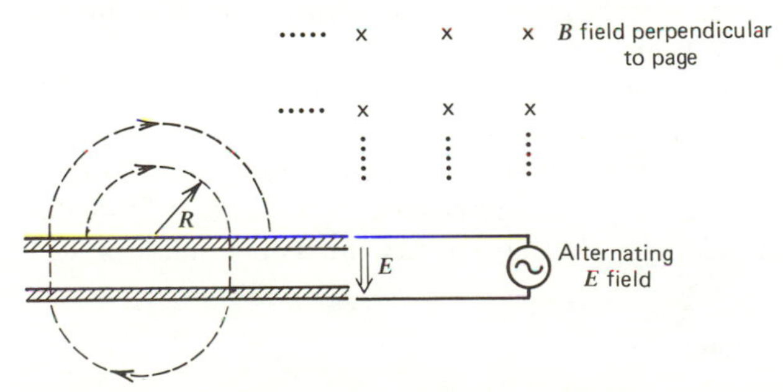

Fig. 1.17 A particle orbiting by means of the Lorentz force is repeatedly accelerated by an electric field E.

rotation frequency ω of the protons given by (1.19) (hence the name synchrocyclotron). Only protons entering the cavity at the right moment are accelerated. Thus, we are forced to accelerate "bunches" of protons rather than a continuous beam. In the CERN machine, 4600 proton bunches rotate with a revolution time of 23 μsec. The machine operates therefore at 200 MHz. To create the necessary 5 million volts per revolution requires a radio-frequency power of 2 MW. During acceleration, the particles circulate roughly 10^5 times around the machine; this takes a total of 2 sec. They cover 500,000 km but are nevertheless kept in their orbit to a precision of 1 mm. To achieve this accuracy, quadrupole focusing magnets are interspaced with the rf cavities and bending magnets.

An increasingly popular technique is to accelerate two beams of particles which have opposite charges inside the same circular accelerator. The beams, circulating in opposite directions, are made to intersect at specific interaction regions, leading to violent head-on collisions (see Exercise 3.3). The products of a collision can be viewed with detectors surrounding the interaction region. Electron–positron colliders with beam energies of up to 20 GeV are in operation at Hamburg (West Germany) and at SLAC (Stanford, California, U.S.A.). The same technique is used in a proton–antiproton collider at CERN where the beams circulate with an energy of 270 GeV. Designs to reach higher energies are under study at CERN and Fermilab, and a new generation of accelerators is forthcoming, with Europe, the United States, Japan, and the Soviet Union all engaged in their construction.

Electron or proton beams can also be extracted from accelerators and directed onto external hydrogen or nuclear targets to study particle–nucleon interactions. Alternatively, we can focus the charged pions or kaons or antiprotons produced in these interactions into secondary beams, which in turn can be used to bombard a nuclear target, and so study their interactions with nucleons.

In a secondary π^+-beam, some particles will decay in flight via $\pi^+ \rightarrow \mu^+ + \nu$. Therefore, the beam becomes "contaminated" with muons and neutrinos. By sending the beam through shielding material, the π^+-component in the beam will be absorbed, because the π^+-particles, unlike muons and neutrinos, interact strongly with the absorber. In so doing, we have constructed a μ^+-beam "contaminated" with neutrinos. By increasing further the amount of absorber, eventually only the weakly interacting neutrinos survive. Our π^+-beam has thus become a neutrino beam!

We should not forget that interactions of cosmic rays with the nitrogen and oxygen nuclei in the atmosphere have been observed with collision energies exceeding those reached in the laboratory by more than five orders of magnitude. The low flux of these high-energy cosmic rays makes a systematic study of the interactions a challenging, but also a very intriguing, mission.

1.10 Particle Detectors

Detection of charged particles is based on a simple physics principle: matter is ionized when traversed by charged particles. The electric field associated with a charged particle moving through matter accelerates the outer electrons of nearby atoms and so ionizes them. The charged particle will thus leave behind a trail of ionized atoms which can be used to infer the trajectory it has followed. A common technique is to place a dielectric, usually a gas, between condensor plates. A potential close to breakdown is applied across the plates. Then, when a charged particle enters the detector, it ionizes the dielectric and the condensor will discharge. This is the principle on which the Geiger counter is based. *Spark*, *streamer*, or *flash* chambers used in high-energy physics experiments are basically a large collection of such counters. By observing the sequence of discharges in the multicomponent detector, we can reconstruct the particle track. The same technique is used in *proportional* or *drift* chambers. They are ionization counters operated in the "sub-Geiger" mode; that is, the voltage is maintained below breakdown of the dielectric. The ionization along the particle's trajectory is observed via electric pulses on the anode wires which collect the electrons that result from the ionization (see Fig. 1.18). The positive ions will drift toward cathode planes and also contribute to the detected current. It is not unusual to construct counters of 5 m × 5 m using large numbers of anode wires and cathode strips.

The *bubble chamber* illustrates an alternative method of displaying the ionization which signals the passage of a charged particle. A similar method is also used in *cloud chambers* and *photographic emulsions*. In a large vessel a few meters in diameter, a liquid is kept under 5 ~ 20 atmospheres of overpressure. Through a sudden decompression, the liquid is superheated and the boiling will start with the formation of bubbles along the particle track. It is the ionized atoms which catalyze the formation of these bubbles. The bubbles are allowed to grow for 10 msec and are then recorded by stereo cameras.

In some materials the atoms are excited, rather than ionized, by high-energy charged particles. The excited levels decay with the emission of light which can be observed by photomultiplier tubes. Such detectors, called *scintillation counters*,

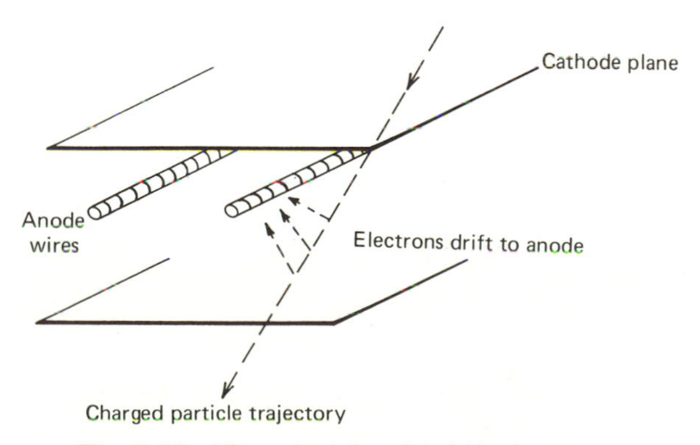

Fig. 1.18 The principle of a drift chamber.

can be built from an organic crystal scintillator, or organic materials where the particle excites the molecular levels.

Detectors exist which can discriminate between π-mesons, K-mesons, and nucleons by measuring their mass. In these experiments, particles move close to the speed of light. Their velocity can nevertheless be measured by sending the particles through a dielectric and observing the Čerenkov light emitted by excited atoms. Čerenkov light is emitted when a particle's velocity exceeds the speed of light in the dielectric medium ($v_{th} = c/n$, where n is the refractive index). The phenomenon is similar to the sonic boom radiated by an aircraft flying faster than the speed of sound in air. This *velocity* threshold can be used to distinguish particles with the same *momentum* but different masses.

Clearly, neutral particles cannot be observed by these techniques. However, we can detect their presence by observing their charged decay products, for example, $\pi^0 \to \gamma\gamma$ followed by $\gamma \to e^- e^+$. The last reaction is forbidden by energy conservation, but it is possible in the electric field of a nucleus $\gamma N \to e^- e^+ N'$. When a photon or electron enters a high-Z material, repeated bremsstrahlung and pair production will create an avalanche of such particles. Their number increases exponentially with depth inside the detector until eventually the cascade has used up the total energy of the incident particle. Such detectors are called *calorimeters*.

A modern particle detector is usually a hybrid system made up from several of the above detectors. It will usually contain a magnet which deflects the particle tracks and allows a determination of their momenta. Its iron will absorb the hadrons and so identify surviving muons. Detectors often consist of over 1000 tons of magnetized iron. A detailed discussion of these multipurpose detectors and of the sophisticated electronics required to collect and digest the information from its various components is outside the scope of this book. For further reading, see, for example, Chapter 2 of Perkins (1982) and Kleinknecht (1982).

2

Symmetries and Quarks

SYMMETRIES AND GROUPS

2.1 Symmetries in Physics: An Example

A glance at a table of particle masses shows that the proton and neutron masses are amazingly close in value. Nuclear physicists took this as a hint that they are in fact two manifestations of one and the same particle called the "nucleon," in very much the same way as the two states of an electron with spin up and down are thought of as one, not two, particles. Indeed, the mathematical structure used to discuss the similarity of the neutron and the proton is almost a carbon copy of spin, and is called "*isospin*" (see Fig. 2.1).

This concept is very useful. As a simple illustration, consider the description of the two-nucleon system. Each nucleon has spin $\frac{1}{2}$ (with spin states $\uparrow$ and $\downarrow$), and so, following the rules for the addition of angular momenta, the composite system may have total spin $S = 1$ or $S = 0$. We are using units with $\hbar = 1$ (see Section 1.4). The composition of these spin triplet and spin singlet states is

$$\begin{cases} |S = 1, M_S = 1\rangle = \uparrow\uparrow \\ |S = 1, M_S = 0\rangle = \sqrt{\tfrac{1}{2}}\,(\uparrow\downarrow + \downarrow\uparrow) \\ |S = 1, M_S = -1\rangle = \downarrow\downarrow \end{cases} \tag{2.1}$$

$$|S = 0, M_S = 0\rangle = \sqrt{\tfrac{1}{2}}\,(\uparrow\downarrow - \downarrow\uparrow).$$

Each nucleon is similarly postulated to have isospin $I = \frac{1}{2}$, with $I_3 = \pm\frac{1}{2}$ for protons and neutrons, respectively. $I = 1$ and $I = 0$ states of the nucleon–nucleon system can be constructed in exact analogy to spin:

$$\begin{cases} |I = 1, I_3 = 1\rangle = \mathrm{pp} \\ |I = 1, I_3 = 0\rangle = \sqrt{\tfrac{1}{2}}\,(\mathrm{pn} + \mathrm{np}) \\ |I = 1, I_3 = -1\rangle = \mathrm{nn} \end{cases} \tag{2.2}$$

$$|I = 0, I_3 = 0\rangle = \sqrt{\tfrac{1}{2}}\,(\mathrm{pn} - \mathrm{np}).$$

EXERCISE 2.1 Justify the decomposition shown in (2.1) by either (1) considering the symmetry of the states under interchange of the nucleons or (2) using the angular momentum "step-down" operator.

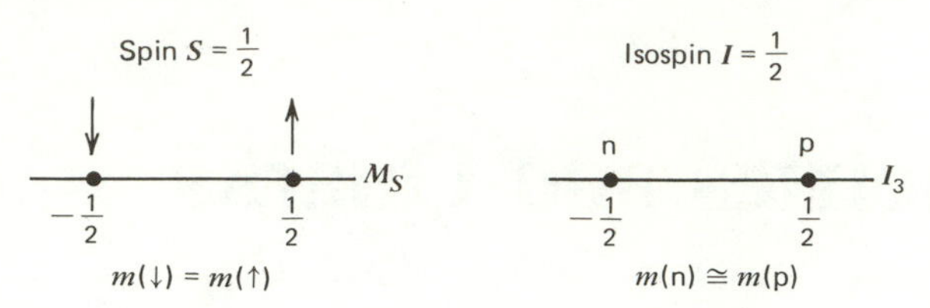

Fig. 2.1 Spin and isospin doublets.

EXERCISE 2.2 If the nucleons are in a state of relative orbital angular momentum $L = 0$, use the Pauli exclusion principle to show that $S + I$ must be an odd integer.

There is much evidence to show that the nuclear force is invariant under isospin transformations [for example, that it is independent of the value of I_3 in the $I = 1$ multiplet of (2.2)]. For instance, consider the three nuclei ^{6}He, ^{6}Li, and ^{6}Be, which can be regarded respectively as an nn, np, and pp system attached to a ^{4}He core of $I = 0$. After correcting for the Coulomb repulsion between the protons and for the neutron–proton mass difference, the observed nuclear masses are as sketched in Fig. 2.2. Furthermore, isospin invariance requires that the same nuclear physics should be obtained for each of the three $I = 1$ states ($I_3 = -1, 0, 1$), just as rotational invariance ensures that the $2J + 1$ substates of an isolated system of total angular momentum J describe exactly equivalent physical systems.

EXERCISE 2.3 Use isospin invariance to show that the reaction cross sections σ must satisfy

$$\frac{\sigma(\text{pp} \to \pi^+\text{d})}{\sigma(\text{np} \to \pi^0\text{d})} = 2,$$

given that the deuteron d has isospin $I = 0$ and the π has isospin $I = 1$.

Hint You may assume that the reaction rate is

$$\sigma \sim |\text{amplitude}|^2 \sim \sum_I \left|\langle I', I_3'|A|I, I_3\rangle\right|^2$$

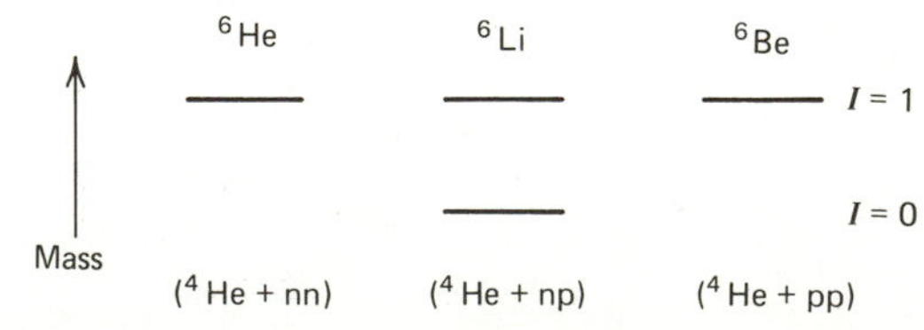

Fig. 2.2 Nuclear energy levels. The states reflect those of eq. (2.2). The two states of ^{6}Li are separated by about 2 MeV. After correcting for electromagnetic effects, the excited state of ^{6}Li and the ground states of ^{6}He and ^{6}Be are found to be degenerate in mass.

where I and I' are the total isospin quantum numbers of the initial and final states, respectively, and $I = I'$ and $I_3 = I_3'$.

In Sections 2.2 to 2.9, we amplify these ideas and take the opportunity to introduce some useful group theory concepts. You may prefer to skip these sections and, instead, refer to the appropriate results as and when necessary.

2.2 Symmetries and Groups: A Brief Introduction

Group theory is the branch of mathematics that underlies the treatment of symmetry. Although we shall not need the formal machinery of group theory, it is useful to introduce some of the concepts and the terminology which belongs to the jargon of particle physics. We take the rotation group as an illustrative example.

The set of rotations of a system form a group, each rotation being an element of the group. Two successive rotations R_1 followed by R_2 (written as the "product" $R_2 R_1$) are equivalent to a single rotation (that is, to another group element). The set of rotations is closed under "multiplication." There is an identity element (no rotation), and every rotation has an inverse (rotate back again). The "product" is not necessarily commutative, $R_1 R_2 \neq R_2 R_1$, but the associative law $R_3(R_2 R_1) = (R_3 R_2) R_1$ always holds. The rotation group is a continuous group in that each rotation can be labeled by a set of continuously varying parameters $(\alpha_1, \alpha_2, \alpha_3)$. These can be regarded as the components of a vector $\boldsymbol{\alpha}$ directed along the axis of rotation with magnitude given by the angle of rotation.

The rotation group is a Lie group. The crucial property here is that every rotation can be expressed as the product of a succession of infinitesimal rotations (rotations arbitrarily close to the identity). The group is then completely defined by the "neighborhood of the identity."

We do not want an experimental result to depend on the specific laboratory orientation of the system we are measuring. Rotations must therefore form a symmetry group of a system. They are a subset of the Lorentz transformations that can be performed on a system, namely, those transformations that leave it at rest. By definition, the physics is unchanged by a symmetry operation. In particular, these operations leave the transition probabilities of the system invariant. For example, suppose that under a rotation R the states of a system transform as

$$|\psi\rangle \rightarrow |\psi'\rangle = U|\psi\rangle. \tag{2.3}$$

The probability that a system described by $|\psi\rangle$ will be found in state $|\phi\rangle$ must be unchanged by R,

$$|\langle\phi|\psi\rangle|^2 = |\langle\phi'|\psi'\rangle|^2 = |\langle\phi|U^\dagger U|\psi\rangle|^2, \tag{2.4}$$

and so U must be a unitary operator. The operators $U(R_1)$, $U(R_2), \ldots,$ form a group with exactly the same structure as the original group $R_1, R_2, \ldots$; they are said to form a unitary representation of the rotation group.

Moreover, the Hamiltonian is unchanged by a symmetry operation R of the system, and the matrix elements are preserved:

$$\langle \phi' | H | \psi' \rangle = \langle \phi | U^\dagger H U | \psi \rangle = \langle \phi | H | \psi \rangle,$$

so that

$$H = U^\dagger H U \qquad \text{or} \qquad [U, H] \equiv UH - HU = 0. \tag{2.5}$$

The transformation U has no explicit time dependence, and the equation of motion,

$$i \frac{d}{dt} | \psi(t) \rangle = H | \psi(t) \rangle, \tag{2.6}$$

is unchanged by the symmetry operation. As a consequence, the expectation value of U is a constant of the motion:

$$i \frac{d}{dt} \langle \psi(t) | U | \psi(t) \rangle = \langle \psi(t) | UH - HU | \psi(t) \rangle = 0. \tag{2.7}$$

All the group properties follow from considering infinitesimal rotations in the neighborhood of the identity. As an example, consider a rotation through an infinitesimal angle ε about the 3- (or z) axis. We may write, to first order in ε,

$$U = 1 - i\varepsilon J_3. \tag{2.8}$$

The operator J_3 is called the generator of rotations about the 3-axis. Now,

$$\begin{aligned} 1 = U^\dagger U &= \left(1 + i\varepsilon J_3^\dagger\right)\left(1 - i\varepsilon J_3\right) \\ &= 1 + i\varepsilon\left(J_3^\dagger - J_3\right) + 0(\varepsilon^2). \end{aligned}$$

Therefore, J_3 is hermitian and hence is a (quantum mechanical) observable. The i was introduced in (2.8) to make this so.

To identify the observable J_3, we consider the effect of a rotation on the wave function $\psi(\mathbf{r})$ describing the system. First, we must distinguish between two points of view. Either we may rotate the axes and keep the physical system fixed (the passive viewpoint) or we can keep the axes fixed and rotate the system (the active viewpoint). The viewpoints are equivalent; a rotation of the axes through an angle θ is the same as a rotation of the physical system by $-\theta$. We adopt the active viewpoint and rotate the physical system. The wave function ψ' describing the rotated state at $\mathbf{r}$ is then equal to the original function ψ at the point $R^{-1}\mathbf{r}$, which is transformed into $\mathbf{r}$ under the rotation R, that is,

$$\psi'(\mathbf{r}) = \psi(R^{-1}\mathbf{r}). \tag{2.9}$$

This specifies the one-to-one correspondence between ψ' and ψ, which we have written [cf. (2.3)]

$$\psi' = U\psi. \tag{2.10}$$

For an infinitesimal rotation ε about the z axis, (2.9) and (2.10) give

$$\begin{aligned} U\psi(x, y, z) &= \psi(R^{-1}\mathbf{r}) \simeq \psi(x + \varepsilon y, y - \varepsilon x, z) \\ &\simeq \psi(x, y, z) + \varepsilon\left(y\frac{\partial\psi}{\partial x} - x\frac{\partial\psi}{\partial y}\right) \\ &= \left(1 - i\varepsilon(xp_y - yp_x)\right)\psi. \end{aligned} \tag{2.11}$$

Comparing (2.11) with (2.8), that is, with

$$U\psi = (1 - i\varepsilon J_3)\psi,$$

we identify the *generator*, J_3, of rotations about the 3- (or z) axis with the third-component of the *angular momentum operator*.

From (2.7), we see that the eigenvalues of the observable J_3 are constants of the motion. They are conserved quantum numbers. A symmetry of the system has led to a conservation law. The fact that experiments performed with different orientations of an apparatus give the same physics results (rotational symmetry) has led to the conservation of angular momentum.

A rotation through a finite angle θ may be built up from a succession of n infinitesimal rotations

$$U(\theta) = (U(\varepsilon))^n = \left(1 - i\frac{\theta}{n}J_3\right)^n \underset{n\to\infty}{\rightarrow} e^{-i\theta J_3}. \tag{2.12}$$

We may introduce similar hermitian generators of rotations about the 1- and 2-axes, J_1 and J_2, respectively. The commutator algebra of the generators is (see Exercise 2.4)

$$[J_j, J_k] = i\varepsilon_{jkl}J_l \tag{2.13}$$

where $\varepsilon_{jkl} = +1(-1)$ if jkl are a cyclic (anticyclic) permutation of 1 2 3 and $\varepsilon_{jkl} = 0$ otherwise. Relations (2.13) completely define the group properties; the ε_{jkl} coefficients are called the structure constants of the group. The J's are said to form a Lie algebra. Since no two J's commute with each other, only the eigenvalues of one generator, say J_3, are useful quantum numbers.

EXERCISE 2.4 Show that the four successive infinitesimal rotations (ε about the 1-axis, followed by η about the 2-axis, then $-\varepsilon$ about the 1-axis, and finally $-\eta$ about the 2-axis) are equivalent to the second-order rotation $\varepsilon\eta$ about the 3-axis. Hence, show that the generators satisfy

$$[J_1, J_2] = iJ_3.$$

Nonlinear functions of the generators which commute with all the generators are called invariants or Casimir operators. For the rotation group,

$$J^2 = J_1^2 + J_2^2 + J_3^2 \tag{2.14}$$

is the only Casimir operator,

$$[J^2, J_i] = 0 \qquad \text{with } i = 1, 2, 3. \tag{2.15}$$

It follows that we can construct simultaneous eigenstates $|jm\rangle$ of J^2 and one of the generators, say J_3. Using only (2.13), it is possible to show that

$$\begin{aligned} J^2|jm\rangle &= j(j+1)|jm\rangle \\ J_3|jm\rangle &= m|jm\rangle \end{aligned} \tag{2.16}$$

with $m = -j, -j+1, \ldots, j$, and where j can take one of the values $0, \frac{1}{2}, 1, \frac{3}{2}, \ldots$.

EXERCISE 2.5 Verify (2.16). To do this, it is useful to form the so-called "step-up" and "step-down" operators

$$J_{\pm} = J_1 \pm iJ_2. \tag{2.17}$$

First, show that

$$J_{\pm}|jm\rangle = (C - m(m \pm 1))^{1/2}|j, m \pm 1\rangle, \tag{2.18}$$

that is, $J_{\pm}$ step m up and down by one unit, respectively. Show that $C = j(j+1)$.

A state $|jm\rangle$ is transformed under a rotation through an angle θ about the 2-axis into a linear combination of the $2j+1$ states $|jm'\rangle$, with $m' = -j, -j+1, \ldots, j$:

$$e^{-i\theta J_2}|jm\rangle = \sum_{m'} d^j_{m'm}(\theta)|jm'\rangle, \tag{2.19}$$

where the coefficients $d^j_{m'm}$ are written in conventional notation and are frequently called rotation matrices. From (2.19), we see the states having the same j but all possible m values transform among themselves under rotations. In fact, all the $2j+1$ states are mixed by rotations. They form the basis of a $(2j+1)$-dimensional irreducible representation of the rotation group. The set of states is called a multiplet.

EXERCISE 2.6 Show that the rotation matrices

$$d^j_{m'm}(\theta) = \langle jm'|e^{-i\theta J_2}|jm\rangle$$

for $j = \frac{1}{2}$ and $j = 1$ are

$$j = \tfrac{1}{2} \begin{cases} d_{++} = d_{--} = \cos\frac{1}{2}\theta \\ d_{-+} = -d_{+-} = \sin\frac{1}{2}\theta \end{cases} \tag{2.20}$$

where $\pm$ denote $m = \pm\frac{1}{2}$, respectively, and

$$j = 1 \begin{cases} d_{01} = -d_{10} = -d_{0-1} = d_{-10} = \sqrt{\frac{1}{2}}\sin\theta \\ d_{11} = d_{-1-1} = \frac{1}{2}(1 + \cos\theta) \\ d_{-11} = d_{1-1} = \frac{1}{2}(1 - \cos\theta) \\ d_{00} = \cos\theta. \end{cases} \tag{2.21}$$

2.3 The Group $SU(2)$

In the lowest-dimension nontrivial representation of the rotation group ($j = \frac{1}{2}$), the generators may be written

$$J_i = \tfrac{1}{2}\sigma_i \qquad \text{with } i = 1, 2, 3, \tag{2.22}$$

where σ_i are the Pauli matrices

$$\sigma_1 = \begin{pmatrix} 0 & 1 \\ 1 & 0 \end{pmatrix}, \qquad \sigma_2 = \begin{pmatrix} 0 & -i \\ i & 0 \end{pmatrix}, \qquad \sigma_3 = \begin{pmatrix} 1 & 0 \\ 0 & -1 \end{pmatrix}. \tag{2.23}$$

The basis (or set of base states) for this representation is conventionally chosen to be the eigenvectors of σ_3, that is, the column vectors

$$\begin{pmatrix} 1 \\ 0 \end{pmatrix} \quad \text{and} \quad \begin{pmatrix} 0 \\ 1 \end{pmatrix}$$

describing a spin-$\frac{1}{2}$ particle of spin projection up ($m = +\frac{1}{2}$ or $\uparrow$) and spin projection down ($m = -\frac{1}{2}$ or $\downarrow$) along the 3-axis, respectively.

The Pauli matrices σ_i are hermitian, and the transformation matrices

$$U(\theta_i) = e^{-i\theta_i\sigma_i/2} \tag{2.24}$$

are unitary. The set of all unitary 2×2 matrices is known as the group $U(2)$. However, $U(2)$ is larger than the group of matrices $U(\theta_i)$, since the generators σ_i all have zero trace. Now, for any hermitian traceless matrix σ, we can show that

$$\det(e^{i\sigma}) = e^{iTr(\sigma)} = 1. \tag{2.25}$$

Since the unit determinant is preserved in matrix multiplication, the set of traceless unitary 2×2 matrices form a subgroup, $SU(2)$, of $U(2)$. $SU(2)$ denotes the special unitary group in two dimensions. The set of transformation matrices $U(\theta_i)$ therefore form an $SU(2)$ group. The $SU(2)$ algebra is just the algebra of the generators J_i, relations (2.13). There are thus $1, 2, 3, 4, \ldots$ dimensional representations of $SU(2)$ corresponding to $j = 0, \frac{1}{2}, 1, \frac{3}{2}, \ldots$, respectively. The two-dimensional representation is, of course, just the σ-matrices themselves. It is called the fundamental representation of $SU(2)$, the representation from which all other representations can be built, as we will now show.

EXERCISE 2.7 Show that the rotation of a spin-$\frac{1}{2}$ system through a finite angle θ about the 2-axis corresponds to the unitary transformation

$$e^{-i\theta\sigma_2/2} = \cos\frac{\theta}{2} - i\sigma_2 \sin\frac{\theta}{2}. \tag{2.26}$$

2.4 Combining Representations

A composite system formed from two systems having angular momentum j_A and j_B may be described in terms of the basis

$$|j_A j_B m_A m_B\rangle \equiv |j_A m_A\rangle |j_B m_B\rangle.$$

However, the combined operator

$$\mathbf{J} = \mathbf{J}_A + \mathbf{J}_B \tag{2.27}$$

also satisfies the Lie algebra of (2.13), and it is the eigenvalues $J(J+1)$, M of J^2, J_3 which are the conserved quantum numbers. In fact, the "product" of the two irreducible representations of dimension $2j_A + 1$ and $2j_B + 1$ may be decomposed into the sum of irreducible representations of dimension $2J + 1$ with

$$J = |j_A - j_B|, |j_A - j_B| + 1, \ldots, j_A + j_B, \tag{2.28}$$

with basis $|j_A j_B J M\rangle$, where

$$M = m_A + m_B. \tag{2.29}$$

The last equality follows directly from the third component of (2.27). One basis may be expressed in terms of the other by

$$|j_A j_B J M\rangle = \sum_{m_A, m_B} C(m_A m_B; JM) |j_A j_B m_A m_B\rangle, \tag{2.30}$$

where the coefficients C are called Clebsch–Gordan coefficients; they have been tabulated, for example, in the *Review of Particle Properties* (1982). These coefficients are readily calculated by repeatedly applying the step-down operator [cf. (2.17)]

$$J_- = (J_A)_- + (J_B)_-$$

to the "fully stretched" state

$$|j_A j_B J, M = J\rangle = |j_A j_B, m_A = j_A, m_B = j_B\rangle$$

and using orthogonality when necessary.

Equation (2.1) is a simple example of (2.30). The composite system of two spin-$\frac{1}{2}$ particles $j_A = j_B = \frac{1}{2}$ may have spin $J = 1$ or 0. We may write the decomposition symbolically as

$$2 \otimes 2 = 3 \oplus 1, \tag{2.31}$$

using the dimensions (that is, the size of the multiplet) to label the irreducible representations. We can readily extend this procedure. Combining a third spin-$\frac{1}{2}$

particle, we have

$$(2 \otimes 2) \otimes 2 = (3 \otimes 2) \oplus (1 \otimes 2)$$
$$= 4 \oplus 2 \oplus 2. \tag{2.32}$$

That is, three spin-$\frac{1}{2}$ particles group together into a quartet of spin $\frac{3}{2}$ and two doublets of spin $\frac{1}{2}$.

2.5 Finite Symmetry Groups: *P* and *C*

A finite group is one which contains only a finite number of elements. In particle physics, we encounter a very simple symmetry group containing just two elements, the identity e and an element g satisfying $g^2 = e$. For example, g may be space inversion or particle–antiparticle conjugation. Invariance of the physics under g means that g is represented by a unitary (or antiunitary) operator $U(g)$ which satisfies [see (2.5)]

$$[U, H] = 0, \tag{2.33}$$

Time-reversal invariance is the only symmetry requiring an antiunitary operator [see, for example, Schiff (1968), Martin and Spearman (1970), and Messiah (1962)], and so here we take U to be unitary. For our two-element group, we have

$$U^2 = 1, \tag{2.34}$$

and, since U is unitary, it must also be hermitian. Thus, U itself is an observable conserved quantity [cf. (2.7)], and its eigenvalues are conserved quantum numbers. If p is an eigenvalue of U corresponding to eigenvector $|p\rangle$, then

$$U^2|p\rangle = p^2|p\rangle. \tag{2.35}$$

From (2.34), $p^2 = 1$, and so the allowed eigenvalues are $p = \pm 1$. Invariance of the system under the symmetry operation g (for example, space inversion or particle–antiparticle conjugation) means that if the system is in an eigenstate of U (with $U = P$ or C), then transitions can only occur to eigenstates with the same eigenvalue. We see that the eigenvalues of U are *multiplicative* quantum numbers. By contrast, the eigenvalues of the commuting generators of $SU(n)$ are *additive* quantum numbers.

Strong and electromagnetic interactions are invariant under both P and C, whereas weak interactions do not respect these symmetries. However, to a good approximation, weak interactions are invariant under the product transformation CP (see Chapter 12).

2.6 *SU*(2) of Isospin

Isospin arises because the nucleon may be viewed as having an internal degree of freedom with two allowed states, the proton and the neutron, which the nuclear interaction does not distinguish. We therefore have an $SU(2)$ symmetry in which the (n, p) form the fundamental representation. It is a mathematical copy of spin in that the isospin generators satisfy

$$\left[I_j, I_k\right] = i\varepsilon_{jkl}I_l \tag{2.36}$$

[cf. (2.13)]. In the fundamental representation, the generators are denoted $I_i \equiv \frac{1}{2}\tau_i$, where

$$\tau_1 = \begin{pmatrix} 0 & 1 \\ 1 & 0 \end{pmatrix}, \quad \tau_2 = \begin{pmatrix} 0 & -i \\ i & 0 \end{pmatrix}, \quad \tau_3 = \begin{pmatrix} 1 & 0 \\ 0 & -1 \end{pmatrix} \tag{2.37}$$

are the isospin version of the Pauli matrices (2.23). They act on the proton and neutron states represented by

$$\mathrm{p} = \begin{pmatrix} 1 \\ 0 \end{pmatrix}, \quad \mathrm{n} = \begin{pmatrix} 0 \\ 1 \end{pmatrix}.$$

In general, the most positively charged particle is chosen to have the maximum value of I_3.

2.7 Isospin for Antiparticles

The construction of antiparticle isospin multiplets requires care. It is well illustrated by a simple example. Consider a particular isospin transformation of the nucleon doublet, a rotation through π about the 2-axis. We obtain [see (2.26)]

$$\begin{pmatrix} \mathrm{p}' \\ \mathrm{n}' \end{pmatrix} = e^{-i\pi(\tau_2/2)}\begin{pmatrix} \mathrm{p} \\ \mathrm{n} \end{pmatrix} = -i\tau_2\begin{pmatrix} \mathrm{p} \\ \mathrm{n} \end{pmatrix} = \begin{pmatrix} 0 & -1 \\ 1 & 0 \end{pmatrix}\begin{pmatrix} \mathrm{p} \\ \mathrm{n} \end{pmatrix}. \tag{2.38}$$

We define antinucleon states using the particle–antiparticle conjugation operator C,

$$C\mathrm{p} = \bar{\mathrm{p}}, \qquad C\mathrm{n} = \bar{\mathrm{n}}. \tag{2.39}$$

Applying C to (2.38) therefore gives

$$\begin{pmatrix} \bar{\mathrm{p}}' \\ \bar{\mathrm{n}}' \end{pmatrix} = \begin{pmatrix} 0 & -1 \\ 1 & 0 \end{pmatrix}\begin{pmatrix} \bar{\mathrm{p}} \\ \bar{\mathrm{n}} \end{pmatrix}. \tag{2.40}$$

However, we want the antiparticle doublet to transform in exactly the same way as the particle doublet, so that we can combine particle and antiparticle states using the same Clebsch–Gordan coefficients, and so on. We must therefore make two changes. First, we must reorder the doublet so that the most positively charged particle has $I_3 = +\frac{1}{2}$, and then we must introduce a minus sign to keep the matrix transformation identical to (2.38). We obtain

$$\begin{pmatrix} -\bar{\mathrm{n}}' \\ \bar{\mathrm{p}}' \end{pmatrix} = \begin{pmatrix} 0 & -1 \\ 1 & 0 \end{pmatrix}\begin{pmatrix} -\bar{\mathrm{n}} \\ \bar{\mathrm{p}} \end{pmatrix}. \tag{2.41}$$

That is, the antiparticle doublet $(-\bar{\mathrm{n}}, \bar{\mathrm{p}})$ transforms exactly as the particle doublet (p, n). This is a special property of $SU(2)$; it is not possible, for example, to arrange an $SU(3)$ triplet of antiparticles so that it transforms as the particle triplet.

A composite system of a nucleon–antinucleon pair has isospin states [compare (2.2)]:

$$\begin{cases} |I=1, I_3=1\rangle = -\mathrm{p}\bar{\mathrm{n}} \\ |I=1, I_3=0\rangle = \sqrt{\tfrac{1}{2}}\,(\mathrm{p}\bar{\mathrm{p}} - \mathrm{n}\bar{\mathrm{n}}) \\ |I=1, I_3=-1\rangle = \mathrm{n}\bar{\mathrm{p}} \end{cases} \tag{2.42}$$

$$|I=0, I_3=0\rangle = \sqrt{\tfrac{1}{2}}\,(\mathrm{p}\bar{\mathrm{p}} + \mathrm{n}\bar{\mathrm{n}}).$$

2.8 The Group $SU(3)$

The set of unitary 3×3 matrices with $\det U = 1$ form the group $SU(3)$. The generators may be taken to be any $3^2 - 1 = 8$ linearly independent traceless hermitian 3×3 matrices. Since it is possible to have only two of these traceless matrices diagonal, this is the maximum number of mutually commuting generators. This number is called the rank of the group, so that $SU(3)$ has rank 2 and $SU(2)$ has rank 1. It can be shown that the number of Casimir operators is equal to the rank of the group.

The fundamental representation of $SU(3)$ is a triplet. The three color charges of a quark, R, G, and B of Section 1.2, form the fundamental representation of an $SU(3)$ symmetry group. In this representation, the generators are 3×3 matrices. They are traditionally denoted λ_i, with $i = 1, \ldots, 8$, and the diagonal matrices are taken to be

$$\lambda_3 = \begin{pmatrix} 1 & & \\ & -1 & \\ & & 0 \end{pmatrix}, \qquad \lambda_8 = \sqrt{\tfrac{1}{3}} \begin{pmatrix} 1 & & \\ & 1 & \\ & & -2 \end{pmatrix} \tag{2.43}$$

with simultaneous eigenvectors

$$\mathrm{R} = \begin{pmatrix} 1 \\ 0 \\ 0 \end{pmatrix}, \qquad \mathrm{G} = \begin{pmatrix} 0 \\ 1 \\ 0 \end{pmatrix}, \qquad \mathrm{B} = \begin{pmatrix} 0 \\ 0 \\ 1 \end{pmatrix}.$$

These base states R, G, B are plotted in Fig. 2.3 in terms of their λ_3, λ_8 eigenvalues. The figure also shows how the remaining six generators give the analogues of the "step-up" and "step-down" operators of $SU(2)$. With this numbering of the λ_i matrices, $\lambda_1, \lambda_2, \lambda_3$ correspond to the three Pauli matrices and thus they exhibit explicitly one $SU(2)$ subgroup of $SU(3)$. The λ_i are known as the Gell-Mann matrices.

EXERCISE 2.8 Obtain the matrix representations of the λ_i of Fig. 2.3. Show that

$$\left[\frac{\lambda_i}{2}, \frac{\lambda_j}{2}\right] = i \sum_k f_{ijk} \frac{\lambda_k}{2} \tag{2.44}$$

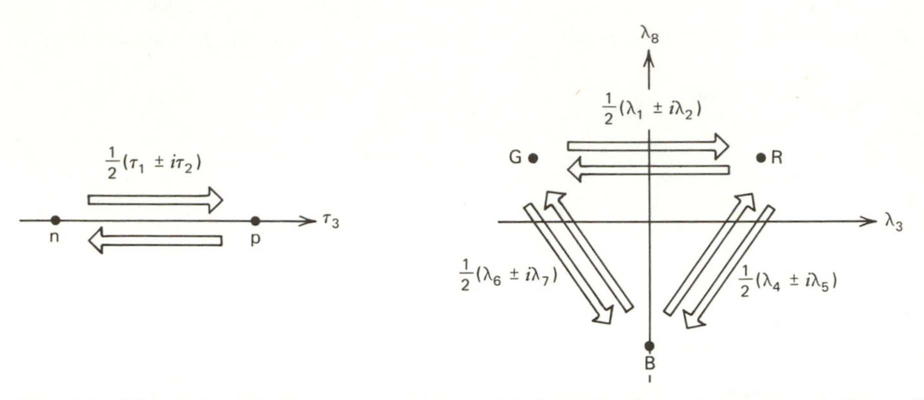

Fig. 2.3 The action of the generators (τ_i and λ_i) on fundamental representations of $SU(2)$ of isospin and $SU(3)$ of color, respectively.

where the $SU(3)$ structure constants f_{ijk} are fully antisymmetric under interchange of any pair of indices (see exercise 14.9) and the nonvanishing values are permutations of

$$f_{123} = 1, \qquad f_{458} = f_{678} = \sqrt{3}/2,$$

$$f_{147} = f_{165} = f_{246} = f_{257} = f_{345} = f_{376} = \tfrac{1}{2}.$$

2.9 Another Example of an $SU(3)$ Group: Isospin and Strangeness

In 1947, the pion was discovered, and from that date on, the nucleon lost its unique role in particle physics. Subsequently, many more strongly interacting particles (hadrons) have been identified. Some of the new particles were surprisingly long-lived on the time scale of strong interactions, despite being massive enough to decay into lighter objects without violating the conservation of charge or baryon number. For instance, a Σ^- is readily produced by the strong interaction $\pi^- p \rightarrow K^+ \Sigma^-$ and yet decays only weakly via $\Sigma^- \rightarrow n\pi^-$. The contrast to a typical strong decay, $\Delta \rightarrow n\pi$, was emphasized in Section 1.6. Gell-Mann and, independently, Nishijima, took this as a manifestation of a new additive quantum number, which was called "strangeness," S. They assigned to each hadron an integer value of strangeness,

$$\begin{aligned} S = 0: &\quad \pi, N, \Delta, \ldots, \\ S = 1: &\quad K^+, \ldots, \\ S = -1: &\quad \Lambda, \Sigma, \ldots. \end{aligned} \tag{2.45}$$

with $-S$ for their antiparticles, and they asserted that strong and electromagnetic interactions are forbidden unless S is conserved by the reaction. Gell-Mann and Nishijima's proposal immediately accounts for the strong production and the weak decay of the Σ. Indeed, in the reaction $\pi^- p \to K^+ \Sigma^-$ both the initial and final states have a total strangeness $S = 0$ [see (2.45)]. The Σ-particle can therefore be produced by the strong interaction. It could also decay via the strong interaction $\Sigma^- \to \Lambda\pi^-$ if it were not for the fact that the Λ is too heavy so that the strangeness conserving decay is kinematically forbidden. The Σ^- can only decay by the strangeness-violating weak interaction $\Sigma^- \to n\pi^-$, thus "explaining" its long lifetime (see Section 1.6). The Gell-Mann and Nishijima scheme was confirmed by observations of the properties of the large number of strange particles that were subsequently discovered.

With the existence of a second additive quantum number S, in addition to I_3, it was natural to attempt to enlarge isospin symmetry to a larger group, namely, a group of rank 2. This new symmetry group had to naturally fit the hadrons with similar properties into its multiplet representations. This task was relatively easy for the $SU(2)$ group of isospin: the neutron and proton, which are almost identical in mass, are nicely accommodated in an $SU(2)$ doublet. However, no strange particles exist that are close in mass to the nucleon, so the appropriate grouping is difficult to identify and the choice of group far from obvious. $SU(3)$ was originally proposed in 1961: we shall see (Fig. 2.8) that it groups the $n, p, \Sigma^+, \Sigma^0, \Sigma^-, \Lambda, \Xi^0$, and Ξ^-, with a mass spread of nearly 400 MeV, into an $SU(3)$ octet representation. Moreover, the lightest mesons are also fitted into an octet, with the K-meson belonging to the same representation as the much lighter π ($m_K > 3m_\pi$!). It is clear that the extra symmetry linking strange and non-strange particles is much more approximate than is isospin. Not until 1964 was $SU(3)$ symmetry firmly established. The $SU(3)$ multiplet structure of the so-called "elementary" particles was reminiscent of the grouping of chemical elements in Mendeleev's table. Like the periodic table, the $SU(3)$ classification strongly hinted at the existence of a substructure. The role of the $SU(3)$ group of isospin and strangeness is mostly historical: it set the scene for the entry of quarks into particle physics.

With hindsight, we now realize that the success of $SU(2)$ isospin symmetry is due to the essentially equal mass of the u, d quark constituents. However, $SU(3)$ incorporating the heavier s quark is not such a good symmetry; rather, we use it to enumerate the hadronic states. We call this "flavor $SU(3)$," u, d, s being the three lightest flavors of quark. It is completely unrelated to "color $SU(3)$," which is believed to be an exact symmetry of fundamental origin (see Chapter 14).

QUARK "ATOMS"

According to the quark model, all hadrons are made up of a small variety of more basic entities, called quarks, bound together in different ways. The fundamental representation of $SU(3)$, the multiplet from which all other multiplets can be built, is a triplet. This basic quark multiplet is given in Fig. 2.4; also shown is the antiquark multiplet in which the signs of the additive quantum numbers are

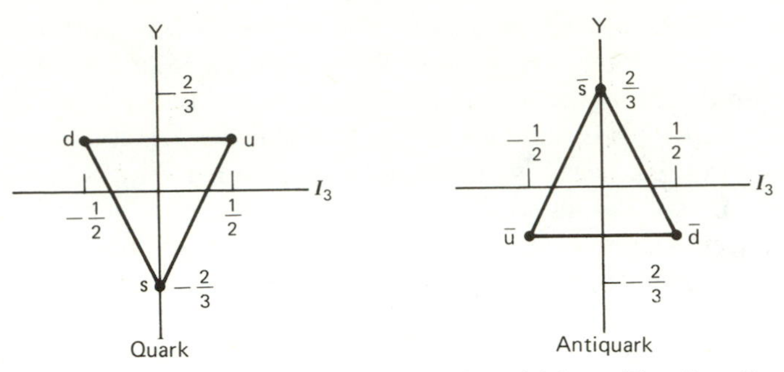

Fig. 2.4 $SU(3)$ quark and antiquark multiplets; $Y \equiv B + S$.

reversed. Each quark is assigned spin $\frac{1}{2}$ and baryon number $B = \frac{1}{3}$. Baryons are made of three quarks (qqq) and the mesons of a quark–antiquark pair ($q\bar{q}$). In Fig. 2.4, the new additive quantum number is shown as the "hypercharge,"

$$Y \equiv B + S, \tag{2.46}$$

rather than the strangeness S. This choice has no physical significance; it simply centers the multiplets on the origin. The charge, Qe, is

$$Q = I_3 + \frac{Y}{2}.$$

The quantum numbers of the quarks are listed in Table 2.1. Baryon conservation means it is impossible to destroy or to make a single quark, but we can annihilate or create a quark–antiquark pair (a meson). Moreover, quarks retain their identity under strong or electromagnetic transitions; that is, transmutations such as s → u + leptons, s → u + $d\bar{u}$ occur only by weak interactions.

2.10 Quark–Antiquark States: Mesons

In the quark model, mesons are made of a quark and an antiquark bound together. Let us start with two flavors, q = u or d. The $q\bar{q}$ bound-state wave functions are readily obtained by making the substitutions p → u and n → d in

TABLE 2.1
Quantum Numbers of the Quarks ($Y = B + S$, $Q = I_3 + Y/2$)[a]

Quark	Spin	B	Q	I_3	S	Y
u	$\frac{1}{2}$	$\frac{1}{3}$	$\frac{2}{3}$	$\frac{1}{2}$	0	$\frac{1}{3}$
d	$\frac{1}{2}$	$\frac{1}{3}$	$-\frac{1}{3}$	$-\frac{1}{2}$	0	$\frac{1}{3}$
s	$\frac{1}{2}$	$\frac{1}{3}$	$-\frac{1}{3}$	0	-1	$-\frac{2}{3}$

[a] Here, S denotes the strangeness.

(2.42). We thus obtain an isotriplet and an isosinglet of mesons

$$\begin{cases} |I = 1, I_3 = 1\rangle = -\mathrm{u\bar{d}} \\ |I = 1, I_3 = 0\rangle = \sqrt{\tfrac{1}{2}}\,(\mathrm{u\bar{u}} - \mathrm{d\bar{d}}) \\ |I = 1, I_3 = -1\rangle = \mathrm{d\bar{u}}, \end{cases} \tag{2.47}$$

$$|I = 0, I_3 = 0\rangle = \sqrt{\tfrac{1}{2}}\,(\mathrm{u\bar{u}} + \mathrm{d\bar{d}}).$$

For three flavors of quarks, q = u, d, or s, there are nine possible $\mathrm{q\bar{q}}$ combinations. The resulting multiplet structure is shown in Fig. 2.5b and is readily obtained by superimposing the center of gravity of the antiquark multiplet on top of every site of the quark multiplet, Fig. 2.5a. The nine states divide into an $SU(3)$ octet and an $SU(3)$ singlet; that is, under operations of the $SU(3)$ group, the eight states transform among themselves but do not mix with the singlet state. This is the extension to $SU(3)$ of the familiar separation (2.31) or (2.47) of $SU(2)$.

Of the nine $\mathrm{q\bar{q}}$ states, we note that three, labeled A, B, and C on Fig. 2.5, have $I_3 = Y = 0$. These are linear combinations of $\mathrm{u\bar{u}}$, $\mathrm{d\bar{d}}$, and $\mathrm{s\bar{s}}$ states. The singlet combination, C, must contain each quark flavor on an equal footing; and so, after normalization, we have

$$\mathrm{C} = \sqrt{\tfrac{1}{3}}\,(\mathrm{u\bar{u}} + \mathrm{d\bar{d}} + \mathrm{s\bar{s}}). \tag{2.48}$$

State A is taken to be a member of the isospin triplet ($\mathrm{d\bar{u}}$, A, $-\mathrm{u\bar{d}}$) and so

$$\mathrm{A} = \sqrt{\tfrac{1}{2}}\,(\mathrm{u\bar{u}} - \mathrm{d\bar{d}}), \tag{2.49}$$

see (2.47). By requiring orthogonality to both A and C, the isospin singlet state B is found to be

$$\mathrm{B} = \sqrt{\tfrac{1}{6}}\,(\mathrm{u\bar{u}} + \mathrm{d\bar{d}} - 2\mathrm{s\bar{s}}). \tag{2.50}$$

Like any quantum-mechanical bound system, the $\mathrm{q\bar{q}}$ pair will have a discrete energy level spectrum corresponding to the different modes of $\mathrm{q\bar{q}}$ excitations,

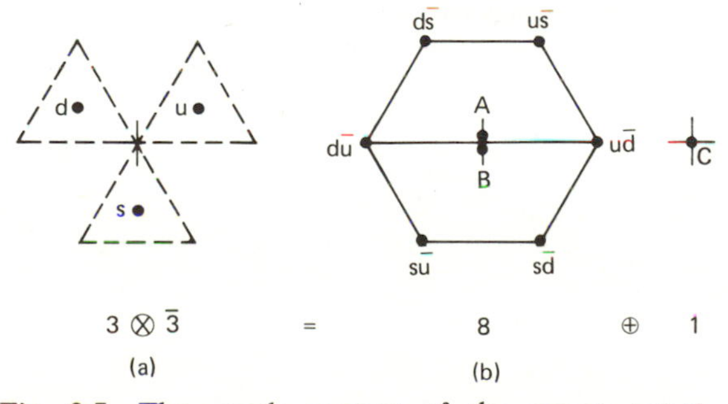

Fig. 2.5 The quark content of the meson nonet, showing the $SU(3)$ decomposition in the I_3, Y plane.

rotations, vibrations, and so on. These must correspond to the observed meson states. Even in the absence of knowledge about the potential which binds the quark to the antiquark, the model is very predictive. Recall that the quark has spin $\frac{1}{2}$, and so the total intrinsic spin of the $q\bar{q}$ pair can be either $S = 0$ or 1. The spin J of the composite meson is the vector sum of this spin S and the relative orbital angular momentum L of the q and $\bar{q}$. Moreover, the parity of the meson is

$$P = -(-1)^L \tag{2.51}$$

where the minus sign arises because the q and $\bar{q}$ have opposite intrinsic parity [cf. (5.63)], and $(-1)^L$ arises from the space inversion replacements $\theta \to \pi - \theta$, $\phi \to \phi + \pi$ in the angular part of the $q\bar{q}$ wavefunction $Y_{LM}(\theta, \phi)$. A neutral $q\bar{q}$ system is an eigenstate of the particle–antiparticle conjugation operator C. The value of C can be deduced by $q \leftrightarrow \bar{q}$ and then interchanging their positions and spins. The combined operation gives

$$C = -(-1)^{S+1}(-1)^L = (-1)^{L+S} \tag{2.52}$$

where the minus sign arises from interchanging fermions, the $(-1)^{S+1}$ from the symmetry of the $q\bar{q}$ spin states [see (2.1)], and the $(-1)^L$ is as before. Here S is the total intrinsic spin of the $q\bar{q}$ pair.

The allowed sets of quantum numbers for the ground ($L = 0$) and first ($L = 1$) excited states are shown in Table 2.2, together with the observed candidate meson states. In each nonet, there are two isospin doublets (see Fig. 2.5). For example, in the $J^P = 0^-$ nonet, we have

$$\begin{aligned} &K^0(d\bar{s}), \quad K^+(u\bar{s}) \quad \text{with } Y = 1 \\ &K^-(s\bar{u}), \quad \bar{K}^0(s\bar{d}) \quad \text{with } Y = -1. \end{aligned} \tag{2.53}$$

These pseudoscalar mesons form an octet along with the $Y = 0$ isotriplet (the π^+, π^0, π^- states) and the $I = 0$ state (the η meson). The $SU(3)$ singlet state is identified with η' meson, see Table 2.2.

TABLE 2.2
Quantum Numbers of Observed Mesons Composed of u, d, and s Quarks

$q\bar{q}$ Orbital Ang. Mom.	$q\bar{q}$ Spin	J^{PC}	Observed Nonet $I = 1$	$I = \frac{1}{2}$	$I = 0$	Typical Mass (MeV)
$L = 0$	$S = 0$	0^{-+}	π	K	η, η'	500
	$S = 1$	1^{--}	ρ	K*	ω, ϕ	800
$L = 1$	$S = 0$	1^{+-}	B	Q_2	H, ?	1250
		2^{++}	A_2	K*	f, f′	1400
	$S = 1$	1^{++}	A_1	Q_1	D, ?	1300
		0^{++}	δ	κ	ε, S*	1150

In the 1^- and 2^+ nonets, the two neutral octet and singlet $I = 0$ states, (2.49) and (2.48), are found to mix with one another so that to a good approximation the physical particles are those made entirely from strange and nonstrange quarks. For example, in the vector (1^-) nonet,

$$\phi \approx s\bar{s}, \qquad \omega \approx \sqrt{\tfrac{1}{2}}\,(u\bar{u} + d\bar{d}). \tag{2.54}$$

Note that, in general, the physical neutral $q\bar{q}$ states correspond to orthogonal quantum-mechanical superpositions of the singlet and the $I = 0$ neutral octet states, which have indeed identical quantum numbers. For example, although we said above that the observed η is an octet state, there is in fact a small singlet admixture.

With reference to Table 2.2, we should add that we would not have expected L and the spin S to be good quantum numbers. However, parity conservation forbids the mixing of even and odd L states, and then C conservation requires the spin S to be unique. This leaves the possibility of mixing only for $S = 1$ states for which L differs by two units.

The $L = 1$ states of Table 2.2 are examples of orbital excitations. The typical excitation energy is about 600 MeV. Just as in positronium, we would also expect radial excitations; for example, a repeat of the $L = 0$ nonets at a higher mass.

The random names attributed to the observed states of Table 2.2 are relics of the past, when experimenters had the difficult task of identifying mesons and determining their quantum numbers. However, there is no doubt that the success of the quark model predictions is impressive; all established mesons lie within the expected $q\bar{q}$ multiplets. Before about 1971, when the first good data with high enough resolution to provide direct evidence for quarks became available, tests of this type were the principal basis for accepting the quark hypothesis.

EXERCISE 2.9 Make use of (2.54) and (2.53) to predict the decay modes and branching ratios of the ϕ-meson (mass 1020 MeV). Comment on the width of the resonance.

EXERCISE 2.10 Explain why a particularly good way of identifying mesons coupled to the $\pi\pi$-channel is to study the reaction $\pi N \to (\pi\pi)N$ at high energies. Show that $I + J$ must be an even integer for these mesons.

In passing, we should note that particles that decay by strong interactions do not live long enough to leave tracks in an experimentalist's detector. Rather, they are identified by tracking their decay products. The mass of the decaying particle is the total energy of these products as measured in its rest frame. Due to its short lifetime, the uncertainty in its mass $(\sim \hbar/\Delta t)$ is sufficiently large to be directly observable. For example, the Δ is formed and rapidly decays in πN scattering, $\pi N \to \Delta \to \pi N$. Such an unstable particle decays according to the exponential law

$$|\psi(t)|^2 = |\psi(0)|^2 e^{-\Gamma t}, \tag{2.55}$$

where $\tau \equiv 1/\Gamma$ is called the lifetime of the state. Thus, the time dependence of $\psi(t)$ for an unstable state must include the decay factor $\Gamma/2$; that is,

$$\psi(t) \sim e^{-iMt} e^{-\Gamma t/2}$$

where M is the rest mass energy of the state. As a function of the center-of-mass energy E of the πN system, the state is described by the Fourier transform

$$\begin{aligned}\chi(E) &= \int \psi(t) e^{iEt}\, dt \\ &\sim \frac{1}{E - M + (i\Gamma/2)}.\end{aligned} \tag{2.56}$$

The experimenter thus sees a πN reaction rate of the form

$$|\chi(E)|^2 = \frac{A}{(E-M)^2 + (\Gamma/2)^2}. \tag{2.57}$$

This function has a sharp peak centered at M with a width determined by Γ. Equation (2.57) is called a Breit–Wigner resonance form, and M and Γ are known as the mass and width of the resonance, respectively. In a detailed resonance analysis, the form (2.55) will include kinematic factors. For instance, resonance production and decay near threshold are inhibited by phase space; its observed width is suppressed by kinematic factors.

2.11 Three-Quark States: Baryons

The flavor $SU(3)$ decomposition of the 27 possible qqq combinations is more involved than that for mesons; nevertheless, the quark content of baryons can be readily obtained using the same techniques. We first combine two of the quarks. Figure 2.6 shows that the nine qq combinations arrange themselves into two $SU(3)$ multiplets,

$$3 \otimes 3 = 6 \oplus \bar{3}, \tag{2.58}$$

where the 6 is symmetric and the $\bar{3}$ is antisymmetric under interchange of the two

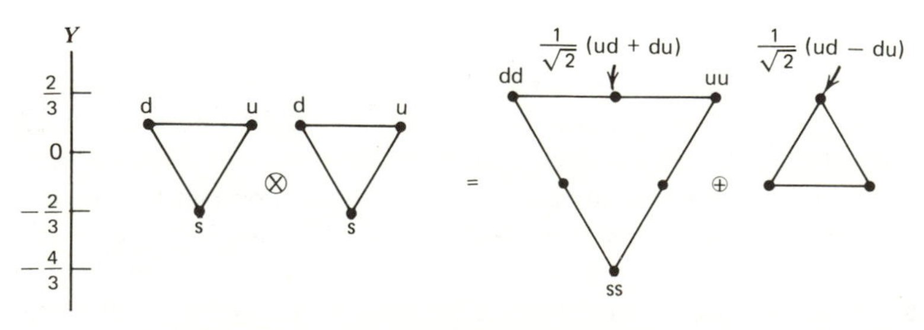

Fig. 2.6 The qq $SU(3)$ multiplets; $3 \otimes 3 = 6 \oplus \bar{3}$.

quarks. The quark content of the nonstrange sector is found by combining the two (d, u) I-spin doublets. It is explicitly shown on the figure and simply repeats eqs. (2.2), with the substitution n → d, p → u. The quark content of the other states is found in exactly the same way. For example, we combine two (s, d) doublets to obtain the states in the s, d sector. The state on the line linking the dd and ss states in Fig. 2.6 is $(\mathrm{ds} + \mathrm{sd})/\sqrt{2}$. We talk of combining U-spin doublets, and in the u, s sector we speak of V-spin. The mathematics of I, U, and V spin are identical, all based on the $SU(2)$ group, which also underlies the description of ordinary spin. In fact, we shall see that almost all the $SU(3)$ structure that we require can be obtained simply by successive application of $SU(2)$.

We are now ready to add the third quark triplet. The final decomposition,

$$\begin{aligned} 3 \otimes 3 \otimes 3 &= (6 \otimes 3) \oplus (\overline{3} \otimes 3) \\ &= 10 \oplus 8 \oplus 8 \oplus 1, \end{aligned} \tag{2.59}$$

is displayed in Fig. 2.7. As an example, we form the three "uud" combinations which are denoted Δ, p_S, and p_A on the figure. Combining the nonstrange member of the $\overline{3}$ (see Fig. 2.6) with the u quark of the 3, we have immediately

$$\mathrm{p}_A = \sqrt{\tfrac{1}{2}}\,(\mathrm{ud} - \mathrm{du})\mathrm{u}. \tag{2.60}$$

The decuplet states are totally symmetric under interchange of quarks, as evidenced by the uuu, ddd, and sss members. The symmetric combination of "uud" is

$$\Delta = \sqrt{\tfrac{1}{3}}\,[\mathrm{uud} + (\mathrm{ud} + \mathrm{du})\mathrm{u}]. \tag{2.61}$$

Requiring orthogonality of the remaining "uud" state to both p_A and Δ gives

$$\mathrm{p}_S = \sqrt{\tfrac{1}{6}}\,[(\mathrm{ud} + \mathrm{du})\mathrm{u} - 2\mathrm{uud}]. \tag{2.62}$$

The states p_S and p_A have mixed symmetry; however, the subscripts are to remind us that they have symmetry and antisymmetry, respectively, under interchange of the first two quarks. The quark structure of the other states can be readily obtained in a similar way (by application of either U or V spin).

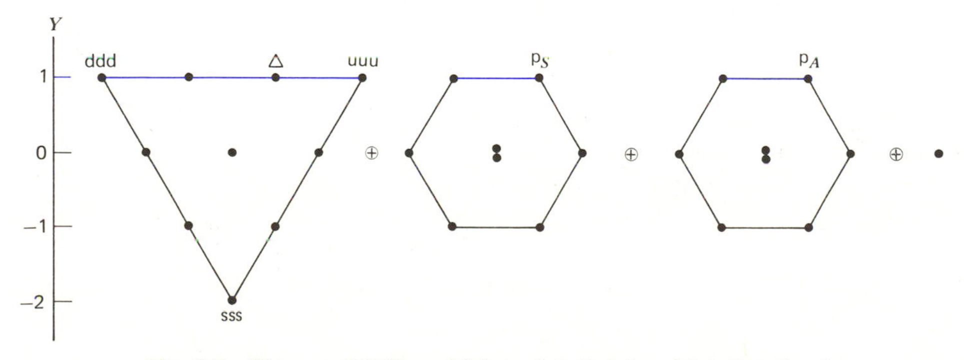

Fig. 2.7 The qqq $SU(3)$ multiplets; $3 \otimes 3 \otimes 3 = 10 \oplus 8 \oplus 8 \oplus 1$.

EXERCISE 2.11 Write down the quark composition of the three "dds" states.

EXERCISE 2.12 Determine the structure of the six "uds" states. In particular, show that the $SU(3)$ singlet state is the completely antisymmetric combination

$$(\text{qqq})_{\text{singlet}} = \sqrt{\tfrac{1}{6}}\,(\text{uds} - \text{usd} + \text{sud} - \text{sdu} + \text{dsu} - \text{dus}). \tag{2.63}$$

In the ground state, the baryon spin is found simply by the addition of three spin-$\frac{1}{2}$ angular momenta. Writing the decomposition in terms of the multiplicities of the spin states, we found [see (2.32)]

$$\underset{}{2} \otimes 2 \otimes 2 = (\underset{S}{3} \oplus \underset{A}{1}) \otimes 2 = \underset{S}{4} \oplus \underset{M_S}{2} \oplus \underset{M_A}{2}, \tag{2.64}$$

or, in other words, baryon spin multiplets with $S = \frac{3}{2}, \frac{1}{2}, \frac{1}{2}$. The subscripts on the mixed symmetry doublets (M_S, M_A) indicate that the spin states are symmetric or antisymmetric under interchange of the first two quarks. The four $S = \frac{3}{2}$ spin states are totally symmetric.

Note that in deriving eqs. (2.60)–(2.62), we were working in the $SU(2)$ isospin sector of $SU(3)$. We can therefore apply these results directly to $SU(2)$ spin if we make the replacements u $\rightarrow \uparrow$ and d $\rightarrow \downarrow$. Using this analogy, we immediately obtain the composition of the spin "up" state belonging to each of the three spin multiplets

$$\begin{aligned} \chi(S) &= \sqrt{\tfrac{1}{3}}\,(\uparrow\uparrow\downarrow + \uparrow\downarrow\uparrow + \downarrow\uparrow\uparrow) \\ \chi(M_S) &= \sqrt{\tfrac{1}{6}}\,(\uparrow\downarrow\uparrow + \downarrow\uparrow\uparrow - 2\uparrow\uparrow\downarrow) \\ \chi(M_A) &= \sqrt{\tfrac{1}{2}}\,(\uparrow\downarrow\uparrow - \downarrow\uparrow\uparrow). \end{aligned} \tag{2.65}$$

To enumerate the baryons expected in the quark model, we must combine the $SU(3)$ flavor decomposition of (2.59) with the $SU(2)$ spin decomposition of (2.64),

$$(\underset{S}{10} + \underset{M_S}{8} + \underset{M_A}{8} + \underset{A}{1}), \qquad (\underset{S}{4} + \underset{M_S}{2} + \underset{M_A}{2}). \tag{2.66}$$

Considering the product symmetries, we are led to assign the ($SU(3)$, $SU(2)$) multiplets to the following categories:

$$\begin{aligned} S&: (10, 4) + (8, 2) \\ M_S&: (10, 2) + (8, 4) + (8, 2) + (1, 2) \\ M_A&: (10, 2) + (8, 4) + (8, 2) + (1, 2) \\ A&: (1, 4) + (8, 2) \end{aligned} \tag{2.67}$$

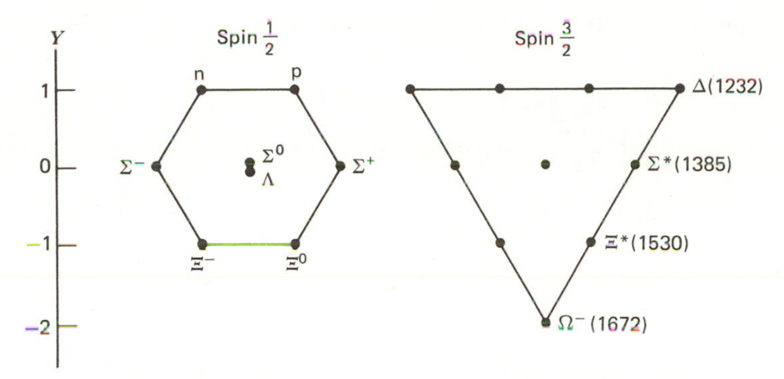

Fig. 2.8 Ground-state baryons: (8, 2) + (10, 4).

where, for example, the totally symmetric (S) octet arises from the combination

$$\sqrt{\tfrac{1}{2}}\,[\underset{M_S,\,M_S}{(8,2)} + \underset{M_A,\,M_A}{(8,2)}]. \tag{2.68}$$

The lowest-mass baryons fit neatly into the symmetric spin-$\frac{3}{2}$ decuplet (10, 4) and the spin-$\frac{1}{2}$ octet (8, 2) (see Fig. 2.8).

This symmetry of the ground state poses a problem, however. For example, a Δ^{++} of $J_3 = \frac{3}{2}$ is described by the symmetric wave function

$$\mathrm{u}\!\uparrow \mathrm{u}\!\uparrow \mathrm{u}\!\uparrow, \tag{2.69}$$

whereas we expect antisymmetry under the exchange of identical fermion quarks. As noted in Chapter 1, the explanation is that the quarks possess an additional attribute, called color, which can take three possible values, R, G, or B. The quarks form a fundamental triplet of an $SU(3)$ color symmetry which, unlike $SU(3)$ flavor, is believed to be exact. All hadrons are postulated to be colorless; that is, they belong to singlet representations of the $SU(3)$ color group. The color wavefunction for a baryon is therefore [compare (2.63)]

$$(\mathrm{qqq})_{\text{col. singlet}} = \sqrt{\tfrac{1}{6}}\,(\mathrm{RGB} - \mathrm{RBG} + \mathrm{BRG} - \mathrm{BGR} + \mathrm{GBR} - \mathrm{GRB}). \tag{2.70}$$

The required antisymmetric character of the total wavefunction is achieved; it is overall symmetric in space, spin, and flavor structure and antisymmetric in color. As the color structure of (2.70) is common to all baryons, we suppress it from now on, but remember to select only overall symmetric representations of space × spin × flavor.

A relevant example of an explicit quark model wavefunction is that for a spin-up proton. From (2.68),

$$|p\uparrow\rangle = \sqrt{\tfrac{1}{2}}\,(\mathrm{p}_S\chi(M_S) + \mathrm{p}_A\chi(M_A)),$$

where the flavor and spin components are given by (2.60), (2.62), and (2.65). Thus, we have

$$\begin{aligned} |p\uparrow\rangle &= \sqrt{\tfrac{1}{18}}\,[\mathrm{uud}(\uparrow\downarrow\uparrow + \downarrow\uparrow\uparrow - 2\uparrow\uparrow\downarrow) + \mathrm{udu}(\uparrow\uparrow\downarrow + \downarrow\uparrow\uparrow - 2\uparrow\downarrow\uparrow) \\ &\quad + \mathrm{duu}(\uparrow\downarrow\uparrow + \uparrow\uparrow\downarrow - 2\downarrow\uparrow\uparrow)] \\ &= \sqrt{\tfrac{1}{18}}\,[\mathrm{u}\uparrow\mathrm{u}\downarrow\mathrm{d}\uparrow + \mathrm{u}\downarrow\mathrm{u}\uparrow\mathrm{d}\uparrow - 2\mathrm{u}\uparrow\mathrm{u}\uparrow\mathrm{d}\downarrow + \text{permutations}]. \end{aligned} \tag{2.71}$$

EXERCISE 2.13 Construct the quark model wavefunctions $|\mathrm{p}\downarrow\rangle$, $|\mathrm{n}\uparrow\rangle$, and $|\mathrm{n}\downarrow\rangle$. The charge operator is defined as $Q = \sum_i Q_i$, where Q_i are the charges of the quarks in units of the proton charge e. The sum is over the constituent quarks of the hadron. Show that

$$\langle \mathrm{p}\uparrow|Q|\mathrm{p}\uparrow\rangle = \langle \mathrm{p}\downarrow|Q|\mathrm{p}\downarrow\rangle = 1$$

$$\langle \mathrm{n}\uparrow|Q|\mathrm{n}\uparrow\rangle = \langle \mathrm{n}\downarrow|Q|\mathrm{n}\downarrow\rangle = 0.$$

EXERCISE 2.14 Express the π^+-wavefunction in terms of the spin, flavor, and color of the component quarks.

EXERCISE 2.15 Convince yourself that the photon is a U-spin scalar; that is, $U = 0$. By inspection of Fig. 2.8, show that if $SU(3)$ flavor symmetry were exact, the electromagnetic decay $\Sigma^*(1385)^- \to \Sigma^-\gamma$ is forbidden, whereas $\Sigma^*(1385)^+ \to \Sigma^+\gamma$ is allowed.

The three quarks have zero orbital angular momentum in ground-state baryons. That is, $l = l' = 0$ in Fig. 2.9, and so the parity of the state, $(-1)^{l+l'}$, is therefore positive. The first excited state has either $l = 1$, $l' = 0$, or $l = 0$, $l' = 1$; in fact, it is a combination of the two which, when combined with the mixed symmetry multiplets of (2.67), gives a totally symmetric space, spin, and flavor wavefunction. That is, the first excited state is predicted to contain (1 + 8 + 10) flavor multiplets of $S = \frac{1}{2}$ baryons and an octet of $S = \frac{3}{2}$ baryons [see (2.67)]. These spins combine with $L = 1$ to give

$$\text{Multiplets } 1, 8, 10 \text{ with } J^P = \tfrac{1}{2}^- \quad \text{and} \quad J^P = \tfrac{3}{2}^-$$

$$\text{Three octets with } J^P = \tfrac{1}{2}^-, \tfrac{3}{2}^-, \tfrac{5}{2}^-.$$

Fig. 2.9 Orbital angular momenta, l and l', of qq and (qq)q systems, respectively.

Impressive agreement with the observed baryons with masses around 1600 MeV is again found.

EXERCISE 2.16 The $Y = 1$ baryons are most easily identified as resonances in πN elastic scattering. Show that the relative orbital angular momentum, L', between the π and the N is a good quantum number, and that it is even for resonances of negative parity. Use the quark model to list the isospin, spin, and L' of the πN states expected in the first excited level. Identify these resonances in the particle data tables.

2.12 Magnetic Moments

The calculation of the charge in Exercise 2.13 can be repeated for the magnetic moments of the hadrons. The magnetic moment operator is $\Sigma_i\, \mu_i(\sigma_3)_i$, where the summation is again over the constituent quarks. According to the accepted convention, the operator is evaluated between states with $M_J = +J$. Now the magnetic moment of a point-like spin-$\frac{1}{2}$ particle of charge e is $e/2m$ (see Chapter 5). Thus, a structureless quark of charge $Q_i e$ and mass m_i has magnetic moment

$$\mu_i = Q_i\left(\frac{e}{2m_i}\right) \tag{2.72}$$

Hence, in the nonrelativistic approximation, we may write the magnetic moment of the proton as

$$\mu_p = \sum_{i=1}^{3} \langle p\uparrow | \mu_i(\sigma_3)_i | p\uparrow \rangle.$$

Using the explicit wavefunction (2.71), we obtain

$$\mu_p = \tfrac{1}{18}\{(\mu_u - \mu_u + \mu_d) + (-\mu_u + \mu_u + \mu_d) + 4(2\mu_u - \mu_d)\} \times 3,$$

where the factor 3 takes care of the "permutations." The magnetic moment of the proton in terms of the component moments is therefore

$$\mu_p = \tfrac{1}{3}(4\mu_u - \mu_d). \tag{2.73}$$

The neutron magnetic moment is obtained by the interchange of u and d,

$$\mu_n = \tfrac{1}{3}(4\mu_d - \mu_u).$$

In the limit that $m_u = m_d$, we have, from (2.72),

$$\mu_u = -2\mu_d \tag{2.74}$$

and so the quark model prediction is

$$\frac{\mu_n}{\mu_p} = -\frac{2}{3}. \tag{2.75}$$

This agrees quite well with experiment:

$$\frac{\mu_n}{\mu_p} = -0.68497945 \pm 0.00000058.$$

EXERCISE 2.17 Determine the magnetic moments of the other members of the $J^P = \frac{1}{2}^+$ baryon octet in terms of μ_p and compare with the measured values.

EXERCISE 2.18 The spin-flavor wavefunctions of the ground-state baryons are symmetric, and color was invoked to recover the required antisymmetric character. You should notice, however, and some people did, that we can construct a totally antisymmetric proton wavefunction, for example,

$$|p\uparrow\rangle = \sqrt{\tfrac{1}{2}}\,[p_A\,\chi(M_S) - p_S\,\chi(M_A)],$$

and forget about color! Write this function in an explicit form, comparable to (2.71). Obtain $|n\uparrow\rangle$, and hence show that

$$\frac{\mu_n}{\mu_p} = -2.$$

So this option is ruled out by experiment. In fact, glancing at your derivation, you will notice that μ_p is negative. It is measured to be positive. Long live color.

EXERCISE 2.19 Prove that the quark model relations for the magnetic moments of the $\rho^\pm$ mesons are

$$\mu_{\rho^+} = -\mu_{\rho^-} = \mu_p.$$

EXERCISE 2.20 Use the quark model to calculate the amplitude for the radiative decay $\omega \to \pi^0\gamma$. The ω and π^0 belong to the $J^P = 1^-$, $S = 1$ and

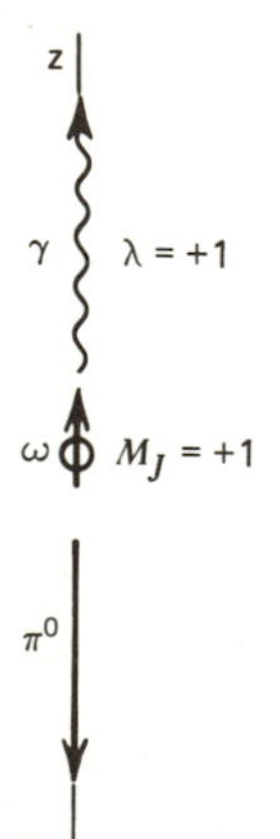

Fig. 2.10 The radiative decay $\omega \to \pi^0\gamma$.

the $J^P = 0^-$, $S = 0$ nonets, respectively. We therefore require a quark spin flip (magnetic dipole) transition. This will involve the quark magnetic moment operator.

First, assume (2.54) and obtain the spin-flavor wavefunctions for an ω with $M_J = 1$ and for a π^0. If the z axis is chosen as in Fig. 2.10, show that the required amplitude is

$$\sum_{i=1,2} \langle \pi^0 | \mu_i \boldsymbol{\sigma}_i \cdot \boldsymbol{\varepsilon}_R^* | \omega(M_J = 1) \rangle = -\sqrt{2} \sum_{i=1,2} \langle \pi^0 | \mu_i (\sigma_-)_i | \omega(M_J = 1) \rangle$$

$$= \mu_d - \mu_u$$

where $\boldsymbol{\varepsilon}_R \equiv -\sqrt{\frac{1}{2}}(1, i, 0)$ is the polarization vector of the emitted (helicity-one) photon and $\sigma_- \equiv \frac{1}{2}(\sigma_1 - i\sigma_2)$ is the operator which "steps down" or "flips" the quark spin.

EXERCISE 2.21 Assuming (2.54), show that the quark model forbids the decay $\phi \to \pi^0\gamma$, and predicts that

$$\frac{\text{Rate}(\omega \to \pi^0\gamma)}{\text{Rate}(\rho \to \pi^0\gamma)} = \left(\frac{\mu_d - \mu_u}{\mu_d + \mu_u}\right)^2 = 9.$$

2.13 Heavy Quarks: Charm and Beyond

The discovery in November 1974 of a very narrow resonance, called ψ, in e^+e^- annihilation near a center-of-mass energy of 3.1 GeV, followed two weeks later by the appearance of a second narrow resonance, ψ', at 3.7 GeV, can rightly be called the "November revolution." Independently, a group of experimentalists discovered the ψ-particle by producing it in proton–proton collisions. They called it J, and it is therefore often referred to in the literature as the J/ψ particle.

The ψ and ψ' were immediately interpreted as the lowest bound states of a new quark and its antiquark, $c\bar{c}$. This new charmed quark, c, had been much heralded. As we shall see in Chapter 12, the existence of another quark of charge $+\frac{2}{3}$ had long been needed in the theory of the weak interactions of hadrons (cf. the GIM mechanism).

As the total energy of the colliding e^+ and e^- beams was increased beyond 3.7 GeV, the cross section for $e^+e^- \to$ hadrons showed complicated resonance structure. A peep ahead at the column marked e^+e^- in Fig. 2.13 shows the four resonances identified below about 4 GeV. Above 3.7 GeV, the widths of the resonances become larger and more typical of hadronic decays. This phenomenon is a replay of ϕ-decay (see Exercise 2.9). The narrow width (4 MeV) of the ϕ-meson arises because it is an $s\bar{s}$ bound state just above the $K\bar{K}$ threshold, K being the lightest strange meson. The decay

$$\phi(s\bar{s}) \to K(q\bar{s}) + \bar{K}(\bar{q}s) \tag{2.76}$$

with q = u, d, is inhibited by lack of phase space, while $\phi \to \pi\pi\pi$ has plenty of phase space but requires annihilation of the $s\bar{s}$ pair. Nevertheless, the dominant decay of the ϕ is through the $K\bar{K}$ mode. This result is pictured in terms of quark lines in Fig. 2.11 and is an example of the Zweig or OZI rule which asserts that disconnected quark-line diagrams (Fig. 2.11a) are highly suppressed relative to connected ones (Fig. 2.11b). Similarly, in the case of the $\psi(3.1)$ and $\psi'(3.68)$, the very small widths (69 and 225 keV, respectively) arise because they are $c\bar{c}$ bound states below the $D\bar{D}$ threshold, where D is the lightest charmed meson. Hence, their hadronic decays $\psi \to \pi\pi\pi$, and so forth, require annihilation of the $c\bar{c}$ pair (Fig. 2.11c). The larger hadronic-type widths of the higher ψ-states are attributed to the allowed decay

$$\psi(c\bar{c}) \to D(\bar{q}c) + \bar{D}(q\bar{c}) \tag{2.77}$$

with q = u, d (see Fig. 2.11d). Thus, we expect the kinematic threshold for $D\bar{D}$ production to lie somewhere between the $\psi'(3.68)$ and the $\psi''(3.77)$. We therefore predict the mass of the D meson to be $m(D) \gtrsim 3.7/2 = 1.85$ GeV.

Just like the introduction of strangeness S, we assign an additive quantum number $C = \pm 1$ to the c, $\bar{c}$ quarks, respectively, and $C = 0$ to the lighter quarks. The c quark has charge $Q = +\frac{2}{3}$ and isospin $I = 0$, and so we should update the relations of Table 2.1 to read

$$Y = B + S + C, \qquad Q = I_3 + \tfrac{1}{2}Y. \tag{2.78}$$

We show the basic quark multiplet in Fig. 2.12, together with the antiquark tetrahedron. Now that we have basic building blocks, we can repeat our proce-

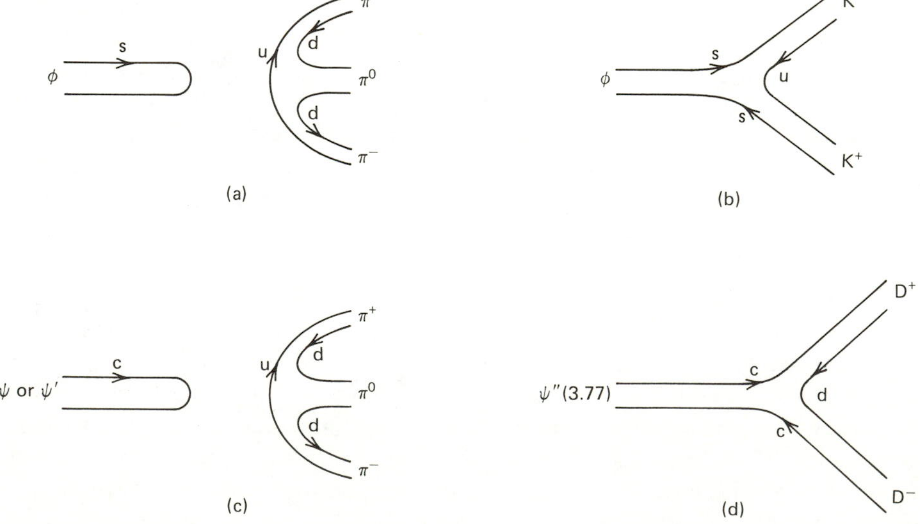

Fig. 2.11 Suppressed decay modes $\phi(s\bar{s}), \psi(c\bar{c}) \to \pi\pi\pi$ and allowed decay modes $\phi \to K\bar{K}$, $\psi'' \to D\bar{D}$.

dure of combining them to form hadrons. The $q\bar{q}$ states, the mesons, are constructed as in Fig. 2.12; the mesons shown in parentheses are members of the lowest-lying ($J^P = 0^-$) multiplet. Charmed members have been observed, with masses

$$m(\text{D}) = 1.86 \text{ GeV}, \qquad m(\text{F}) = 2.03 \text{ GeV}. \tag{2.79}$$

Charmed mesons are also required to complete the other multiplets listed in Table 2.2. The charmed states of the $J^P = 1^-$ multiplet are, not surprisingly, called D* and F*. The observed masses are

$$m(\text{D}^*) = 2.01 \text{ GeV}, \qquad m(\text{F}^*) = 2.14 \text{ GeV}.$$

EXERCISE 2.22 According to weak interaction theory, the dominant hadronic weak decay proceeds via the quark transmutations $c \to s$ and/or $u \leftrightarrow d$ (see Chapter 12). For example, an allowed charmed meson decay is $c\bar{u} \to s\bar{d}(u\bar{u})$.

Assuming that these, and only these, transmutations can occur, show that

$$\text{D}^0 \to \text{K}^-\pi^+ \qquad \text{and} \qquad \text{K}^-\pi^+\pi^+\pi^-$$

are allowed decay modes, but that

$$\text{D}^0 \to \pi^+\pi^-, \text{K}^+\text{K}^-, \text{K}^+\pi^-, \qquad \text{and} \qquad \text{K}^+\pi^-\pi^+\pi^-$$

are all forbidden. Further, show that $\text{D}^+ \to \text{K}^-\pi^+\pi^+$ is an allowed weak decay, but that $\text{D}^+ \to \text{K}^+\pi^+\pi^-$ is forbidden. This distinctive feature of D^+ decays was in fact convincing evidence in the first ever observation of a charmed particle in 1976, some 18 months after the revolutionary discovery of the "hidden" charm state $\psi(c\bar{c})$.

Each meson multiplet contains a state, $c\bar{c}$, of "hidden" charm. For the $J^P = 0^-$ and 1^- multiplets, it is $\eta_c(2.98)$ and the original $\psi(3.1)$, respectively. The states of the bound $c\bar{c}$ system can be compared with those of positronium e^+e^-. We speak of "charmonium." It is a particularly clean system to study and has revo-

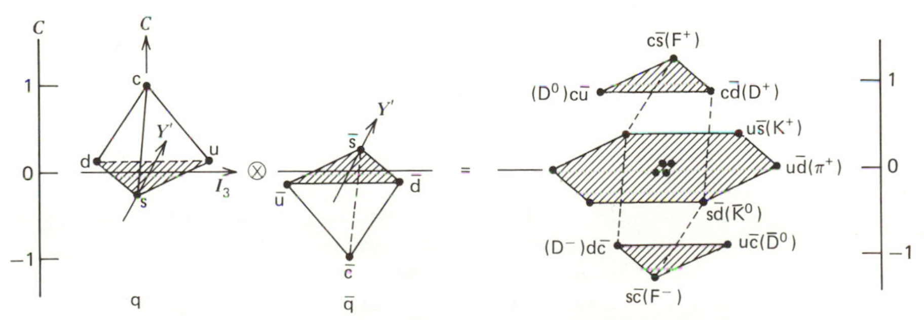

Fig. 2.12 The 16 meson states made from u, d, s, c quarks, plotted in (I_3, Y', C) space with $Y' = Y - \frac{4}{3}C$. Some members of the $J^P = 0^-$ multiplet are indicated.

lutionized meson spectroscopy. States with $J^{PC} = 1^{--}$ can be directly produced ($e^+e^- \rightarrow$ virtual $\gamma \rightarrow c\bar{c}$); and, via their decays, other charmonium states can be identified. The observed states are shown in Fig. 2.13, labeled in the conventional spectroscopic manner $^{2S+1}L_J$, where S, L, and J are, respectively, the total intrinsic spin, orbital angular momentum, and total angular momentum of the $c\bar{c}$ system. This is, of course, a nonrelativistic classification; it is the heavy mass of the c quark which makes it possible to use a nonrelativistic picture. We also show the J^{PC} values of the states and note that the observations coincide with quark model expectations. The six J^{PC} values listed in Table 2.2 are reproduced, except that the 1^{+-} (or $^1P_1$) state still awaits discovery. As in positronium, radial as well as orbital excitations are expected. In fact, the 2 ^3S and 3 ^3S excitations are seen directly as resonances in the cross section for $e^+e^- \rightarrow$ hadrons (see Fig. 2.13).

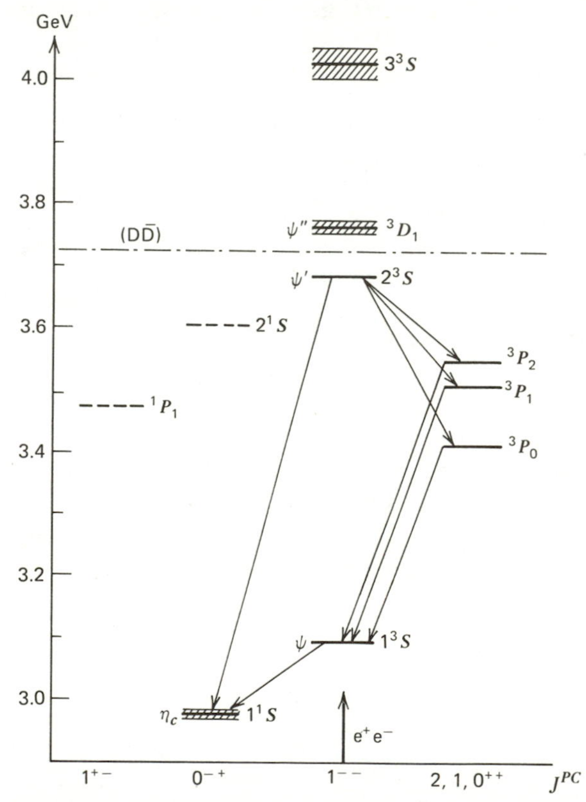

Fig. 2.13 The observed charmonium spectrum. The transitions shown have all been observed. The $^1P_1$ and 2^1S states await discovery. The particle widths are shown by shaded bands. The dot-dash line shows the D$\bar{\text{D}}$ threshold; states below this line cannot decay into charmed mesons. The states with $J^{PC} = 1^{--}$ can be directly produced by e^+e^- collisions.

EXERCISE 2.23 The decay

$$\psi'(3.7) \to \psi(3.1) + \text{hadrons}$$

is observed. What are the hadrons?

EXERCISE 2.24 Mark on Fig. 2.13 the expected radiative transitions between the levels, indicating which are electric and which are magnetic dipole transitions.

Justify that the rates for the radiative transitions from $\psi'(3.7)$ to the three χ-levels, ^3P_J, with $J = 2, 1, 0$, are proportional to $(2J + 1)k^3$, where k is the momentum of the emitted photon. Hence, show that the branching ratios of these decay modes of ψ' are approximately equal.

EXERCISE 2.25 The leptonic decay of neutral vector ($J^{PC} = 1^{--}$) mesons can be pictured as proceeding via a virtual photon,

$$\mathrm{V}(\mathrm{q\bar{q}}) \to \gamma \to \mathrm{e^+e^-}. \tag{2.80}$$

The technique for calculating such amplitudes will be explained in succeeding chapters. Here, it suffices to note that the V–γ coupling is proportional to the charge of the quark q. Neglecting a possible dependence on the vector meson mass, show that the leptonic decay widths are in the ratios

$$\rho : \omega : \phi : \psi = 9 : 1 : 2 : 8.$$

EXERCISE 2.26 Comment on the rate you would expect for the $\mathrm{e^+e^-}$ decay mode of the $^3D_1$ state as compared to the $\psi'(3.7)$ state. Can these two states mix?

EXERCISE 2.27 The hadronic decay widths of η_c and $\psi(3.1)$ are estimated using

$$\eta_c(\mathrm{c\bar{c}}) \to n\mathrm{g} \to \text{hadrons}$$

$$\psi(\mathrm{c\bar{c}}) \to n'\mathrm{g} \to \text{hadrons}, \tag{2.81}$$

where g is a gluon and n and n' are integers. These are QCD analogues of the QED process of (2.80). Show that the minimum values of n and n' are 2 and 3, respectively.

Properties of the potential between the c and $\bar{\mathrm{c}}$ can be inferred from the charmonium spectrum. In Chapter 1, we noted that at small c and $\bar{\mathrm{c}}$ separations, QCD predicts a Coulomb-type potential $-\alpha_s/r$, but that at large separation r, we expect a confining potential which increases with r. A glance at the $1S$, $2S$, and the "center of gravity" of the P levels of Fig. 2.13 shows that the potential is in fact somewhere between Coulomb (which has $2S$ and P degenerate) and an oscillator potential $V \sim r^2$ (which has the P level halfway between $1S$ and $2S$). A naive potential, which is phenomenologically rather satisfactory, is

$$V(r) = -\frac{4}{3}\frac{\alpha_s}{r} + ar \tag{2.82}$$

where a is a constant parameter and $\frac{4}{3}$ is the color factor associated with the quark–gluon coupling α_s [see (2.98)].

Let us now repeat the steps of constructing baryons, but this time include the c quark. Combining three basic quark multiplets, we find that the analogue of (2.59) is

$$\underset{}{4 \otimes 4 \otimes 4} = \underset{S}{20} \oplus \underset{M_S}{20} \oplus \underset{M_A}{20} \oplus \underset{A}{\bar{4}}. \tag{2.83}$$

Rather than to derive this decomposition, it is better at this stage to use the elegant techniques of group theory (Young tableaux); see, for example, Close (1979). Including spin, (2.64), we can as before form the required symmetric spin-flavor ground state in two ways: either the symmetric 20 with a symmetric spin $\frac{3}{2}$ or a mixed-symmetry 20 with spin $\frac{1}{2}$ constructed in exact analogy to (2.68). Extracting the flavor multiplets from a superposition of three basic (quark) tetrahedra leads to the ground-state baryons of Fig. 2.14a.

The spin-$\frac{1}{2}$ multiplet can be viewed as three $SU(3)$ octets propping each other up and based on the edges of a fourth $SU(3)$ octet. In fact, we do not need the elegance of group theory to enumerate the states. For example, the $C = 1$ spin-$\frac{1}{2}$ baryons are cqq composites with q = u, d, or s. The qq decomposition is given in (2.58), namely,

$$3 \otimes 3 = 6 \oplus \bar{3},$$

and the states are shown in Fig. 2.14b. The lightest charmed baryons are the Σ_c isospin triplet and Λ_c^+. The observed masses are

$$m(\Lambda_c) = 2.28 \text{ GeV}, \qquad m(\Sigma_c) = 2.44 \text{ GeV}. \tag{2.84}$$

EXERCISE 2.28 Determine the flavor wavefunctions of the Λ_c and Σ_c baryons. Give an observable decay sequence of Σ_c^{++}.

The c quark was desired theoretically. The same cannot be said of the b quark. Evidence for this fifth quark came in a replay of the charmonium phenomenon in the e^+e^- energy region around 10 GeV. Four e^+e^- resonances were quickly identified: $\Upsilon(1S)$, $\Upsilon(2S)$, $\Upsilon(3S)$, and $\Upsilon(4S)$ with masses of 9.46, 10.02, 10.35, and 10.57 GeV, respectively. The first three states are narrow and the fourth is much wider. The lightest meson ($b\bar{u}$ or $b\bar{d}$) with explicit beauty is therefore expected to have mass $m(D_b) \approx 10.4/2 = 5.2$ GeV [cf. (2.77)].

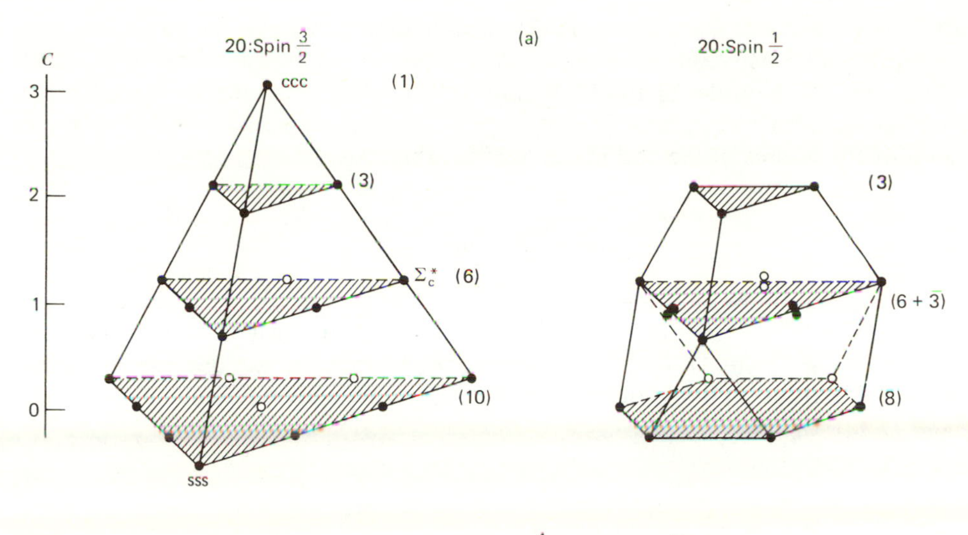

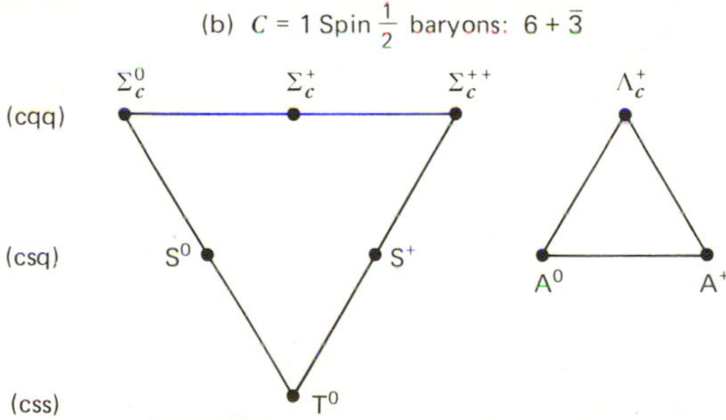

Fig. 2.14 (a) The spin-$\frac{1}{2}$ and spin-$\frac{3}{2}$ ground-state baryons made from u, d, s, c quarks, with the $SU(3)$ multiplicities of the $C = 0, 1, 2, 3$ states shown in parentheses. (b) The $C = 1$ spin-$\frac{1}{2}$ baryons $(6 + \bar{3})$ and their quark content with q = u, d.

2.14 Hadron Masses

If $SU(4)$ flavor symmetry were exact, all members of a given $SU(4)$ multiplet would have the same mass. This is manifestly not the case. For example, within the 1^- meson multiplet, we have

$$
\begin{aligned}
m_\omega \approx m_\rho(\mathrm{u\bar{u}}) &= 0.78 \text{ GeV} \\
m_\phi(\mathrm{s\bar{s}}) &= 1.02 \text{ GeV} \\
m_{K^*}(\mathrm{s\bar{u}}) &= 0.89 \text{ GeV} \\
m_{D^*}(\mathrm{c\bar{u}}) &= 2.01 \text{ GeV} \\
m_{F^*}(\mathrm{c\bar{s}}) &= 2.14 \text{ GeV} \\
m_\psi(\mathrm{c\bar{c}}) &= 3.1 \text{ GeV}.
\end{aligned}
\tag{2.85}
$$

It is true that members of an $SU(2)$ isospin multiplet have the same mass to within about 5 MeV. However, u, d, s $SU(3)$ flavor symmetry is broken by mass differences of the order of 100 MeV, and $SU(4)$ flavor symmetry by considerably greater than 1 GeV. Indeed, if we attribute the mass of a hadron to simply the sum of the masses of the constituent quarks, then eqs. (2.85) imply

$$\begin{aligned} m_u &\approx m_d \approx 0.39 \text{ GeV} \\ m_s &\approx 0.51 \text{ GeV} \\ m_c &\approx 1.6 \text{ GeV}. \end{aligned} \tag{2.86}$$

We may use the observed magnetic moments of the baryons to obtain an alternative estimate of the quark masses. For example, we use the magnetic moment of the proton and the Λ to estimate m_u and m_s, respectively. From (2.73) and (2.74), we have

$$\mu_u = \frac{2}{3}\mu_p = \frac{2}{3}(2.79)\frac{e}{2m_p}$$

and so, noting (2.72),

$$m_u = \frac{m_p}{2.79} = 0.34 \text{ GeV}. \tag{2.87}$$

Similarly, the quark model prediction

$$\mu_s = \mu_\Lambda = -0.61\frac{e}{2m_p}$$

gives $m_s = 0.51$ GeV. The agreement with (2.86) increases our confidence that quarks are indeed point-like constituents with Dirac magnetic moments.

Equations (2.86) are to be regarded as the effective masses of quarks bound within (color singlet) hadrons. We speak of constituent quark masses. It is useful to think of the constituent masses of the quark and antiquark as their zero-point energy when they are bound by some potential like (2.82) with an energy spectrum that corresponds to the masses of the observed mesons. For charm, and heavier quarks, it appears that the total zero-point energy is not much different from the masses of the lowest-lying meson states. These $c\bar{c}$ or $b\bar{b}$ states can therefore be regarded as essentially nonrelativistic bound states of the quark and antiquark.

The success of simple quark counting in explaining the gross features of the baryon and meson masses leads us to attempt to understand more detailed properties of the mass spectra. Why, for instance, is the Δ (spin $\frac{3}{2}$) heavier than the N (spin $\frac{1}{2}$), and the ρ (spin 1) heavier than the π (spin 0), despite having the same quark content? How do we account for the different masses of the neutral Λ, Σ (spin $\frac{1}{2}$), and Σ^* (spin $\frac{3}{2}$) baryons, even though they are each made of uds quarks? The differing spin configurations of the quarks offer a clue. In QED, we know that the forces are spin dependent. Should we therefore not expect an analogous result in QCD?

We begin by recalling (see, for example, Bethe and Salpeter, 1957) that the spin–spin, or magnetic moment, interaction leads to hyperfine splitting of the ground-state level of the hydrogen atom (or of positronium):

$$\Delta E_{hf} = -\frac{2}{3}\boldsymbol{\mu}_1\cdot\boldsymbol{\mu}_2|\psi(0)|^2 = \frac{2\pi\alpha}{3}\frac{\boldsymbol{\sigma}_1\cdot\boldsymbol{\sigma}_2}{m_1 m_2}|\psi(0)|^2, \tag{2.88}$$

where the magnetic moment $\boldsymbol{\mu}_i = e_i\boldsymbol{\sigma}/2m_i$ and $e_1 e_2 = -e^2 = -4\pi\alpha$. This is a contact interaction; it involves the square of the relative wavefunction evaluated at zero separation and so only applies to $L = 0$ states. For the hydrogen atom, it is truly a hyperfine splitting; but for positronium, we see that it is enhanced by a factor m_p/m_e.

EXERCISE 2.29 Verify that the spin 1 level ($^3S_1$) is higher than the spin 0 level ($^1S_0$).

The QED result, (2.88), can be taken over directly to QCD, provided we replace the electromagnetic coupling $e_1 e_2$ by the product of color charges. For mesons and baryons, the substitutions are

$$-\alpha \rightarrow \begin{cases} -\frac{4}{3}\alpha_s & \text{for } (q\bar{q}) \quad (2.89) \\ -\frac{2}{3}\alpha_s & \text{for } (qqq) \quad (2.90) \end{cases}$$

where $\frac{4}{3}$ and $\frac{2}{3}$ are the appropriate color factors. We show how to compute these factors in a moment.

We can now make a model for the ground-state hadron masses. We assume (1) that quark confinement, which is operative at large separations, is independent of the spins and of the masses of the quarks; (2) that at near-separation, α_s is small enough for QCD hyperfine splitting to be relevant; and (3) that the only symmetry breaking arises from the different constituent masses assigned to the quarks of different flavors. In this scheme, the meson and baryon masses are therefore

$$m(q_1\bar{q}_2) = m_1 + m_2 + [a(\boldsymbol{\sigma}_1\cdot\boldsymbol{\sigma}_2)/m_1 m_2] \tag{2.91}$$

$$m(q_1 q_2 q_3) = m_1 + m_2 + m_3 + \left[\frac{a'}{2}\sum_{i>j}(\boldsymbol{\sigma}_i\cdot\boldsymbol{\sigma}_j)/m_i m_j\right] \tag{2.92}$$

where a and a' are positive constants [see (2.88)–(2.90)].

EXERCISE 2.30 For the π (spin 0) and the K* (spin 1), show that (2.91) gives

$$m(\pi) = m_u + m_d - (3a/m_u m_d)$$

$$m(\mathrm{K}^*) = m_u + m_s + (a/m_u m_s).$$

Calculate the masses of all the members of the 0^- and 1^- meson multiplets (Fig. 2.12) using

$$m_u = m_d = 0.31, \qquad m_s = 0.48, \qquad m_c = 1.65, \qquad a/m_u^2 = 0.16,$$

all in units of GeV. Compare your predictions with the meson masses listed in the particle data tables.

Check that

$$(\rho - \pi) = \frac{m_s}{m_u}(\mathrm{K}^* - \mathrm{K}) = \frac{m_c}{m_u}(\mathrm{D}^* - \mathrm{D}) = \frac{m_c m_s}{m_u^2}(\mathrm{F}^* - \mathrm{F}),$$

where the meson names are used to denote their masses.

EXERCISE 2.31 Show that the model gives the Δ heavier than the nucleon. Further, show that if $a = a'$ in (2.91) and (2.92), then

$$m(\Delta) - m(\mathrm{N}) = \tfrac{1}{2}[m(\rho) - m(\pi)].$$

EXERCISE 2.32 Use (2.92) to study the relative masses of the Λ, Σ, Σ^*, Λ_c, Σ_c, and Σ_c^* baryons of Figs. 2.8 and 2.14. Each baryon is a qqQ composite, where q = u or d and Q = s or c. For the Λ and Σ baryons, show that the qq have isospin $I = 0$ and $I = 1$, respectively, and hence spin 0 and spin 1, respectively. Use this result to evaluate $\mathbf{S}_1 \cdot \mathbf{S}_2$, where $\mathbf{S}_i \equiv \frac{1}{2}\boldsymbol{\sigma}_i$. By considering $(\mathbf{S}_1 + \mathbf{S}_2 + \mathbf{S}_3)^2$, show that

$$(\mathbf{S}_1 + \mathbf{S}_2) \cdot \mathbf{S}_3 = 0, \ -1, \text{ and } +\tfrac{1}{2} \qquad \text{for } \Lambda, \Sigma, \text{ and } \Sigma^*, \text{ respectively.}$$

Thus, confirm that (2.92) gives

$$m(\Lambda_Q) = m_0 - \frac{3a'}{2m_u^2}$$

$$m(\Sigma_Q) = m_0 + \frac{2a'}{m_u^2}\left(\frac{1}{4} - \frac{m_u}{m_Q}\right)$$

$$m(\Sigma_Q^*) = m_0 + \frac{a'}{m_u^2}\left(\frac{1}{2} + \frac{m_u}{m_Q}\right)$$

where $m_0 = 2m_u + m_Q$. Verify that [see (2.84)]

$$[m(\Sigma_c) - m(\Lambda_c)] = \frac{m_s}{m_c}\frac{(m_c - m_u)}{(m_s - m_u)}[m(\Sigma) - m(\Lambda)] \simeq 0.16 \text{ GeV},$$

The masses of the other $\frac{1}{2}^+$ and $\frac{3}{2}^+$ baryons can also be calculated from (2.92) in terms of a' and the quark masses.

Considering the crude nature of the model, the quantitative agreement between the predictions and the observed masses is impressive. Indeed, all the observed features are reproduced. It is straightforward to enlarge the calculation to include hadrons containing b quarks.

2.15 Color Factors

We have already mentioned some of the evidence for color. We saw that there are good reasons to believe that each of the N flavors (u, d, ...) of quark comes in three colors which we called R, G, and B. To be precise, the quarks are assigned to a triplet of an $SU(3)$ color group (see Fig. 2.3). Unlike $SU(\mathrm{N})$ flavor symmetry, $SU(3)$ color symmetry is expected to be exactly conserved. A glance back at Fig. 1.4 reminds us that the gluons, which mediate the QCD force between color charges, come in eight different color combinations:

$$\mathrm{R\overline{G}, R\overline{B}, G\overline{R}, G\overline{B}, B\overline{R}, B\overline{G}}, \sqrt{\tfrac{1}{2}}\,(\mathrm{R\overline{R} - G\overline{G}}), \sqrt{\tfrac{1}{6}}\,(\mathrm{R\overline{R} + G\overline{G} - 2B\overline{B}}). \tag{2.93}$$

In other words, the gluons belong to an $SU(3)$ color octet [recall the $SU(3)$ flavor analogy of (2.49), (2.50) and Fig. 2.5]. The remaining combination, the $SU(3)$ color singlet,

$$\sqrt{\tfrac{1}{3}}\,(\mathrm{R\overline{R} + G\overline{G} + B\overline{B}}), \tag{2.94}$$

does not carry color and cannot mediate between color charges.

In QED, the strength of the electromagnetic coupling between two quarks is given by $e_1 e_2 \alpha$, where e_i is the electric charge in units of e (that is, $e_i = +\frac{2}{3}$ or $-\frac{1}{3}$) and α is the fine structure constant. Similarly, in QCD, the strength of the (strong) coupling for single-gluon exchange between two color charges is $\frac{1}{2} c_1 c_2 \alpha_s$, where c_1 and c_2 are the color coefficients associated with the vertices. It has become conventional to call

$$C_F \equiv \tfrac{1}{2}|c_1 c_2| \tag{2.95}$$

the color factor (although, in fact, it would have been more natural to absorb the factor $\frac{1}{2}$ in a redefinition of the strong coupling α_s and to just let the product $|c_1 c_2|$ be known as the color factor).

As a first example, we calculate the color factor for the interaction between two quarks of the same color, say B. Out of the eight gluons, only the one containing the $\mathrm{B\overline{B}}$ combination can be exchanged. The product $c_1 c_2$ is therefore $\frac{2}{3}$ (see Fig. 2.15a). On the other hand, the interaction between colored R quarks can be mediated by two different gluons (see Fig. 2.15b). Nevertheless, the total $c_1 c_2 = \frac{1}{6} + \frac{1}{2} = \frac{2}{3}$ is the same, as indeed it has to be from color symmetry.

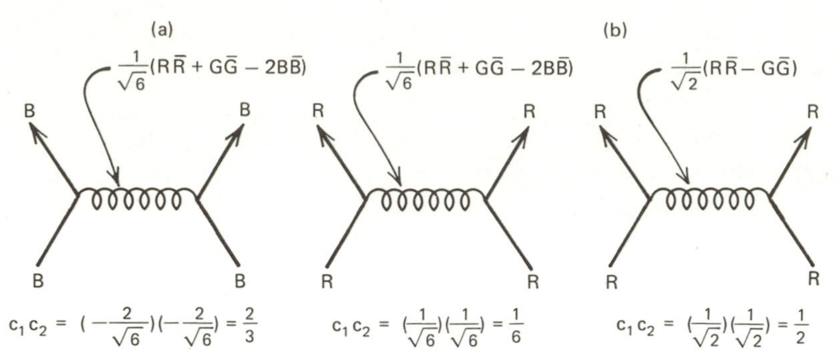

Fig. 2.15 The product of color couplings for (a) the B–B quark interaction and (b) the R–R quark interaction.

What about the interaction between two quarks of different color, say R and B? Here again, two different gluons are allowed with $c_1c_2 = -\frac{1}{3}$ and $+1$ (see Fig. 2.16). Do we add or subtract these two (indistinguishable) amplitudes? The answer depends on the symmetry of the color wavefunction under interchange of the quarks. For a symmetric (antisymmetric) state, we sum (subtract) to give a factor $+\frac{2}{3}$ $(-\frac{4}{3})$. Indeed, we already followed this prescription when we added the two amplitudes describing the R–R interaction.

All the results so far can be concisely summarized by

$$c_1c_2 = P - \tfrac{1}{3} \tag{2.96}$$

where $P = \pm 1$ according to whether the two quarks are in a color symmetric or antisymmetric state.

It is relevant to compute the color factor for gluon exchange between a quark and an antiquark in the color singlet state

$$\sqrt{\tfrac{1}{3}}\,(\mathrm{R\bar{R}} + \mathrm{G\bar{G}} + \mathrm{B\bar{B}}), \tag{2.97}$$

that is, between a $q\bar{q}$ pair in a meson. Here, all colors occur on an equal footing, and so it is sufficient to consider, say, the B–B̄ interaction. There are three possible diagrams (see Fig. 2.17). In computing c_1c_2, we insert a minus sign at the antiparticle vertex, just as in QED, where the antiparticle has opposite charge to

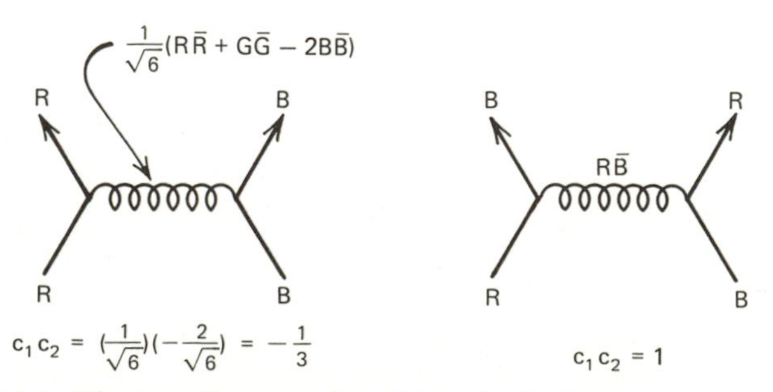

Fig. 2.16 The two diagrams describing the R–B quark interaction.

the particle. In a color singlet meson, each initial and final state of Fig. 2.17 has a factor $\sqrt{\frac{1}{3}}$ [see (2.97)]. The total factor for the q$\bar{\text{q}}$ interaction, via single-gluon exchange, in a meson is therefore

$$c_1 c_2 = 3\sqrt{\tfrac{1}{3}}\sqrt{\tfrac{1}{3}}\left(-\tfrac{2}{3} - 1 - 1\right) = -\tfrac{8}{3},$$

where the first factor of 3 allows for the contributions when R$\bar{\text{R}}$ and G$\bar{\text{G}}$ are the initial states. That is, the color factor

$$C_F = \tfrac{4}{3}, \tag{2.98}$$

a result we used in (2.89).

It is instructive to also derive (2.90), that is, to calculate the color factor for two quarks exchanging a gluon within a baryon. Now remembering that $3 \otimes 3 = 6 \oplus \bar{3}$, we see that every quark pair in a baryon is in a color $\bar{3}$. The reason is that the pair must be coupled to a third quark (a color 3) to give an overall color singlet. The alternative, $6 \otimes 3$, does not contain a singlet. The $\bar{3}$ is an antisymmetric color state [compare (2.58)], and so we must use (2.96) with $P = -1$. The required color factor, (2.90), is therefore

$$C_F = \tfrac{2}{3}. \tag{2.99}$$

The message of this chapter is that all the observed strongly interacting particles (hadrons) are bound states of quarks. Historically, it was flavor $SU(3)$ that led to the discovery of this fundamental fact, but its role in particle physics is now superseded by the quark model. Later, we develop a dynamical theory for the interactions of the (constituent) quarks which is based on color $SU(3)$ (gauge) symmetry.

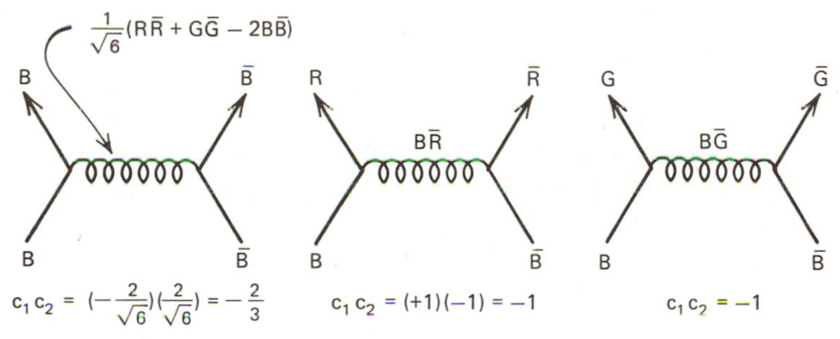

Fig. 2.17 Diagrams describing the B–$\bar{\text{B}}$ interaction.

3
Antiparticles

Quantum electrodynamics (QED) is the theory describing the electromagnetic interaction of quarks and leptons. Some of the high-energy physicist's favorite QED processes are $e^+e^- \to \mu^+\mu^-$, $eq \to eq$, $\gamma q \to (e^+e^-)q$, and so on, where q denotes a quark. The list reveals the technical problems to be faced in any computation of their transition rates: (1) we are concerned with a many-particle situation, (2) we are dealing with a relativistic problem. Indeed, not only are experiments routinely performed using beams of highly relativistic particles, ruling out any nonrelativistic approach, but also antiparticles occur. They are of course not required by nonrelativistic theory.

The problem is not as formidable as it looks; perturbation theory will save us. We obtain the solutions of the one-particle wave equations for *free* leptons (or quarks) and then study the scattering of one particle by another by treating the interaction as a perturbation.

At first sight, it is very surprising that single-particle wave equations can be used to describe interactions in which particles can be created and annihilated. The crucial observation is that relativistic wave equations have negative energy solutions which can be exploited so as to introduce antiparticles into the theory.

The final formalism is a covariant copy of nonrelativistic perturbation theory using only solutions to single-particle wave equations. As a calculational scheme, it is most readily implemented by summing the relevant "Feynman diagrams" that can be drawn for the process under study, where the diagrams are evaluated using a set of well-established rules: the Feynman rules. This heuristic, but very intuitive, approach is due to Feynman and has the practical advantage that we can calculate transition rates and cross sections at an early stage in the development of the formalism. We thus avoid having to develop the formal machinery of quantum field theory, which eventually yields the Feynman rules from a Lagrangian. (For an introduction to quantum field theory, see, for example, Mandl (1966) or Sakurai (1967).)

The spin of the quarks and leptons complicates to some extent the essential simplicity of Feynman's approach. We therefore introduce the calculational scheme using the unphysical example of "spinless leptons" (Chapters 3 and 4) and subsequently introduce their spin as a technical complication (Chapters 5 and 6). Chapter 7 offers a glimpse of the accuracy that can be achieved in QED calculations and describes computations which are verified by experiment to, for example, one part per million in the case of the magnetic moment of the electron.

3.1 Nonrelativistic Quantum Mechanics

We begin by recalling that a prescription for obtaining the Schrödinger equation for a free particle of mass m is to substitute into the classical energy momentum relation

$$E = \frac{p^2}{2m} \tag{3.1}$$

the differential operators

$$E \to i\hbar\frac{\partial}{\partial t}, \qquad \mathbf{p} \to i\hbar\nabla. \tag{3.2}$$

The resulting operator equation is understood to act on a (complex) wavefunction $\psi(\mathbf{x}, t)$. That is (with $\hbar \equiv 1$),

$$i\frac{\partial\psi}{\partial t} + \frac{1}{2m}\nabla^2\psi = 0, \tag{3.3}$$

where we interpret

$$\rho = |\psi|^2$$

as the probability density ($|\psi|^2 d^3x$ gives the probability of finding the particle in a volume element d^3x).

We are often concerned with moving particles, for example, the collision of one particle with another. We therefore need to be able to calculate the density flux of a beam of particles, $\mathbf{j}$. Now from the conservation of probability, the rate of decrease of the number of particles in a given volume is equal to the total flux of particles out of that volume, that is,

$$-\frac{\partial}{\partial t}\int_V \rho\, dV = \int_S \mathbf{j}\cdot\hat{\mathbf{n}}\, dS = \int_V \nabla\cdot\mathbf{j}\, dV$$

where the last equality is Gauss's theorem and $\hat{\mathbf{n}}$ is a unit vector along the outward normal to the element dS of the surface S enclosing volume V. The probability and the flux densities are therefore related by the "continuity" equation

$$\frac{\partial\rho}{\partial t} + \nabla\cdot\mathbf{j} = 0. \tag{3.4}$$

To determine the flux, we first form $\partial\rho/\partial t$ by subtracting the wave equation, (3.3), multiplied by $-i\psi^*$ from the complex conjugate equation multiplied by $-i\psi$. We then obtain

$$\frac{\partial\rho}{\partial t} - \frac{i}{2m}(\psi^*\nabla^2\psi - \psi\nabla^2\psi^*) = 0. \tag{3.5}$$

Comparing this with (3.4), we identify the probability flux density as

$$\mathbf{j} = -\frac{i}{2m}(\psi^* \nabla\psi - \psi \nabla\psi^*). \tag{3.6}$$

For example, a solution of (3.3),

$$\psi = N e^{i\mathbf{p}\cdot\mathbf{x} - iEt}, \tag{3.7}$$

which describes a free particle of energy E and momentum $\mathbf{p}$, has

$$\rho = |N|^2, \qquad \mathbf{j} = \frac{\mathbf{p}}{m}|N|^2. \tag{3.8}$$

3.2 Lorentz Covariance and Four-Vector Notation

A cornerstone of modern physics is that the fundamental laws have the same form in all Lorentz frames; that is, in reference frames which have a uniform relative velocity. The fundamental equations are said to be *Lorentz covariant*. Recall that the theory of special relativity is based on the premise that the velocity of light, c, is the same in all Lorentz frames. A Lorentz transformation relates the coordinates in two such frames. The basic invariant is $c^2t^2 - \mathbf{x}^2$.

> ***EXERCISE 3.1*** Consider a Lorentz transformation in which the new frame (primed coordinates) moves with velocity v along the z axis of the original frame (unprimed coordinates). For such a Lorentz "boost", show that
>
> $$ct' = \cosh\theta\, ct - \sinh\theta\, z,$$
>
> $$z' = -\sinh\theta\, ct + \cosh\theta\, z,$$
>
> with x and y unchanged; here, $\tanh\theta = v/c$. As $\cos i\theta = \cosh\theta$ and $\sin i\theta = i\sinh\theta$, we see that the Lorentz transformation may be regarded as a rotation through an imaginary angle $i\theta$ in the ict–z plane.

By definition, any set of four quantities which transform like $(ct, \mathbf{x})$ under Lorentz transformations is called a *four-vector*. We use the notation

$$(ct, \mathbf{x}) \equiv (x^0, x^1, x^2, x^3) \equiv x^\mu. \tag{3.9}$$

According to the theory of special relativity, the total energy E and the momentum $\mathbf{p}$ of an isolated system transform as the components of a four-vector

$$\left(\frac{E}{c}, \mathbf{p}\right) \equiv (p^0, p^1, p^2, p^3) = p^\mu$$

with the basic invariant $(E^2/c^2) - \mathbf{p}^2$. The simplest system is a free particle, for which

$$\frac{E^2}{c^2} - \mathbf{p}^2 = m^2c^2, \tag{3.10}$$

where m is the rest mass of the particle. From now on, we revert back to the use of natural units with $c \equiv 1$ (see Section 1.4).

Just as in three-dimensional space, we may introduce the scalar product of two four-vectors $A^\mu \equiv (A^0, \mathbf{A})$ and $B^\mu \equiv (B^0, \mathbf{B})$

$$A \cdot B \equiv A^0 B^0 - \mathbf{A} \cdot \mathbf{B},$$

which is left invariant under Lorentz transformations. Due to the minus sign, it is convenient to introduce a new type of four-vector, $A_\mu \equiv (A^0, -\mathbf{A})$, so that the scalar product is

$$A \cdot B = A_\mu B^\mu = A^\mu B_\mu = g_{\mu\nu} A^\mu B^\nu = g^{\mu\nu} A_\mu B_\nu . \tag{3.11}$$

Here, we have introduced the (metric) tensor $g_{\mu\nu}$, which is defined by

$$g_{00} = 1, \qquad g_{11} = g_{22} = g_{33} = -1, \qquad \text{other components} = 0$$

(and similarly for $g^{\mu\nu}$). A summation over repeated indices is implied in (3.11). Upper (lower) index vectors are called contravariant (covariant) vectors. The rule for forming Lorentz invariants is to make the upper indices balance the lower indices. If an equation is Lorentz covariant, we must ensure that all unrepeated indices (upper and lower separately) balance on either side of the equation, and that all repeated indices appear once as an upper and once as a lower index.

EXERCISE 3.2 Show that $g_{\mu\nu} g^{\mu\nu} = 4$.

Examples of scalar products are

$$p^\mu x_\mu \equiv p \cdot x = Et - \mathbf{p} \cdot \mathbf{x}$$

$$p^\mu p_\mu \equiv p \cdot p \equiv p^2 = E^2 - \mathbf{p}^2.$$

These quantities are Lorentz invariants. For a free particle, we have $p^2 = m^2$, see (3.10). We say that the particle is on its mass shell.

EXERCISE 3.3 The collision of two particles, each of mass M, is viewed in a Lorentz frame in which they hit head-on with momenta equal in magnitude but opposite in direction. We speak of this as the "center-of-mass" frame (though the name "center-of-momentum" would be more appropriate). The total energy of the system is E_{cm}. Show that the Lorentz invariant

$$s \equiv (p_1 + p_2)_\mu (p_1 + p_2)^\mu \equiv (p_1 + p_2)^2 = E_{cm}^2. \tag{3.12}$$

If the collision is viewed in the "laboratory" frame where one of the particles is at rest, then show, by evaluating the invariant s, that the other has energy

$$E_{lab} = \frac{E_{cm}^2}{2M} - M.$$

We can see from this result that colliding-beam accelerators have an enormous advantage over fixed-target accelerators in achieving a given total center-of-mass energy $\sqrt{s}$. List some advantages of fixed-target accelerators.

Note that the space-like components of A^μ and A_μ are $\mathbf{A}$ and $-\mathbf{A}$, respectively. The exception is

$$\partial^\mu = \left(\frac{\partial}{\partial t}, -\nabla\right) \quad \text{and} \quad \partial_\mu = \left(\frac{\partial}{\partial t}, \nabla\right), \tag{3.13}$$

which can be shown to transform like $x^\mu = (t, \mathbf{x})$ and $x_\mu = (t, -\mathbf{x})$, respectively. Thus, the covariant form of (3.2) is

$$p^\mu \to i\partial^\mu. \tag{3.14}$$

From ∂_μ and ∂^μ we can form the invariant (D'Alembertian) operator

$$\square^2 \equiv \partial_\mu \partial^\mu. \tag{3.15}$$

3.3 The Klein–Gordon Equation

Wave equation (3.3) violates Lorentz covariance and is not suitable for a particle moving relativistically. It is tempting to repeat the steps of Section 3.2, but starting from the relativistic energy–momentum relation, (3.10),

$$E^2 = \mathbf{p}^2 + m^2.$$

Making the operator substitutions (3.2), we obtain

$$-\frac{\partial^2 \phi}{\partial t^2} + \nabla^2 \phi = m^2 \phi, \tag{3.16}$$

which is known as the Klein–Gordon equation (but could, more correctly, have been called the relativistic Schrödinger equation). Multiplying the Klein–Gordon equation by $-i\phi^*$ and the complex conjugate equation by $-i\phi$, and subtracting, gives the relativistic analogue of (3.5)

$$\frac{\partial}{\partial t} \underbrace{\left[i\left(\phi^* \frac{\partial \phi}{\partial t} - \phi \frac{\partial \phi^*}{\partial t} \right) \right]}_{\rho} + \nabla \cdot \underbrace{\left[-i(\phi^* \nabla \phi - \phi \nabla \phi^*) \right]}_{j} = 0. \tag{3.17}$$

By comparison with (3.4), we identify the probability and the flux densities with the terms in square brackets. For example, for a free particle of energy E and momentum $\mathbf{p}$, described by the Klein–Gordon solution

$$\phi = N e^{i\mathbf{p}\cdot\mathbf{x} - iEt},$$

we find from (3.17) that [see (3.8)].

$$\begin{aligned} \rho &= i(-2iE)|N|^2 = 2E|N|^2 \\ \mathbf{j} &= -i(2i\mathbf{p})|N|^2 = 2\mathbf{p}|N|^2. \end{aligned} \tag{3.18}$$

We see that the probability density is proportional to E, the relativistic energy of the particle. (We defer the explanation of this for a moment.)

It is advantageous to express these results in four-vector notation. Not only are they then more concise, but also the covariance becomes explicit. Using the D'Alembertian operator, (3.15), the Klein–Gordon equation becomes

$$(\square^2 + m^2)\phi = 0. \tag{3.19}$$

Moreover, the probability and the flux densities form a four-vector

$$j^\mu = (\rho, \mathbf{j}) = i(\phi^* \, \partial^\mu \phi - \phi \, \partial^\mu \phi^*) \tag{3.20}$$

which satisfies the (covariant) continuity relation

$$\partial_\mu j^\mu = 0. \tag{3.21}$$

Taking the free particle solution

$$\phi = N e^{-ip \cdot x}, \tag{3.22}$$

we have [see (3.18)]

$$j^\mu = 2 p^\mu |N|^2. \tag{3.23}$$

We noted that the probability density ρ is the time-like component of a four-vector; ρ is proportional to E. This result may be anticipated since under a Lorentz boost of velocity $\mathbf{v}$, a volume element suffers a Lorentz contraction $d^3x \to d^3x\sqrt{1 - v^2}$; and so, to keep $\rho \, d^3x$ invariant, we require ρ to transform as the time-like component of a four-vector $\rho \to \rho/\sqrt{1 - v^2}$.

So far, so good; but what are the energy eigenvalues of the Klein–Gordon equation? Substitution of (3.22) into (3.19) gives

$$E = \pm(\mathbf{p}^2 + m^2)^{1/2}. \tag{3.24}$$

Thus, in addition to the acceptable $E > 0$ solutions, we have negative energy solutions. This looks at first like a total disaster, because transitions can occur to lower and lower (more negative) energies. A second problem is that the $E < 0$ solutions are associated with a negative probability density from (3.18). To summarize, the difficulties are

$$\boxed{E < 0 \text{ solutions with } \rho < 0.}$$

It is clear that this problem cannot be simply ignored. We cannot simply discard the negative energy solutions as we have to work with a complete set of states, and this set inevitably includes the unwanted states.

3.4 Historical Interlude

In 1927, in an attempt to avoid these problems, Dirac devised a relativistic wave equation linear in $\partial/\partial t$ and ∇. He succeeded in overcoming the problem of the negative probability density, with the unexpected bonus that the equation de-

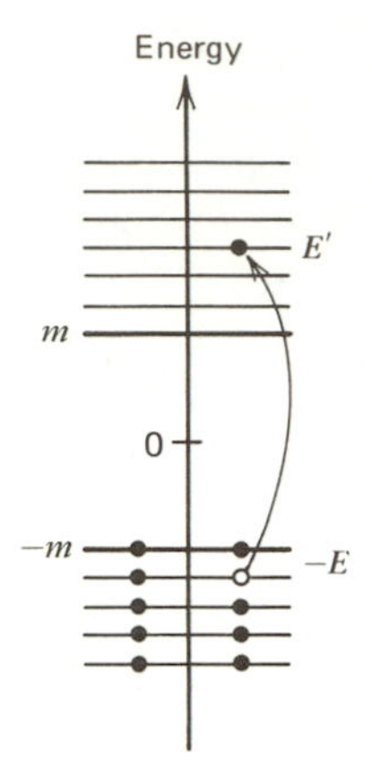

Fig. 3.1 Energy level spectrum for the electron. Dirac's picture of the vacuum has all the negative energy states occupied. We show two states per level to account for the two spin states of the electron.

scribed spin-$\frac{1}{2}$ particles. However, $E < 0$ solutions still occurred, as can be seen in the energy spectrum for a free Dirac electron sketched in Fig. 3.1. Dirac sidestepped the negative energy solutions by invoking the exclusion principle. He postulated that all the negative energy states are occupied and regarded the vacuum as an infinite sea of $E < 0$ electrons. Now the positive energy electrons cannot collapse into the lower (negative) energy levels, as this is prevented by the exclusion principle. One can, however, create a "hole" in the sea by excitation of an electron from a negative energy $(-E)$ state to a positive energy (E') state, as shown. The absence of an electron of charge $-e$ and energy $-E$ is interpreted as the presence of an antiparticle (a positron) of charge $+e$ and energy $+E$. Thus, the net effect of this excitation is the production of a pair of particles

$$e^-(E') + e^+(E),$$

which clearly requires energy $E + E' \geq 2m$ (see diagram). Until 1934, the Dirac equation was considered to be the only acceptable relativistic wave equation.

In 1934, Pauli and Weisskopf revived the Klein–Gordon equation by inserting the charge $-e$ into j^μ and interpreting it as the charge-current density of the electron,

$$j^\mu = -ie(\phi^* \partial^\mu \phi - \phi \partial^\mu \phi^*). \tag{3.25}$$

Now, $\rho = j^0$ represents a charge density, not a probability density, and so the fact that it can be negative is no longer objectionable. In some sense, which we shall make clear in a moment, the $E < 0$ solutions may then be regarded as $E > 0$ solutions for particles of opposite charge (antiparticles). Unlike "hole theory," this interpretation is applicable to bosons as well as fermions. One cannot fill up the Dirac sea with bosons, as there is no exclusion principle operative to stack the particles. To develop the antiparticle idea and to introduce Feynman diagrams, it is useful to first ignore the complications due to the spin of the electrons. We therefore begin by obtaining the Feynman rules for "spinless" electrons and use them to calculate the scattering amplitudes and cross sections for interacting particles. Not until then do we turn to the Dirac equation and to the Feynman rules for the physically realistic case of electromagnetic interactions of spin-$\frac{1}{2}$ electrons.

3.5 The Feynman–Stückelberg Interpretation of $E < 0$ Solutions

The prescription we use for handling negative energy states was proposed by Stückelberg (1941) and by Feynman (1948). Expressed most simply, the idea is that a negative energy solution describes a particle which propagates backward in time or, equivalently, a positive energy antiparticle propagating forward in time. It is crucial to master this idea, as it lies at the heart of our approach to Feynman diagrams. We try to make it plausible in the following way.

Consider an electron of energy E, three-momentum $\mathbf{p}$, and charge $-e$. From (3.25) and (3.22), we know that the electromagnetic four-vector current is

$$j^{\mu}(e^{-}) = -2e|N|^{2}(E,\mathbf{p}). \tag{3.26}$$

Now take an antiparticle, a positron, with the same $E,\mathbf{p}$. Since its charge is $+e$,

$$\begin{aligned} j^{\mu}(e^{+}) &= +2e|N|^{2}(E,\mathbf{p}) \\ &= -2e|N|^{2}(-E,-\mathbf{p}), \end{aligned} \tag{3.27}$$

which is exactly the same as the current j^{μ} for an electron with $-E$, $-\mathbf{p}$. Thus, as far as a system is concerned, the emission of a positron with energy E is the same as the absorption of an electron of energy $-E$. Pictorially, we have

$$\left.\begin{matrix} \uparrow & e^{+} \\ & E > 0 \end{matrix}\right. \equiv \left.\begin{matrix} \downarrow & e^{-} \\ & (-E) < 0 \end{matrix}\right. \qquad \uparrow \text{time} \tag{3.28}$$

In other words, negative-energy *particle* solutions going backward in time describe positive-energy *antiparticle* solutions going forward in time. Of course, the reason why this identification can be made is simply because

$$e^{-i(-E)(-t)} = e^{-iEt}.$$

The single-particle (e^{-}) wavefunction formalism not only handles antiparticles but can even describe many-particle situations. As an example, we consider the double scattering of an electron in a potential. We picture this in the space-time (Feynman) diagrams of Fig. 3.2. The crucial observation is that there are two pictures corresponding to the same observation. There are two different time orderings of the two interactions with the potential that lead to the same observable event. Indeed, note that the (observable) path of the electron before and after the double scattering is the same in the two diagrams. The second picture is only possible because of the antiparticle prescription. At time t_2, the electron scatters backward in time (with $E < 0$). This electron is interpreted as a

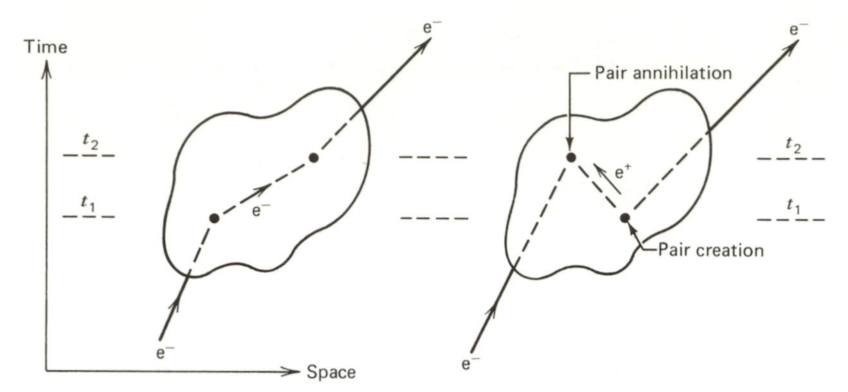

Fig. 3.2 Different time orderings of the double scattering of an electron.

positron (with $E > 0$) going forward in time. Then, the events in the diagram can be viewed as follows: first, at time t_1, an e^-e^+ pair is created; then, at a later time t_2, the e^+ is annihilated by the incident e^-. Therefore, between t_1 and t_2, the electron trajectory drawn in the second diagram actually describes three particles: the initial and final electrons and a positron! As both double scatterings lead to the same observed final electron, they both have to be included in computing the probability of this event. Note that, just as in hole theory, the vacuum has become a very complex environment: e^-e^+ pairs can pop out of it and disappear into it as a result of the antiparticle prescription!

All possible processes can be described with the interpretation of a single-particle (e^-) wavefunction; the antiparticle (e^+) states are never used. For example, for a single e^+ scattering, Fig. 3.3, we use negative-energy e^- solutions with the exit and entrance states interchanged.

Our objective is to calculate transition rates and cross sections. Yet, so far, we have only the wavefunctions for free particles. How are interactions to be included? As implied by the above discussion, with its mention of "single" and "double" scattering, perturbation theory will be the method that we shall use to calculate scattering amplitudes. It is therefore appropriate to recall the main results of perturbation theory we shall need.

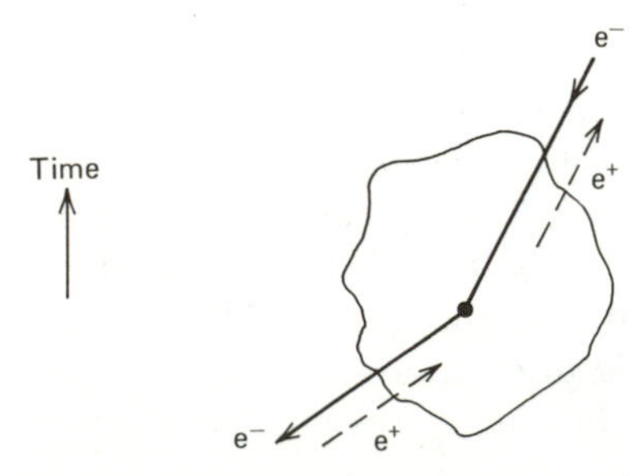

Fig. 3.3 Positron scattering by a potential.

3.6 Nonrelativistic Perturbation Theory

Suppose we know the solutions to the free-particle Schrödinger equation

$$H_0\phi_n = E_n\phi_n \qquad \text{with} \int_V \phi_m^*\phi_n\, d^3x = \delta_{mn} \tag{3.29}$$

where H_0, the Hamiltonian, is time independent. For simplicity, we have normalized the solutions to one particle in a box of volume V. The objective is to solve Schrödinger's equation

$$(H_0 + V(\mathbf{x}, t))\psi = i\frac{\partial\psi}{\partial t} \tag{3.30}$$

for a particle moving in the presence of an interaction potential $V(\mathbf{x}, t)$.

Any solution of (3.30) can be expressed in the form

$$\psi = \sum_n a_n(t)\,\phi_n(\mathbf{x})e^{-iE_nt}. \tag{3.31}$$

Now, to find the unknown coefficients $a_n(t)$, we substitute (3.31) into (3.30) and obtain

$$i\sum_n \frac{da_n}{dt}\phi_n(\mathbf{x})e^{-iE_nt} = \sum_n V(\mathbf{x}, t)\, a_n\, \phi_n(\mathbf{x})e^{-iE_nt}.$$

Multiplying by ϕ_f^*, integrating over the volume, and using the orthonormality relation (3.29) leads to the following coupled linear differential equations for the a_n coefficients:

$$\frac{da_f}{dt} = -i\sum_n a_n(t)\int \phi_f^* V\phi_n\, d^3x\, e^{i(E_f - E_n)t}. \tag{3.32}$$

Suppose that before the potential V acts the particle is in an eigenstate i of the unperturbed Hamiltonian, that is, at time $t = -T/2$:

$$\begin{aligned} a_i(-T/2) &= 1, \\ a_n(-T/2) &= 0 \qquad \text{for } n \neq i, \end{aligned} \tag{3.33}$$

and

$$\frac{da_f}{dt} = -i\int d^3x\, \phi_f^* V\phi_i\, e^{i(E_f - E_i)t}. \tag{3.34}$$

Now, provided that the potential is small and transient, we can, as a first approximation, assume that these initial conditions remain true at all times. Then, integrating (3.34), we obtain

$$a_f(t) = -i\int_{-T/2}^{t} dt' \int d^3x\, \phi_f^* V\phi_i\, e^{i(E_f - E_i)t'} \tag{3.35}$$

and, in particular, at time $t = +T/2$ after the interaction has ceased,

$$T_{fi} \equiv a_f(T/2) = -i\int_{-T/2}^{T/2} dt \int d^3x\left[\phi_f(\mathbf{x})\, e^{-iE_ft}\right]^* V(\mathbf{x}, t)\left[\phi_i(\mathbf{x})e^{-iE_it}\right] \tag{3.36}$$

which we may write in the covariant form

$$\boxed{T_{fi} = -i\int d^4x\, \phi_f^*(x)\, V(x)\, \phi_i(x).} \tag{3.37}$$

Of course, the expression for $a_f(t)$ is only a good approximation if $a_f(t) \ll 1$, as this has been assumed in obtaining the result.

We are tempted to interpret $|T_{fi}|^2$ as the probability that the particle is scattered from an initial state i to a final state f. Is this interpretation valid? Consider the case when $V(\mathbf{x}, t) = V(\mathbf{x})$ is time independent; then, (3.36) can be written as

$$\begin{aligned} T_{fi} &= -iV_{fi}\int_{-\infty}^{\infty} dt\, e^{-i(E_f - E_i)t} \\ &= -2\pi i V_{fi}\, \delta(E_f - E_i) \end{aligned} \tag{3.38}$$

with

$$V_{fi} \equiv \int d^3x\, \phi_f^*(\mathbf{x})\, V(\mathbf{x})\, \phi_i(\mathbf{x}). \tag{3.39}$$

The δ-function in (3.38) expresses the fact that the energy of the particle is conserved in the transition $i \to f$. By the uncertainty principle, this means that an infinite time separates the states i and f, and $|T_{fi}|^2$ is therefore not a meaningful quantity. We define instead a transition probability per unit time

$$W = \lim_{T\to\infty} \frac{|T_{fi}|^2}{T}. \tag{3.40}$$

Squaring (3.38),

$$\begin{aligned} W &= \lim_{T\to\infty} 2\pi \frac{|V_{fi}|^2}{T} \delta(E_f - E_i) \int_{-T/2}^{+T/2} dt\, e^{i(E_f - E_i)t} \\ &= \lim_{T\to\infty} 2\pi \frac{|V_{fi}|^2}{T} \delta(E_f - E_i) \int_{-T/2}^{+T/2} dt \\ &= 2\pi |V_{fi}|^2 \delta(E_f - E_i). \end{aligned} \tag{3.41}$$

This equation can only be given physical meaning after integrating over a set of initial and final states. In particle physics, we usually deal with situations where we start with a specified initial state and end up in one of a set of final states. Let $\rho(E_f)$ be the density of final states; that is, $\rho(E_f)\, dE_f$ is the number of states in the energy interval E_f to $E_f + dE_f$. We integrate over this density, imposing energy conservation, and obtain the transition rate

$$\begin{aligned} W_{fi} &= 2\pi \int dE_f \rho(E_f) |V_{fi}|^2\, \delta(E_f - E_i) \\ &= 2\pi |V_{fi}|^2\, \rho(E_i). \end{aligned} \tag{3.42}$$

This is Fermi's Golden Rule.

Clearly, we can improve on the above approximation by inserting the result for $a_n(t)$, (3.35), in the right-hand side of (3.32):

$$\frac{da_f}{dt} = \ldots + (-i)^2 \left[\sum_{n \neq i} V_{ni} \int_{-T/2}^{t} dt' \, e^{i(E_n - E_i)t'} \right] V_{fn} \, e^{i(E_f - E_n)t} \tag{3.43}$$

where the dots represent the first-order result. The correction to T_{fi} is

$$T_{fi} = \ldots - \sum_{n \neq i} V_{fn} V_{ni} \int_{-\infty}^{\infty} dt \, e^{i(E_f - E_n)t} \int_{-\infty}^{t} dt' \, e^{i(E_n - E_i)t'}.$$

To make the integral over dt' meaningful, we must include a term in the exponent involving a small positive quantity ε which we let go to zero after integration

$$\int_{-\infty}^{t} dt' \, e^{i(E_n - E_i - i\varepsilon)t} = i \frac{e^{i(E_n - E_i - i\varepsilon)t}}{E_i - E_n + i\varepsilon}.$$

The second-order correction to T_{fi} is therefore

$$T_{fi} = \cdots - 2\pi i \sum_{n \neq i} \frac{V_{fn} V_{ni}}{E_i - E_n + i\varepsilon} \delta(E_f - E_i). \tag{3.44}$$

EXERCISE 3.4 Show that the rate for the $i \to f$ transition is given by (3.42) with the replacement

$$V_{fi} \to V_{fi} + \sum_{n \neq i} V_{fn} \frac{1}{E_i - E_n + i\varepsilon} V_{ni} + \cdots. \tag{3.45}$$

Obtain the form in the next correction.

Equation (3.45) is the perturbation series for the amplitude with terms to first, second, ... order in V. The Feynman diagrams of Fig. 3.4 represent the first two terms in the nonrelativistic perturbation series. For each interaction vertex, we

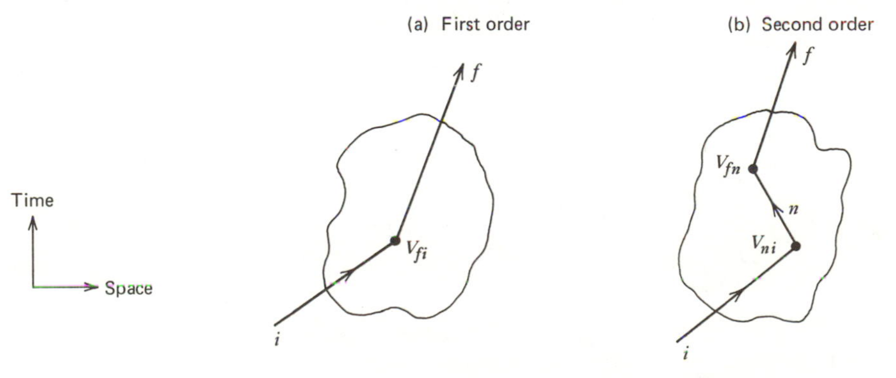

Fig. 3.4 First- and second-order contributions to the $i \to f$ transition.

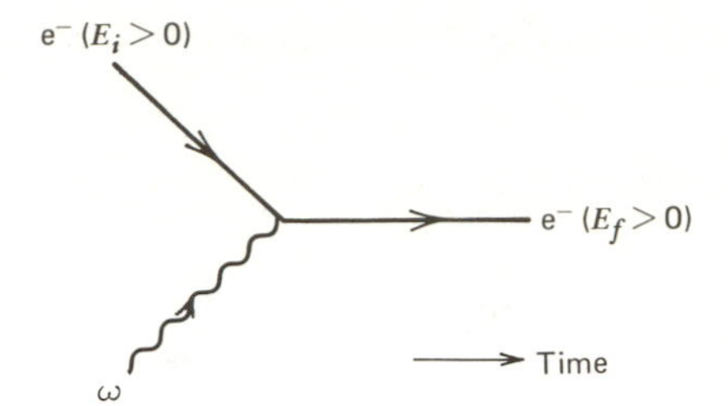

Fig. 3.5 e⁻ Scattering, with time increasing from left to right (rather than upward).

have a factor like V_{ni}, and for the propagation of each intermediate state, we have a "propagator" factor like $1/(E_i - E_n)$. The intermediate states are "virtual" in the sense that energy is not conserved, $E_n \neq E_i$, but there is of course energy conservation between the initial and final states, $E_f = E_i$, as indicated by the delta function $\delta(E_f - E_i)$. The central problem is to generalize this scheme to handle relativistic particles, including their antiparticles. This is the topic of the next chapter.

3.7 Rules for Scattering Amplitudes in the Feynman–Stückelberg Approach

How are we to form scattering amplitudes T_{fi} involving antiparticles if antiparticles are to be regarded as negative-energy particle solutions going backward in time? Clearly, our antiparticle prescription will have to be consistent with energy conservation.

The diagrams of Fig. 3.4 represent a noncovariant situation; they refer to scattering from a fixed, static potential. However, we are interested in the scattering of one particle by another, and to do this we will take one particle to be moving in the electromagnetic potential V due to the other. In Chapter 1, we described how the electromagnetic interaction between electrons is due to the emission and absorption of photons. Consider now energy conservation at the vertex of Fig. 3.5, in which a photon is absorbed by an electron. The form of the (covariant) interaction V is derived in the next chapter, but it is clear that V

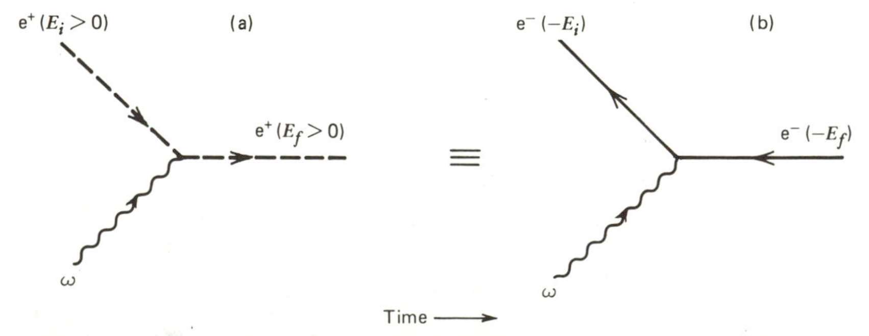

Fig. 3.6 e⁺ Scattering, pictured as negative-energy e⁻ scattering backward in time.

has a time dependence $e^{-i\omega t}$ for an incoming photon of energy ω. Thus, the transition amplitude T_{fi}, (3.36), is proportional to

$$\int (e^{-iE_f t})^* e^{-i\omega t} e^{-iE_i t}\, dt = 2\pi\,\delta(E_f - \omega - E_i)$$

so that $E_f = E_i + \omega$.

Obviously, exactly the same argument will follow for antiparticle (positron) scattering in Fig. 3.6a. However, as before, we wish to formulate the matrix element in terms of electron states alone, as shown in Fig. 3.6b. Now, the ingoing state describes an electron of (negative) energy $-E_f$, and the transition amplitude contains the factor

$$\int (e^{-i(-E_i)t})^* e^{-i\omega t} e^{-i(-E_f)t}\, dt = 2\pi\,\delta(-E_i - \omega + E_f)$$

so again, $E_f = E_i + \omega$, as required.

The *rule* is to form the matrix element

$$\int \phi^*_{\text{outgoing}} V \phi_{\text{ingoing}}\, d^4x,$$

where ingoing and outgoing always refer to the arrows on the *particle* (electron) lines.

EXERCISE 3.5 Check that the rule satisfies the conservation of energy for (a) e^-e^+ pair creation, and (b) e^-e^+ annihilation of Fig. 3.7. Following the same idea, use the space part of the matrix element to show that the expected three-momentum conservation laws are obtained.

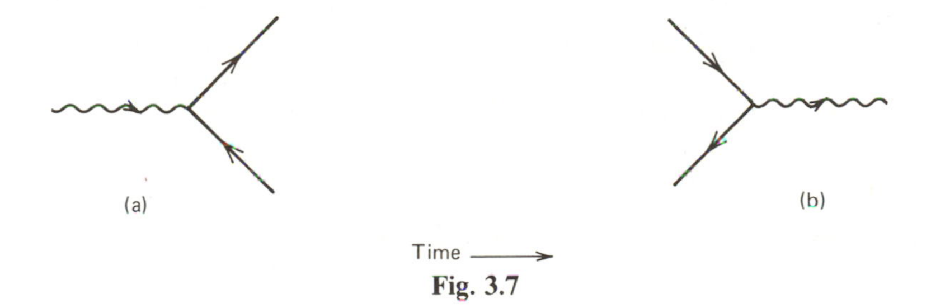

Fig. 3.7

We have now set up a formalism based on perturbation theory which can handle interactions of particles and antiparticles. It can even describe multiparticle situations. The next task is to cast it in a relativistically covariant form. Note that (3.37) is already covariant.

4

Electrodynamics of Spinless Particles

The title of this chapter requires some explanation. No spinless quark or lepton has ever been observed in an experiment. Spinless hadrons exist (e.g., the π-meson), but they are complicated composite structures of spin-$\frac{1}{2}$ quarks and spin-1 gluons. The spin-zero leptons, that is, leptons satisfying the Klein–Gordon equation, which appear throughout this chapter, are completely fictitious objects. The goal of this chapter is to find out how to use perturbation theory in a covariant way. To illustrate this, we have to choose particles and an interaction. For simplicity, we choose the particles to be "spinless" charged leptons. Clearly, it is desirable to begin by avoiding the complications of their spin. For the interaction, we choose the electromagnetic force. Electromagnetic interactions are of fundamental importance in particle physics. Quantum electrodynamics is the simplest example of a gauge theory in the sense that it has only one gauge particle, the photon. Gauge theories are now thought to be capable of describing all interactions. Chromodynamics and weak interactions are described by gauge theories that copy QED. This is demonstrated in Chapters 14 and 15. So, although the "spinless" leptons in this chapter are fictitious, the interaction is not. Understanding this is essential for further progress. The complications that arise from the spin of the leptons are revealed in Chapters 5 and 6.

4.1 An "Electron" in an Electromagnetic Field A^μ

A free "spinless" electron satisfies the Klein–Gordon equation, (3.19):

$$\left(\partial_\mu \partial^\mu + m^2\right)\phi = 0.$$

In classical electrodynamics, the motion of a particle of charge $-e$ in an electromagnetic potential $A^\mu = (A^0, \mathbf{A})$ is obtained by the substitution

$$p^\mu \to p^\mu + eA^\mu \tag{4.1}$$

(see a standard text such as Goldstein or Jackson). The corresponding quantum-mechanical substitution is therefore

$$\boxed{i\partial^\mu \to i\partial^\mu + eA^\mu,} \tag{4.2}$$

see (3.14); and the Klein–Gordon equation becomes

$$\left(\partial_\mu \partial^\mu + m^2\right)\phi = -V\phi, \tag{4.3}$$

where the (electromagnetic) perturbation is

$$V = -ie\left(\partial_\mu A^\mu + A^\mu \partial_\mu\right) - e^2 A^2. \tag{4.4}$$

The sign of V in (4.3) is chosen to be in accord with the relative sign of the kinetic and potential energy terms of the Schrödinger equation.

Everything is happening in (4.2)! This *is* quantum electrodynamics. (In Chapter 14, we shall see that this fundamental prescription emerges naturally from insisting that physics is unchanged under gauge, or phase, transformations.)

The potential, (4.4), is characterized by the parameter e, which (in natural units) is related to the fine structure constant α by

$$\alpha = \frac{e^2}{4\pi} \simeq \frac{1}{137}, \tag{4.5}$$

recall (1.3). The smallness of the electromagnetic coupling means that it is sensible to make a perturbation expansion of V in powers of α. The lowest-order (in α) contribution to a scattering amplitude should be a good approximation.

Working to lowest order, we omit the e^2A^2 term in (4.4). The amplitude, (3.37), for the scattering of a "spinless" electron from a state ϕ_i to ϕ_f off an electromagnetic potential A_μ, which we represent by Fig. 4.1, is

$$\begin{aligned} T_{fi} &= -i\int \phi_f^*(x)\, V(x)\, \phi_i(x)\, d^4x \\ &= i\int \phi_f^* ie\left(A^\mu \partial_\mu + \partial_\mu A^\mu\right)\phi_i\, d^4x. \end{aligned} \tag{4.6}$$

The derivative, in the second term, which acts on both A^μ and ϕ_i, can be turned around by integration by parts, so that it acts on ϕ_f^*:

$$\int \phi_f^*\, \partial_\mu\left(A^\mu \phi_i\right) d^4x = -\int \partial_\mu\left(\phi_f^*\right) A^\mu \phi_i\, d^4x, \tag{4.7}$$

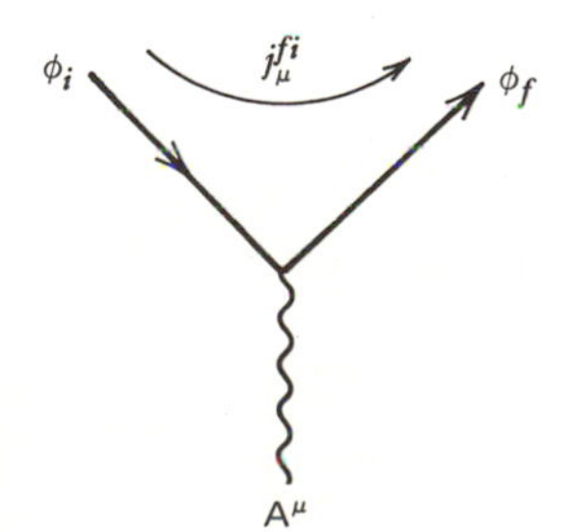

Fig. 4.1 A "spinless" electron interacting with A^μ.

where we have omitted the surface term as the potential is taken to vanish as $|\mathbf{x}|$, $t \to \pm\infty$. We may therefore rewrite the amplitude T_{fi} in the suggestive form

$$T_{fi} = -i\int j_\mu^{fi} A^\mu \, d^4x, \tag{4.8}$$

where

$$j_\mu^{fi}(x) \equiv -ie\big(\phi_f^*(\partial_\mu\phi_i) - (\partial_\mu\phi_f^*)\phi_i\big) \tag{4.9}$$

which, by comparison with (3.25), can be regarded as the electromagnetic current for the $i \to f$ electron transition. If the ingoing spinless electron has four-momentum p_i, we have

$$\phi_i(x) = N_i e^{-ip_i \cdot x}, \tag{4.10}$$

where N_i is the normalization constant. Using a similar expression for ϕ_f, it follows that

$$j_\mu^{fi} = -eN_i N_f(p_i + p_f)_\mu \, e^{i(p_f - p_i)\cdot x}. \tag{4.11}$$

4.2 "Spinless" Electron–Muon Scattering

Using the results for the scattering of an electron off an electromagnetic potential A^μ, shown in Fig. 4.1, we are able to calculate the scattering of the same electron off another charged particle, say, another electron or a muon. Let us choose a muon to avoid dealing with identical particles. The Feynman diagram corresponding to the process is shown in Fig. 4.2. It suggests how to approach the problem. The calculation is an extension of the previous one; we just have to identify the electromagnetic potential A^μ with its source, the charged "spinless" muon. This identification is done using Maxwell's equations

$$\square^2 A^\mu = j^\mu_{(2)} \tag{4.12}$$

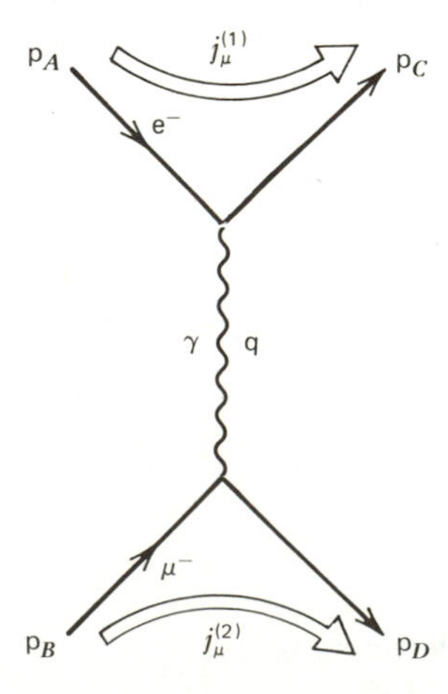

Fig. 4.2 Electron–muon scattering showing the particle four-momenta.

which determine the electromagnetic field A^μ associated with the current $j^\mu_{(2)}$ of the muon. (If you are unfamiliar with (4.12), that is, with the covariant form of Maxwell's equations in the Lorentz gauge, work through Exercises 6.9 and 6.10.) Now what do we take for the current in (4.12)? Again, Fig. 4.2 suggests the answer. The current associated with a spinless muon has the same form as that for the electron, which is given by (4.11). Thus, we have

$$j^\mu_{(2)} = -eN_B N_D (p_D + p_B)^\mu e^{i(p_D - p_B)\cdot x}, \tag{4.13}$$

where the momenta are defined in Fig. 4.2. Since

$$\square^2 e^{iq\cdot x} = -q^2 e^{iq\cdot x}, \tag{4.14}$$

the solution of (4.12) is

$$A^\mu = -\frac{1}{q^2} j^\mu_{(2)} \qquad \text{with } q = p_D - p_B. \tag{4.15}$$

Inserting this field due to the muon into (4.8), we find that the (lowest-order) amplitude for electron–muon scattering is

$$T_{fi} = -i \int j^{(1)}_\mu (x) \left(-\frac{1}{q^2} \right) j^\mu_{(2)}(x)\, d^4x. \tag{4.16}$$

Inserting (4.13), together with the corresponding expression for the electron current, (4.11), and carrying out the x integration, we find

$$T_{fi} = -iN_A N_B N_C N_D (2\pi)^4 \delta^{(4)}(p_D + p_C - p_B - p_A)\mathfrak{M} \tag{4.17}$$

with

$$-i\mathfrak{M} = \left(ie(p_A + p_C)^\mu \right) \left(-i\frac{g_{\mu\nu}}{q^2} \right) \left(ie(p_B + p_D)^\nu \right). \tag{4.18}$$

A consistency check on result (4.18) is that we get the same amplitude if we take the muon to be moving in the field A_μ produced by the electron. $\mathfrak{M}$, as defined by (4.17), is known as the *invariant* amplitude. The delta function expresses energy–momentum conservation for the process.

In order to catalogue the different terms in the perturbative expansion of T_{fi} in nonrelativistic perturbation theory, we drew pictures like Fig. 3.4. Furthermore, the different factors in T_{fi} as given by (3.44) were associated with interaction vertices and particle propagators in Fig. 3.4b. It is useful to draw the same pictures for the covariant form of the perturbation series. For example, Fig. 4.3 represents spinless electron–muon scattering to order e^2 (or α), and the amplitude is given by (4.17) and (4.18). This is the lowest-order Feynman diagram. The wavy line represents a photon exchanged between the leptons, and the associated factor $-ig_{\mu\nu}/q^2$ is called the *photon propagator*; it carries Lorentz indices because the photon is a spin-1 particle (see Chapter 6). The four-momentum q of the photon is determined by four-momentum conservation at the vertices. We see that

$q^2 \neq 0$, and we say the photon is "virtual" or "off-mass shell." At each of the vertices of the diagram, we associate the factor shown. Each *vertex factor* contains the electromagnetic coupling e and a four-vector index to connect with the photon index. The particular distribution of minus signs and factors i has been made to give the correct results for higher-order diagrams. Note that the multiplication of the three factors gives $-i\mathfrak{M}$.

Whenever the same vertex or internal line occurs in a Feynman diagram, the corresponding factor will contribute multiplicatively to the amplitude $-i\mathfrak{M}$ for that diagram. We may thus start to draw up a table of Feynman rules for quantum electrodynamics, which will, when complete, allow us to quickly write down the expression for the amplitude, $-i\mathfrak{M}$, for any Feynman diagram. The table is given in Section 6.17.

4.3 The Cross Section in Terms of the Invariant Amplitude $\mathfrak{M}$

To relate these calculations to experimental observables, we need to fix the normalization N of our free particle wavefunctions:

$$\phi = Ne^{-ip\cdot x}. \tag{4.19}$$

Recall from (3.18) that the probability density ρ of particles described by ϕ is

$$\rho = 2E|N|^2.$$

The proportionality of ρ to E was just what we needed to compensate for the Lorentz contraction of the volume element d^3x and to keep the number of particles $\rho\, d^3x$ unchanged. Let us therefore work in a volume V and normalize to $2E$ particles in V,

$$\int_V \rho\, dV = 2E. \tag{4.20}$$

That is, we adopt a covariant normalization

$$N = \frac{1}{\sqrt{V}}. \tag{4.21}$$

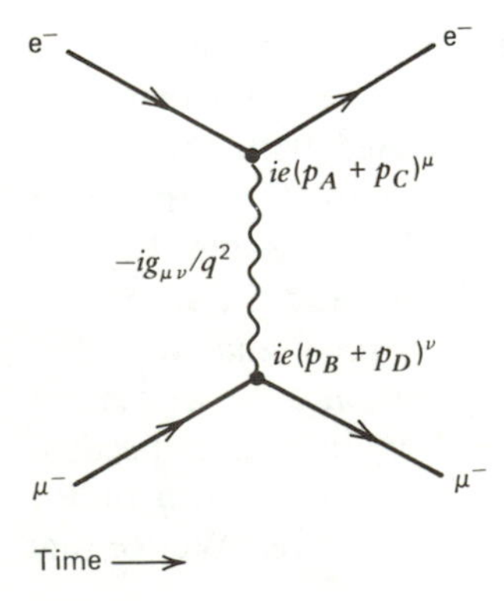

Fig. 4.3 The vertex factors and propagator for "spinless" electron–muon scattering.

Now, the A + B → C + D transition rate per unit volume is

$$W_{fi} = \frac{|T_{fi}|^2}{TV}$$

where T is the time interval of the interaction, and the transition amplitude is

$$T_{fi} = -iN_A N_B N_C N_D (2\pi)^4 \delta^{(4)}(p_C + p_D - p_A - p_B)\mathfrak{M},$$

see (4.17). On squaring, one delta function remains, and $(2\pi)^4$ times the other gives TV [the calculation is identical to the one that led to (3.41)]. Thus, making use of (4.21), we obtain

$$W_{fi} = (2\pi)^4 \frac{\delta^{(4)}(p_C + p_D - p_A - p_B)|\mathfrak{M}|^2}{V^4}. \tag{4.22}$$

Experimental results on AB → CD scattering are usually quoted in the form of a "cross section." It is related to the transition rate by

$$\text{Cross section} = \frac{W_{fi}}{(\text{initial flux})}(\text{number of final states}), \tag{4.23}$$

where the factors in brackets allow for the "density" of the incoming and outgoing states. We first carefully define these factors and then show how the cross section, so defined, may be regarded as the effective area over which particles A, B interact to produce C, D.

For a single particle, quantum theory restricts the *number of final states* in a volume V with momenta in element d^3p to be $V d^3p/(2\pi)^3$ (see Exercise 4.1). But we have $2E$ particles in V, and so

$$\text{No. of final states/particle} = \frac{V d^3p}{(2\pi)^3 2E}. \tag{4.24}$$

Thus, for particles C, D scattered into momentum elements d^3p_C, d^3p_D,

$$\text{No. of available final states} = \frac{V d^3p_C}{(2\pi)^3 2E_C} \frac{V d^3p_D}{(2\pi)^3 2E_D}. \tag{4.25}$$

EXERCISE 4.1 Working in a box of volume $V = L^3$, show that the number of allowed states of momentum with x component in the range p_x to $p_x + dp_x$ is $(L/2\pi)\,dp_x$. Convince yourself that you need to impose periodic boundary conditions on the wavefunction and its derivative to ensure no net particle flow out of the volume.

Turning now to the *initial flux*, we find that it is easiest to calculate it in the laboratory frame. The number of beam particles passing through unit area per unit time is $|\mathbf{v}_A| 2E_A/V$, and the number of target particles per unit volume is $2E_B/V$. To obtain a normalization-independent measure of the ingoing "density," we therefore take

$$\text{Initial flux} = |\mathbf{v}_A| \frac{2E_A}{V} \frac{2E_B}{V}. \tag{4.26}$$

Inserting (4.22), (4.25), and (4.26) into (4.23), we arrive at a differential cross section $d\sigma$ for scattering into $d^3p_C\, d^3p_D$:

$$d\sigma = \frac{V^2}{|\mathbf{v}_A| 2E_A 2E_B} \frac{1}{V^4} |\mathfrak{M}|^2 \frac{(2\pi)^4}{(2\pi)^6} \delta^{(4)}(p_C + p_D - p_A - p_B) \frac{d^3p_C}{2E_C} \frac{d^3p_D}{2E_D} V^2. \tag{4.27}$$

The arbitrary normalization volume V cancels, as indeed it must. From now on, we drop V and work in unit volume. That is, we normalize to $2E$ particles/unit volume, and the normalization factor (4.21) of the wavefunction is

$$N = 1. \tag{4.28}$$

This is the origin of the multiplicative factors $N = 1$ which are associated with the external lines of spinless particles (see the table of Feynman rules in Section 6.17).

What is the physical interpretation of the cross section as defined by (4.23) [and (4.27)]? The appearance of the number of final states in conjunction with the transition rate W_{fi} is familiar, see (3.42); it represents the number of scatters, n_s, per unit time. The flux is introduced in (4.23) to make the rate independent of the number of particles present in the beam or target used in a particular experimental setup. That is, we want the cross section to represent an intrinsic scattering probability, that is, the intrinsic strength of the AB → CD interaction. We therefore divide in (4.23) by the number of particles in the target (n_t) and the flux of the beam ($n_b v_b$) which counts the number of beam particles traversing a unit area perpendicular to the beam velocity per unit time (v_b is the relative velocity of beam and target if the latter is not stationary). Thus, (4.23) can be symbolically written as

$$n_s = (n_b v_b) n_t \sigma.$$

The counting rate n_s is always proportional to the (beam flux $\times\, n_t$); it is the proportionality constant σ which contains the physics, that is, the intrinsic scattering probability. It has the units of area. This can be checked by realizing that the left-hand side has units $(\text{time})^{-1}$ and that $(n_b v_b) n_t$ has the units of flux, namely, $(\text{area} \times \text{time})^{-1}$. We can interpret σ intuitively as the effective area of the beam seen by a target particle, or the area over which A, B interact to produce C, D.

We may write the differential cross section, (4.27), in the symbolic form

$$d\sigma = \frac{|\mathfrak{M}|^2}{F} dQ, \tag{4.29}$$

where dQ is the Lorentz invariant phase space factor (sometimes written as dLips)

$$dQ = (2\pi)^4 \delta^{(4)}(p_C + p_D - p_A - p_B)\frac{d^3p_C}{(2\pi)^3 2E_C}\frac{d^3p_D}{(2\pi)^3 2E_D} \tag{4.30}$$

(recall d^3p/E is a Lorentz invariant quantity), and the incident flux in the laboratory is

$$F = |\mathbf{v}_A| 2E_A \cdot 2E_B, \tag{4.31}$$

with $\mathbf{v}_A = \mathbf{p}_A/E_A$. For a general collinear collision between A and B,

$$\begin{aligned} F &= |\mathbf{v}_A - \mathbf{v}_B| \cdot 2E_A \cdot 2E_B \\ &= 4(|\mathbf{p}_A|E_B + |\mathbf{p}_B|E_A) \\ &= 4\left((p_A \cdot p_B)^2 - m_A^2 m_B^2\right)^{1/2}, \end{aligned} \tag{4.32}$$

which is manifestly invariant.

Equation (4.29) is the final result. We see that the physics resides in the invariant amplitude $\mathfrak{M}$. To relate the observable rate $d\sigma$ to this universal measure of the interaction, we need to include the "bookkeeping" factors dQ and F.

EXERCISE 4.2 In the center-of-mass frame for the process AB → CD, show that

$$dQ = \frac{1}{4\pi^2}\frac{p_f}{4\sqrt{s}}\,d\Omega \tag{4.33}$$

$$F = 4p_i\sqrt{s}, \tag{4.34}$$

and hence that the differential cross section is

$$\boxed{\left.\frac{d\sigma}{d\Omega}\right|_{cm} = \frac{1}{64\pi^2 s}\frac{p_f}{p_i}|\mathfrak{M}|^2}, \tag{4.35}$$

where $d\Omega$ is the element of solid angle about $\mathbf{p}_C$, $s = (E_A + E_B)^2$, $|\mathbf{p}_A| = |\mathbf{p}_B| = p_i$ and $|\mathbf{p}_C| = |\mathbf{p}_D| = p_f$.

EXERCISE 4.3 Use (4.18) to show that for very high-energy "spinless" electron–muon scattering,

$$\left.\frac{d\sigma}{d\Omega}\right|_{cm} = \frac{\alpha^2}{4s} \times \left(\frac{3 + \cos\theta}{1 - \cos\theta}\right)^2$$

where θ is the scattering angle and $\alpha = e^2/4\pi$. Neglect the particle masses.

4.4 The Decay Rate in Terms of $\mathfrak{M}$

The derivation of the formula for particle decay rates proceeds along similar lines. The differential rate for the decay A $\to 1 + 2 + \cdots n$ into momentum elements $d^3p_1, \ldots, d^3p_n$ of the final state particles is

$$d\Gamma = \frac{1}{2E_A}|\mathfrak{M}|^2 \frac{d^3p_1}{(2\pi)^3 2E_1} \cdots \frac{d^3p_n}{(2\pi)^3 2E_n} (2\pi)^4 \delta^{(4)}(p_A - p_1 - \cdots p_n). \tag{4.36}$$

The formula has the form of (4.29) and (4.30). $2E_A$ is the number of decaying particles per unit volume and $\mathfrak{M}$ is the invariant amplitude which has been computed from the relevant Feynman diagram. One common application is the calculation of the integrated rate for the decay mode A $\to 1 + 2$, that is, we integrate (4.36) over all possible momenta $\mathbf{p}_1, \mathbf{p}_2$. In the rest frame of A, we find, using (4.33),

$$\Gamma(\mathrm{A} \to 1 + 2) = \frac{p_f}{32\pi^2 m_A^2} \int |\mathfrak{M}|^2 \, d\Omega. \tag{4.37}$$

The total decay rate, Γ, is the sum of the rates for all the decay channels. Clearly, the rate

$$\Gamma = -\frac{dN_A}{dt} / N_A, \tag{4.38}$$

which leads to the exponential decay law for the number of A particles

$$N_A(t) = N_A(0) e^{-\Gamma t}. \tag{4.39}$$

We say Γ^{-1} is the lifetime of particle A.

4.5 "Spinless" Electron–Electon Scattering

We return to the application of the Feynman rules to some sample processes. For electron–electron scattering, the new feature is that we have identical particles in the initial and the final states, and so the amplitude should be symmetric under interchange of particle labels C $\leftrightarrow$ D (and A $\leftrightarrow$ B). Consequently, in addition to the Feynman diagram of Fig. 4.4a, we have a second diagram, Fig. 4.4b, which, to maintain the order of A, B, C, and D, is drawn as Fig. 4.4c.

There is no way to experimentally distinguish whether electron C came from A or B, so we must add amplitudes (rather than probabilities). Thus, the invariant amplitude for the scattering of spinless electrons is, to lowest order, the sum of the amplitudes for diagrams (a) and (c):

$$-i\mathfrak{M}_{e^-e^-} = -i\left(-\frac{e^2(p_A + p_C)_\mu (p_B + p_D)^\mu}{(p_D - p_B)^2} - \frac{e^2(p_A + p_D)_\mu (p_B + p_C)^\mu}{(p_C - p_B)^2} \right). \tag{4.40}$$

The first term follows directly from (4.18), and the second is simply that with

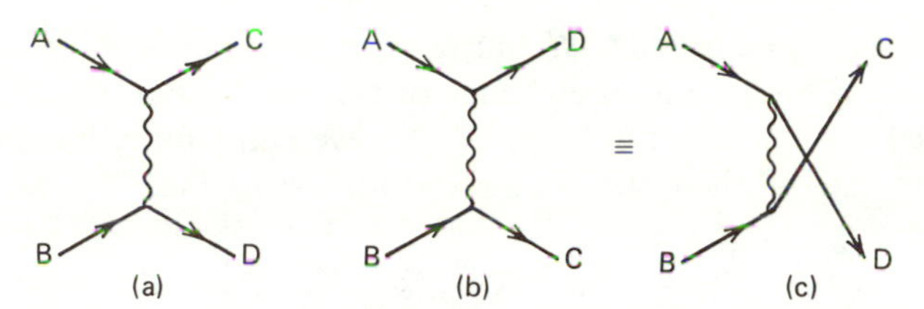

Fig. 4.4 The two (lowest-order) Feynman diagrams for electron–electron scattering.

$p_C \leftrightarrow p_D$. Note that symmetry under $p_C \leftrightarrow p_D$ ensures that $\mathfrak{M}$ is also symmetric under $p_A \leftrightarrow p_B$.

4.6 Electron–Positron Scattering: An Application of Crossing

Again, here we have two possible Feynman diagrams, Figs. 4.5a and 4.5c. We are working only in terms of particle (electron) states, and so we must use the antiparticle prescription, (3.28), to translate these to diagrams (b) and (d), respectively. We can use the Feynman rules (obtained above and collected in Section 6.17) to calculate the (lowest-order) $e^-e^+ \to e^-e^+$ amplitude

$$-i\mathfrak{M}_{e^-e^+} = -i\left(-e^2\frac{(p_A+p_C)_\mu(-p_D-p_B)^\mu}{(p_D-p_B)^2} - e^2\frac{(p_A-p_B)_\mu(-p_D+p_C)^\mu}{(p_C+p_D)^2}\right). \tag{4.41}$$

Note that, for example, the factor at the B, D vertex of diagram (b) has simply changed sign from that of Fig. 4.4(a), which is just what we expect in going from

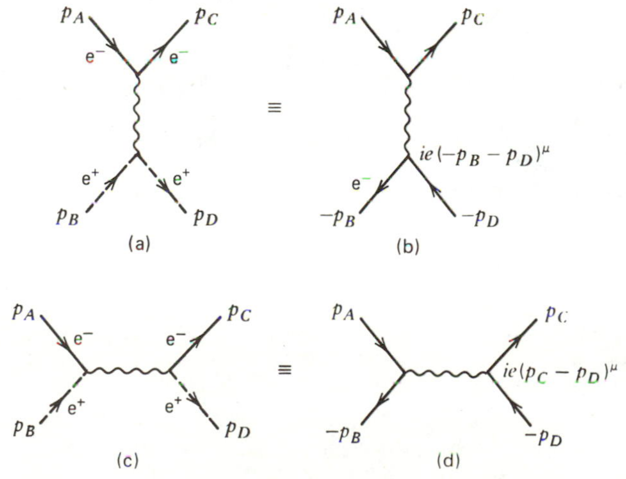

Fig. 4.5 The two (lowest-order) Feynman diagrams, (b) and (d), for spinless $e^-e^+ \to e^-e^+$ scattering.

a charge $-e$ to a $+e$ vertex. We also observe that $\mathfrak{M}$ is symmetric under $p_C \leftrightarrow -p_B$, that is, under the interchange of the two "outgoing" electrons.

But we need not do this; to obtain $\mathfrak{M}_{e^-e^+}$, we can simply use the antiparticle prescription to "cross" the result we derived for $\mathfrak{M}_{e^-e^-}$ (see Fig. 4.6). In this way, we obtain

$$\mathfrak{M}_{e^-e^+\to e^-e^+}(p_A, p_B, p_C, p_D) = \mathfrak{M}_{e^-e^-\to e^-e^-}(p_A, -p_D, p_C, -p_B). \quad (4.42)$$

Indeed, we see that (4.41) is simply (4.40) with $p_D \leftrightarrow -p_B$.

EXERCISE 4.4 Use the Feynman rules to obtain the invariant amplitudes for the "spinless" processes $e^-\mu^+ \to e^-\mu^+$ and $e^-e^+ \to \mu^-\mu^+$. Check your answers by appropriately crossing the $e^-\mu^- \to e^-\mu^-$ amplitude of Section 4.2.

4.7 Invariant Variables

For a scattering process of the form AB → CD, we expect two independent kinematic variables; for example, the incident energy and the scattering angle. It is however possible, and desirable, to express the invariant amplitude $\mathfrak{M}$ as a function of variables invariant under Lorentz transformations. We have at our disposal the particle four-momenta, and so possible invariant variables are the scalar products $p_A \cdot p_B$, $p_A \cdot p_C$, $p_A \cdot p_D$. Since $p_i^2 = m_i^2$ [see (3.10)], and since $p_A + p_B = p_C + p_D$ due to energy–momentum conservation, only two of the three variables are independent. Rather than these, it is conventional to use the related (Mandelstam) variables

$$\begin{aligned} s &= (p_A + p_B)^2, \\ t &= (p_A - p_C)^2, \\ u &= (p_A - p_D)^2. \end{aligned} \quad (4.43)$$

EXERCISE 4.5 Show that

$$s + t + u = m_A^2 + m_B^2 + m_C^2 + m_D^2, \quad (4.44)$$

where m_i is the rest mass of particle i.

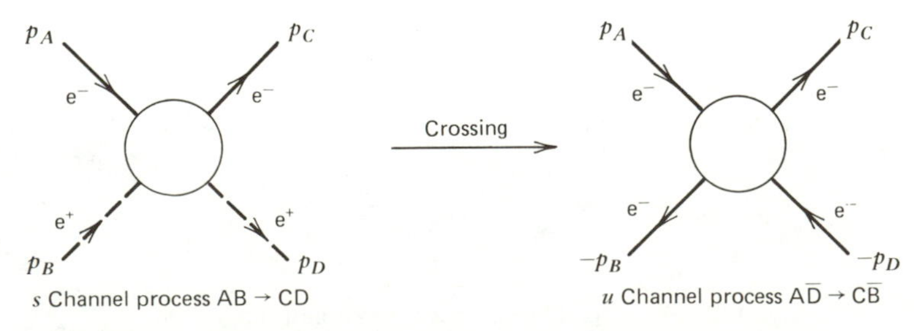

Fig. 4.6 The crossing (or interchange) of particles B and D.

To display the kinematic (or physical) regions of processes related by crossing, we construct a two-dimensional plot which maintains the symmetry of s, t, u. The three axes $s, t, u = 0$ are drawn (Fig. 4.7) to form an equilateral triangle of height Σm_i^2. From any point inside and also outside (if attention is paid to the signs of s, t, u) the triangle, the sum of the perpendicular distances to the axes is equal to the height of the triangle [see (4.44)].

It is easy to show that s is the square of the total center-of-mass energy of the process $\mathrm{AB} \to \mathrm{CD}$ [see (3.12) or Exercise 4.6]. It is conventional to take the reaction under study to be the s channel process. In our last example, this was $e^-e^+ \to e^-e^+$ scattering. The crossed reactions $\mathrm{A\overline{D}} \to \mathrm{C\overline{B}}$ and $\mathrm{\overline{D}B} \to \mathrm{C\overline{A}}$ are called the u and t channels, respectively, since u and t are equal to the square of the total center-of-mass energy in the respective channels (see Exercise 4.7).

EXERCISE 4.6 Taking $e^-e^+ \to e^-e^+$ to be the s channel process, verify that

$$\begin{aligned} s &= 4(k^2 + m^2) \\ t &= -2k^2(1 - \cos\theta) \\ u &= -2k^2(1 + \cos\theta) \end{aligned} \tag{4.45}$$

where θ is the center-of-mass scattering angle and $k = |\mathbf{k}_i| = |\mathbf{k}_f|$, where $\mathbf{k}_i$ and $\mathbf{k}_f$ are, respectively, the momenta of the incident and scattered electrons

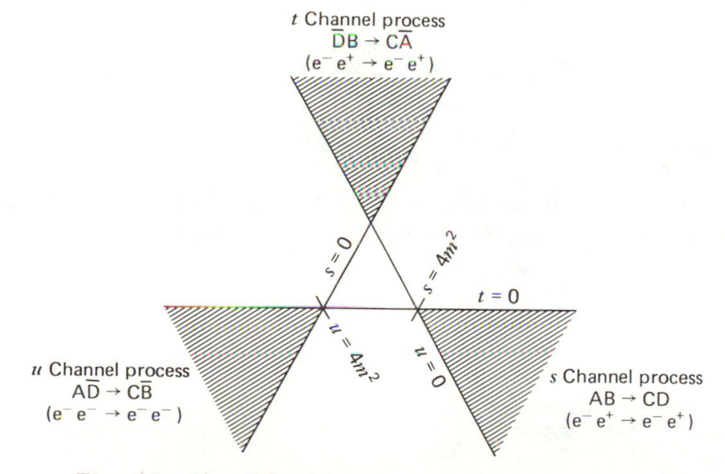

Fig. 4.7 The (Mandelstam) s, t, u plot showing the physical regions for $e^-e^+ \to e^-e^+$ and the crossed reactions. For scattering between particles of unequal masses, the boundaries of the physical regions are more complicated, but the general result of three nonoverlapping regions holds true (see Exercise 4.8).

in the center-of-mass frame. Show that the process is physically allowed provided $s \geq 4m^2$, $t \leq 0$, and $u \leq 0$. The physical region is shown shaded on Fig. 4.7. Note that $t = 0$ ($u = 0$) corresponds to forward (backward) scattering.

EXERCISE 4.7 For the crossed reaction $A\overline{D} \to C\overline{B}$ ($e^-e^- \to e^-e^-$), show that u becomes the square of the total center-of-mass energy and that this process would become physical in a different kinematic region: $u \geq 4m^2$, $t \leq 0$, and $s \leq 0$. (Note that, for example, $-p_D = (E, \mathbf{p})$, where E and $\mathbf{p}$ refer to the incoming $\overline{D}$).

EXERCISE 4.8 If the s channel process is $e^-\mu^- \to e^-\mu^-$, show that the boundaries of the physical regions of this and the crossed channel reactions are given by

$$t = 0, \qquad su = (M^2 - m^2)^2$$

where m and M are the electron and muon masses, respectively. Construct the Mandelstam plot.

EXERCISE 4.9 Verify that crossing relation (4.42) is of the form

$$\mathcal{M}_{e^-e^+}(s, t, u) = \mathcal{M}_{e^-e^-}(u, t, s) \tag{4.46}$$

EXERCISE 4.10 Show that the invariant amplitude, (4.41), for "spinless" electron–positron scattering can be written as

$$\mathcal{M}_{e^-e^+}(s, t, u) = e^2\left(\frac{s-u}{t} + \frac{t-u}{s}\right). \tag{4.47}$$

Comment on the symmetry of $\mathcal{M}$ under $s \leftrightarrow t$.

Let us look back at the amplitude for "spinless" electron–electron scattering. The amplitude (4.40) is derived taking, of course, the process to be AB → CD,

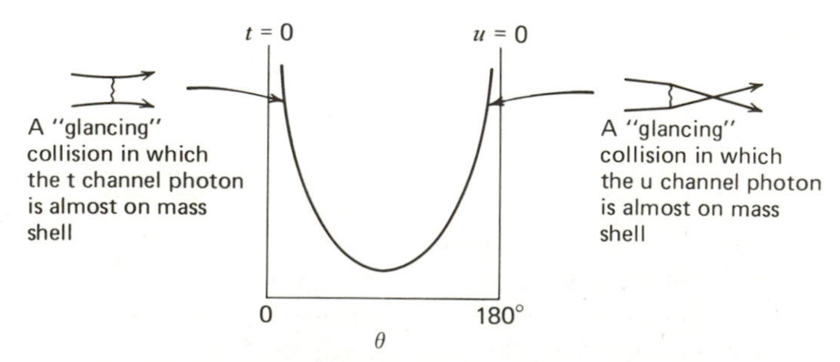

Fig. 4.8 The differential cross section, $d\sigma/d\Omega$, for electron–electron scattering.

that is, the s channel process. In terms of invariant variables, (4.40) becomes

$$\mathcal{M}_{e^-e^-} = e^2\left(\frac{u-s}{t} + \frac{t-s}{u}\right).$$

The resulting cross section is sketched in Fig. 4.8, and the origin of the forward and backward peaks is identified; $-t$ and $-u$ are the squares of the three-momentum transferred in Figs. 4.4a and 4.4c, respectively, that is, of the momentum carried by the virtual photon. When the photon has a very small momentum squared $(-q^2)$, that is, almost on its mass shell, then by the uncertainty principle the range of the interaction is very large. Interactions with small deflections therefore occur with large cross sections.

4.8 The Origin of the Propagator

We saw earlier (Section 4.2) that a virtual photon line in a Feynman diagram corresponds to a propagator $1/q^2$, where q is the four-momentum carried by the virtual photon. For example, the photon propagator in the annihilation process $e^-e^+ \to \gamma \to e^-e^+$ of Fig. 4.9 is of the form $1/q^2$, where $q = p_A + p_B$ is given by four-momentum conservation.

In general (aside from complications of spin), a particle of mass m has a propagator $1/(p^2 - m^2)$. Feynman (1962) has given a nice explanation of why this is so. For instance, for the photon propagator of Fig. 4.9, the argument goes as follows. There are two interaction vertices, and so we must be able to interpret the result as the relativistic generalization of the second-order term in perturbation series (3.44),

$$T_{fi}^{(2)} = -i\sum_{n\neq i} V_{fn}\frac{1}{E_i - E_n}V_{ni}2\pi\,\delta(E_f - E_i). \tag{4.48}$$

Here, we are not interested in the precise details [a detailed discussion is given by Aitchison (1972)], but rather to simply see how the propagator generalizes:

$$\frac{1}{E_i - E_n} \to \frac{1}{(p_A + p_B)^2}? \tag{4.49}$$

We take the energy difference to refer to relativistic energies.

A Feynman diagram is the sum of all possible time-ordered diagrams. There are two possible time-ordered diagrams corresponding to Fig. 4.9. These are

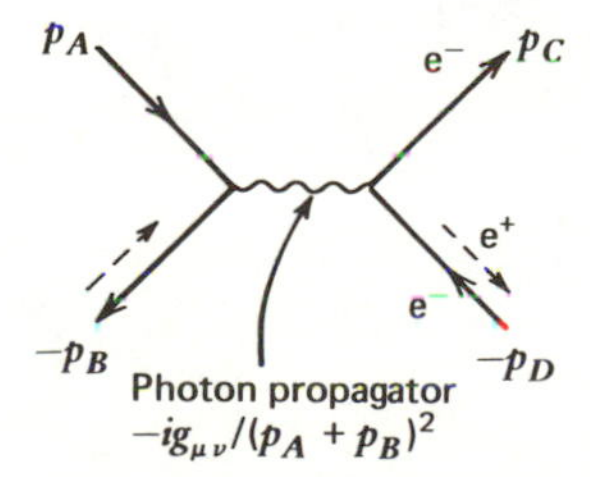

Fig. 4.9 Annihilation diagram $e^-e^+ \to \gamma \to e^-e^+$, drawn using only particle (electron) lines.

shown in Fig. 4.10. The resulting amplitude is thus of the form

$$\mathfrak{M} \sim V_{fn}\frac{1}{E_i - E_\gamma}V_{ni} + V_{fn}\frac{1}{E_i - 2E_i - E_\gamma}V_{ni} = V_{fn}\frac{2E_\gamma}{E_i^2 - E_\gamma^2}V_{ni}, \tag{4.50}$$

where the factor $2E_\gamma$ has to do with normalization. This method of calculating the amplitude is now often referred to as old-fashioned perturbation theory, OFPT. In OFPT, three-momentum is conserved at a vertex, but not energy, as pointed out in Chapter 3 (clearly a noninvariant situation); moreover, particles stay on mass shell. To determine the propagator, we calculate

$$E_i^2 = (p_A + p_B)^2 + (\mathbf{p}_A + \mathbf{p}_B)^2,$$
$$E_\gamma^2 = m_\gamma^2 + \mathbf{p}^2.$$

Now, since $\mathbf{p} = \mathbf{p}_A + \mathbf{p}_B$, we obtain

$$\frac{1}{E_i^2 - E_\gamma^2} = \frac{1}{(p_A + p_B)^2 - m_\gamma^2} = \frac{1}{q^2}. \tag{4.51}$$

We have displayed the photon mass ($m_\gamma = 0$) until the last equality so as to be able to compare with propagators for $m \neq 0$ particles. The relativistic generalization of the propagator for a spinless particle of mass m is [see (4.51)]

$$\frac{1}{(p_A + p_B)^2 - m^2} = \frac{1}{p^2 - m^2}. \tag{4.52}$$

Each of the two time-ordered diagrams of Fig. 4.10 (considered separately) is not invariant; but by including the second term together with the standard nonrelativistic result, we have obtained an invariant expression.

Another example is the electron propagator in the process $\gamma e^- \to \gamma e^-$. In this case, we make a pictorial comparison of OFPT and covariant perturbation theory.

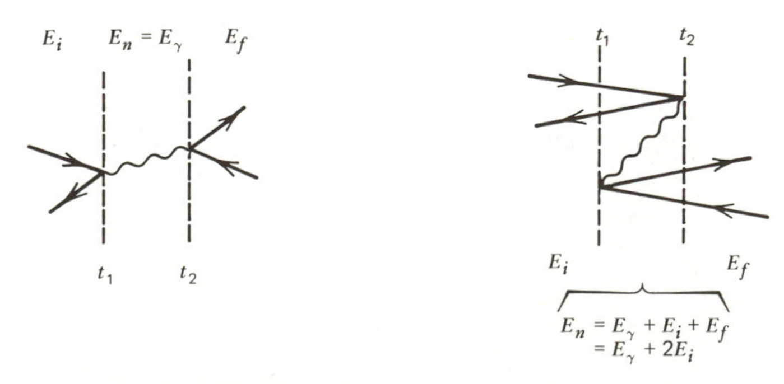

Fig. 4.10 Time-ordered diagrams for $e^-e^+ \to \gamma \to e^-e^+$.

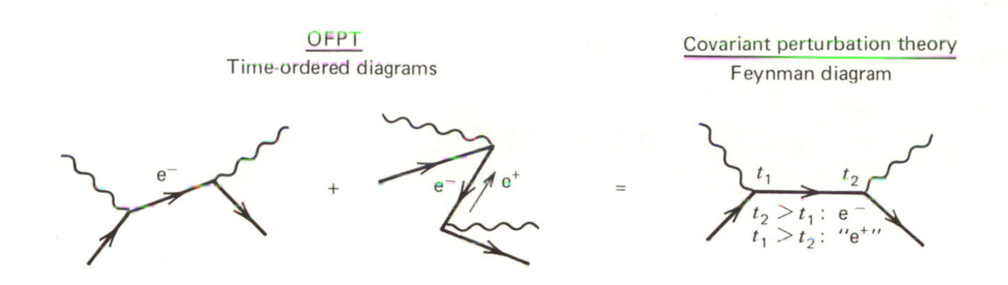

Clearly not an invariant separation, but the sum is invariant

Three-momentum is conserved at each vertex, but not energy. The intermediate particle is on mass shell, $p^2 = m^2$.

Four-momentum is conserved at each vertex, see (4.15). The intermediate particle is not on mass shell, $p^2 \neq m^2$.

The single Feynman diagram embodies the possibility that the intermediate particle is an electron ($t_2 > t_1$) or a positron ($t_1 > t_2$). How this "magic" comes about is explained in Section 6.16; there, we also discuss how to handle the singularity in the propagator at $p^2 = m^2$.

4.9 Summary

We now have shown how to do nonrelativistic perturbation theory in a covariant way. The crucial step was to exploit the fact that the expression (3.37) for T_{fi} was already covariant. The invariant amplitude $\mathfrak{M}$ [related to T_{fi} by (4.22)] is calculated by identifying the covariant replacements for the vertex factors V_{ni} and the propagators $1/(E_i - E_n)$. We explicitly derived their form for the electromagnetic interaction of spinless charged particles. An important difference with the nonrelativistic formalism should be remembered: energy, as well as three-momentum, is conserved at each vertex. Finally, the procedure for relating the invariant amplitude to observables was discussed.

This in fact concludes the discussion of relativistic perturbation theory and Feynman diagrams. The rest of the work is purely technical in nature. Electrons, muons, and quarks have spin $\frac{1}{2}$. They therefore satisfy the Dirac equation, not the Klein–Gordon equation. Repeating the procedure described in this chapter for particles satisfying the Dirac equation will clearly lead to modified results for vertex factors and propagators. The calculational procedure, however, is the same.

5
The Dirac Equation

At this point, we would have mastered the computational techniques to attack the particle physics problems presented in the rest of the book if it were not for the fact that quarks and leptons are spin-$\frac{1}{2}$ particles. We constructed the Feynman rules for particles (and antiparticles) described by wavefunctions ϕ that satisfy the Klein–Gordon equation. These wavefunctions do not have the required two-component structure to accommodate, for instance, the spins of the electron and positron. We are looking for a relativistic equation with solutions that have two-component structure for both particle and antiparticle. For some time, it was thought that the Klein–Gordon equation was the only relativistic generalization of the Schrödinger equation until Dirac discovered an alternative one. His goal was to write an equation which, unlike the Klein–Gordon equation, was linear in $\partial/\partial t$. In order to be covariant, it must then also be linear in ∇ and has therefore the general form

$$H\psi = (\boldsymbol{\alpha}\cdot\mathbf{P} + \beta m)\psi. \tag{5.1}$$

The four coefficients β and α_i ($i = 1, 2, 3$) are determined by the requirement that a free particle must satisfy the relativistic energy–momentum relation (3.10),

$$H^2\psi = (\mathbf{P}^2 + m^2)\psi. \tag{5.2}$$

Equations (5.1) and (5.2) represent the Dirac equation. We will show that its solutions have sufficiently rich structure to describe spin-$\frac{1}{2}$ particles and antiparticles.

The historical impact of Dirac's suggestion is most profound and goes far beyond providing us with a relativistic equation to describe fermions, which is our current interest. Its study led to developments ranging from quantum field theory to semiconductors and beyond. Note, however, that Dirac's original motivation for linearizing the Klein–Gordon equation in $\partial/\partial t$ was not to explain spin but to remove negative probability densities. The appearance of $\partial/\partial t$ in the probability (3.17) is indeed at the root of this "problem." However, for us this feature of the Klein–Gordon equation is no longer a problem. It is a bonus that allows the correct treatment of antiparticles, at least when they have no spin!

Let us forget history and study how (5.1) describes leptons (or quarks) with spin. From (5.1), we have

$$\begin{aligned} H^2\psi &= (\alpha_i P_i + \beta m)(\alpha_j P_j + \beta m)\psi \\ &= (\underset{\downarrow\atop 1}{\alpha_i^2} P_i^2 + \underbrace{(\alpha_i\alpha_j + \alpha_j\alpha_i)}_{0} P_i P_j + \underbrace{(\alpha_i\beta + \beta\alpha_i)}_{0} P_i m + \underset{\downarrow\atop 1}{\beta^2} m^2)\psi, \end{aligned}$$

where we sum over repeated indices, with the condition $i > j$ on the second term. Comparing with (5.2), we see that

- $\alpha_1, \alpha_2, \alpha_3, \beta$ all anticommute with each other,
- $\alpha_1^2 = \alpha_2^2 = \alpha_3^2 = \beta^2 = 1.$ (5.3)

Since the coefficients α_i and β do not commute, they cannot simply be numbers, and we are led to consider matrices operating on a wavefunction ψ, which is a multicomponent column vector.

EXERCISE 5.1 Prove that the α_i and β are hermitian, traceless matrices of even dimensionality, with eigenvalues ± 1.

The lowest dimensionality matrices satisfying all these requirements are 4×4. The choice of the four matrices $(\boldsymbol{\alpha}, \beta)$ is not unique. The Dirac–Pauli representation is most frequently used:

$$\boldsymbol{\alpha} = \begin{pmatrix} 0 & \boldsymbol{\sigma} \\ \boldsymbol{\sigma} & 0 \end{pmatrix}, \qquad \beta = \begin{pmatrix} I & 0 \\ 0 & -I \end{pmatrix} \tag{5.4}$$

where I denotes the unit 2×2 matrix (which is frequently written as 1) and where $\boldsymbol{\sigma}$ are the Pauli matrices:

$$\sigma_1 = \begin{pmatrix} 0 & 1 \\ 1 & 0 \end{pmatrix}, \qquad \sigma_2 = \begin{pmatrix} 0 & -i \\ i & 0 \end{pmatrix}, \qquad \sigma_3 = \begin{pmatrix} 1 & 0 \\ 0 & -1 \end{pmatrix}. \tag{5.5}$$

Another possible representation, the Weyl representation, is

$$\boldsymbol{\alpha} = \begin{pmatrix} -\boldsymbol{\sigma} & 0 \\ 0 & \boldsymbol{\sigma} \end{pmatrix}, \qquad \beta = \begin{pmatrix} 0 & I \\ I & 0 \end{pmatrix}. \tag{5.6}$$

Most of the results are independent of the choice of representation. Certainly, all the physics depends only on the properties listed in (5.3). In fact, not until we exhibit explicit solutions of the Dirac equation in Section 5.3 will we use a particular representation. Unless stated otherwise, we shall always choose the Dirac–Pauli representation, (5.4).

A four-component column vector ψ which satisfies the Dirac equation (5.1) is called a Dirac spinor. We might have anticipated two independent solutions (particles and antiparticles), but instead we have four!

Maybe the surprise should not have been total. We know at least one other example where a field with more components appears when linearizing the equation. The covariant Maxwell equations $\square^2 A_\mu = 0$ are second-order but can be written in a linear form $\partial_\mu F^{\mu\nu} = 0$ by introducing the field strength $F_{\mu\nu}$, which has more components than A_μ.

5.1 Covariant Form of the Dirac Equation. Dirac γ-Matrices

On multiplying Dirac's equation, (5.1), by β from the left, we obtain

$$i\beta \frac{\partial \psi}{\partial t} = -i\beta \boldsymbol{\alpha} \cdot \nabla \psi + m\psi,$$

which may be rewritten

$$\boxed{(i\gamma^\mu \partial_\mu - m)\psi = 0,} \tag{5.7}$$

where we have introduced four Dirac γ-matrices

$$\gamma^\mu \equiv (\beta, \beta\boldsymbol{\alpha}). \tag{5.8}$$

Equation (5.7) is called the covariant form of the Dirac equation. We must wait until Section 5.2 to understand the sense in which the four 4×4 matrices $\gamma^0, \gamma^1, \gamma^2, \gamma^3$ are to be regarded as a four-vector. The Dirac equation is really four differential equations which couple the four components of a single column vector ψ:

$$\sum_{k=1}^{4} \left[\sum_\mu i(\gamma^\mu)_{jk} \partial_\mu - m\delta_{jk} \right] \psi_k = 0.$$

Using (5.3) and (5.8), it is straightforward to show that the Dirac γ-matrices satisfy the anticommutation relations

$$\gamma^\mu \gamma^\nu + \gamma^\nu \gamma^\mu = 2g^{\mu\nu}. \tag{5.9}$$

Moreover, since $\gamma^0 = \beta$, we have

$$\gamma^{0\dagger} = \gamma^0, \qquad (\gamma^0)^2 = I \tag{5.10}$$

and

$$\left.\begin{aligned} \gamma^{k\dagger} &= (\beta\alpha^k)^\dagger = \alpha^k\beta = -\gamma^k \\ (\gamma^k)^2 &= \beta\alpha^k\beta\alpha^k = -I \end{aligned}\right\} \quad k = 1, 2, 3. \tag{5.11}$$

Note that the hermitian conjugation results can be summarized by

$$\gamma^{\mu\dagger} = \gamma^0 \gamma^\mu \gamma^0.$$

5.2 Conserved Current and the Adjoint Equation

In order to construct the currents, we proceed as for the Klein–Gordon equation, see (3.17), except that, as we now have a matrix equation, we must consider the hermitian, rather than the complex, conjugate equation. The hermitian conjugate of Dirac's equation,

$$i\gamma^0 \frac{\partial \psi}{\partial t} + i\gamma^k \frac{\partial \psi}{\partial x^k} - m\psi = 0 \tag{5.12}$$

where $k = 1, 2, 3$, is

$$-i\frac{\partial \psi^\dagger}{\partial t}\gamma^0 - i\frac{\partial \psi^\dagger}{\partial x^k}(-\gamma^k) - m\psi^\dagger = 0. \tag{5.13}$$

To restore the covariant form, we need to remove the minus sign of $-\gamma^k$ while leaving the first term unchanged. Since $\gamma^0\gamma^k = -\gamma^k\gamma^0$, this can be done by multiplying (5.13) from the right by γ^0. Introducing the adjoint (row) spinor

$$\bar{\psi} \equiv \psi^\dagger\gamma^0, \tag{5.14}$$

we obtain

$$\boxed{i\partial_\mu\bar{\psi}\gamma^\mu + m\bar{\psi} = 0.} \tag{5.15}$$

We can now derive a continuity equation, $\partial_\mu j^\mu = 0$, by multiplying (5.7) from the left by $\bar{\psi}$ and (5.15) from the right by ψ, and adding. We find

$$\bar{\psi}\gamma^\mu\partial_\mu\psi + \left(\partial_\mu\bar{\psi}\right)\gamma^\mu\psi = \partial_\mu\left(\bar{\psi}\gamma^\mu\psi\right) = 0.$$

Thus, we see that

$$j^\mu = \bar{\psi}\gamma^\mu\psi$$

satisfies the continuity equation, which suggests that we should identify j^μ with the probability and flux densities, ρ and $\mathbf{j}$. The probability density

$$\rho \equiv j^0 = \bar{\psi}\gamma^0\psi = \psi^\dagger\psi = \sum_{i=1}^{4} |\psi_i|^2 \tag{5.16}$$

is now positive definite. As previously remarked, this result historically motivated Dirac's work.

However, from the Pauli–Weisskopf prescription in Chapter 3, we saw that $j^\mu = (\rho, \mathbf{j})$ should be identified with the charge current density. We therefore insert the charge $-e$ in j^μ,

$$\boxed{j^\mu = -e\bar{\psi}\gamma^\mu\psi}, \tag{5.17}$$

and from now on regard j^μ as the electron (four-vector) current density. Recall that the reason for $-e$ is that the electron (rather than the positron) is regarded as the particle.

For the covariance of the continuity equation $\partial_\mu j^\mu = 0$, it is necessary that j^μ transforms as a four-vector. This can indeed be proved; that is, using the four Dirac matrices γ^μ, we can form a four-vector $\bar{\psi}\gamma^\mu\psi$ (cf. Section 5.6).

5.3 Free-Particle Spinors

EXERCISE 5.2 Operate on (5.7) with $\gamma^\nu \partial_\nu$ and show that each of the four components ψ_i satisfies the Klein–Gordon equation

$$(\square^2 + m^2)\psi_i = 0.$$

For a free particle, we can therefore seek four-momentum eigensolutions of Dirac's equation of the form

$$\psi = u(\mathbf{p})e^{-ip\cdot x}, \tag{5.18}$$

where u is a four-component spinor independent of x. Substituting in (5.7), we have

$$(\gamma^\mu p_\mu - m)u(\mathbf{p}) = 0 \tag{5.19}$$

or, using the abbreviated notation $\not{A} \equiv \gamma^\mu A_\mu$ for any four-vector A_μ,

$$\boxed{(\not{p} - m)u = 0}. \tag{5.20}$$

Since we are seeking energy eigenvectors, it is easier to use the original form, (5.1),

$$Hu = (\boldsymbol{\alpha}\cdot\mathbf{p} + \beta m)u = Eu. \tag{5.21}$$

There are four independent solutions of this equation, two with $E > 0$ and two with $E < 0$. This is particularly easy to see in the Dirac–Pauli representation of $\boldsymbol{\alpha}$ and β. First, take the particle at rest, $\mathbf{p} = 0$. Using (5.4), we have

$$Hu = \beta m u = \begin{pmatrix} mI & 0 \\ 0 & -mI \end{pmatrix} u$$

with eigenvalues $E = m, m, -m, -m$, and eigenvectors

$$\begin{pmatrix} 1 \\ 0 \\ 0 \\ 0 \end{pmatrix}, \quad \begin{pmatrix} 0 \\ 1 \\ 0 \\ 0 \end{pmatrix}, \quad \begin{pmatrix} 0 \\ 0 \\ 1 \\ 0 \end{pmatrix}, \quad \begin{pmatrix} 0 \\ 0 \\ 0 \\ 1 \end{pmatrix}. \tag{5.22}$$

As we have just mentioned, the electron, charge $-e$, is regarded as the particle. The first two solutions therefore describe an $E > 0$ electron. The $E < 0$ particle solutions are to be interpreted, as before, as describing an $E > 0$ antiparticle (positron) (see Section 3.6).

For $\mathbf{p} \neq 0$, (5.21) becomes, using (5.4),

$$Hu = \begin{pmatrix} m & \boldsymbol{\sigma}\cdot\mathbf{p} \\ \boldsymbol{\sigma}\cdot\mathbf{p} & -m \end{pmatrix}\begin{pmatrix} u_A \\ u_B \end{pmatrix} = E\begin{pmatrix} u_A \\ u_B \end{pmatrix}, \tag{5.23}$$

where u has been divided into two two-component spinors, u_A and u_B. This reduces to

$$\boldsymbol{\sigma}\cdot\mathbf{p}u_B = (E-m)u_A,$$
$$\boldsymbol{\sigma}\cdot\mathbf{p}u_A = (E+m)u_B. \tag{5.24}$$

For the two $E > 0$ solutions, we may take $u_A^{(s)} = \chi^{(s)}$, where

$$\chi^{(1)} = \begin{pmatrix}1\\0\end{pmatrix}, \qquad \chi^{(2)} = \begin{pmatrix}0\\1\end{pmatrix}. \tag{5.25}$$

The corresponding lower components of u are then specified by using the second equation of (5.24),

$$u_B^{(s)} = \frac{\boldsymbol{\sigma}\cdot\mathbf{p}}{E+m}\chi^{(s)}, \tag{5.26}$$

and so the positive-energy four-spinor solutions of Dirac's equation are

$$u^{(s)} = N\begin{pmatrix}\chi^{(s)}\\ \dfrac{\boldsymbol{\sigma}\cdot\mathbf{p}}{E+m}\chi^{(s)}\end{pmatrix}, \qquad E > 0 \tag{5.27}$$

with $s = 1, 2$, where N is the normalization constant. For the $E < 0$ solutions, we take $u_B^{(s)} = \chi^{(s)}$, and so, from (5.24),

$$u_A^{(s)} = \frac{\boldsymbol{\sigma}\cdot\mathbf{p}}{E-m}u_B^{(s)} = -\frac{\boldsymbol{\sigma}\cdot\mathbf{p}}{|E|+m}\chi^{(s)}. \tag{5.28}$$

Hence, we obtain

$$u^{(s+2)} = N\begin{pmatrix}\dfrac{-\boldsymbol{\sigma}\cdot\mathbf{p}}{|E|+m}\chi^{(s)}\\ \chi^{(s)}\end{pmatrix}, \qquad E < 0. \tag{5.29}$$

For an electron of given momentum $\mathbf{p}$, we have four solutions: $u^{(1,2)}$, corresponding to positive energy, and $u^{(3,4)}$, corresponding to negative energy. We can readily verify that the four solutions are orthogonal:

$$u^{(r)\dagger}u^{(s)} = 0 \qquad r \neq s,$$

with $r, s = 1, 2, 3, 4$.

The bonus embodied in the Dirac equation, for instance, (5.23), is the extra twofold degeneracy. This means that there must be another observable which commutes with H and $\mathbf{P}$, whose eigenvalues can be taken to distinguish the states. On inspection, we see that the operator

$$\boldsymbol{\Sigma}\cdot\hat{\mathbf{p}} \equiv \begin{pmatrix}\boldsymbol{\sigma}\cdot\hat{\mathbf{p}} & 0\\ 0 & \boldsymbol{\sigma}\cdot\hat{\mathbf{p}}\end{pmatrix} \tag{5.30}$$

commutes with H and $\mathbf{P}$; $\hat{\mathbf{p}}$ is the unit vector pointing in the direction of the momentum, $\mathbf{p}/|\mathbf{p}|$. The "spin" component in the direction of motion, $\frac{1}{2}\boldsymbol{\sigma}\cdot\hat{\mathbf{p}}$, is therefore a "good" quantum number and can be used to label the solutions. We call this quantum number the *helicity* of the state. The possible eigenvalues λ of the helicity operator $\frac{1}{2}\boldsymbol{\sigma}\cdot\hat{\mathbf{p}}$ are

$$\lambda = \begin{cases} +\frac{1}{2} & \text{positive helicity,} \\ -\frac{1}{2} & \text{negative helicity.} \end{cases}$$

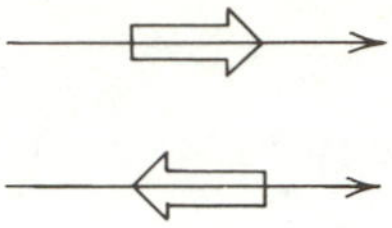

We see that no other component of $\boldsymbol{\sigma}$ has eigenvalues which are good quantum numbers.

With the above choice (5.25), of the spinors $\chi^{(s)}$, it is appropriate to choose $\mathbf{p}$ along the z axis, $\mathbf{p} = (0, 0, p)$. Then

$$\tfrac{1}{2}\boldsymbol{\sigma}\cdot\hat{\mathbf{p}}\chi^{(s)} = \tfrac{1}{2}\sigma_3\chi^{(s)} = \lambda\chi^{(s)}$$

with $\lambda = \pm\frac{1}{2}$ corresponding to $s = 1, 2$, respectively.

EXERCISE 5.3 Calculate the $\lambda = +\frac{1}{2}$ helicity eigenspinor of an electron of momentum $\mathbf{p}' = (p\sin\theta, 0, p\cos\theta)$.

EXERCISE 5.4 Confirm the desired result that the Dirac equation describes "intrinsic" angular momentum ($\equiv$ spin)- $\frac{1}{2}$ particles.

Hint We are clearly interested in angular momentum, so first we should explore the commutation of the orbital angular momentum $\mathbf{L} = \mathbf{r}\times\mathbf{P}$ with the Hamiltonian. Use $[x_i, P_j] = i\delta_{ij}$ to show that

$$[H, \mathbf{L}] = -i(\boldsymbol{\alpha}\times\mathbf{P}).$$

So $\mathbf{L}$ is not conserved! There must be some other angular momentum. Show that

$$[H, \boldsymbol{\Sigma}] = +2i(\boldsymbol{\alpha}\times\mathbf{P}), \qquad \text{where } \boldsymbol{\Sigma} \equiv \begin{pmatrix} \boldsymbol{\sigma} & 0 \\ 0 & \boldsymbol{\sigma} \end{pmatrix}.$$

Clearly, neither $\mathbf{L}$ nor $\boldsymbol{\Sigma}$ are conserved. The combination

$$\mathbf{J} = \mathbf{L} + \tfrac{1}{2}\boldsymbol{\Sigma},$$

which is nothing other than the total angular momentum, is however conserved, as now

$$[H, \mathbf{J}] = 0.$$

The eigenvalues of $\frac{1}{2}\boldsymbol{\Sigma}$ are $\pm\frac{1}{2}$.

EXERCISE 5.5 For a nonrelativistic electron of velocity v, use (5.24) to show that u_A is larger than u_B by a factor of the order v/c. In nonrelativistic problems, ψ_A and ψ_B are referred to as the "large" and "small" components of the electron wavefunction ψ.

In the nonrelativistic limit, show that the Dirac equation for an electron (charge $-e$) in an electromagnetic field $A^\mu = (A^0, \mathbf{A})$ reduces to the Schrödinger–Pauli equation

$$\left(\frac{1}{2m}(\mathbf{P} + e\mathbf{A})^2 + \frac{e}{2m}\boldsymbol{\sigma}\cdot\mathbf{B} - eA^0\right)\psi_A = E_{NR}\psi_A, \tag{5.31}$$

where the magnetic field $\mathbf{B} = \nabla \times \mathbf{A}$ and $E_{NR} = E - m$. Assume $|eA^0| \ll m$.

Hint Make the substitution $P^\mu \to P^\mu + eA^\mu$ in eqs. (5.24) written in terms of $\psi_{A,B}$, eliminate ψ_B, and use

$$\mathbf{P} \times \mathbf{A} - \mathbf{A} \times \mathbf{P} = -i\nabla \times \mathbf{A}$$

where $\mathbf{P} = -i\nabla$.

We see from the form (5.31) of the nonrelativistic reduction of the Dirac equation for an electron in an electromagnetic field that we may associate with the electron an intrinsic (or spin) magnetic moment

$$\boldsymbol{\mu} = -\frac{e}{2m}\boldsymbol{\sigma} \equiv -g\frac{e}{2m}\mathbf{S}, \tag{5.32}$$

where the gyromagnetic ratio g is 2 and the spin angular momentum operator $\mathbf{S}$ is $\frac{1}{2}\boldsymbol{\sigma}$. We speak of the Dirac moment $-e/2m$ of the electron. Experimentally, $g = 2.00232$. The prediction $g = 2$ is a triumph of the Dirac equation. The small difference from 2 is discussed in Chapter 7.

5.4 Antiparticles

The first two solutions of the Dirac equation,

$$u^{(1,2)}(\mathbf{p})e^{-ip\cdot x},$$

clearly describe a free electron of energy E and momentum $\mathbf{p}$. The two negative-energy electron solutions $u^{(3,4)}$ are to be associated with the antiparticle, the positron. Indeed, using the antiparticle prescription of Section 3.6, a positron of energy E, momentum $\mathbf{p}$ is described by one of the $-E$, $-\mathbf{p}$ electron solutions, namely,

$$u^{(3,4)}(-\mathbf{p})e^{-i[-p]\cdot x} \equiv v^{(2,1)}(\mathbf{p})e^{ip\cdot x}, \tag{5.33}$$

where $p^0 \equiv E > 0$. The "positron" spinors, v, are introduced for notational convenience. Recall that the Dirac equation for $u(\mathbf{p})$ is [see (5.20)]

$$(\not{p} - m)\,u(\mathbf{p}) = 0.$$

What does this imply for $v(\mathbf{p})$? For an electron of energy $-E$ and momentum

$-\mathbf{p}$, we have

$$(-\not{p} - m)\, u(-\mathbf{p}) = 0,$$

and so

$$\boxed{(\not{p} + m) v(\mathbf{p}) = 0.} \tag{5.34}$$

We emphasize that here that $p^0 \equiv E > 0$.

As before, we continue to draw the Feynman diagrams entirely in terms of particle (electron) states. For example, an incoming positron of energy E is drawn as an outgoing electron of energy $-E$ [see (5.35)]. The only new feature in the antiparticle–particle correspondence is particle spin. Note the identification of the spinor labels 1, 2 with the negative energy 4, 3 states in (5.33). One way to anticipate this reverse order is to observe that, in the rest frame, the absence of spin up along a certain axis is equivalent to the presence of spin down along that axis. The detailed connection is made in Exercise 5.6. As both spin and momentum are reversed, the helicity, $\frac{1}{2}\boldsymbol{\sigma} \cdot \hat{\mathbf{p}}$, is unchanged. We may summarize these results by

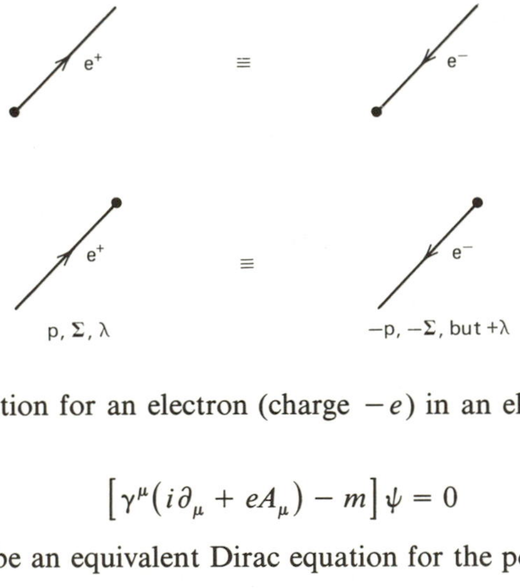

(5.35)

The Dirac equation for an electron (charge $-e$) in an electromagnetic field is [see (4.2)]

$$\left[\gamma^\mu\left(i\partial_\mu + eA_\mu\right) - m\right]\psi = 0 \tag{5.36}$$

Now, there must be an equivalent Dirac equation for the positron $(+e)$:

$$\left[\gamma^\mu\left(i\partial_\mu - eA_\mu\right) - m\right]\psi_C = 0. \tag{5.37}$$

There must be a one-to-one correspondence between ψ_C and ψ. To relate ψ_C to ψ, we first take the complex conjugate of (5.36),

$$\left[-\gamma^{\mu *}\left(i\partial_\mu - eA_\mu\right) - m\right]\psi^* = 0. \tag{5.38}$$

Thus, if we can find a matrix, denoted $(C\gamma^0)$, which satisfies

$$-(C\gamma^0)\gamma^{\mu *} = \gamma^\mu(C\gamma^0),$$

then (5.38) can be written in the form of (5.37), namely,

$$\left[\gamma^{\mu}(i\partial_{\mu} - eA_{\mu}) - m\right](C\gamma^{0}\psi^{*}) = 0,$$

and we can satisfy

$$\psi_{C} = C\gamma^{0}\psi^{*} = C\bar{\psi}^{T},$$

where T denotes the transpose of a matrix.

EXERCISE 5.6 In representation (5.4) and (5.8) of the γ-matrices, show that a possible choice of C is

$$C\gamma^{0} = i\gamma^{2} = \begin{pmatrix} & & & 1 \\ & & -1 & \\ & -1 & & \\ 1 & & & \end{pmatrix}.$$

Try this operation out on a particular spinor, and show, for instance, that

$$\psi_{C}^{(1)} = i\gamma^{2}\left[u^{(1)}(\mathbf{p})\, e^{-ip\cdot x}\right]^{*} = u^{(4)}(-\mathbf{p})\, e^{ip\cdot x} = v^{(1)}(\mathbf{p})\, e^{ip\cdot x}.$$

Further, show that in this representation

$$\begin{aligned} C^{-1}\gamma^{\mu}C &= (-\gamma^{\mu})^{T}, \\ C &= -C^{-1} = -C^{\dagger} = -C^{T}, \\ \bar{\psi}_{C} &= -\psi^{T}C^{-1}. \end{aligned} \tag{5.39}$$

We have seen that the electron current is

$$j^{\mu} = -e\bar{\psi}\gamma^{\mu}\psi.$$

The current associated with the charge conjugate field is therefore

$$\begin{aligned} j_{C}^{\mu} &= -e\bar{\psi}_{C}\gamma^{\mu}\psi_{C} \\ &= +e\psi^{T}C^{-1}\gamma^{\mu}C\bar{\psi}^{T} \\ &= -e\psi^{T}(\gamma^{\mu})^{T}\bar{\psi}^{T} \\ &= -(-)e\bar{\psi}\gamma^{\mu}\psi \end{aligned} \tag{5.40}$$

[see (5.39)]. The origin of the extra minus sign introduced in the last line is subtle but important. It is clearly necessary for a physically meaningful result, that is, if j_{C}^{μ} is to be the positron current. The minus sign is related to the connection between spin and statistics; in field theory, it occurs because of the antisymmetric nature of the fermion fields. In field theory, the charge conjugation operator C changes a positive-energy electron into a positive-energy positron, and the formalism is completely $e^+ \leftrightarrow e^-$ symmetric. However, in a single-particle (electron) theory, positron states are not allowed; rather, C changes a positive-energy electron state into a negative-energy electron state. As a result, we can show that we must add to our Feynman rules the requirement that we insert by hand an extra minus sign for every negative-energy electron in the final state of the process. The C invariance of electromagnetic interactions then follows:

$$j_\mu^C (A^\mu)^C = (-j_\mu)(-A^\mu) = j_\mu A^\mu.$$

5.5 Normalization of Spinors and the Completeness Relations

For fermions, we choose the covariant normalization in which we have $2E$ particles/unit volume, just as we did for bosons. That is,

$$\int_{\text{unit vol.}} \rho \, dV = \int \psi^\dagger \psi \, dV = u^\dagger u = 2E,$$

where we have used (5.16) and (5.18). Thus, we have the orthogonality relations

$$u^{(r)\dagger} u^{(s)} = 2E\delta_{rs}, \qquad v^{(r)\dagger} v^{(s)} = 2E\delta_{rs}, \tag{5.41}$$

with $r, s = 1, 2$. Now, using (5.27), we calculate

$$u^{(s)\dagger} u^{(s)} = |N|^2 \left[1 + \left(\frac{\boldsymbol{\sigma} \cdot \mathbf{p}}{E + m} \right)^2 \right] = |N|^2 \frac{2E}{E + m},$$

and so we may take the normalization constant to be

$$N = \sqrt{E + m}, \tag{5.42}$$

and the same for $v^{(s)}$.

To obtain the Dirac equation for $\bar{u} \equiv u^\dagger \gamma^0$, we need the hermitian conjugate of (5.19):

$$u^\dagger \gamma^{\mu\dagger} p_\mu - m u^\dagger = 0.$$

If we multiply from the right by γ^0 and note from (5.9)–(5.11) that

$$\gamma^{\mu\dagger} \gamma^0 = \gamma^0 \gamma^\mu, \tag{5.43}$$

this reduces to

$$\bar{u}(\not{p} - m) = 0. \tag{5.44}$$

Similarly, (5.35) gives

$$\bar{v}(\not{p} + m) = 0. \tag{5.45}$$

EXERCISE 5.7 Use (5.41) to show that

$$\bar{u}^{(s)} u^{(s)} = 2m, \qquad \bar{v}^{(s)} v^{(s)} = -2m. \tag{5.46}$$

EXERCISE 5.8 Show that $(\boldsymbol{\sigma}\cdot\mathbf{p})^2 = |\mathbf{p}|^2$.

EXERCISE 5.9 Derive the completeness relations

$$\boxed{\begin{aligned}\sum_{s=1,2} u^{(s)}(p)\,\bar{u}^{(s)}(p) &= \not{p} + m,\\ \sum_{s=1,2} v^{(s)}(p)\,\bar{v}^{(s)}(p) &= \not{p} - m.\end{aligned}} \tag{5.47}$$

These 4×4 matrix relations are used extensively in the evaluation of Feynman diagrams (see Chapter 6).

EXERCISE 5.10 Show that $\not{p}\not{p} = p^2$.

EXERCISE 5.11 Show that

$$\Lambda_+ = \frac{\not{p}+m}{2m}, \qquad \Lambda_- = \frac{-\not{p}+m}{2m} \tag{5.48}$$

project over positive and negative energy states, respectively. Recall that the projection operators must satisfy

$$\Lambda_\pm^2 = \Lambda_\pm \qquad \text{and} \qquad \Lambda_+ + \Lambda_- = 1.$$

5.6 Bilinear Covariants

In order to construct the most general form of currents consistent with Lorentz covariance, we need to tabulate bilinear quantities of the form

$$(\bar{\psi})(4\times 4)(\psi)$$

which have definite properties under Lorentz transformations, where the 4×4 matrix is a product of γ-matrices. To simplify the notation, we introduce

$$\gamma^5 \equiv i\gamma^0\gamma^1\gamma^2\gamma^3. \tag{5.49}$$

It follows that

$$\gamma^{5\dagger} = \gamma^5, \qquad \left(\gamma^5\right)^2 = I, \qquad \gamma^5\gamma^\mu + \gamma^\mu\gamma^5 = 0. \tag{5.50}$$

In the Dirac–Pauli representation, (5.4),

$$\gamma^0 = \begin{pmatrix} I & 0\\ 0 & -I\end{pmatrix}, \qquad \boldsymbol{\gamma} = \begin{pmatrix} 0 & \boldsymbol{\sigma}\\ -\boldsymbol{\sigma} & 0\end{pmatrix}, \qquad \gamma^5 = \begin{pmatrix} 0 & I\\ I & 0\end{pmatrix}. \tag{5.51}$$

We are interested in the behavior of bilinear quantities under proper Lorentz transformations (that is, rotations, boosts) and under space inversion (the parity operation). An exhaustive list of the possibilities is

		No. of Compts.	Space Inversion, P	
Scalar	$\bar{\psi}\psi$	1	+ under P	
Vector	$\bar{\psi}\gamma^\mu\psi$	4	Space compts.: − under P	
Tensor	$\bar{\psi}\sigma^{\mu\nu}\psi$	6		(5.52)
Axial Vector	$\bar{\psi}\gamma^5\gamma^\mu\psi$	4	Space compts.: + under P	
Pseudoscalar	$\bar{\psi}\gamma^5\psi$	1	− under P	

Due to the anticommutation relations, (5.9), the tensor is antisymmetric:

$$\sigma^{\mu\nu} = \frac{i}{2}(\gamma^\mu\gamma^\nu - \gamma^\nu\gamma^\mu). \tag{5.53}$$

The list is arranged in increasing order of the number of γ^μ matrices that are sandwiched between $\bar{\psi}$ and ψ. The pseudoscalar is the product of four matrices [see (5.49)]. If five matrices were used, at least two would be the same, in which case the product would reduce to three and be already included in the axial vector.

It is useful to see how the above properties are established. To do this, we must consider Dirac's equation in two frames (x and x') related by a Lorentz transformation Λ. From (5.7), we have

$$i\gamma^\mu \frac{\partial\psi(x)}{\partial x^\mu} - m\,\psi(x) = 0, \tag{5.54}$$

$$i\gamma^\mu \frac{\partial\psi'(x')}{\partial x'^\mu} - m\,\psi'(x') = 0, \tag{5.55}$$

where $x' = \Lambda x$. There must exist a relation

$$\psi'(x') = S\,\psi(x). \tag{5.56}$$

If we recall (5.18), it is clear that S is independent of x and acts only on the spinor u. Substituting (5.56) into (5.55) and demanding consistency with (5.54), we obtain

$$S^{-1}\gamma^\mu S = \Lambda^\mu{}_\nu \gamma^\nu, \tag{5.57}$$

where we have used $\partial/\partial x^\mu = \Lambda^\nu{}_\mu \partial/\partial x'^\nu$.

EXERCISE 5.12 For a proper infinitesimal Lorentz transformation

$$\Lambda^\nu{}_\mu = \delta^\nu{}_\mu + \varepsilon^\nu{}_\mu, \tag{5.58}$$

show that an S which satisfies (5.57) is

$$S_L = 1 - \frac{i}{4}\sigma_{\mu\nu}\varepsilon^{\mu\nu}. \tag{5.59}$$

Hence, show that

$$S_L^{-1} = \gamma^0 S_L^\dagger \gamma^0 \tag{5.60}$$

$$\gamma^5 S_L = S_L \gamma^5. \tag{5.61}$$

For space inversion, or the parity operation,

$$\Lambda^\nu{}_\mu = \begin{pmatrix} 1 & & & \\ & -1 & & \\ & & -1 & \\ & & & -1 \end{pmatrix}.$$

Then, (5.57) becomes

$$S_P^{-1} \gamma^0 S_P = \gamma^0,$$

$$S_P^{-1} \gamma^k S_P = -\gamma^k \qquad \text{for } k = 1, 2, 3,$$

which is satisfied by

$$S_P = \gamma^0. \tag{5.62}$$

In the Dirac–Pauli representation of γ^0, (5.51), the behavior of the four components of ψ under parity is therefore

$$\psi'_{1,2} = \psi_{1,2} \qquad \text{and} \qquad \psi'_{3,4} = -\psi_{3,4}. \tag{5.63}$$

The "at rest" states, (5.22), are therefore eigenstates of parity, with the positive and negative energy states (that is, the electron and the positron) having opposite intrinsic parities.

Armed with S_L and S_P, we can now check the claimed properties of the bilinear covariants. First, we note that

$$\begin{aligned} \bar{\psi}' &= \psi'^\dagger \gamma^0 = \psi^\dagger S^\dagger \gamma^0 = \psi^\dagger \gamma^0 S^{-1} \\ &= \bar{\psi} S^{-1}, \end{aligned} \tag{5.64}$$

where we have used (5.56) and (5.60). As an example, let us establish the character of $\bar{\psi}\gamma^\mu\psi$. Under Lorentz transformations,

$$\bar{\psi}'\gamma^\mu\psi' = \bar{\psi} S_L^{-1} \gamma^\mu S_L \psi = \Lambda^\mu{}_\nu (\bar{\psi}\gamma^\nu\psi), \tag{5.65}$$

using (5.64) and (5.57), while under the parity operation,

$$\bar{\psi}'\gamma^\mu\psi' = \bar{\psi} S_P^{-1} \gamma^\mu S_P \psi = \begin{cases} \bar{\psi}\gamma^0\psi, \\ -\bar{\psi}\gamma^k\psi. \end{cases} \tag{5.66}$$

These transformation properties are precisely what we expect for a Lorentz four-vector.

From (5.56) and (5.64), if follows immediately that $\bar{\psi}\psi$ is a Lorentz scalar. The probability density $\rho = \psi^\dagger\psi$ is not a scalar, but is the time-like component of the four-vector $\bar{\psi}\gamma^\mu\psi$. Since

$$\gamma^5 S_P = -S_P\gamma^5, \tag{5.67}$$

the presence of γ^5 gives rise to the pseudo-nature of the axial vector and pseudoscalar. For instance, a pseudoscalar is a scalar under proper Lorentz transformations but, unlike a scalar, changes sign under parity.

5.7 Zero-Mass Fermions: The Two-Component Neutrino

We return to the Dirac equation, (5.1),

$$H\psi = (\boldsymbol{\alpha}\cdot\mathbf{p} + \beta m)\psi \tag{5.68}$$

and the discussion at the opening of this chapter. We derived algebraic relations which were demanded of the α_i's and β, and found that they could be satisfied by 4×4 matrices. However, note that β is not involved in the case of zero-mass particles and that we need only satisfy

$$\alpha_i\alpha_j + \alpha_j\alpha_i = 2\delta_{ij}, \qquad \alpha_i = \alpha_i^\dagger \tag{5.69}$$

[see (5.3)]. These relations can be realized by the 2×2 Pauli matrices. We can take $\alpha_i = -\sigma_i$ and $\alpha_i = \sigma_i$, and the massless Dirac equation divides into two *decoupled* equations for two-component spinors $\chi(\mathbf{p})$ and $\phi(\mathbf{p})$:

$$E\chi = -\boldsymbol{\sigma}\cdot\mathbf{p}\,\chi, \tag{5.70}$$

$$E\phi = +\boldsymbol{\sigma}\cdot\mathbf{p}\,\phi. \tag{5.71}$$

Each equation is based on the relativistic energy–momentum relation, $E^2 = \mathbf{p}^2$, and so has one positive and one negative energy solution.

Suppose (5.70) is the wave equation for a massless fermion, a neutrino. The positive energy solution has $E = |\mathbf{p}|$ and so satisfies

$$\boldsymbol{\sigma}\cdot\hat{\mathbf{p}}\,\chi = -\chi. \tag{5.72}$$

That is, χ describes a left-handed neutrino (helicity $\lambda = -\frac{1}{2}$) of energy E and momentum $\mathbf{p}$. The remaining solution has negative energy. To interpret this, we consider a neutrino solution with energy $-E$ and momentum $-\mathbf{p}$. It satisfies

$$\boldsymbol{\sigma}\cdot(-\hat{\mathbf{p}})\chi = \chi \tag{5.73}$$

with positive helicity, and hence describes a right-handed antineutrino ($\lambda = +\frac{1}{2}$) of energy E and momentum $\mathbf{p}$. Symbolically, we say (5.70) describes ν_L and $\bar{\nu}_R$. Such a wave equation was first proposed by Weyl in 1929 but was rejected because of noninvariance under the parity operation P, which takes $\nu_L \to \nu_R$. For massless neutrinos, this is no longer an objection as weak interactions do not respect parity conservation (see Chapter 12). The second equation, (5.71), describes the other helicity states ν_R and $\bar{\nu}_L$.

Translating these results to four-component form

$$u = \begin{pmatrix} \chi \\ \phi \end{pmatrix}, \qquad \boldsymbol{\alpha} = \begin{pmatrix} -\boldsymbol{\sigma} & 0 \\ 0 & \boldsymbol{\sigma} \end{pmatrix}, \tag{5.74}$$

we see that we are here working in the Weyl (or chiral) representation, (5.6). In this representation,

$$\boldsymbol{\gamma} = \begin{pmatrix} 0 & \boldsymbol{\sigma} \\ -\boldsymbol{\sigma} & 0 \end{pmatrix}, \qquad \gamma^0 = \begin{pmatrix} 0 & I \\ I & 0 \end{pmatrix}, \qquad \gamma^5 = \begin{pmatrix} -I & 0 \\ 0 & I \end{pmatrix}. \tag{5.75}$$

At this stage, it is appropriate to peep ahead at weak interactions, which are discussed in Chapter 12. A vast number of different experiments indicate that leptons enter the "charged-current" weak interactions in a special combination of two of the bilinear covariants. For example, for the electron and its neutrino,

$$J^\mu = \bar{\psi}_e \gamma^\mu \tfrac{1}{2}(1 - \gamma^5)\psi_\nu. \tag{5.76}$$

We speak of the V–A form of the weak current J^μ, in contrast to the V form of the electromagnetic current (5.17). Our concern here is the presence of the $\frac{1}{2}(1 - \gamma^5)$; this mixture of vector (V) and axial vector (A) ensures that parity is violated, and violated maximally. Indeed, from (5.75),

$$\tfrac{1}{2}(1 - \gamma^5)u_\nu = \begin{pmatrix} I & 0 \\ 0 & 0 \end{pmatrix}\begin{pmatrix} \chi \\ \phi \end{pmatrix} = \begin{pmatrix} \chi \\ 0 \end{pmatrix} \tag{5.77}$$

and so projects out just ν_L (and $\bar{\nu}_R$). That is, only left-handed neutrinos (and right-handed antineutrinos) are coupled to charged leptons by the weak interactions. As far as we know, neutrinos have only the weak interaction, and hence this is the only way we can observe them. There is thus no empirical evidence for the existence of ν_R (and $\bar{\nu}_L$), and it could well be that they do not exist in nature. Of course, the assertion that only ν_L (and $\bar{\nu}_R$) occur can only be made if the mass is strictly zero. Otherwise, we could perform a Lorentz transformation which would change a ν_L into a ν_R.

EXERCISE 5.13 Show that the operators

$$P_R \equiv \tfrac{1}{2}(1 + \gamma^5), \qquad P_L \equiv \tfrac{1}{2}(1 - \gamma^5)$$

have the appropriate properties to be (right- and left-hand) projection operators, that is,

$$P_i^2 = P_i, \qquad P_L + P_R = 1, \qquad P_R P_L = 0.$$

Here, γ^5 is called the chirality operator.

For a massive fermion, we define the projections $\frac{1}{2}(1 \pm \gamma^5)u$ to be right- and left-handed components of u.

EXERCISE 5.14 For a massive fermion, show that handedness is not a good quantum number. That is, show that γ^5 does not commute with the Hamiltonian. However, verify that helicity is conserved but is frame dependent. In particular, show that the helicity is reversed by "overtaking" the particle concerned.

EXERCISE 5.15 Working in the Dirac–Pauli representation of γ-matrices, (5.51), show that at high energies

$$\gamma^5 u^{(s)} \simeq \begin{pmatrix} \boldsymbol{\sigma} \cdot \hat{\mathbf{p}} & 0 \\ 0 & \boldsymbol{\sigma} \cdot \hat{\mathbf{p}} \end{pmatrix} u^{(s)}, \tag{5.78}$$

where $u^{(s)}$ is the electron spinor of (5.27). That is, show that in the extreme relativistic limit, the chirality operator (γ^5) is equal to the helicity operator; and so, for example, $\frac{1}{2}(1 - \gamma^5)u = u_L$ corresponds to an electron of negative helicity.

Of course, the fact that $\frac{1}{2}(1 - \gamma^5)$ projects out negative helicity fermions at high energies does not depend on the choice of representation. We need only choose a representation if we wish to show explicit spinors. The particular advantage of the Dirac–Pauli representation is that it diagonalizes the energy in the nonrelativistic limit (γ^0 is diagonal), whereas the Weyl representation diagonalizes the helicity in the extreme relativistic limit (γ^5 is diagonal).

Finally suppose, for the moment, that a nonzero mass was established for one of the neutrinos (ν_e, ν_μ, ν_τ and so on). Remembering exercise (5.14), how can we have a massive neutrino and still ensure that weak interactions couple only to ν_L and $\bar{\nu}_R$? Majorana neutrinos accomplish this. They are formed by making the neutrino its own antiparticle. Thus, we may identify ν_L and $\bar{\nu}_R$ as two helicity components of a four-component spinor. The other two components, ν_R and $\bar{\nu}_L$ (if these exist), can then be a Majorana fermion of different mass. Clearly this is a different structure to the four-component Dirac spinor for, say, $e^\pm$. A detailed discussion of Majorana spinors is given by P. Roman in *Theory of Elementary Particles*, North Holland (1961), page 306.

6

Electrodynamics of Spin-$\frac{1}{2}$ Particles

In this chapter, we are going to reach the goal set in Chapter 3: compute cross sections for the electromagnetic interactions of leptons and photons. This is quantum electrodynamics. We proceed by retracing the steps of Chapter 4 for particles now described by the Dirac equation. The result will be Feynman rules for the electromagnetic interactions of spin-$\frac{1}{2}$ leptons and quarks. Not only are the interactions completely physical but so, now, are the particles. They are no longer "pedagogical constructs" of spin zero. We shall finally confront experimental measurements.

6.1 An Electron Interacting with an Electromagnetic Field A^{μ}

We have seen that a free electron of four-momentum p^{μ} is described by a four-component wavefunction

$$\psi = u(\mathbf{p})\, e^{-ip\cdot x}$$

which satisfies the Dirac equation, (5.19),

$$(\gamma_{\mu} p^{\mu} - m)\psi = 0.$$

The equation for an electron in an electromagnetic field A^{μ} is obtained by the substitution [see (4.1)]

$$p^{\mu} \to p^{\mu} + eA^{\mu}, \tag{6.1}$$

where we have again taken $-e$ to be the charge of the electron. We find

$$(\gamma_{\mu} p^{\mu} - m)\psi = \gamma^0 V\psi, \tag{6.2}$$

where the perturbation is given by

$$\gamma^0 V = -e\gamma_{\mu} A^{\mu}. \tag{6.3}$$

The introduction of the γ^0 is to make (6.2) of the form $(E + \cdots)\psi = V\psi$, so that the potential energy enters in the same way as in the Schrödinger equation [see, for example, the $-eA^0$ term in the Schrödinger–Pauli equation, (5.31)].

Using first-order perturbation theory, (3.37), the amplitude for the scattering of an electron from state ψ_i to state ψ_f is

$$\begin{aligned} T_{fi} &= -i\int \psi_f^\dagger(x)\, V(x)\, \psi_i(x)\, d^4x \\ &= ie\int \overline{\psi}_f \gamma_\mu A^\mu \psi_i\, d^4x \\ &= -i\int j_\mu^{fi} A^\mu\, d^4x, \end{aligned} \tag{6.4}$$

where

$$j_\mu^{fi} \equiv -e\overline{\psi}_f \gamma_\mu \psi_i \tag{6.5}$$

$$= -e\overline{u}_f \gamma_\mu u_i e^{i(p_f - p_i)\cdot x}. \tag{6.6}$$

Comparing this result with (5.17), we see that j_μ^{fi} can be regarded as the electromagnetic transition current between electron states i and f.

Recall that the analogous transition current we obtained for a "spinless" electron,

$$j_\mu^{fi} = -e(p_f + p_i)_\mu e^{i(p_f - p_i)\cdot x},$$

resulted in the Feynman rules of Fig. 6.1a. In Fig. 6.1b, we show the corresponding rules that follow for the (physical) spin-$\frac{1}{2}$ electron. The vertex factor is now a 4×4 matrix in spin space. It is sandwiched between column $u^{(s)}(p_i)$ and row $\overline{u}^{(r)}(p_f)$ spinors describing incoming and outgoing electrons of momentum p_i, p_f and in spin states s, r, respectively.

EXERCISE 6.1 A "spinless" electron can interact with A^μ only via its charge; the coupling involves $(p_f + p_i)^\mu$. Show that

$$\overline{u}_f \gamma^\mu u_i = \frac{1}{2m}\overline{u}_f\left((p_f + p_i)^\mu + i\sigma^{\mu\nu}(p_f - p_i)_\nu\right)u_i, \tag{6.7}$$

from which it is possible to establish that the physical spin-$\frac{1}{2}$ electron interacts via both its charge and its magnetic moment; see also Exercise 6.2. Equation (6.7) is known as the Gordon decomposition of the current.

Fig. 6.1 Factors for the Feynman rules listed in Section 6.17.

EXERCISE 6.2 Show that in the nonrelativistic limit, the Gordon decomposition, (6.7), of the electron current, (6.6), separates the electron interaction with an electromagnetic field A_μ into a part arising from its charge, $-e$, and a part due to its magnetic moment, $-e/2m$. Assume that A_μ is independent of t, so that (6.4) becomes

$$T_{fi} = -i2\pi\,\delta(E_f - E_i)\int j_\mu^{fi} A^\mu \, d^3x.$$

To identify the magnetic moment interaction $(-\boldsymbol{\mu}\cdot\mathbf{B})$, it suffices to show that

$$\int\left[-\frac{e}{2m}\bar{\psi}_f i\sigma_{\mu\nu}(p_f - p_i)^\nu \psi_i\right] A^\mu \, d^3x = \int \psi_A^{f\dagger}\left(\frac{e}{2m}\boldsymbol{\sigma}\cdot\mathbf{B}\right)\psi_A^i \, d^3x,$$

where ψ_A denotes the upper two (or "large") components of ψ; compare with eqs. (5.31) and (5.32).

Hint Since $E_i = E_f$, only the space-like components of $(p_f - p_i)$ contribute. Note that $\bar{\psi}(\sigma_{23}, \sigma_{31}, \sigma_{12})\psi \simeq \bar{\psi}_A \boldsymbol{\sigma} \psi_A$.

6.2 Møller Scattering $e^-e^- \to e^-e^-$

As an illustration of how to use the QED vertex factor of Fig. 6.1b, we compute the Feynman diagram of Fig. 6.2 for e^-e^- scattering. Repeating the steps in Section 4.2, the transition amplitude is given by [see (4.16)]

$$\begin{aligned} T_{fi} &= -i\int j_\mu^{(1)}(x)\left(-\frac{1}{q^2}\right) j_{(2)}^\mu(x)\, d^4x \\ &= -i(-e\bar{u}_C\gamma_\mu u_A)\left(-\frac{1}{q^2}\right)(-e\bar{u}_D\gamma^\mu u_B)(2\pi)^4\,\delta^{(4)}(p_A + p_B - p_C - p_D), \end{aligned}$$

with $q = p_A - p_C$, where $(2\pi)^4$ times the delta function arises from the integration over the x dependence of the currents [see (6.6)]. Recall that the invariant

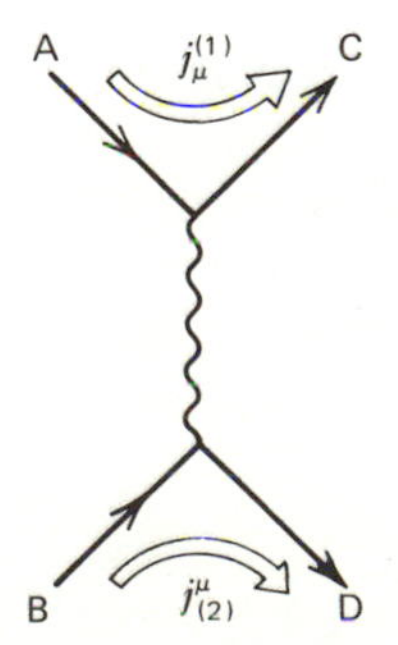

Fig. 6.2 Feynman diagram for $e^-e^- \to e^-e^-$.

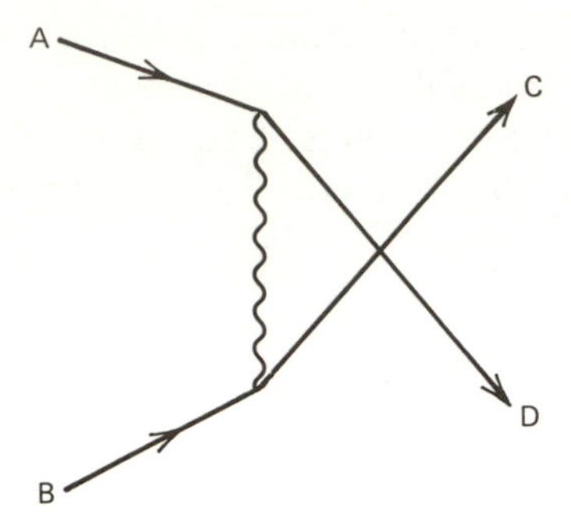

Fig. 6.3 Second diagram for $e^-e^- \to e^-e^-$.

amplitude $\mathfrak{M}$ is defined by

$$T_{fi} = -i(2\pi)^4 \delta^{(4)}(p_A + p_B - p_C - p_D)\mathfrak{M},$$

and so we have

$$-i\mathfrak{M} = (ie\bar{u}_C\gamma^\mu u_A)\left(\frac{-ig_{\mu\nu}}{q^2}\right)(ie\bar{u}_D\gamma^\nu u_B), \tag{6.8}$$

in agreement with the factors assigned to Fig. 6.1b.

For e^-e^- scattering, there is a second Feynman diagram, Fig. 6.3. The amplitude is obtained from (6.8) with the interchange C $\leftrightarrow$ D, but with a relative minus sign on account of the interchange of identical fermions. Thus, the complete (lowest-order) amplitude for Møller scattering is

$$\mathfrak{M} = -e^2\frac{(\bar{u}_C\gamma^\mu u_A)(\bar{u}_D\gamma_\mu u_B)}{(p_A - p_C)^2} + e^2\frac{(\bar{u}_D\gamma^\mu u_A)(\bar{u}_C\gamma_\mu u_B)}{(p_A - p_D)^2}. \tag{6.9}$$

To calculate the unpolarized cross section, we must amend the cross section formulae of Section 4.3. By unpolarized we mean that no information about the electron spins is recorded in the experiment. To allow for scattering in all possible spin configurations, we therefore have to make the replacement

$$|\mathfrak{M}|^2 \to \overline{|\mathfrak{M}|^2} \equiv \frac{1}{(2s_A + 1)(2s_B + 1)} \sum_{\substack{\text{all spin}\\ \text{states}}} |\mathfrak{M}|^2, \tag{6.10}$$

where s_A, s_B are the spins of the incoming particles. That is, we average over the spins of the incoming particles and we sum over the spins of the particles in the final state. Clearly, to carry out this calculation is a nontrivial task.

To get some feel for what is involved, we first evaluate the unpolarized cross section for Møller scattering in the *nonrelativistic limit*. In this case, the spin sums become relatively simple. In the limit $|\mathbf{p}| \to 0$, (5.27) and (5.42) give

$$\begin{aligned} &\text{Ingoing } e^-: u^{(s)} = \sqrt{2m}\begin{pmatrix} \chi^{(s)} \\ 0 \end{pmatrix}, \\ &\text{Outgoing } e^-: \bar{u}^{(s)} = \sqrt{2m}\begin{pmatrix} \chi^{(s)} & 0 \end{pmatrix}, \end{aligned} \tag{6.11}$$

where $s = 1, 2$ correspond to spin up, down along the z axis [see (5.25)]. Using representation (5.4) in which

$$\gamma^0 = \begin{pmatrix} I & \\ & -I \end{pmatrix}, \qquad \boldsymbol{\gamma} = \begin{pmatrix} & \boldsymbol{\sigma} \\ -\boldsymbol{\sigma} & \end{pmatrix}, \tag{6.12}$$

we find

$$\begin{aligned} \bar{u}^{(s)}\gamma^\mu u^{(s)} &= \begin{cases} 2m & \text{if } \mu = 0, \\ 0 & \text{if } \mu \neq 0, \end{cases} \\ \bar{u}^{(s)}\gamma^\mu u^{(s')} &= 0 \qquad \text{for all } \mu, \text{ if } s \neq s'. \end{aligned} \tag{6.13}$$

In other words, the spin direction does not change in the scattering of nonrelativistic electrons. This is to be expected, as the electrons interact dominantly via an electric field which cannot change their spin direction. At higher energies (velocities), it is the magnetic field which flips the spins. Inserting (6.13) into (6.9) and labeling the six nonzero amplitudes by the electron spins gives

$$\mathfrak{M}(\uparrow\uparrow \to \uparrow\uparrow) = \mathfrak{M}(\downarrow\downarrow \to \downarrow\downarrow) = -e^2 4m^2\left(\frac{1}{t} - \frac{1}{u}\right)$$

$$\mathfrak{M}(\uparrow\downarrow \to \uparrow\downarrow) = \mathfrak{M}(\downarrow\uparrow \to \downarrow\uparrow) = -e^2 4m^2 \frac{1}{t}$$

$$\mathfrak{M}(\uparrow\downarrow \to \downarrow\uparrow) = \mathfrak{M}(\downarrow\uparrow \to \uparrow\downarrow) = e^2 4m^2 \frac{1}{u},$$

where $1/t$, $1/u$, defined by (4.43), are the photon propagators. Arrows label the spin-up (down) state of each particle. We can now perform the spin summation (6.10), and find

$$\overline{|\mathfrak{M}|^2} = \frac{1}{4}(4m^2e^2)^2 2\left[\left(\frac{1}{t} - \frac{1}{u}\right)^2 + \frac{1}{t^2} + \frac{1}{u^2}\right]. \tag{6.14}$$

In the center-of-mass frame [see (4.45)],

$$\begin{aligned} t &= -2p^2(1 - \cos\theta) = -4p^2\sin^2\frac{\theta}{2}, \\ u &= -2p^2(1 + \cos\theta) = -4p^2\cos^2\frac{\theta}{2}, \end{aligned} \tag{6.15}$$

where θ is the scattering angle and $p = |\mathbf{p}_i|$ with $i = \text{A, B, C, D}$. Using this, and inserting (6.14) into (4.35), we find that the e^-e^- differential cross section is

$$\left.\frac{d\sigma}{d\Omega}\right|_{cm} = \frac{m^2\alpha^2}{16p^4}\left(\frac{1}{\sin^4\frac{\theta}{2}} + \frac{1}{\cos^4\frac{\theta}{2}} - \frac{1}{\sin^2\frac{\theta}{2}\cos^2\frac{\theta}{2}}\right) \tag{6.16}$$

is the nonrelativistic limit, where $\alpha \equiv e^2/4\pi$; see (4.5).

6.3 The Process $e^-\mu^- \to e^-\mu^-$

How are these spin summations done without the benefit of the simplifications introduced by the nonrelativistic approximation? We use the example of $e^-\mu^-$

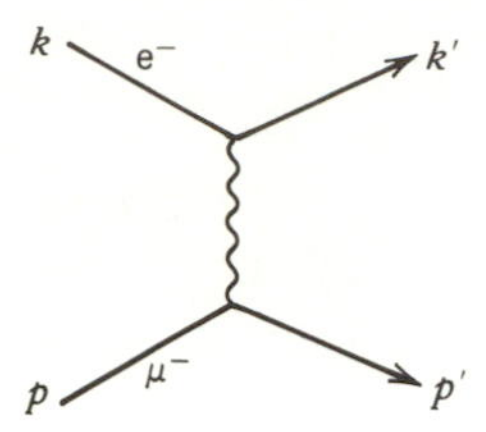

Fig. 6.4 Feynman diagram for electron–muon scattering.

scattering to illustrate the general technique for summing over spins since it has only one (lowest-order) Feynman diagram (Fig. 6.4). The invariant amplitude follows from the Feynman rules [see (6.8)]:

$$\mathfrak{M} = -e^2 \bar{u}(k')\gamma^\mu u(k)\frac{1}{q^2}\bar{u}(p')\gamma_\mu u(p). \tag{6.17}$$

The momenta are defined in Fig. 6.4 and $q = k - k'$. To obtain the (unpolarized) cross section, we have to take the square of the modulus of $\mathfrak{M}$ and then carry out the spin sums. It is convenient to separate the sums over the electron and muon spins by writing (6.10) as

$$\overline{|\mathfrak{M}|^2} = \frac{e^4}{q^4} L_e^{\mu\nu} L_{\mu\nu}^{\text{muon}}, \tag{6.18}$$

where the tensor associated with the electron vertex is

$$L_e^{\mu\nu} \equiv \frac{1}{2} \sum_{(e \text{ spins})} [\bar{u}(k')\gamma^\mu u(k)][\bar{u}(k')\gamma^\nu u(k)]^*, \tag{6.19}$$

and with a similar expression for $L_{\mu\nu}^{\text{muon}}$.

The spin summations look a forbidding task. Fortunately, well-established trace techniques considerably simplify such calculations. To begin, we note that the second square bracket of (6.19) (a 1×1 matrix for which the complex and hermitian conjugates are the same) is equal to

$$\begin{aligned}\left[u^\dagger(k')\gamma^0\gamma^\nu u(k)\right]^\dagger &= \left[u^\dagger(k)\gamma^{\nu\dagger}\gamma^0 u(k')\right] \\ &= \left[\bar{u}(k)\gamma^\nu u(k')\right],\end{aligned}$$

where we have used $\gamma^{\nu\dagger}\gamma^0 = \gamma^0\gamma^\nu$ [see (5.43)]. That is, the complex conjugation in (6.19) simply reverses the order of the matrix product. We now write the complete product in (6.19) explicitly in terms of individual matrix elements (labeled $\alpha, \beta \ldots$, with a summation over repeated indices implied)

$$L_e^{\mu\nu} = \frac{1}{2} \underbrace{\sum_{s'} \bar{u}_\alpha^{(s')}(k')}_{(\not{k}' + m)_{\delta\alpha}} \gamma_{\alpha\beta}^\mu \underbrace{\sum_s u_\beta^{(s)}(k)\bar{u}_\gamma^{(s)}(k)}_{(\not{k} + m)_{\beta\gamma}} \gamma_{\gamma\delta}^\nu u_\delta^{(s')}(k').$$

The completeness relation (5.47) allows the sums over both the initial and the final electron spins to be performed. The repositioning of u_δ makes this evident; it can be moved as the matrix character is recorded by the component. Thus, L becomes the trace of the product of four 4×4 matrices,

$$L_e^{\mu\nu} = \tfrac{1}{2}\mathrm{Tr}((\not{k}' + m)\gamma^\mu(\not{k} + m)\gamma^\nu), \tag{6.20}$$

where m is the mass of the electron. To evaluate L, we use the trace theorems.

6.4 Trace Theorems and Properties of γ-Matrices

Recall that the Dirac γ-matrices satisfy the commutation algebra

$$\gamma^\mu\gamma^\nu + \gamma^\nu\gamma^\mu = 2g^{\mu\nu}. \tag{6.21}$$

As a consequence, it is straightforward to show that the trace of a product of γ-matrices can be evaluated without ever explicitly calculating a matrix product. The trace theorems are (using again the notation $\not{a} = \gamma_\mu a^\mu$):

$$\begin{aligned}
&\mathrm{Tr}\,1 = 4,\\
&\text{Trace of an odd number of } \gamma_\mu\text{'s vanishes,}\\
&\mathrm{Tr}(\not{a}\not{b}) = 4a \cdot b,\\
&\mathrm{Tr}(\not{a}\not{b}\not{c}\not{d}) = 4[(a \cdot b)(c \cdot d) - (a \cdot c)(b \cdot d) + (a \cdot d)(b \cdot c)],
\end{aligned} \tag{6.22}$$

$$\begin{aligned}
&\mathrm{Tr}\,\gamma_5 = 0,\\
&\mathrm{Tr}(\gamma_5\not{a}\not{b}) = 0,\\
&\mathrm{Tr}(\gamma_5\not{a}\not{b}\not{c}\not{d}) = 4i\varepsilon_{\mu\nu\lambda\sigma}a^\mu b^\nu c^\lambda d^\sigma,
\end{aligned} \tag{6.23}$$

where $\varepsilon_{\mu\nu\lambda\sigma} = +1\ (-1)$ for $\mu, \nu, \lambda, \sigma$ an even (odd) permutation of 0, 1, 2, 3; and 0 if two indices are the same.

Other useful results for simplifying trace calculations are:

$$\begin{aligned}
\gamma_\mu\gamma^\mu &= 4\\
\gamma_\mu\not{a}\gamma^\mu &= -2\not{a}\\
\gamma_\mu\not{a}\not{b}\gamma^\mu &= 4a \cdot b\\
\gamma_\mu\not{a}\not{b}\not{c}\gamma^\mu &= -2\not{c}\not{b}\not{a}
\end{aligned} \tag{6.24}$$

EXERCISE 6.3 Making use of (6.21), prove the trace theorems and the identities (6.24).

6.5 $e^-\mu^-$ Scattering and the Process $e^+e^- \to \mu^+\mu^-$

It is straightforward to evaluate the tensor associated with the electron vertex, (6.20), using the trace theorems listed in (6.22). We have

$$\begin{aligned}
L_e^{\mu\nu} &= \tfrac{1}{2}\,\mathrm{Tr}(\not{k}'\gamma^\mu\not{k}\gamma^\nu) + \tfrac{1}{2}m^2\,\mathrm{Tr}(\gamma^\mu\gamma^\nu),\\
&= 2(k'^\mu k^\nu + k'^\nu k^\mu - (k' \cdot k - m^2)g^{\mu\nu}).
\end{aligned} \tag{6.25}$$

The evaluation of $L_{\mu\nu}^{\text{muon}}$ of (6.18) is identical. We find

$$L_{\mu\nu}^{\text{muon}} = 2\left(p'_\mu p_\nu + p'_\nu p_\mu - (p' \cdot p - M^2) g_{\mu\nu} \right), \tag{6.26}$$

where M is the mass of the muon. Forming the product of (6.25) and (6.26), we finally arrive at the following "exact" form for the spin-averaged $e^-\mu^- \to e^-\mu^-$ amplitude, (6.18):

$$\overline{|\mathfrak{M}|^2} = \frac{8e^4}{q^4}\left[(k' \cdot p')(k \cdot p) + (k' \cdot p)(k \cdot p') \right.$$
$$\left. - m^2 p' \cdot p - M^2 k' \cdot k + 2m^2 M^2 \right]. \tag{6.27}$$

In the extreme relativistic limit, we may neglect the terms involving m^2 and M^2; therefore,

$$\overline{|\mathfrak{M}|^2} = \frac{8e^4}{(k - k')^4}\left[(k' \cdot p')(k \cdot p) + (k' \cdot p)(k \cdot p')\right]. \tag{6.28}$$

Moreover, in this limit, the (Mandelstam) variables of (4.43) become

$$\begin{aligned} s &\equiv (k + p)^2 \simeq 2k \cdot p \simeq 2k' \cdot p', \\ t &\equiv (k - k')^2 \simeq -2k \cdot k' \simeq -2p \cdot p', \\ u &\equiv (k - p')^2 \simeq -2k \cdot p' \simeq -2k' \cdot p. \end{aligned} \tag{6.29}$$

Thus, at high energies, unpolarized $e^-\mu^-$ scattering, (6.28), is given by

$$\boxed{\overline{|\mathfrak{M}|^2} = 2e^4 \frac{s^2 + u^2}{t^2}.} \tag{6.30}$$

We may also obtain the amplitude for $e^-e^+ \to \mu^+\mu^-$ by "crossing" the above result for $e^-\mu^- \to e^-\mu^-$ (see Section 4.6). The required interchange is $k' \leftrightarrow -p$, that is, $s \leftrightarrow t$ in (6.30), and we obtain

$$\overline{|\mathfrak{M}|^2} = 2e^4 \frac{t^2 + u^2}{s^2} \tag{6.31}$$

where now $e^-e^+ \to \mu^+\mu^-$ is the s-channel process. The corresponding diagram is drawn in Fig. 6.5. This result for the square of the amplitude can be translated into a differential cross section for $e^+e^- \to \mu^+\mu^-$ scattering using (4.35). In the

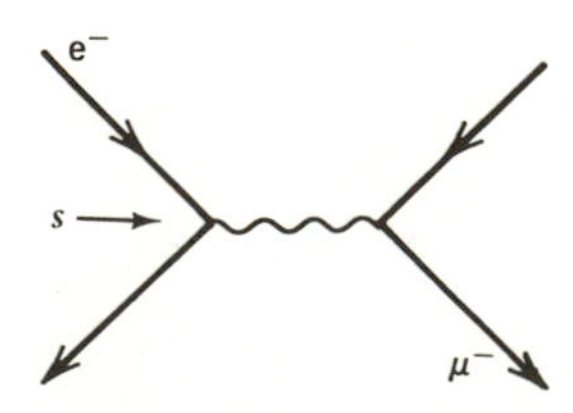

Fig. 6.5 The Feynman diagram for $e^-e^+ \to \mu^+\mu^-$. As always, the antiparticles are drawn using only particle (e^-, μ^-) lines.

center-of-mass frame, we have

$$\left.\frac{d\sigma}{d\Omega}\right|_{cm} = \frac{1}{64\pi^2 s} 2e^4\left[\tfrac{1}{2}(1+\cos^2\theta)\right],$$

where the quantity in square brackets is $(t^2+u^2)/s^2$; see (4.45). Using $\alpha = e^2/4\pi$, this becomes

$$\boxed{\left.\frac{d\sigma}{d\Omega}\right|_{cm} = \frac{\alpha^2}{4s}(1+\cos^2\theta).} \tag{6.32}$$

To obtain the reaction cross section, we integrate over θ, ϕ:

$$\boxed{\sigma(e^+e^- \to \mu^+\mu^-) = \frac{4\pi\alpha^2}{3s}.} \tag{6.33}$$

A comparison of this result with PETRA data is shown in Fig. 6.6. The PETRA accelerator consists of a ring of magnets which simultaneously accelerate an electron and positron beam circulating in opposite directions. In selected spots, these beams are crossed, resulting in e^+e^- interactions with center-of-mass energy $\sqrt{s} = 2E_b$, where E_b is the energy of each beam. Equation (6.33) can be written in

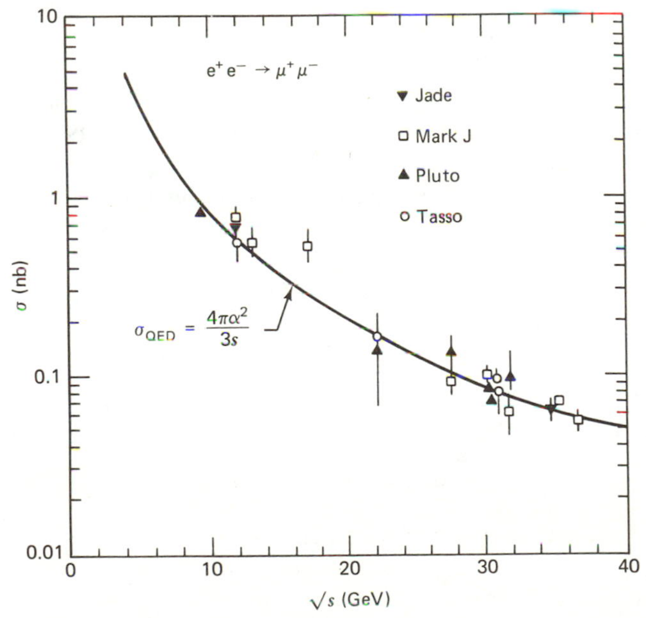

Fig. 6.6 The total cross section for $e^-e^+ \to \mu^-\mu^+$ measured at PETRA versus the center-of-mass energy.

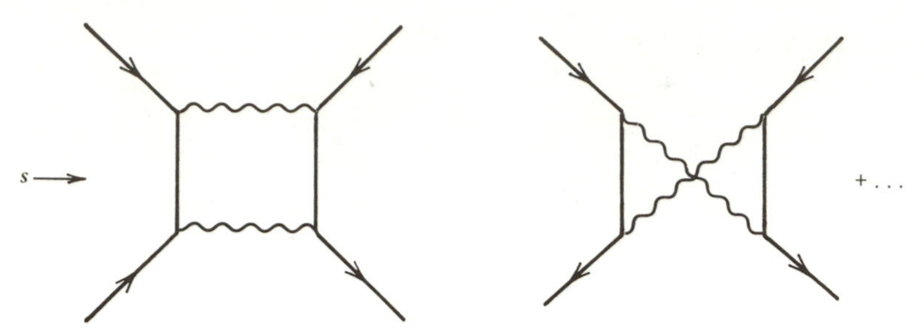

Fig. 6.7 Some higher-order diagrams for $e^-e^+ \to \mu^+\mu^-$.

numerical form as

$$\sigma(e^+e^- \to \mu^+\mu^-) \simeq \frac{20\ (\text{nb})}{E_b^2\ (\text{in GeV}^2)}.$$

This quadratic dependence of the annihilation cross section on beam energy predicted by our calculations can be checked by varying the energy of the beams, as displayed in Fig. 6.6. The above results are the contribution from the lowest-order Feynman diagram, Fig. 6.5. There are, of course, corrections of order $\alpha^3, \alpha^4, \ldots,$ arising due to interference with, or directly from, the amplitudes of higher-order diagrams, such as those of Fig. 6.7.

6.6 Helicity Conservation at High Energies

We can obtain further physical insight into the structure of results like (6.30) and (6.31) by looking at the helicity of the particles. As we will be often interested in highly relativistic interactions, it is useful to study the structure of the electromagnetic current in this limit. We start with (5.78) and note that for a fermion of energy $E \gg m$,

$$\begin{aligned} \tfrac{1}{2}(1 - \gamma^5)u &= u_L \\ \tfrac{1}{2}(1 + \gamma^5)u &= u_R. \end{aligned} \tag{6.34}$$

That is, $\frac{1}{2}(1 \pm \gamma^5)$ project out the helicity $\lambda = \pm\frac{1}{2}$ components of a spinor, respectively. We can use this observation to show that

$$\bar{u}\gamma^\mu u \equiv (\bar{u}_L + \bar{u}_R)\gamma^\mu(u_L + u_R) = \bar{u}_L\gamma^\mu u_L + \bar{u}_R\gamma^\mu u_R, \tag{6.35}$$

and so at high energies the electromagnetic interaction [see (6.4) and (6.6)] conserves the helicity of the scattered fermion.

The proof is as follows. First note that

$$\bar{u}_L = u_L^\dagger\gamma^0 = u^\dagger\tfrac{1}{2}(1 - \gamma^5)\gamma^0 = \bar{u}\tfrac{1}{2}(1 + \gamma^5), \tag{6.36}$$

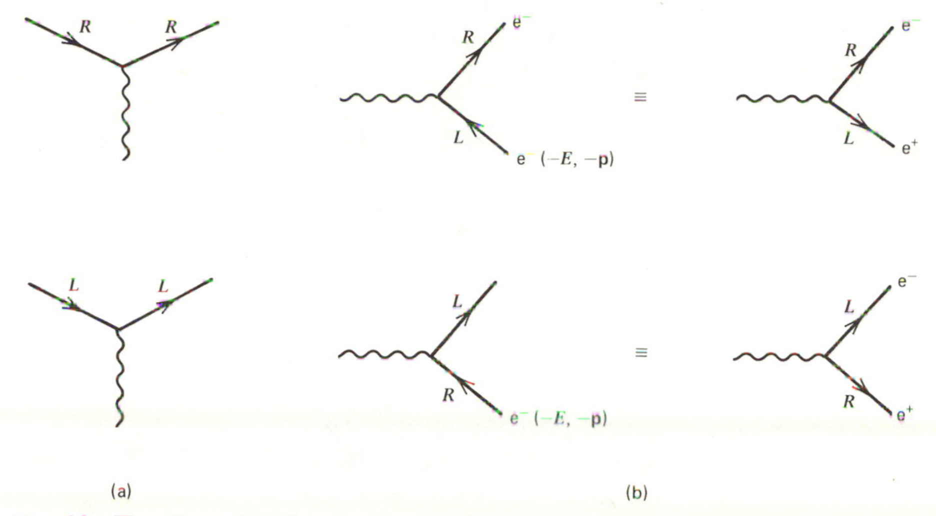

Fig. 6.8 The allowed vertices to $O(m/E)$ for (a) fermion scattering and (b) the crossed or annihilation channel. In all diagrams time increases from left to right.

since $\gamma^5 = \gamma^{5\dagger}$ and $\gamma^5\gamma^0 = -\gamma^0\gamma^5$. Hence,

$$\begin{aligned}\bar{u}_L\gamma^\mu u_R &= \tfrac{1}{4}\bar{u}(1+\gamma^5)\gamma^\mu(1+\gamma^5)u \\ &= \tfrac{1}{4}\bar{u}\gamma^\mu(1-\gamma^5)(1+\gamma^5)u \\ &= 0,\end{aligned} \tag{6.37}$$

where we have used $\gamma^5\gamma^\mu = -\gamma^\mu\gamma^5$ and $(\gamma^5)^2 = 1$. Helicity conservation clearly holds for any vector (or, for that matter, axial vector) interaction at high energies.

The allowed vertices for high-energy fermion scattering are shown in Fig. 6.8a. On the other hand, in an annihilation channel such as e^-e^+, the vertices of Fig. 6.8b dominate to order m/E, that is, the outgoing fermions have opposite helicity. We shall see throughout the book how the application of these helicity conservation rules provides valuable insights into electromagnetic and weak interactions.

EXERCISE 6.4 Assuming a vector–axial vector form of the weak interaction, explain why the electron emitted in the μ^--decay process, $\mu^- \rightarrow e^-\bar{\nu}_e\nu_\mu$, must be left-handed. What is the helicity of e^+ from μ^+ decay?

Let us consider e^+e^- annihilation in more detail. Helicity conservation requires that the incoming e^- and e^+ have opposite helicities, see Fig. 6.8b. The same is true for the μ^- and μ^+ in the final state. Thus, in the center-of-mass frame, scattering proceeds from an initial state with $J_z = +1$ or -1 to a final state with $J_{z'} = +1$ or -1, where the z, z' axes are along the ingoing e^- and outgoing μ^- directions, respectively. One of the four possibilities is sketched in Fig. 6.9. The reaction proceeds via an intermediate photon of spin $j = 1$, and therefore the

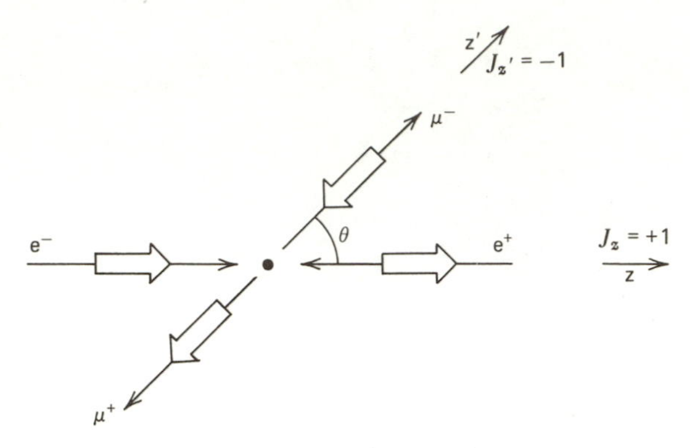

Fig. 6.9 The process $e^-e^+ \to \mu^-\mu^+$ in a particular helicity configuration: $\mathfrak{M}(e_R^- e_L^+ \to \mu_L^- \mu_R^+) \sim d^1_{-11}(\theta)$.

amplitudes are proportional to the rotation matrices

$$d^j_{\lambda'\lambda}(\theta) \equiv \langle j\lambda' | e^{-i\theta J_y} | j\lambda \rangle, \tag{6.38}$$

where y is perpendicular to the reaction plane, θ is the center-of-mass scattering angle, and λ, λ' are the net helicities along the z, z' axes. We adopt the conventional notation for the rotation matrices. They can be readily worked out from angular momentum theory and are tabulated in many places (e.g., Martin and Spearman, 1970). The four possible helicity amplitudes have the same vertex factors and are thus proportional to [see (4.45) and (2.21)]

$$\begin{aligned} d^1_{11}(\theta) = d^1_{-1-1}(\theta) &= \tfrac{1}{2}(1 + \cos\theta) \simeq -\frac{u}{s} \\ d^1_{1-1}(\theta) = d^1_{-11}(\theta) &= \tfrac{1}{2}(1 - \cos\theta) \simeq -\frac{t}{s}. \end{aligned} \tag{6.39}$$

Spin averaging these amplitudes gives the desired result [see (6.31)]

$$\overline{|\mathfrak{M}|^2} \propto \frac{u^2 + t^2}{s^2}. \tag{6.40}$$

It is a straightforward consequence of angular momentum conservation and embodies, for example, the requirement that the amplitude for the process of Fig. 6.9 must vanish in the forward direction, because here the net helicity is not conserved.

EXERCISE 6.5 Use rotation matrix arguments to show that for "spinless" electrons and muons

$$\mathfrak{M}(e^-e^+ \to \mu^-\mu^+) \propto \frac{t - u}{s}. \tag{6.41}$$

Compare the s-channel photon contribution of (4.47).

TABLE 6.1
Leading Order Contributions to Representative QED Processes

	Feynman Diagrams		$\overline{\lvert\mathcal{M}\rvert^2}/2e^4$		
	Forward peak	Backward peak	Forward	Interference	Backward
Møller scattering $e^-e^- \to e^-e^-$		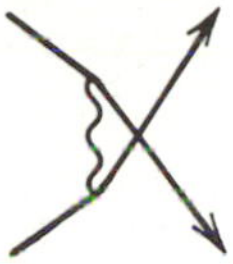	$\frac{s^2+u^2}{t^2}$	$+\ \frac{2s^2}{tu}$	$+\ \frac{s^2+t^2}{u^2}$
			($u \leftrightarrow t$ symmetric)		
(Crossing $s \leftrightarrow u$) ↓	Forward	"Time-like"	Forward	Interference	Time-like
Bhabha scattering $e^-e^+ \to e^-e^+$			$\frac{s^2+u^2}{t^2}$	$+\ \frac{2u^2}{ts}$	$+\ \frac{u^2+t^2}{s^2}$
$e^-\mu^- \to e^-\mu^-$	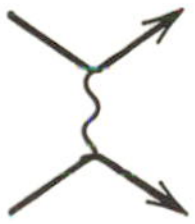		$\frac{s^2+u^2}{t^2}$		
(Crossing ↓ $s \leftrightarrow t$)					
$e^-e^+ \to \mu^-\mu^+$		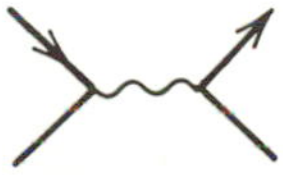			$\frac{u^2+t^2}{s^2}$

6.7 Survey of $e^-e^+ \to e^-e^+, \mu^-\mu^+$

We have calculated above some typical QED processes in the extreme relativistic limit. It is instructive to summarize these and related results in Table 6.1. Apart from the interference terms, all the contributions listed in the table follow simply from the last entry, which itself is easily obtained by helicity arguments.

Similar results are found in QCD for the "strong" $qq \to qq$, $q\bar{q} \to q\bar{q}$ interactions via single gluon exchange. In fact, the results are identical except that we must average (sum) over the colors of the initial (final) quarks, in addition to their spins, and make the replacement $\alpha \to \alpha_s$, where α_s is the quark–gluon coupling introduced in Chapters 1 and 2.

EXERCISE 6.6 Show that the spin-averaged interference term between the two Feynman diagrams for electron–electron scattering is that shown in the table.

Wherever an internal photon line occurs in a Feynman diagram, we shall see in Chapter 13 that we also have the possibility of a contribution from a massive neutral weak boson Z^0. Although the Feynman rules for the weak bosons are given in detail in Chapter 13, it is worth quickly anticipating the possible effect of γ–Z^0 interference. Processes such as $e^-e^+ \to e^-e^+$, $\mu^-\mu^+$ with s-channel or "time-like" Feynman diagrams should increasingly feel the effects of any such Z bosons as the center-of-mass energy approaches its mass value M_Z. From Table 6.1, it is clear that it is preferable to study $e^-e^+ \to \mu^-\mu^+$ rather than Bhabha scattering, since in the latter process, Z effects will be swamped by t-channel photon exchange. The details are given in Section 13.6.

6.8 $e^-\mu^- \to e^-\mu^-$ in the Laboratory Frame. Kinematics Relevant to the Parton Model

Before we leave fermion scattering, it is useful to introduce laboratory frame kinematics, that is, the frame where the initial μ is at rest. We can then directly apply these results to electron–quark scattering when probing the structure of hadrons in Chapter 8 and onward.

We return to the "exact" formula (6.27) for $e^-(k) + \mu^-(p) \to e^-(k') + \mu^-(p')$ and neglect only the terms involving the electron mass m,

$$\overline{|\mathfrak{M}|^2} = \frac{8e^4}{q^4}\left[(k'\cdot p')(k\cdot p) + (k'\cdot p)(k\cdot p') - M^2 k'\cdot k\right]$$

$$= \frac{8e^4}{q^4}\left[-\frac{1}{2}q^2(k\cdot p - k'\cdot p) + 2(k'\cdot p)(k\cdot p) + \frac{1}{2}M^2q^2\right] \quad (6.42)$$

where $q = k - k'$. To obtain the last line, we have used $p' = k - k' + p$, $k^2 = k'^2 \simeq 0$ and $q^2 \simeq -2k\cdot k'$.

We wish to evaluate the cross section in the laboratory frame, the frame in which the muon is initially at rest, $p = (M, \mathbf{0})$. The particle momenta in this frame are shown in Fig. 6.10.

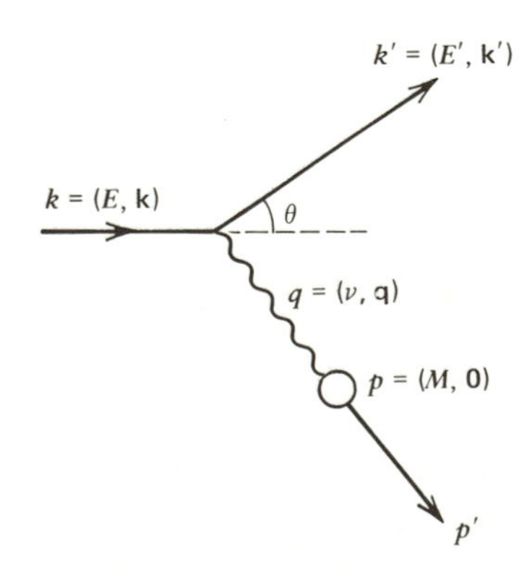

Fig. 6.10 The process $e^-\mu^- \to e^-\mu^-$ in the laboratory frame.

Evaluating (6.42) in the laboratory frame, we find

$$\overline{|\mathfrak{M}|^2} = \frac{8e^4}{q^4}\left(-\frac{1}{2}q^2M(E-E') + 2EE'M^2 + \frac{1}{2}M^2q^2\right)$$

$$= \frac{8e^4}{q^4}2M^2E'E\left\{1 + \frac{q^2}{4EE'} - \frac{q^2}{2M^2}\frac{M(E-E')}{2EE'}\right\}$$

$$= \frac{8e^4}{q^4}2M^2E'E\left\{\cos^2\frac{\theta}{2} - \frac{q^2}{2M^2}\sin^2\frac{\theta}{2}\right\}. \tag{6.43}$$

To reach the last line, we have used the following kinematic relations:

$$q^2 \simeq -2k\cdot k' \simeq -2EE'(1-\cos\theta) = -4EE'\sin^2\frac{\theta}{2}. \tag{6.44}$$

Also, squaring $q + p = p'$, we obtain

$$q^2 = -2p\cdot q = -2\nu M \qquad \text{so} \qquad \nu \equiv E - E' = -\frac{q^2}{2M}. \tag{6.45}$$

To calculate the $e^-\mu^- \to e^-\mu^-$ cross section, we make use of (4.27):

$$d\sigma = \frac{1}{(2E)(2M)}\frac{\overline{|\mathfrak{M}|^2}}{4\pi^2}\frac{d^3k'}{2E'}\frac{d^3p'}{2p_0'}\delta^{(4)}(p + k - p' - k')$$

$$= \frac{1}{4ME}\frac{\overline{|\mathfrak{M}|^2}}{4\pi^2}\frac{1}{2}E'\,dE'\,d\Omega\frac{d^3p'}{2p_0'}\delta^{(4)}(p + q - p'). \tag{6.46}$$

The flux is the product of beam and target densities $(2E)(2M)$ multiplied by relative velocity which is 1 (i.e., the speed of light) in the limit where the electron mass has been neglected.

EXERCISE 6.7 Justify the following relations:

$$\int \frac{d^3p'}{2p_0'}\delta^{(4)}(p + q - p') = \int d^3p'\,dp_0'\,\delta^{(4)}(p + q - p')\,\theta(p_0')\,\delta(p'^2 - M^2)$$

$$= \frac{1}{2M}\delta\left(\nu + \frac{q^2}{2M}\right) \tag{6.47}$$

$$= \frac{1}{2MA}\delta(E' - E/A), \tag{6.48}$$

where $A = 1 + (2E/M)\sin^2\theta$, and the step function $\theta(x)$ is 1 if $x > 0$ and 0 otherwise.

Inserting (6.43) into (6.46) and using (6.47), we obtain

$$\frac{d\sigma}{dE'\,d\Omega} = \frac{(2\alpha E')^2}{q^4}\left\{\cos^2\frac{\theta}{2} - \frac{q^2}{2M^2}\sin^2\frac{\theta}{2}\right\}\delta\left(\nu + \frac{q^2}{2M}\right). \tag{6.49}$$

Using (6.48), we may perform the dE' integration and, replacing q^2 by (6.44), we finally arrive at the following formula for the differential cross section for $e^-\mu^-$ scattering in the laboratory frame:

$$\left.\frac{d\sigma}{d\Omega}\right|_{\text{lab.}} = \left(\frac{\alpha^2}{4E^2\sin^4\dfrac{\theta}{2}}\right)\frac{E'}{E}\left\{\cos^2\frac{\theta}{2} - \frac{q^2}{2M^2}\sin^2\frac{\theta}{2}\right\}. \tag{6.50}$$

A powerful technique for exploring the internal structure of a target is to bombard it with a beam of high-energy electrons and to observe the angular distribution and energy of the scattered electrons. Such experiments have repeatedly led to major advances in our understanding of the structure of matter. Starting in Chapter 8, we describe how this method has revealed the structure of the proton. Equation (6.50) plays a central role in the story.

EXERCISE 6.8 Show that the cross section for elastic scattering of unpolarized electrons from spinless point-like particles is

$$\left.\frac{d\sigma}{d\Omega}\right|_{\text{lab.}} = \left(\frac{\alpha^2}{4E^2\sin^4\dfrac{\theta}{2}}\right)\frac{E'}{E}\cos^2\frac{\theta}{2}, \tag{6.51}$$

where as before we neglect the mass of the electron. Justify using (6.18) with $L_{\mu\nu}^{\text{muon}}$ replaced by $(p+p')_\mu(p+p')_\nu$. Comparing the cross section with that for $e^-\mu^- \to e^-\mu^-$, we see that the $\sin^2(\theta/2)$ in (6.50) is due to scattering from the magnetic moment of the muon.

6.9 Photons. Polarization Vectors

We have already noted that, in the presence of a current density j, the electromagnetic field A^μ satisfies

$$\square^2 A^\mu = j^\mu. \tag{6.52}$$

The following two exercises recall how this equation arises from Maxwell's equations.

EXERCISE 6.9 Maxwell's equations of classical electrodynamics are, *in vacuo*,

$$\begin{aligned} \nabla\cdot\mathbf{E} &= \rho, \qquad \nabla\times\mathbf{E} + \frac{\partial\mathbf{B}}{\partial t} = 0 \\ \nabla\cdot\mathbf{B} &= 0, \qquad \nabla\times\mathbf{B} - \frac{\partial\mathbf{E}}{\partial t} = \mathbf{j} \end{aligned} \tag{6.53}$$

(where we are using Heaviside–Lorentz rationalized units, see Appendix C of Aitchison and Hey). Show that these equations are equivalent to the following covariant equation for A^μ:

$$\Box^2 A^\mu - \partial^\mu(\partial_\nu A^\nu) = j^\mu, \tag{6.54}$$

with $j^\mu = (\rho, \mathbf{j})$, and where $A^\mu = (\phi, \mathbf{A})$, the four-vector potential, is related to the electric and magnetic fields by

$$\mathbf{E} = -\frac{\partial\mathbf{A}}{\partial t} - \nabla\phi, \qquad \mathbf{B} = \nabla\times\mathbf{A}. \tag{6.55}$$

Further, show that in terms of the antisymmetric field strength tensor

$$F^{\mu\nu} \equiv \partial^\mu A^\nu - \partial^\nu A^\mu \tag{6.56}$$

Maxwell's equations take the compact form

$$\partial_\mu F^{\mu\nu} = j^\nu, \tag{6.57}$$

and that current conservation, $\partial_\nu j^\nu = 0$, follows as a natural compatibility condition. [Note that $\nabla\times(\nabla\times\mathbf{A}) = -\nabla^2\mathbf{A} + \nabla(\nabla\cdot\mathbf{A})$.]

EXERCISE 6.10 Verify that **E** and **B** in (6.55) are unchanged by the gauge transformation

$$A_\mu \to A'_\mu = A_\mu + \partial_\mu\chi, \tag{6.58}$$

where χ can be any function of x. Use this freedom to write Maxwell's equations in the form

$$\Box^2 A^\mu = j^\mu \qquad \text{with } \partial_\mu A^\mu = 0. \tag{6.59}$$

The requirement $\partial_\mu A^\mu = 0$ is known as the Lorentz condition. However, even after imposing this, there is still some residual freedom in the choice of the potential A^μ. We can still make another gauge transformation,

$$A_\mu \to A'_\mu = A_\mu + \partial_\mu \Lambda \tag{6.60}$$

where Λ is any function that satisfies

$$\square^2 \Lambda = 0. \tag{6.61}$$

This last equation ensures that the Lorentz condition is still satisfied.

Turning now from classical to quantum mechanics, we see that the wavefunction A^μ for a free photon satisfies the equation

$$\square^2 A^\mu = 0, \tag{6.62}$$

which has solutions

$$A^\mu = \varepsilon^\mu(\mathbf{q}) e^{-iq\cdot x}. \tag{6.63}$$

The four-vector ε^μ is called the polarization vector of the photon. On substituting into the equation, we find that q, the four-momentum of the photon, satisfies

$$q^2 = 0, \qquad \text{that is, } m_\gamma = 0. \tag{6.64}$$

The polarization vector has four components and yet it describes a spin-1 particle. How does this come about? First, the Lorentz condition, $\partial_\mu A^\mu = 0$, gives

$$q_\mu \varepsilon^\mu = 0, \tag{6.65}$$

and this reduces the number of independent components of ε^μ to three. Moreover, we have to explore the consequences of the additional gauge freedom (6.60). Choose a gauge parameter

$$\Lambda = iae^{-iq\cdot x}$$

with a constant so that (6.61) is satisfied. Substituting this, together with (6.63), into (6.60) shows that the physics is unchanged by the replacement

$$\varepsilon_\mu \to \varepsilon'_\mu = \varepsilon_\mu + aq_\mu. \tag{6.66}$$

In other words, two polarization vectors (ε_μ, ε'_μ) which differ by a multiple of q_μ describe the same photon. We may use this freedom to ensure that the time component of ε^μ vanishes, $\varepsilon^0 \equiv 0$; and then the Lorentz condition (6.65) reduces to

$$\boldsymbol{\varepsilon}\cdot\mathbf{q} = 0. \tag{6.67}$$

This (noncovariant) choice of gauge is known as the Coulomb gauge.

From (6.67), we see that there are only *two* independent polarization vectors and that they are both transverse to the three-momentum of the photon. For example, for a photon traveling along the z axis, we may take

$$\boldsymbol{\varepsilon}_1 = (1,0,0), \qquad \boldsymbol{\varepsilon}_2 = (0,1,0). \tag{6.68}$$

A free photon is thus described by its momentum q and a polarization vector $\boldsymbol{\varepsilon}_i$. Since $\boldsymbol{\varepsilon}_i$ transforms as a vector, we anticipate that it is associated with a particle of spin 1.

EXERCISE 6.11 Determine how the linear combinations

$$\boldsymbol{\varepsilon}_R = -\sqrt{\tfrac{1}{2}}\,(\boldsymbol{\varepsilon}_1 + i\boldsymbol{\varepsilon}_2)$$
$$\boldsymbol{\varepsilon}_L = \sqrt{\tfrac{1}{2}}\,(\boldsymbol{\varepsilon}_1 - i\boldsymbol{\varepsilon}_2) \tag{6.69}$$

transform under a rotation θ about the z axis. Hence, show that $\boldsymbol{\varepsilon}_R$ and $\boldsymbol{\varepsilon}_L$ describe a photon of helicity $+1$ and -1, respectively; $\boldsymbol{\varepsilon}_{R,L}$ are called circular polarization vectors.

EXERCISE 6.12 Show that (in the transverse gauge) the completeness relation is

$$\sum_{\lambda=R,L} (\varepsilon_\lambda)_i^*(\varepsilon_\lambda)_j = \delta_{ij} - \hat{q}_i\hat{q}_j \tag{6.70}$$

If $\boldsymbol{\varepsilon}$ were along $\mathbf{q}$, it would be associated with a helicity-zero photon. This state is missing because of the transversality condition, $\mathbf{q}\cdot\boldsymbol{\varepsilon} = 0$. It can only be absent because the photon is massless. We return to a further discussion of photon polarization vectors in Section 6.13.

6.10 More on Propagators. The Electron Propagator

Here, we wish to consolidate our earlier discussion of propagators and also introduce the propagator for the electron.

First, let us recall the nonrelativistic perturbation expansion of the transition amplitude [see (3.44) and (4.48)]

$$T_{fi} = -i2\pi\,\delta(E_f - E_i)\left(\langle f|V|i\rangle + \sum_{n\neq i}\langle f|V|n\rangle\frac{1}{E_i - E_n}\langle n|V|i\rangle + \cdots\right). \tag{6.71}$$

Recall also that we associated factors such as $\langle f|V|n\rangle$ with the vertices and identified $1/(E_i - E_n)$ as the propagator (see Fig. 3.4 and Section 4.8). The state vectors are eigenstates of the Hamiltonian in the absence of V [see (3.29)]

$$H_0|n\rangle = E_n|n\rangle.$$

Formally, we may therefore rewrite (6.71) as

$$T_{fi} = 2\pi\,\delta(E_f - E_i)\langle f|(-iV) + (-iV)\frac{i}{E_i - H_0}(-iV) + \cdots\,|i\rangle, \tag{6.72}$$

where we have made use of the completeness relation $\Sigma|n\rangle\langle n| = 1$. (The prescription for handling the singularity at $E_n = E_i$ is discussed in Section 6.16.) It is natural to take $-iV$, rather than V, as the perturbation parameter. (The $-i$ arises from the i in $i\partial\psi/\partial t = V\psi$, which leads to a time dependence $\exp(-iVt)$ in the interaction picture.) That is, the vertex factor is $-iV$, and the propagator may thus be regarded as i times the inverse of the Schrödinger operator,

$$-i(E_i - H_0)\psi = -iV\psi, \tag{6.73}$$

acting on the intermediate state.

Let us apply the same technique to the various relativistic wave equations and so deduce the form of the propagators for the corresponding particles.

The Propagator for a Spinless Particle

The form of the Klein–Gordon equation corresponding to (6.73) is

$$i(\square^2 + m^2)\phi = -iV\phi, \tag{6.74}$$

see (4.3). Guided by the relativistic generalization of (6.72), we expect the propagator for a spinless particle to be the inverse of the operator on the left-hand side of (6.74). For an intermediate state of momentum p, this gives

$$\frac{1}{i(-p^2 + m^2)} = \boxed{\frac{i}{p^2 - m^2}}. \tag{6.75}$$

Indeed, we have already discussed how this form arises as the relativistic generalization of the propagator (see Section 4.8).

The Electron Propagator

An electron in an electromagnetic field satisfies

$$(\not{p} - m)\psi = -e\gamma^\mu A_\mu \psi, \tag{6.76}$$

see (6.2) and (6.3). As before, we must multiply by $-i$. Hence, the vertex factor is $ie\gamma^\mu$, The electron propagator is therefore the inverse of $-i$ times the left-hand side of (6.76):

$$\frac{1}{-i(\not{p} - m)} = \frac{i}{\not{p} - m} = \boxed{\frac{i(\not{p} + m)}{p^2 - m^2}} = \frac{i\sum_s u\bar{u}}{p^2 - m^2}, \tag{6.77}$$

where we have used $\not{p}\not{p} = p^2$ and the completeness relation (5.47). The numerator contains the sum over the spin states of the virtual electron; a further discussion is given in Section 6.16.

In summary, the general form of the propagator of a virtual particle is

$$\frac{i\sum_{\text{spins}}}{p^2 - m^2}.$$

The spin sum is the completeness relation; we include all possible spin states of the propagating particle. We would also integrate over the different momentum states that propagate. For the diagrams we have considered so far, this momentum is fixed by the momenta of the external particles.

We discuss the prescription for handling the singularity at $p^2 = m^2$ in Section 6.16.

6.11 The Photon Propagator

The propagator for a photon is not unique, on account of the freedom in the choice of A^μ. Recall that physics is unchanged by the transformation

$$A^\mu \to A^\mu + \partial^\mu \chi$$

which, as will be explained in Chapter 14, is associated with the invariance of QED under phase or "gauge" transformations of the wavefunctions of charged particles.

From (6.54), we see that the wave equation for a photon can be written in the form

$$\left(g^{\nu\lambda}\square^2 - \partial^\nu \partial^\lambda\right) A_\lambda = j^\nu, \tag{6.78}$$

and, in fact, a photon propagator cannot exist until we remove some of the gauge freedom of A_λ, see Exercise 6.13.

EXERCISE 6.13 Verify that the inverse of the "momentum space operator" of (6.78) does not exist.

Hint Attempt to write the inverse in the most general form satisfying Lorentz covariance

$$Aq^2 g_{\mu\nu} + Bq_\mu q_\nu \tag{6.79}$$

where A and B are functions of q^2.

In our discussions so far, we have chosen to work in the Lorentz class of gauges with $\partial_\lambda A^\lambda = 0$. The wave equation, (6.78), then simplifies to

$$g^{\nu\lambda}\square^2 A_\lambda = j^\nu. \tag{6.80}$$

Now, since

$$g_{\mu\nu} g^{\nu\lambda} = \delta_\mu^\lambda, \tag{6.81}$$

where δ_μ^λ equals 1 if $\lambda = \mu$, and 0 otherwise, the propagator (the inverse of the momentum space operator multiplied by $-i$) is

$$\boxed{i\frac{-g_{\mu\nu}}{q^2}}. \tag{6.82}$$

The discussion in Section 6.10 implies that we should associate $-g_{\mu\nu}$ with the sum over the polarization vectors of the virtual photon. This we tackle in Section 6.13. The gauge condition $\partial_\lambda A^\lambda = 0$ has been imposed covariantly, and the resulting covariant propagator, (6.82), is ideal for QED calculations. It is called the *Feynman propagator*, and we say we are working in the Feynman gauge. Indeed, this is the photon propagator we have been using so far, and we have already entered it into our table of Feynman rules, see Section 6.17.

EXERCISE 6.14 The condition $\partial_\lambda A^\lambda = 0$ does not fully define the propagator. We are at liberty to rewrite wave equation (6.78) as

$$\left[g^{\nu\lambda}\square^2 - \left(1 - \frac{1}{\xi}\right)\partial^\nu\partial^\lambda\right]A_\lambda = j^\nu. \tag{6.83}$$

In this case, use (6.79) to show that the propagator is

$$\frac{i}{q^2}\left(-g_{\mu\nu} + (1-\xi)\frac{q_\mu q_\nu}{q^2}\right). \tag{6.84}$$

The Feynman gauge takes $\xi = 1$. But in any case, the extra term in the propagator vanishes in QED calculations in which the virtual photon is coupled to conserved currents which satisfy $q_\mu j^\mu = q_\nu j^\nu = 0$.

6.12 Massive Vector Particles

Massive vector (spin 1) particles, denoted $W^\pm$ and Z^0, play a central role in the theory of weak interactions (see Chapter 12 onward). The wave equation for a spin-1 particle of mass M can be obtained from that for the photon by the replacement $\square^2 \to \square^2 + M^2$; recall the Klein–Gordon operator (3.19). From (6.78), we see that the wavefunction B_λ for a free particle satisfies

$$\left(g^{\nu\lambda}(\square^2 + M^2) - \partial^\nu\partial^\lambda\right)B_\lambda = 0. \tag{6.85}$$

Proceeding exactly as before, we determine the inverse of the momentum space operator by solving

$$\left(g^{\nu\lambda}(-p^2 + M^2) + p^\nu p^\lambda\right)^{-1} = \delta^\mu_\lambda\left(Ag_{\mu\nu} + Bp_\mu p_\nu\right) \tag{6.86}$$

for A and B. The propagator, which is the quantity in brackets on the right-hand side of (6.86) multiplied by i, is found to be

$$\boxed{\frac{i(-g^{\mu\nu} + p^\mu p^\nu/M^2)}{p^2 - M^2}}. \tag{6.87}$$

We can show that the numerator is the sum over the three spin states of the massive particle when taken on-shell $p^2 = M^2$. We first take the divergence, ∂_ν, of (6.85). Two terms cancel, and we find

$$M^2 \partial^\lambda B_\lambda = 0, \qquad \text{that is, } \partial^\lambda B_\lambda = 0. \tag{6.88}$$

Thus, for a massive vector particle, we have no choice but to take $\partial^\lambda B_\lambda = 0$; it is not a gauge condition. As a consequence, the wave equation reduces to

$$(\square^2 + M^2) B_\mu = 0 \tag{6.89}$$

with free-particle solutions

$$B_\mu = \varepsilon_\mu e^{-ip\cdot x}. \tag{6.90}$$

The condition (6.88) demands

$$p^\mu \varepsilon_\mu = 0 \tag{6.91}$$

and so reduces the number of independent polarization vectors from four to three in a covariant fashion.

EXERCISE 6.15 For a vector particle of mass M, energy E, and momentum $\mathbf{p}$ along the z axis, show that states of helicity λ can be described by polarization vectors

$$\begin{aligned} \varepsilon^{(\lambda=\pm 1)} &= \mp(0, 1, \pm i, 0)/\sqrt{2}, \\ \varepsilon^{(\lambda=0)} &= (|\mathbf{p}|, 0, 0, E)/M. \end{aligned} \tag{6.92}$$

EXERCISE 6.16 Show that the completeness relation is

$$\sum_\lambda \varepsilon_\mu^{(\lambda)*} \varepsilon_\nu^{(\lambda)} = -g_{\mu\nu} + \frac{p_\mu p_\nu}{M^2} \tag{6.93}$$

where the sum is over the three polarization states of the massive vector particle.

6.13 Real and Virtual Photons

For a real photon, we saw that there are only *two* polarization states. Indeed, in Section 6.9 we found we could choose $\varepsilon_0^{(\lambda)} \equiv 0$ and $\mathbf{q}\cdot\boldsymbol{\varepsilon}^{(\lambda)} = 0$, so that only transverse polarization states remain. Recall that the completeness relation for these polarization vectors is

$$\sum_{T=R,L} \varepsilon_i^{T*} \varepsilon_j^T = \delta_{ij} - \hat{q}_i \hat{q}_j, \tag{6.94}$$

where T denotes transverse [see (6.70)].

On the other hand, we have associated with a virtual photon the covariant propagator $i(-g_{\mu\nu})/q^2$, where $-g_{\mu\nu}$ implies we are summing over *four* polari-

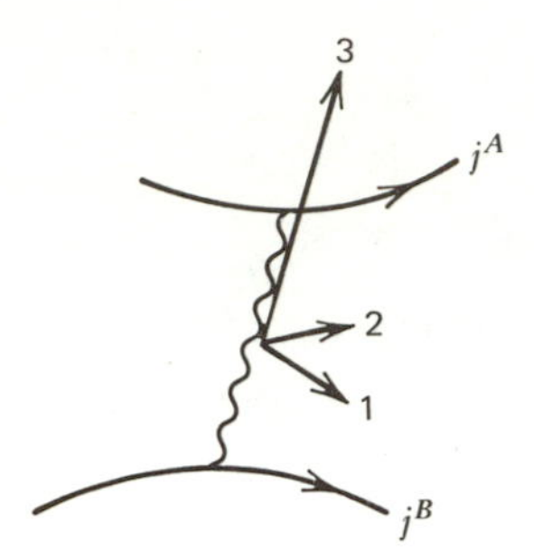

Fig. 6.11 A virtual photon exchanged between two charged particles A and B.

zation states. The completeness relation now reads in an obvious notation

$$-g_{\mu\nu} = \sum_{\lambda=1}^{4} \varepsilon_\mu^{(\lambda)*}\varepsilon_\nu^{(\lambda)} = \sum_T \varepsilon_\mu^{T*}\varepsilon_\nu^T + \varepsilon_\mu^{L*}\varepsilon_\nu^L + \varepsilon_\mu^{S*}\varepsilon_\nu^S$$

$$= \underbrace{\left(\delta_{ij} - \hat{q}_i\hat{q}_j\right)}_{\text{transverse}} + \underbrace{\hat{q}_i\hat{q}_j}_{\text{longitudinal}} + \underbrace{(-g_{\mu 0}g_{\nu 0})}_{\text{scalar}}. \tag{6.95}$$

However, in a sense every photon is virtual, being emitted and then sooner or later being absorbed. How can we reconcile the two descriptions?

Let us look at a typical Feynman diagram, Fig. 6.11, containing a virtual photon exchanged between charged particles. For such diagrams, see, for example, Fig. 6.2, we have found a transition amplitude of the form [see the discussion leading to (6.8)]

$$T_{fi} = -i\int j_\mu^A(x)\left(\frac{-g^{\mu\nu}}{q^2}\right) j_\nu^B(x)\, d^4x$$

$$= -i\int \left(\underbrace{\frac{j_1^A j_1^B + j_2^A j_2^B}{q^2}}_{\text{transverse}} + \underbrace{\frac{j_3^A j_3^B - j_0^A j_0^B}{q^2}}_{\text{longitudinal/scalar}}\right) d^4x \tag{6.96}$$

where we have taken the photon four-momentum $q^\mu = (q^0, 0, 0, |\mathbf{q}|)$. That is, we choose the 3-axis to be along $\hat{\mathbf{q}}$. Recall that charge conservation gives rise to the continuity equation $\partial^\mu j_\mu = 0$. For both the A, B currents, this implies

$$q^\mu j_\mu = q^0 j_0 - |\mathbf{q}| j_3 = 0. \tag{6.97}$$

Thus, if the exchanged photon is almost real, $q^0 \approx |\mathbf{q}|$, then $j_3 \approx j_0$ and the longitudinal and scalar contributions cancel each other, leaving only the two transverse contributions. For a real photon, we can therefore make the replacement

$$\sum_T \varepsilon_\mu^{T*}\varepsilon_\nu^T \to -g_{\mu\nu}. \tag{6.98}$$

On the other hand, for a virtual photon the longitudinal and scalar components cannot be neglected. Indeed, they play an important role. If we use (6.97) to substitute for j_3^A and j_3^B in (6.96), we find

$$T_{fi} = -i\int\left(\frac{j_1^A j_1^B + j_2^A j_2^B}{q^2} + \frac{j_0^A j_0^B}{|\mathbf{q}|^2}\right) d^4x. \tag{6.99}$$

The first term describes the propagation of virtual photons in transverse polarization states, whereas the second term, with its $|\mathbf{q}|^2$ denominator, is not associated with propagation. Instead, it represents the instantaneous Coulomb interaction between the charges of the two particles, j_0^A and j_0^B. This becomes clear if we rewrite the second term of (6.99) in the form

$$T_{fi}^{\text{Coul}} = -i\int dt \int d^3x_1 \int d^3x_2 \frac{j_0^A(t, \mathbf{x}_1)\, j_0^B(t, \mathbf{x}_2)}{4\pi|\mathbf{x}_2 - \mathbf{x}_1|}, \tag{6.100}$$

and note that the charges interact without retardation at time t.

EXERCISE 6.17 Verify (6.100) by making use of the Fourier transform

$$\frac{1}{|\mathbf{q}|^2} = \int d^3x\, e^{i\mathbf{q}\cdot\mathbf{x}} \frac{1}{4\pi|\mathbf{x}|}. \tag{6.101}$$

Finally, by inspection of (6.95), we see that the division of $-g^{\mu\nu}/q^2$ into a transverse propagating contribution and a longitudinal/scalar static contribution is not a Lorentz covariant separation. Only the sum forms a covariant photon propagator.

6.14 Compton Scattering $\gamma e^- \to \gamma e^-$

Compton scattering is a useful example to work through in detail. Not only do the Feynman diagrams involve both the electron propagator and external photons, but we shall need the form of the amplitude for the analogous process $\gamma^* q \to gq$ (involving quarks q and gluons g) in the development of QCD in Chapter 10.

The two (lowest-order) Feynman diagrams are shown in Fig. 6.12, and the factors needed to compute the amplitude are shown in detail on the first diagram.

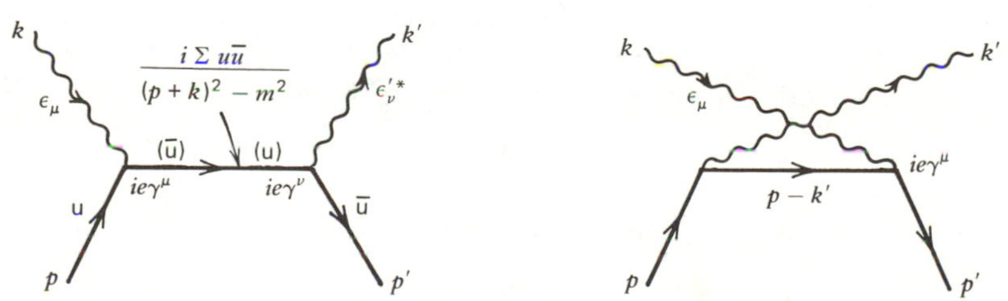

Fig. 6.12 Feynman diagrams for Compton scattering $\gamma e^- \to \gamma e^-$.

A word of explanation is in order. For the incoming photon we have

$$A_\mu = \varepsilon_\mu e^{-ik\cdot x},$$

where ε_μ is one of the two transverse polarization vectors. Referring back to (6.4), we see that the $e^{-ik\cdot x}$ is disposed of by the x integration, which results in momentum conservation at the vertex. We are thus led to include only the factor ε_μ for an incoming photon line in our table of Feynman rules at the end of the chapter. Similarly, for an outgoing photon, $(\varepsilon'_\nu e^{-ik'\cdot x})^*$, we are led to a factor ε'^*_ν. Note that, just as in Fig. 6.1b, the structure of the electron–photon vertex is

$$(\bar{u}\gamma^\mu u)\varepsilon_\mu, \tag{6.102}$$

but that here $\bar{u}$ is already contained in the electron propagator, whereas for electron scattering vertices it is ε_μ that is embodied in the photon propagator.

Using the Feynman rules, we obtain the following amplitudes for the two Feynman diagrams:

$$-i\mathfrak{M}_1 = \bar{u}^{(s')}(p')\left[\varepsilon'^*_\nu(ie\gamma^\nu)\frac{i(\not{p}+\not{k}+m)}{(p+k)^2-m^2}(ie\gamma^\mu)\varepsilon_\mu\right]u^{(s)}(p) \tag{6.103}$$

$$-i\mathfrak{M}_2 = \bar{u}^{(s')}(p')\left[\varepsilon_\mu(ie\gamma^\mu)\frac{i(\not{p}-\not{k}'+m)}{(p-k')^2-m^2}(ie\gamma^\nu)\varepsilon'^*_\nu\right]u^{(s)}(p), \tag{6.104}$$

where p, s and p', s' are the momentum, spin state of the ingoing and outgoing electrons, respectively. Similarly, k, ε and k', ε' are the momentum, polarization vector of the ingoing and outgoing photons, respectively. Note that the invariant amplitude for Compton scattering $(\mathfrak{M}_1 + \mathfrak{M}_2)$ is symmetric under the interchange (or crossing) of the two photons

$$k, \varepsilon \leftrightarrow -k', \varepsilon'^*. \tag{6.105}$$

This is another example of crossing symmetry.

What does gauge invariance have to say about Compton scattering? Provided we impose the Lorentz condition $\partial_\mu A^\mu = 0$, we saw that physics is unchanged by the replacement

$$\varepsilon_\mu \to \varepsilon_\mu + ak_\mu, \tag{6.106}$$

where ε_μ and k_μ are the polarization vector and momentum of the photon, respectively, and a is an arbitrary constant, see (6.66). It therefore follows that if we write the amplitude for Compton scattering in the form

$$\mathfrak{M} = \varepsilon'^*_\nu \varepsilon_\mu T^{\mu\nu}, \tag{6.107}$$

then $\mathfrak{M}$ is unchanged by substitution $\varepsilon_\mu \to ak_\mu$ or $\varepsilon'_\nu \to ak'_\nu$. Thus, gauge invariance requires

$$k_\mu T^{\nu\mu} = k'_\nu T^{\nu\mu} = 0. \tag{6.108}$$

EXERCISE 6.18 Show that, individually, the amplitudes $\mathfrak{M}_1$ and $\mathfrak{M}_2$ are not gauge invariant but that their sum indeed satisfies (6.108).

It is instructive to work through the calculation of the Compton scattering amplitude in detail. For simplicity, we neglect the mass of the electron, and so the invariant variables for $\gamma(k) + e(p) \rightarrow \gamma(k') + e(p')$ are

$$\begin{aligned} s &= (k+p)^2 = 2k \cdot p = 2k' \cdot p' \\ t &= (k-k')^2 = -2k \cdot k' = -2p \cdot p' \\ u &= (k-p')^2 = -2k \cdot p' = -2p \cdot k'. \end{aligned} \tag{6.109}$$

The two invariant amplitudes, (6.103) and (6.104), are

$$\begin{aligned} \mathfrak{M}_1 &= \varepsilon_\nu'^* \varepsilon_\mu e^2 \, \bar{u}(p') \, \gamma^\nu (\not{p} + \not{k}) \gamma^\mu \, u(p)/s, \\ \mathfrak{M}_2 &= \varepsilon_\nu'^* \varepsilon_\mu e^2 \, \bar{u}(p') \, \gamma^\mu (\not{p} - \not{k}') \gamma^\nu \, u(p)/u. \end{aligned} \tag{6.110}$$

To obtain the unpolarized cross section, we must average/sum $|\mathfrak{M}_1 + \mathfrak{M}_2|^2$ over the initial/final electron and photon spins. Fortunately, this is not as difficult as it first appears. For physical photons, (6.98) applies, and we can make the replacement

$$\sum_T \varepsilon_\mu^{T*} \varepsilon_{\mu'}^T \rightarrow -g_{\mu\mu'} \tag{6.111}$$

where T denotes transverse. We have a similar completeness relation for the outgoing photon states, ε'. Thus, for example,

$$\overline{|\mathfrak{M}_1|^2} = \frac{e^4}{4s^2} \sum_{s,\,s'} \left(\bar{u}^{(s')} \gamma^\nu (\not{p} + \not{k}) \gamma^\mu u^{(s)} \right) \left(\bar{u}^{(s)} \gamma_\mu (\not{p} + \not{k}) \gamma_\nu u^{(s')} \right).$$

The factor $\frac{1}{4}$ is due to averaging over the initial electron and photon spins. The spinor completeness relation, (5.47), allows the sum over $u\bar{u}$ states to be performed (just as we described for $e^-\mu^-$ scattering in Section 6.3), and we find

$$\begin{aligned} \overline{|\mathfrak{M}_1|^2} &= \frac{e^4}{4s^2} \mathrm{Tr} \big(\underbrace{\not{p}' \gamma^\nu}_{-2\not{p}'} (\not{p} + \not{k}) \underbrace{\gamma^\mu \not{p} \gamma_\mu}_{-2\not{p}} (\not{p} + \not{k}) \gamma_\nu \big) \\ &= \frac{e^4}{s^2} \mathrm{Tr}(\not{p}' \not{k} \not{p} \not{k}) \\ &= \frac{4e^4}{s^2} 2(p' \cdot k)(p \cdot k) \\ &= 2e^4 \left(-\frac{u}{s} \right), \end{aligned} \tag{6.112}$$

where we have made use of (6.24) and (6.22). Similarly, we obtain

$$\begin{aligned} \overline{|\mathfrak{M}_2|^2} &= 2e^4 \left(-\frac{s}{u} \right), \\ \overline{\mathfrak{M}_1 \mathfrak{M}_2^*} &= 0. \end{aligned}$$

Thus, the spin-averaged Compton amplitude is

$$\overline{|\mathfrak{M}|^2} = \overline{|\mathfrak{M}_1 + \mathfrak{M}_2|^2} = 2e^4\left(-\frac{u}{s} - \frac{s}{u}\right). \tag{6.113}$$

EXERCISE 6.19 Repeat the above calculation for an incident virtual photon of mass $k^2 \equiv -Q^2$. Continue to use (6.111). Show that for $\gamma^* e^- \to \gamma e^-$ (where γ^* denotes a virtual photon),

$$\overline{|\mathfrak{M}|^2} = 2e^4\left(-\frac{u}{s} - \frac{s}{u} + \frac{2Q^2 t}{su}\right). \tag{6.114}$$

We shall make use of this result in Chapter 10.

EXERCISE 6.20 Restore the mass m of the electron and show that at high energy, $s \to \infty$, the integrated cross section for Compton scattering is

$$\sigma = \frac{1}{64\pi^2 s}\int \overline{|\mathfrak{M}|^2}\, d\Omega \to \frac{2\pi\alpha^2}{s}\log\left(\frac{s}{m^2}\right). \tag{6.115}$$

Note that at high energy the dominant contribution comes from $\mathfrak{M}_2$, via a glancing collision in which the u-channel electron is almost on mass shell.

EXERCISE 6.21 Show, by using particle helicities, that high-energy Compton scattering via the first diagram of Fig. 6.12 is, in the center-of-mass frame, given by

$$\begin{aligned}\overline{|\mathfrak{M}_1|^2} &\propto \left|d^{1/2}_{++}(\theta)\right|^2 + \left|d^{1/2}_{--}(\theta)\right|^2 \\ &= (1 + \cos\theta) \simeq -\frac{u}{2s},\end{aligned} \tag{6.116}$$

in agreement with (6.112). An example of this type of calculation is described in Section 6.6.

6.15 Pair Annihilation $e^+e^- \to \gamma\gamma$

EXERCISE 6.22 Draw the lowest-order Feynman diagrams for the pair annihilation process

$$e^+(p_1, s_1) + e^-(p_2, s_2) \to \gamma(k_1, \varepsilon_1) + \gamma(k_2, \varepsilon_2).$$

Fig. 6.13 Feynman diagrams for $e^+e^- \to \gamma\gamma$.

Check your answer with Fig. 6.13. Use the Feynman rules to show that

$$-i\mathfrak{M} = (ie)^2 \bar{v}^{(s_1)}(p_1)\left(\not{\varepsilon}_1^* \frac{i}{\not{p}_1 - \not{k}_1 - m}\not{\varepsilon}_2^* + \not{\varepsilon}_2^* \frac{i}{\not{p}_1 - \not{k}_2 - m}\varepsilon_1^*\right) u^{(s_2)}(p_2). \tag{6.117}$$

Verify that, in the high-energy limit, the spin-averaged rate is given by

$$\overline{|\mathfrak{M}|^2} = 2e^4\left(\frac{u}{t} + \frac{t}{u}\right). \tag{6.118}$$

The $e^+e^- \to \gamma\gamma$ cross section has both forward and backward peaks, corresponding to the t- and u-channel exchanged electrons being almost on mass shell. Result (6.117) can also be obtained by crossing the amplitude for Compton scattering. To go from $\gamma e^- \to \gamma e^-$ to $e^+e^- \to \gamma\gamma$, we simple "cross" the ingoing photon with the outgoing electron,

$$k, \varepsilon \to -k_2, \varepsilon_2^*$$

$$p' \to -p_1 \qquad \text{and} \qquad \bar{u}^{(s')}(p') \to \bar{v}^{(s_1)}(p_1).$$

The initial electron and outgoing photon are unaltered:

$$u^{(s)}(p) \equiv u^{(s_2)}(p_2), \qquad k', \varepsilon'^* \to k_1, \varepsilon_1^*.$$

Making these substitutions in (6.103) and (6.104) gives the pair annihilation amplitude (6.117).

6.16 The $+i\varepsilon$ Prescription for Propagators

The heuristic treatment of quantum electrodynamics that we have given so far is based on Feynman's intuitive space-time approach. Our primary aim has been to motivate the Feynman rules and to calculate physical amplitudes. In so doing, we have avoided a detailed discussion of the underlying propagator theory, although we have made free use of the word "propagator." Here, we try to rectify this omission, but we urge those interested to read Feynman's original papers and the chapter on propagator theory in Bjorken and Drell (1964).

Green's Functions

Propagator theory is based on the Green's function method of solving inhomogeneous differential equations. We explain the method in terms of a simple example. Suppose we wish to solve Poisson's equation

$$\nabla^2\phi(\mathbf{x}) = -\rho(\mathbf{x}) \tag{6.119}$$

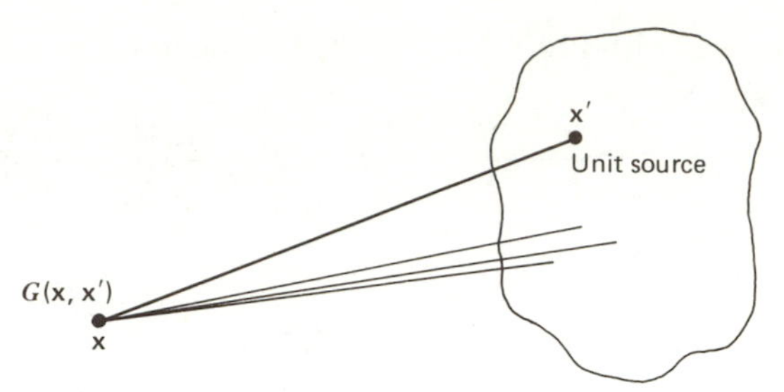

Fig. 6.14 G is the potential at $\mathbf{x}$ due to a unit source at $\mathbf{x}'$. We then use the principle of linear superposition to obtain the cumulative potential at $\mathbf{x}$, (6.121), arising from all possible elemental charges $\rho d^3x'$.

for a known charge distribution $\rho(\mathbf{x})$, subject to some boundary condition. It is easier to first solve the "unit source" problem

$$\nabla^2 G = -\delta^{(3)}(\mathbf{x} - \mathbf{x}') \tag{6.120}$$

where $G(\mathbf{x}, \mathbf{x}')$ is the potential at $\mathbf{x}$ due to a unit source at $\mathbf{x}'$. [For the boundary condition that $G \to 0$ at large distances, it is easy to show that $G = 1/(4\pi|\mathbf{x} - \mathbf{x}'|)$]. We then move this source over the charge distribution and accumulate the total potential at $\mathbf{x}$ from all possible volume elements d^3x':

$$\phi(\mathbf{x}) = \int G(\mathbf{x}, \mathbf{x}')\, \rho(\mathbf{x}')\, d^3x', \tag{6.121}$$

see Fig. 6.14. We can check directly that ϕ is the desired solution of (6.119) by operating with ∇^2 on (6.121).

The Electron Propagator iS_F

We take the electron propagator as our example and use the Green's function method to solve Dirac's equation, (6.76), for an electron in an electromagnetic field:

$$(i\gamma_\mu \partial^\mu - m)\psi = -e\gamma_\mu A^\mu \psi. \tag{6.122}$$

That is, we first solve the unit source problem

$$(i\gamma_\mu \partial^\mu - m) G_F = \delta^{(4)}(x - x'), \tag{6.123}$$

where G_F represents the wave produced at x by a unit source at x'. Once we have found Green's function G_F, we can construct the solution to (6.122):

$$\psi(x) = -e \int d^4x'\, G_F(x, x')\gamma_\mu\, A^\mu(x')\, \psi(x'). \tag{6.124}$$

Note that here ψ appears also on the right-hand side, and so an iterative perturbation series solution in powers of e is obtained.

From translational invariance, $G_F(x, x')$ is a function only of the difference $x - x'$. To solve (6.123), we first Fourier transform to momentum space:

$$G_F(x - x') = \frac{1}{(2\pi)^4} \int S_F(p) e^{-ip\cdot(x-x')}\, d^4p. \tag{6.125}$$

Then, on substituting into (6.123), we obtain

$$\frac{1}{(2\pi)^4} \int (\not{p} - m) S_F(p)\, e^{-ip\cdot(x-x')}\, d^4p = \frac{1}{(2\pi)^4} \int e^{-ip\cdot(x-x')}\, d^4p,$$

where the right-hand side is the Fourier representation of the delta function. In momentum space, (6.123) therefore becomes simply

$$(\not{p} - m)\, S_F(p) = 1.$$

That is,

$$S_F(p) = \frac{1}{\not{p} - m} = \frac{\not{p} + m}{p^2 - m^2}. \tag{6.126}$$

So far, this is just a sophisticated version of the derivation of Section 6.10.

To complete the determination of S_F, we need to know how to treat the singularities at

$$p^2 - m^2 = p_0^2 - (\mathbf{p}^2 + m^2) = (p_0 - E)(p_0 + E) = 0.$$

Since the electron is off mass shell, p_0 and $E = (\mathbf{p}^2 + m^2)^{1/2}$ are independent variables. To obtain the correct prescription for integration over the poles at $p_0 = \pm E$, we need to impose the appropriate boundary conditions on $G_F(x - x')$. From (6.125) and (6.126),

$$\begin{aligned} G_F(x - x') &= \frac{1}{(2\pi)^4} \int \frac{\not{p} + m}{(p_0 - E)(p_0 + E)} e^{-ip\cdot(x-x')}\, d^4p \\ &= \frac{1}{(2\pi)^4} \int d^3p\, e^{i\mathbf{p}\cdot(\mathbf{x}-\mathbf{x}')} \int_{-\infty}^{\infty} dp_0\, \frac{(\gamma_0 p_0 - \boldsymbol{\gamma}\cdot\mathbf{p} + m)}{(p_0 - E)(p_0 + E)} e^{-ip_0(t-t')}. \end{aligned} \tag{6.127}$$

Recall that $G_F(x - x')$ represents the wave produced at x by a unit source at x'. That is, the propagation is from x' to x. Now, we seek an S_F which is associated with the propagation of positive-energy electrons forward in time ($t > t'$) and with negative-energy electrons backward in time ($t < t'$); see Section 3.5. This can be accomplished by performing the p_0 integration along the contour in the complex p_0 plane shown in Fig. 6.15. That is, the required properties of S_F are obtained by choosing the contour along the Re p_0 axis to go below the $p_0 = -E$ pole and above the $p_0 = +E$ pole.

To check that this prescription works, first suppose $t > t'$. Then, from (6.127), we see that to ensure that the contribution from the semicircle vanishes, we must

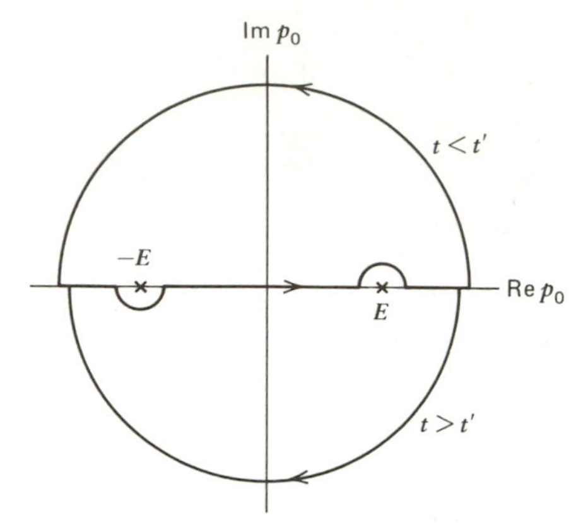

Fig. 6.15 The contours in the complex p_0 plane used to evaluate the dp_0 integral of (6.127).

close the contour in the lower half-plane. We therefore enclose the pole at $p_0 = +E$. Using the Cauchy residue theorem, we obtain

$$G_F(x - x') = \frac{-2\pi i}{(2\pi)^4} \int \frac{d^3p}{2E} e^{-ip\cdot(x-x')}(\gamma_0 E - \boldsymbol{\gamma}\cdot\mathbf{p} + m)$$

$$= \frac{-i}{(2\pi)^3} \int \frac{d^3p}{2E} e^{-ip\cdot(x-x')}(\not{p} + m). \tag{6.128}$$

Here, $\not{p} + m$ is the operator which projects out the positive-energy electron states, (5.48), and so S_F represents the propagation of $+E$ electrons forward in time.

Now, consider propagation backward in time, $t < t'$. In this case, the semicircle contribution will vanish provided we close the contour in the upper half-plane. We now enclose the pole at $p_0 = -E$, and so

$$G_F(x - x') = \frac{2\pi i}{(2\pi)^4} \int \frac{d^3p}{(-2E)} e^{i\mathbf{p}\cdot(\mathbf{x}-\mathbf{x}')} e^{-i(-E)(t-t')}(-\gamma_0 E - \boldsymbol{\gamma}\cdot\mathbf{p} + m).$$

Since we are integrating over all of three-momentum space, G_F is unchanged by the substitution $\mathbf{p} \to -\mathbf{p}$. Therefore,

$$G_F(x - x') = \frac{-i}{(2\pi)^3} \int \frac{d^3p}{2E} e^{ip\cdot(x-x')}(-\not{p} + m), \tag{6.129}$$

where $(-\not{p} + m)$ is the operator which projects out the negative-energy electron states, see (5.48). Thus, S_F represents the propagation of $-E, -\mathbf{p}$ electrons backward in time, which is equivalent to the propagation of $+E, +\mathbf{p}$ positrons forward in time. We see that the origin of the positron states is the pole at $p_0 = -E$, which was not present in nonrelativistic theory. To be really convinced that propagation in both spin states is included, we can recall the completeness relations (5.47).

The required boundary conditions were imposed by displacing the contour round the poles at $p_0 = \pm E$ as shown in Fig. 6.15. An equivalent prescription is

TABLE 6.2
Feynman Rules for $-i\mathcal{M}$

		Multiplicative Factor
• **External Lines**		
Spin 0 boson (or antiboson)		1
Spin $\frac{1}{2}$ fermion (in, out)		$u, \bar{u}$
antifermion (in, out)		$\bar{v}, v$
Spin 1 photon (in, out)	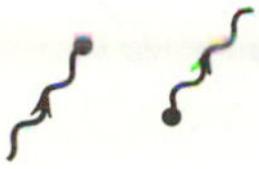	$\varepsilon_\mu, \varepsilon_\mu^*$
• **Internal Lines—Propagators (need $+i\varepsilon$ prescription)**		
Spin 0 boson		$\dfrac{i}{p^2 - m^2}$
Spin $\frac{1}{2}$ fermion		$\dfrac{i(\not{p} + m)}{p^2 - m^2}$
Massive spin 1 boson		$\dfrac{-i\left(g_{\mu\nu} - p_\mu p_\nu/M^2\right)}{p^2 - M^2}$
Massless spin 1 photon (Feynman gauge)		$\dfrac{-ig_{\mu\nu}}{p^2}$
• **Vertex Factors**	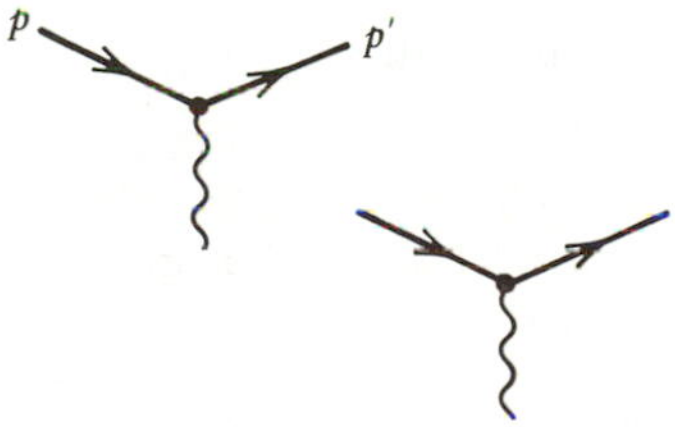 p p'	
Photon—spin 0 (charge $-e$)		$ie(p + p')^\mu$
Photon—spin $\frac{1}{2}$ (charge $-e$)		$ie\gamma^\mu$

Loops: $\int d^4k/(2\pi)^4$ over loop momentum; include -1 if fermion loop and take the trace of associated γ-matrices
Identical Fermions: -1 between diagrams which differ only in $e^- \leftrightarrow e^-$ or initial $e^- \leftrightarrow$ final e^+

to displace the poles slightly off axis and to leave the contour undisturbed. To do this, we write the electron propagator

$$iS_F(p) = i\frac{\not{p} + m}{p^2 - m^2 + i\varepsilon}. \tag{6.130}$$

The introduction of $+i\varepsilon$, with ε infinitesimal and positive, has the effect of displacing the $p_0 = \pm E$ poles slightly below and above the axis, respectively. The same $+i\varepsilon$ prescription is required for the other propagators.

An illuminating way to remember the sign of $i\varepsilon$ is to regard it as a negative-imaginary contribution to the mass, $m \to m - i\varepsilon/2$, so that the time dependence

$$e^{-iEt} \to e^{-i(m - i\varepsilon/2)t} = e^{-imt}e^{-\varepsilon t/2}.$$

Thus, stable particles may be viewed as the limit of unstable particles as the lifetime approaches infinity.

6.17 Summary of the Feynman Rules for QED

The invariant amplitude $\mathfrak{M}$ is obtained by drawing all (topologically distinct and connected) Feynman diagrams for the process and assigning multiplicative factors with the various elements of each diagram. The rules are summarized in Table 6.2.

For a photon–spin 0 interaction, there is also a four-particle vertex; see Fig. 6.16. This originates from the e^2A^2 term in (4.4). No corresponding four-particle photon–spin $\frac{1}{2}$ vertex exists, since no A^2 term occurs in the Dirac equation, (6.2), describing an electron in an electromagnetic field.

> ***EXERCISE 6.23*** Use the Feynman rules to evaluate $\mathfrak{M}(\gamma e^- \to \gamma e^-)$ corresponding to the two Feynman diagrams of Fig. 6.12 with the electron taken to have spin 0. Show that the result is not invariant under the gauge transformation (6.66). Demonstrate that gauge invariance is restored if diagram 6.16 is included with a vertex factor $2ie^2g^{\mu\nu}$.

In these chapters, we have considered only the lowest-order Feynman diagrams. The rules generalize to higher-order graphs. However, new features occur. The diagrams contain closed loops of intermediate particles (see, for example, Fig. 6.7). Even after applying four-momentum conservation at each vertex, there still remains an undetermined four-momentum running round a closed loop. We

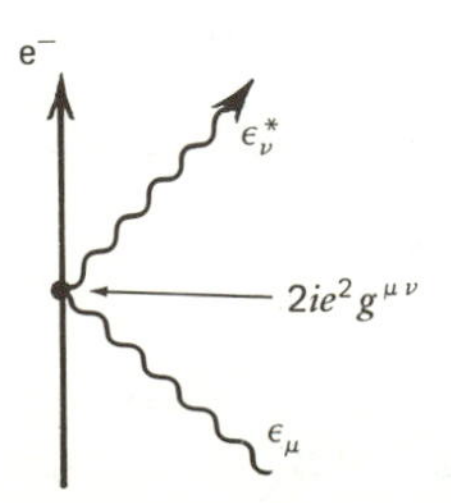

Fig. 6.16 The "sea-gull" diagram for $\gamma e^- \to \gamma e^-$, with "spinless" electrons.

thus need additional Feynman rules to evaluate such diagrams. First, we must integrate over the loop momentum, $\int d^4k/(2\pi)^4$. We must include a factor -1 for each closed fermion loop, and we have to take the trace of the associated γ-matrices. We discuss this in more detail in Chapter 7.

Unfortunately, loop integrations often lead to divergences. However, all the infinities which occur can be removed by well-established techniques. We say that QED is a renormalizable theory. This is the topic of the next chapter.

7

Loops, Renormalization, Running Coupling Constants, and All That

In this chapter, we attempt to give you a glimpse of the beautiful structure of field theory. Field theory is not the main subject of this book, so you can therefore safely skip this chapter; but a successful reading of it will expose you to such inaccessible concepts as loops, renormalization, and running coupling constants in a concise and physical way, we hope. A consequence is that the discussion is rather incomplete, and a few results are not explicitly derived. But only unrevealing algebra is omitted, which can be found in most field theory books.

7.1 Scattering Electrons Off a Static Charge

We use a simple experiment to demonstrate the concepts introduced in this chapter: the scattering of electrons by a static charge. In lowest order, the process is shown in Fig. 7.1a, in which the static charge is represented by a cross. How are the Feynman rules applied to this particular case? The question is best answered by going back to (6.4) and (6.6), where we find that the amplitude for the process of Fig. 7.1a can be written as

$$T_{fi} = -i\int d^4x\, j_\mu^{fi}(x)\, A^\mu(x). \tag{7.1}$$

Here, $j_\mu^{fi}(x)$ is the electron current:

$$j_\mu^{fi} = -e\bar{u}_f\gamma_\mu u_i\, e^{-iq\cdot x}, \tag{7.2}$$

and $A_\mu(x)$ is the four-vector potential associated with the static charge. As before, $q = p_i - p_f$ in terms of the momenta defined in Fig. 7.1a. Equation (7.1) can be written as

$$T_{fi} = ie\bar{u}_f\gamma_\mu u_i A^\mu(q), \tag{7.3}$$

where $A^\mu(q)$ is the Fourier transform:

$$A^\mu(q) = \int d^4x e^{-iq\cdot x}A^\mu(x). \tag{7.4}$$

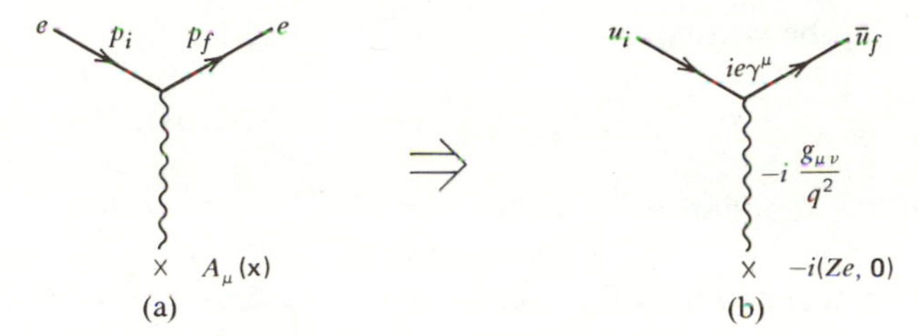

Fig. 7.1 Feynman rules for Rutherford scattering of electrons off a static charge Ze, for example, a nucleus.

For a static source, $A^\mu(x)$ is time independent; therefore

$$A^\mu(q) = \int dt\, e^{-i(E_i - E_f)t} \int d^3x\, e^{i\mathbf{q}\cdot\mathbf{x}} A^\mu(\mathbf{x})$$
$$= 2\pi\, \delta(E_f - E_i)\, A^\mu(\mathbf{q}). \tag{7.5}$$

The three-dimensional Fourier transform, $A^\mu(\mathbf{q})$, is best calculated using Maxwell's equations (6.59). For $A^\mu(x)$ independent of t, we have

$$\nabla^2 A^\mu(\mathbf{x}) = -j^\mu(\mathbf{x}), \tag{7.6}$$

and therefore

$$\int d^3x \left(\nabla^2 A^\mu(x)\right) e^{i\mathbf{q}\cdot\mathbf{x}} = -j^\mu(\mathbf{q}). \tag{7.7}$$

Now, by partial integration the left-hand side equals

$$\int d^3x\, A^\mu(x)\left(\nabla^2 e^{i\mathbf{q}\cdot\mathbf{x}}\right) = -|\mathbf{q}|^2 A^\mu(\mathbf{q}), \tag{7.8}$$

and so, combining (7.7) and (7.8), we have

$$A^\mu(\mathbf{q}) = \frac{1}{|\mathbf{q}|^2} j^\mu(\mathbf{q}). \tag{7.9}$$

We substitute this result into (7.5) and find, from (7.3),

$$T_{fi} = i2\pi\, \delta(E_f - E_i)\, e\bar{u}_f \gamma_\mu u_i \frac{1}{|\mathbf{q}|^2} j^\mu(\mathbf{q}). \tag{7.10}$$

The covariant amplitude $\mathfrak{M}$ is then obtained by removing the δ-function of T_{fi} [see (4.17)]:

$$-i\mathfrak{M} = ie\bar{u}_f \gamma_\mu u_i \frac{1}{|\mathbf{q}|^2} j^\mu(\mathbf{q}). \tag{7.11}$$

The electron recoils off the static charge in Fig. 7.1a, and $\mathbf{p}_i \neq \mathbf{p}_f$, but energy conservation in (7.10) implies $E_i = E_f$ or $q_0 = 0$. Therefore,

$$q^2 = -|\mathbf{q}|^2, \tag{7.12}$$

and so (7.11) can be written

$$-i\mathcal{M} = \left(ie\bar{u}_f\gamma^\mu u_i\right)\left(\frac{-ig_{\mu\nu}}{q^2}\right)\left(-ij^\nu(\mathbf{q})\right). \tag{7.13}$$

We recognize the familiar vertex factor and photon propagator of the Feynman rules for the amplitude $(-i\mathcal{M})$, see Section 6.17. We therefore deduce that the factor $-ij^\nu$ is associated with the source. For a static nucleus of charge Ze,

$$\begin{aligned} j^0(\mathbf{x}) &= \rho(\mathbf{x}) = Ze\,\delta(\mathbf{x}) \\ \mathbf{j}(\mathbf{x}) &= 0, \end{aligned} \tag{7.14}$$

and so

$$-i\mathcal{M} = \left(ie\bar{u}_f\gamma_0 u_i\right)\left(\frac{-i}{q^2}\right)(-iZe). \tag{7.15}$$

The result is recorded in Fig. 7.1b. For a static nucleus, (7.15) just describes Rutherford scattering. We recognize the familiar result for the angular distribution [see (1.6)]:

$$\frac{d\sigma}{d\Omega} \sim |\mathcal{M}|^2 \sim \frac{1}{\sin^4(\theta/2)}, \tag{7.16}$$

where θ is the deflection angle of the electron, shown in Fig. 7.2. This angular distribution is a result of the q^{-4} behavior of $d\sigma/d\Omega$ obtained by inserting (7.15) into (7.16). Indeed,

$$\begin{aligned} q^2 &= (p_i - p_f)^2 \\ &\simeq -2k^2(1-\cos\theta) \\ &\simeq -4k^2\sin^2\frac{\theta}{2}, \end{aligned} \tag{7.17}$$

where we have neglected the electron mass and used

$$k \equiv |\mathbf{p}_i| = |\mathbf{p}_f|.$$

7.2 Higher-Order Corrections

The previous calculation gives the Rutherford cross section to $O(\alpha^2)$ and is therefore an approximate perturbative result. To $O(\alpha^4)$, other Feynman diagrams have to be included, one of which is shown in Fig. 7.3. When the invariant

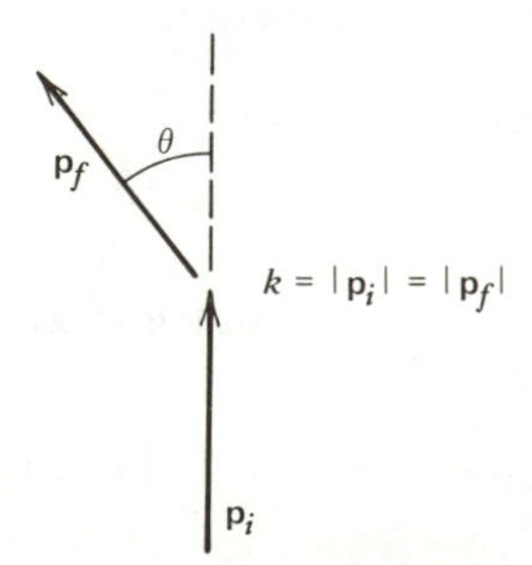

Fig. 7.2 Deflection of an electron by a static charge.

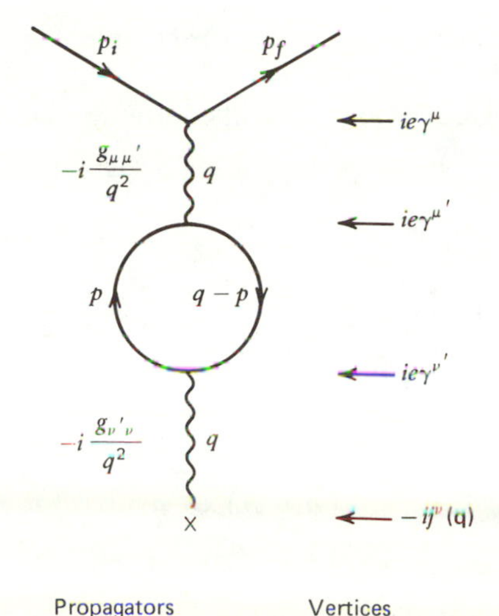

Fig. 7.3 Feynman diagram for Rutherford scattering in which the exchanged photon fluctuates into an e^-e^+ pair.

amplitude $(-i\mathfrak{M})$ of the $O(\alpha^4)$ diagrams is added to (7.15), a more accurate result is obtained for $d\sigma/d\Omega$. In the specific diagram of Fig. 7.3, the exchanged photon spends some time as a virtual e^-e^+ pair; this will lead to a modification of Coulomb's law which results from the lowest-order diagram of Fig. 7.1. We first evaluate the higher-order diagram and then return to discuss this intriguing statement.

By applying the Feynman rules, shown in Fig. 7.3 and Section 6.17, we obtain

$$-i\mathfrak{M} = (-1)^1(ie\bar{u}_f\gamma^\mu u_i)\left(-i\frac{g_{\mu\mu'}}{q^2}\right)$$
$$\times\int\frac{d^4p}{(2\pi)^4}\left[(ie\gamma^{\mu'})_{\alpha\beta}\frac{i(\not{p}+m)_{\beta\lambda}}{p^2-m^2}(ie\gamma^{\nu'})_{\lambda\tau}\frac{i(\not{q}-\not{p}+m)_{\tau\alpha}}{(q-p)^2-m^2}\right]$$
$$\times\left(-i\frac{g_{\nu'\nu}}{q^2}\right)(-ij^\nu(\mathbf{q})). \tag{7.18}$$

The Feynman rules for higher-order diagrams involve some nontrivial extensions of the rules developed in Chapter 6. A factor $(-1)^n$ should be included in an amplitude for a diagram containing n fermion loops [see, for example, the discussion of Mandl (1959) of the Dyson–Wick formalism], hence the factor $(-1)^1$ in (7.18). The only other unfamiliar feature of (7.18) is the $d^4p/(2\pi)^4$ integration. Its origin is easily uncovered. Although four-momentum is conserved at each vertex, the momentum p, which is circulating around the loop, is unrestricted. The magnitude of the loop four-momentum

$$|p| = (p^2)^{1/2} = (p_0^2 - |\mathbf{p}|^2)^{1/2}$$

can be zero or infinite or have any value in between. As p is not observable, we have to sum over all possibilities, hence $\int d^4p$.

The addition of (7.18) to (7.13) can be regarded as a modification to the propagator of the lowest-order result (7.13), namely,

$$-i\frac{g_{\mu\nu}}{q^2} \to -i\frac{g_{\mu\nu}}{q^2} + \left(-i\frac{g_{\mu\mu'}}{q^2}\right) I^{\mu'\nu'} \left(-i\frac{g_{\nu'\nu}}{q^2}\right)$$
$$\to -i\frac{g_{\mu\nu}}{q^2} + \frac{(-i)}{q^2} I_{\mu\nu} \frac{(-i)}{q^2} \tag{7.19}$$

where

$$I_{\mu\nu}(q^2) = (-1)^1 \int \frac{d^4p}{(2\pi)^4} \mathrm{Tr}\left\{ (ie\gamma_\mu) \frac{i(\not{p} + m)}{p^2 - m^2} (ie\gamma_\nu) \frac{i(\not{q} - \not{p} + m)}{(q-p)^2 - m^2} \right\}. \tag{7.20}$$

This $O(\alpha)$ modification to the propagator is shown symbolically in Fig. 7.4. The correction can be calculated once and for all and then substituted into any Feynman diagram.

There is, however, a major problem. $I_{\mu\nu}$ as given by (7.20) apparently has terms of the form $\int |p|^3 d|p| / |p|^2$ for $|p| \to \infty$, and so the correction diverges. Indeed, a rather lengthy but straightforward calculation shows that $I_{\mu\nu}$ can be written as

$$I_{\mu\nu} = -ig_{\mu\nu} q^2 I(q^2) + \cdots \tag{7.21}$$

with

$$I(q^2) = \frac{\alpha}{3\pi} \int_{m^2}^{\infty} \frac{dp^2}{p^2} - \frac{2\alpha}{\pi} \int dz\, z(1-z) \log\left(1 - \frac{q^2 z(1-z)}{m^2}\right), \tag{7.22}$$

where m is the mass of the electron. The dots in (7.21) represent omitted terms which are proportional to $q_\mu q_\nu$ and vanish when the propagator is coupled to external charges or currents. Equation (7.22) divides $I(q^2)$ into logarithmically divergent and finite contributions. We might have expected $I(q^2)$ to diverge quadratically as $\int |p| d|p|$. However, the divergence turns out to be only logarithmic on account of the "conspiratorial" algebra connected with the rest of the integrand. An explicit derivation of (7.21) and (7.22) is given, for example, in Bjorken and Drell (1964), Jauch and Rohrlich (1976), Scadron (1979), or Sakurai (1967) who, in an appendix, gives a collection of tricks for handling loop integrals.

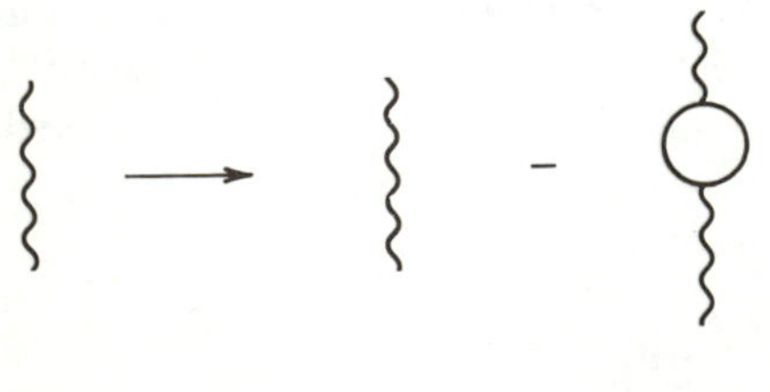

Fig. 7.4 Pictorial representation of (7.19).

Later, we shall study the effects of e^-e^+ loops in the limit of short- or long-range interactions, and so it is useful to evaluate $I(q^2)$ for both large and small values of $(-q^2)$. For $(-q^2)$ small,

$$\log\left(1 - \frac{q^2 z(1-z)}{m^2}\right) \simeq -\frac{q^2 z(1-z)}{m^2},$$

and (7.22) becomes

$$I(q^2) = \frac{\alpha}{3\pi} \log \frac{M^2}{m^2} + \frac{\alpha}{15\pi} \frac{q^2}{m^2}, \tag{7.23}$$

where for the moment we have introduced a cut-off M^2 to replace ∞ as the upper limit of integration in the first term of (7.22). On the other hand, for $(-q^2)$ large,

$$\log\left(1 - \frac{q^2 z(1-z)}{m^2}\right) \simeq \log\left(\frac{-q^2}{m^2}\right)$$

and so, similarly,

$$\begin{aligned} I(q^2) &\simeq \frac{\alpha}{3\pi} \log\left(\frac{M^2}{m^2}\right) - \frac{\alpha}{3\pi} \log\left(\frac{-q^2}{m^2}\right) \\ &= \frac{\alpha}{3\pi} \log\left(\frac{M^2}{-q^2}\right). \end{aligned} \tag{7.24}$$

Unless we can dispose of the infinite part of $I(q^2)$ [which appears as $M^2 \to \infty$ in (7.23) and (7.24)], the result will not be physically meaningful.

The way to proceed is best explained by returning to Rutherford scattering. Including the loop contribution, (7.19), the amplitude (7.15) is

$$-i\mathfrak{M} = (ie\bar{u}\gamma_0 u)\left(-\frac{i}{q^2}\right)\left(1 - \frac{\alpha}{3\pi}\log\left(\frac{M^2}{m^2}\right) - \frac{\alpha}{15\pi}\frac{q^2}{m^2} + O(e^4)\right)(-iZe), \tag{7.25}$$

where we have used (7.21) in the small $-q^2$ limit, (7.23), for $I(q^2)$. We may rewrite (7.25) in the form

$$-i\mathfrak{M} = (ie_R\bar{u}\gamma_0 u)\left(-\frac{i}{q^2}\right)\left(1 - \frac{e_R^2}{60\pi^2}\frac{q^2}{m^2}\right)(-iZe_R), \tag{7.26}$$

with

$$\boxed{e_R \equiv e\left(1 - \frac{e^2}{12\pi^2}\log\frac{M^2}{m^2}\right)^{1/2}.} \tag{7.27}$$

To $O(e^4)$, it is easy to verify that (7.25) and (7.26) are mathematically equivalent. In the previous chapters, we were led to believe that e, the charge appearing in the lowest-order Feynman diagrams, is the charge of the electron as measured in

Thomson scattering or any other long-range Coulomb experiment. We never justified this, and it is, in fact, not true! Suppose that e_R in (7.27) is the electric charge listed in the particle data tables, that is, $e_R^2/4\pi = 1/137$. The invariant amplitude (7.26) is now finite. The infinity associated with the cut-off $M \to \infty$ has been "absorbed" in e_R. This procedure is admittedly bizarre. It is our first contact with "renormalization." We return to this later, but first we explore the physics content of our new amplitude, which we have contrived to be free of infinities.

7.3 The Lamb Shift

As pointed out in Chapters 3 and 4, T_{fi} (or $\mathfrak{M}$) represents the Fourier transform of the potential. The first term in (7.26), which is proportional to $|\mathbf{q}|^{-2}$, is associated with the Coulomb potential, since

$$V_0(r) = -\frac{Ze_R^2}{(2\pi)^3}\int d^3q e^{i\mathbf{q}\cdot\mathbf{r}}\frac{1}{|\mathbf{q}|^2} = -\frac{Ze_R^2}{4\pi r} \tag{7.28}$$

(see Exercise 6.17). The second term, which represents the quantum effect of the virtual e^-e^+ loop in the propagator of the exchanged photon, contains an extra factor $|\mathbf{q}|^2$ relative to the first. In coordinate space $|\mathbf{q}|^2 \to \nabla^2$ and, since

$$\frac{1}{(2\pi)^3}\int d^3q e^{i\mathbf{q}\cdot\mathbf{r}} = \delta(\mathbf{r}), \tag{7.29}$$

(7.26) corresponds to an interaction between the electron and the charge Ze_R of the form

$$V(r) = -\left(1 - \frac{e_R^2}{60\pi^2 m^2}\nabla^2\right)\frac{Ze_R^2}{4\pi r}$$

$$\boxed{V(r) = -\frac{Ze_R^2}{4\pi r} - \frac{Ze_R^4}{60\pi^2 m^2}\delta(\mathbf{r}).} \tag{7.30}$$

The extra interaction (including its sign) was anticipated in the discussion of screening in Chapter 1. When $q^2 \to 0$, the electron probes the static charge Ze_R from a large distance and just interacts via the Coulomb interaction, that is, the first term in (7.30). The charge e_R is by definition the familiar electron charge, the one measured in any long-range electromagnetic interaction, for example, Thomson scattering (see Fig. 1.7). But when the electron comes closer to the nucleus (i.e., $-q^2$ increases), it penetrates the cloud of virtual e^-e^+ pairs which surround it. This leads to an increase in the effective interaction as explained in Fig. 1.6 (note, indeed, that both terms in (7.30) have the same sign), and the second term in (7.30) represents a calculation of this effect to leading order or to "the one-loop level." The presence of the loop thus leads to an additional attractive force between the electron and the nucleus.

This additional interaction can be detected. Loops are not just some graphical construct. Their presence can be experimentally established. For example, if the source in Fig. 7.1 is a proton ($Z = 1$) and the Feynman graph represents the electron–proton interaction of the hydrogen atom, then (7.30) describes the atomic binding, including the additional attraction when the electron ventures inside the cloud of e^-e^+ pairs screening the charge of the nucleus. This effect, represented to lowest order by the $\delta(\mathbf{r})$ potential, contributes to the energy levels E_{nl} of the hydrogen atom. Its magnitude can be computed using ordinary quantum mechanics. Treating the second term in (7.30) as a perturbation, we obtain a contribution to the "Lamb shift"

$$\begin{aligned} \Delta E_{nl} &= -\frac{e_R^4}{60\pi^2 m^2}|\psi_{nl}(0)|^2 \delta_{l0} \\ &= -\frac{8\alpha_R^3}{15\pi n^3} Ry\, \delta_{l0} \end{aligned} \tag{7.31}$$

where the ψ_{nl} are the usual hydrogen atom wavefunctions and $Ry = m\alpha_R^2/2$ is the Rydberg constant. The δ_{l0} in (7.31) arises because the $\delta(\mathbf{r})$ potential can only perturb levels described by wavefunctions which are finite at the origin, namely, those with the $l = 0$. This result can be established experimentally by measuring the Lamb shift between the $2s_{1/2}$ and $2p_{1/2}$ levels. These levels are degenerate if loop contributions are not included. Equation (7.31), which for obvious reasons is called the "vacuum polarization" correction, contributes -27 MHz to the total Lamb shift of $+1057$ MHz between the $2s_{1/2}$ and $2p_{1/2}$ levels. As the Lamb shift can be measured to an accuracy of about 0.01%, the shift (7.31) due to the e^+e^- loop has been verified experimentally. Indeed, this together with other loop contributions exactly reproduces the observed shift (see Section 7.4). We conclude that loop diagrams give real observable effects; but, even more important, our bizarre reinterpretation of the electron charge, (7.27), has received confirmation from experiment.

There is another way to visualize this important result. The hydrogen atom is bound by the exchange of photons between the electron and the proton. The Coulomb force leads to a separation of a Bohr radius on the average. In QED, the electron can deviate from its Bohr orbit as a result of, among other things, the fluctuation of the exchanged photon into e^-e^+ pairs. This quantum screening effect reduces the attraction when the electron is far from the proton and increases the force when it approaches the nucleus. The competing effects do not cancel because the Coulomb force falls with r. The net result is an additional attraction over and above the Coulomb potential $-\alpha_R/r$, given by the second term in (7.30).

7.4 More Loops: The Anomalous Magnetic Moment

The vacuum polarization loop only accounts for a fraction of the splitting of the $2s_{1/2}$–$2p_{1/2}$ levels. Other $O(e^4)$ graphs exist which help to destroy the degeneracy of the levels obtained when calculating to $O(e^2)$ only. The complete set of $O(e^4)$ graphs is shown in Fig. 7.5. We have calculated the contribution of diagram 7.5a. Each of the other diagrams also contains a loop, which, just as before, will diverge for large loop momentum. These divergences can all be hidden in a redefinition of the charge, mass, or wavefunction of the electron, in the same way that we absorbed the divergent part of the e^-e^+ loop occurring in the photon propagator into the charge e_R of (7.27). The "physics" is contained in the finite terms.

We consider the diagram of Fig. 7.5b next. Just as the loop in the propagator affects the attraction between the charges which it connects, so we anticipate that the loop around the vertex will modify the structure of the electron current $-e\bar{u}_f\gamma_\mu u_i$, see Fig. 7.6. Indeed, a computation of the (finite) piece of the diagram gives in the small $(-q^2)$ limit

$$-e\bar{u}_f\gamma_\mu u_i \to -e\bar{u}_f\left\{\gamma_\mu\left[1+\frac{\alpha}{3\pi}\frac{q^2}{m^2}\left(\log\frac{m}{m_\gamma}-\frac{3}{8}\right)\right]-\left[\frac{\alpha}{2\pi}\frac{1}{2m}i\sigma_{\mu\nu}q^\nu\right]\right\}u_i. \tag{7.32}$$

The first square bracket gives an additional contribution to the Lamb shift of similar form to (7.26). A new feature arises here since the loop also diverges for small (infrared) loop momenta $|p|$. In (7.32), we have sidestepped this problem by giving the photon a small fictitious mass m_γ. We shall explain how this takes care of infrared divergences in Chapter 11. The combined effects of (7.26) and (7.32)

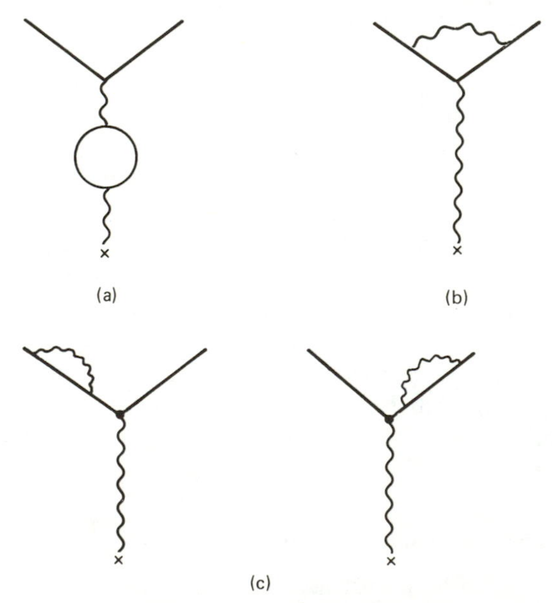

Fig. 7.5 Complete set of $O(\alpha^2)$ Feynman graphs.

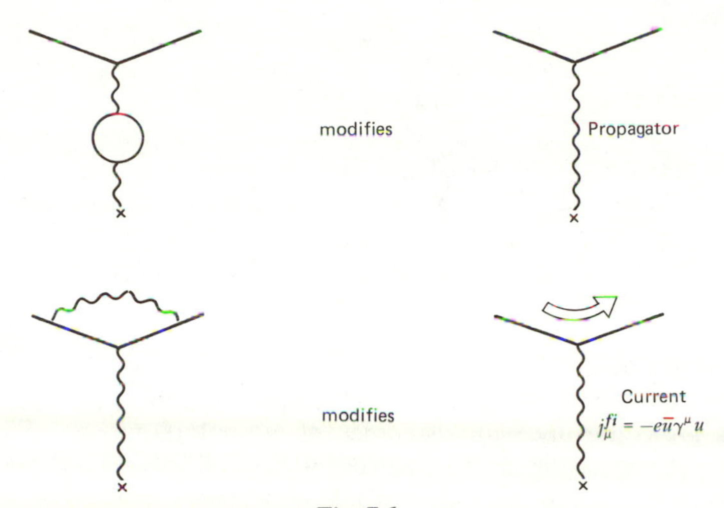

Fig. 7.6

account for the observed value of the Lamb shift. The detailed discussion is rather lengthy and requires a careful treatment of the infrared photons, see Bjorken and Drell (1964) or Jauch and Rohrlich (1976).

An intriguing feature of (7.32) is the second term in square brackets, which modifies the γ_μ Lorentz structure of the electron current. To see the physical implications of this term, we recall the Gordon decomposition of a γ_μ-current, (6.7),

$$-e\bar{u}_f\gamma_\mu u_i = -\frac{e}{2m}\bar{u}_f\big((p_f+p_i)_\mu - i\sigma_{\mu\nu}q^\nu\big)u_i. \tag{7.33}$$

Equation (7.33) exhibits the fact that the electron interacts via both its charge and its magnetic moment. In exercise 6.2, we demonstrated that the $\sigma_{\mu\nu}q^\nu$ term in (7.33) represents a magnetic moment of the electron,

$$\boldsymbol{\mu} = -\frac{e}{2m}\boldsymbol{\sigma}, \tag{7.34}$$

which is often written as

$$\boldsymbol{\mu} = -g\frac{e}{2m}\mathbf{S} \tag{7.35}$$

with $\mathbf{S} = \frac{1}{2}\boldsymbol{\sigma}$ and the gyromagnetic ratio

$$g = 2. \tag{7.36}$$

The second term in (7.32) is therefore just an extra magnetic moment interaction to that already contained in γ_μ via (7.33). In fact, substituting (7.33) in (7.32), we find, using (7.34), that

$$\boldsymbol{\mu} = -\frac{e}{2m}\left(1+\frac{\alpha}{2\pi}\right)\boldsymbol{\sigma}, \tag{7.37}$$

or

$$\boxed{g = 2 + \frac{\alpha}{\pi}.} \tag{7.38}$$

The electron thus has an "anomalous" magnetic moment $\alpha/2\pi$ in addition to its Dirac magnetic moment. To be precise, the anomalous part is given by

$$\begin{aligned} \frac{g-2}{2} &= \frac{1}{2}\left(\frac{\alpha}{\pi}\right) - 0.32848\left(\frac{\alpha}{\pi}\right)^2 + (1.49 \pm 0.2)\left(\frac{\alpha}{\pi}\right)^3 + \cdots \\ &= (1159655.4 \pm 3.3) \times 10^{-9}, \end{aligned} \tag{7.39}$$

where the first term corresponds to our lowest-order result (7.38), while the second and third terms represent the higher-order contributions. The number of diagrams grows rapidly with the order of α, and the error on the $O(\alpha^3)$ contributions hints at the difficult numerical calculations that are involved. The experimental value of the electron's anomalous magnetic moment is

$$\left(\frac{g-2}{2}\right)_{\text{exp}} = (1159657.7 \pm 3.5) \times 10^{-9}, \tag{7.40}$$

in excellent agreement with the prediction (7.39). This triumph of QED has been repeated for the muon magnetic moment, providing further evidence that our strange way of handling the infinities is correct.

7.5 Putting the Loops Together: Ward Identities

Equation (7.27) shows how the infinite part of the loop in the photon propagator is hidden by a redefinition of the electron's charge. When performing the complete $O(e^4)$ calculation, infinite parts of the loops in Fig. 7.5b and 7.5c will also be absorbed into e_R. Suppose we now repeat this calculation for the scattering of a muon, instead of an electron, by a nucleus. Clearly, the first diagram (Fig. 7.5a) contributes an amount to the charge which is independent of the nature of the scattered particle. The result (7.27) is determined by the modified photon propagator and changes the charge of an electron, muon, or any other particle in exactly the same way. But this is not the case for the diagrams of Figs. 7.5b and 7.5c, where the scattered particles are an integral part of the loop. It would appear that we shall get different redefinitions of e for electrons and muons, which would be a serious problem since experimentally the electron and muon charges are equal. It is here that QED displays its full power. A full calculation shows that magically the modification of the charge by the vertex diagram of Fig. 7.5b is exactly canceled by the modification introduced by the diagrams of Fig. 7.5c [see Sakurai (1967), Bjorken and Drell (1964)]. Only the "vacuum polarization" graph of Fig. 7.5a modifies the charge. Equation (7.27) is the full answer, and hence the "renormalized" charge of the electron and muon remain equal.

This cancellation repeats itself in every order of perturbation theory. The electron and muon charges are exactly equal. The conspiracy between the diagrams of Figs. 7.5b and 7.5c reflects a very basic property of (gauge) field theories known as a Ward identity.

7.6 Charge Screening and $e^-\mu^-$ Scattering

Loops in the propagator of the exchanged photon not only modify the interaction of an electron with a static charge but also affect other interactions, for example, electron–muon elastic scattering. The lowest-order $e^-\mu^-$ amplitude is given by (6.50). The $O(e^4)$ vacuum-polarization contribution is readily obtained by replacing the source factor $-ij^\nu = (-iZe, \mathbf{0})$ of (7.18) by the muon current:

$$-ij^\nu = ie\,\bar{u}(p_f')\gamma^\nu u(p_i'),$$

see Fig. 7.7. We require the charge to be the renormalized charge, (7.27), and keep the finite piece of $I(q^2)$ of (7.21). We add this higher-order contribution to the lowest-order result. This is an illustration of the fact that propagator corrections are common to all processes and hence can be calculated once and for all. Of course, there are other $O(e^4)$ contributions to $e^-\mu^-$ scattering which we must also include.

7.7 Renormalization

Despite its phenomenological success, the procedure for treating infinities deserves further consideration. How can we justify perturbation theory in α, when in the next order α is accompanied by an infinite coefficient, namely, $\log(M^2/m^2)$, where M is some arbitrary cutoff? We return to (7.27), which drew attention to a very serious shortcoming of our discussion of relativistic quantum mechanics: the quantity that we called the charge, which appears in the lowest-order Feynman amplitudes of Chapter 6, is changed by higher-order interactions. It is therefore

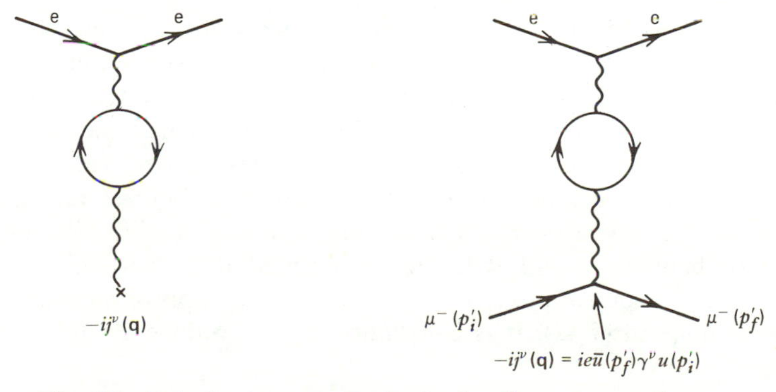

Fig. 7.7 Relation between Rutherford and $e^-\mu^-$ scattering.

not what we thought it was, and it is certainly not the charge the experimentalist measures. We can see this dilemma another way. The charge is associated with the electron–photon coupling, which we symbolically represented as

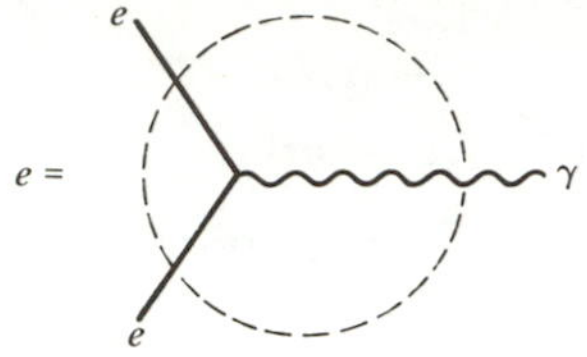

$e =$ (7.41)

But this is absurd, because the charge is also

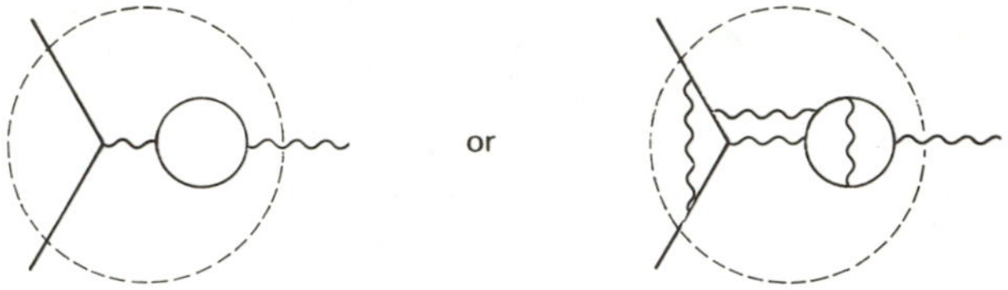

or

In fact, it is all of these things at once, and that is what the experimentalist measures. Therefore, to label (7.41) as e (the quantity measured in Coulomb's experiment giving $\alpha = 1/137$) is simply wrong. Let us therefore call (7.41) the "bare" charge e_0. Here, "bare" refers to the fact that the vertex is stripped of all loops. We can now summarize the situation by writing

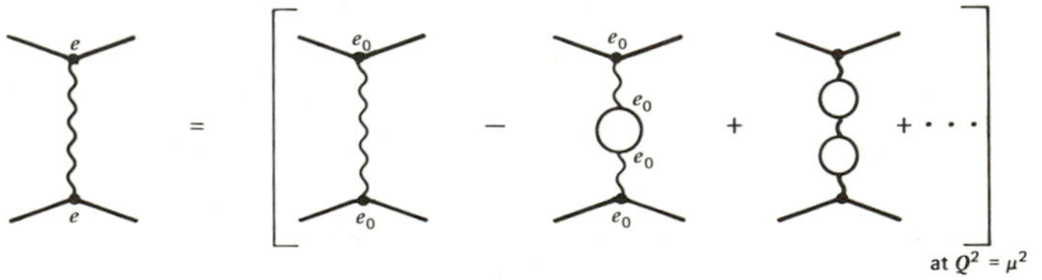

(7.42)

where $\cdots$ stands for diagrams with all possible propagator modifications. We need only consider modifications to the photon propagator because of the Ward identity of Section 7.5. We also explicitly show the negative sign associated with each loop, see (7.18). The charge e in (7.42) is the charge the experimentalist measures when scattering two low-energy electrons or performing a Coulomb experiment, namely, $e^2/4\pi \simeq 1/137$. Its definition recognizes the fact that e_0, appearing in the lowest-order Feynman amplitude, is modified by interactions. The relation between e^2 and e_0^2 has to be specified at the particular value of the virtual photon's momentum, say, $q^2 \equiv -Q^2 = -\mu^2$ appropriate to the experiment, as is done in (7.42). It is conventional to introduce Q^2 for $-q^2$, as this quantity is positive.

To $O(e_0^4)$, we can write the relation between e and the bare charge e_0 as

$$e^2 = e_0^2\left[1 - I\left(q^2 = -\mu^2\right) + O\left(e_0^4\right)\right], \tag{7.43}$$

where $I(q^2)$ is given by (7.19) and (7.21). $I(q^2)$ is $O(e_0^2)$ and represents the result of the one-loop calculation. Indeed, taking the square root of (7.43), we have

$$e = e_0\left[1 - \tfrac{1}{2}I\left(q^2 = -\mu^2\right) + O\left(e_0^4\right)\right], \tag{7.44}$$

which is of the form of (7.27) after expansion of the square root. In the diagrammatic notation of Fig. 7.4, this can be written as

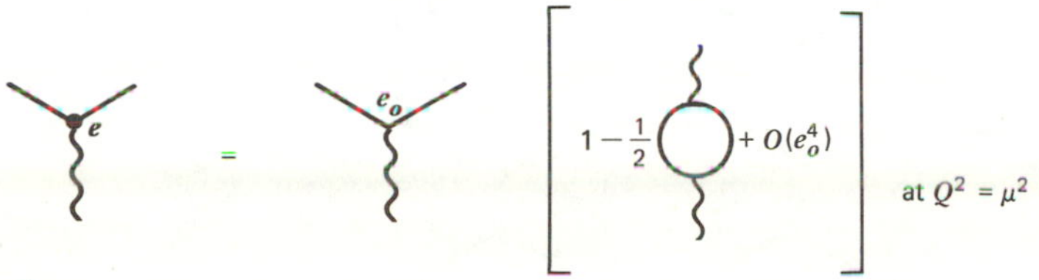

e = e_0 $\left[1 - \frac{1}{2}\,(\text{loop}) + O(e_0^4)\right]_{\text{at } Q^2 = \mu^2}$ (7.44′)

or, to all orders,

$$e = e_0\left(1 + e_0^2 A_1(Q^2) + e_0^4 A_2(Q^2) + \cdots\right)_{\text{at } Q^2 = \mu^2}, \tag{7.45}$$

where $-q^2 \equiv Q^2$. Clearly, $A_1(Q^2)$, which is directly related to $I(-Q^2)$, is an infinite quantity; so are $A_2(Q^2)$ and all subsequent coefficients in (7.45). There is *a priori* nothing wrong with that. It does not matter that a theory is formulated in terms of infinite quantities as long as observable quantities are finite. Extensive use is made of complex quantities in optics, and there is no objection to that as long as the observables are real.

Let us calculate an observable to illustrate this point, for example, $e\mu$ scattering at 90° (see Section 7.6). We fix the angle in order to have an observable $d\sigma/d\Omega(s, t)$ which depends on only one momentum. At 90°, $-t \simeq s/2 \simeq Q^2$. We calculate the invariant amplitude as before, but we now explicitly display the fact that the calculation is in terms of the bare charge e_0;

$-i\mathcal{M}(e_0^2)$ = $\left[\;(e_0,\ e_0) \;-\; (e_0,\ e_0,\ e_0,\ e_0) \;+ \cdots\right]_{\text{at } Q^2}$ (7.46)

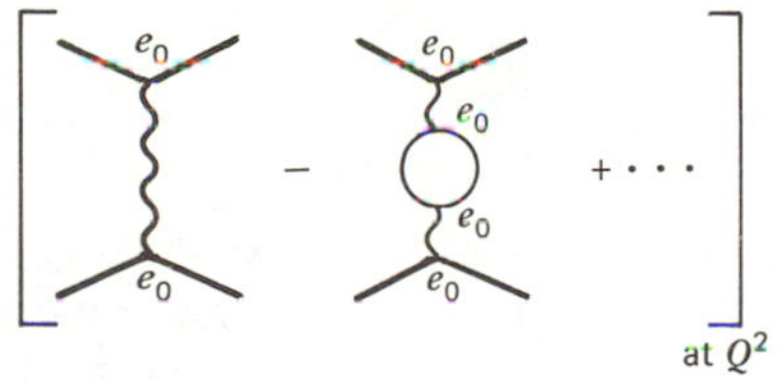

$$= e_0^2\left[F_1(Q^2) + e_0^2 F_2(Q^2) + O\left(e_0^4\right)\right]. \tag{7.46′}$$

Other diagrams are represented by ... in (7.46). To obtain realistic results, they would have to be considered explicitly. Here we just want to demonstrate the techniques by which perturbation theory can manipulate infinite diagrams, for example, the infinity associated with the loop in the e_0^4 term in (7.46). Indeed, all terms in a perturbative calculation in terms of the bare charge e_0, like (7.46), are again infinite. Now comes the crucial step: we reparametrize ("renormalize" in the usual, but unfortunate, terminology) $-i\mathfrak{M}(e_0^2)$ in terms of e^2. To do this, we invert (7.44′) [or (7.44)]:

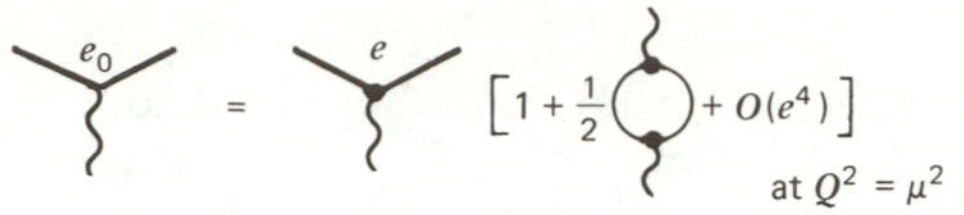

(7.47)

and use this result to replace the e_0 vertices of (7.46). We obtain

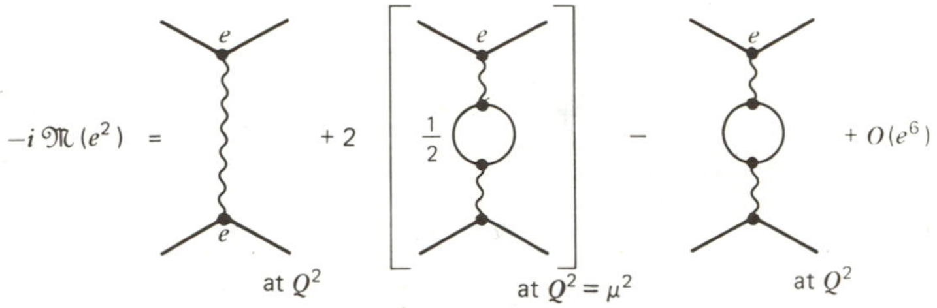

(7.48)

The first two diagrams both come from the first diagram of (7.46); the factor 2 arises because we must replace e_0 by e at each vertex. In the remaining diagram, we can simply write e instead of e_0, as the difference is $O(e^6)$. Equation (7.48) can be written as

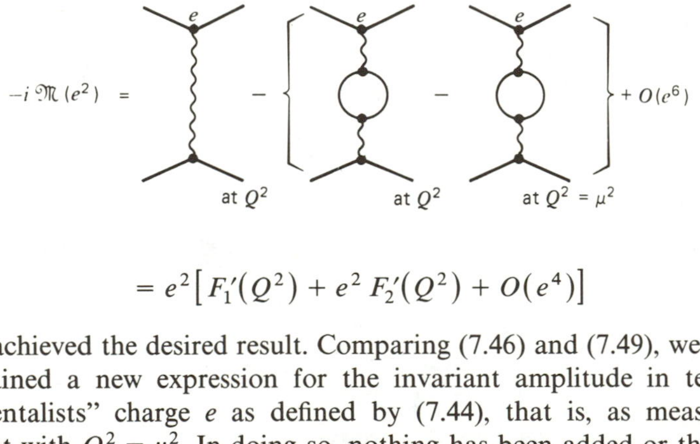

(7.49)

$$= e^2\left[F_1'(Q^2) + e^2 F_2'(Q^2) + O(e^4)\right] \qquad (7.49')$$

We have achieved the desired result. Comparing (7.46) and (7.49), we see that we have obtained a new expression for the invariant amplitude in terms of the "experimentalists" charge e as defined by (7.44), that is, as measured in an experiment with $Q^2 = \mu^2$. In doing so, nothing has been added or thrown away; we have just reparametrized the original calculation (7.46). Therefore, clearly

$$\mathfrak{M}(e^2) = \mathfrak{M}(e_0^2), \qquad (7.50)$$

as indeed it must. So, what have we achieved? The e_0^4 term in (7.46) is infinite, the e^4 term in (7.49) is finite! The e^4 term has been split into two terms, one containing a loop at Q^2 and the other a loop at $Q^2 = \mu^2$. The signs of the two terms are opposite. To see in more detail what happens, take, for example, the result for the loop given by (7.24):

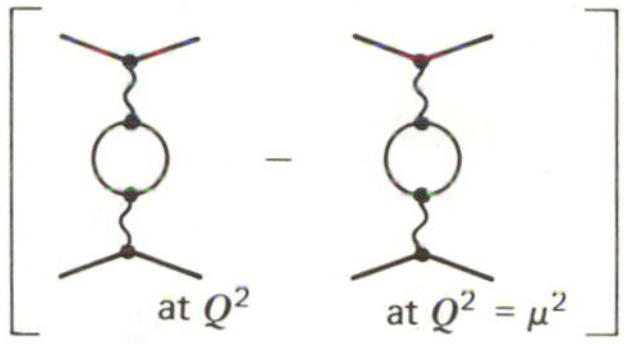

at Q^2 — at $Q^2 = \mu^2$

$$\sim \left(\frac{\alpha}{3\pi}\log\left(\frac{M^2}{Q^2}\right) - \frac{\alpha}{3\pi}\log\left(\frac{M^2}{\mu^2}\right)\right) = \frac{\alpha}{3\pi}\log\left(\frac{\mu^2}{Q^2}\right). \tag{7.51}$$

The difference of the two terms is finite; it does not depend on the *ad hoc* cutoff M^2, which we can now send back off to infinity, where it belongs. We conclude that (7.49), unlike (7.46), defines the observables in terms of finite quantities. The two perturbation expansions are nevertheless equivalent, as we have demonstrated by explicit calculation. The infinite coefficients in the original series, (7.46), arose because e_0 itself is not finite (it is in fact infinitesimal). Once we reorganize the series in terms of the finite quantity e^2, all the coefficients are finite.

Note that a free parameter μ with the dimensions of mass has slipped into the theory via the reparametrization of the charge. Different choices of μ^2, the renormalization mass, will lead to different expansions, (7.49′), of the amplitude. We say we are using different renormalization schemes. But $|\mathfrak{M}|^2$ is an observable and so must be independent of the value chosen for μ. This requirement can be formulated as follows:

$$\mu\frac{d\mathfrak{M}}{d\mu} = \left(\mu\left.\frac{\partial}{\partial\mu}\right|_e + \mu\frac{\partial e}{\partial\mu}\frac{\partial}{\partial e}\right)\mathfrak{M} = 0. \tag{7.52}$$

The dependence of $\mathfrak{M}$ on μ, given by the coefficients $F'(Q^2, \mu^2)$ in (7.49′), must be cancelled by the μ-dependence of $e(\mu^2)$. Equation (7.52) is called the "renormalization group equation." Its importance transcends particle physics; we have done scant justice to it in this passing reference.

7.8 Charge Screening in QED: The Running Coupling Constant

We have seen repeatedly how the charge is modified by the vacuum polarization loop in the photon propagator. We know that the loop will be repeated in higher orders as shown in (7.42). We can rewrite this relation as

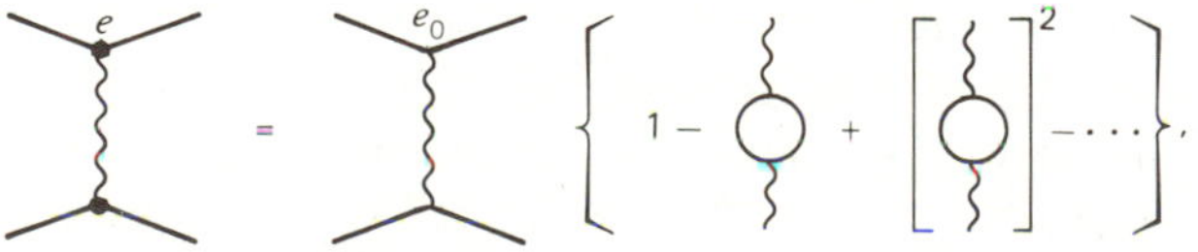

e = e_0 $\left\{1 - \ldots + [\ldots]^2 - \cdots\right\}$. (7.53)

and the geometric series can be summed to give

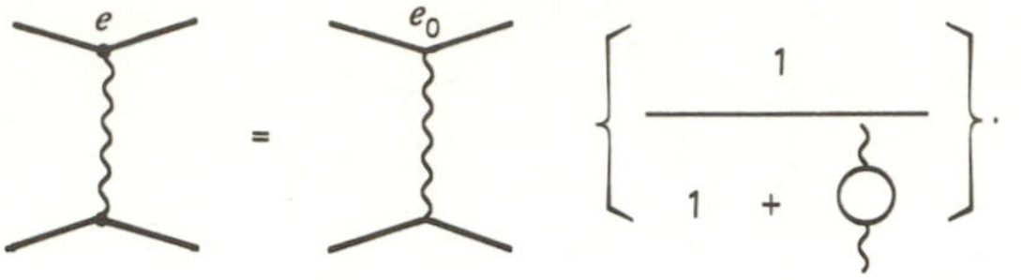

$$\Big\rangle_{e} \; = \; \Big\rangle_{e_0} \left\{ \frac{1}{1 + \text{(loop)}} \right\}. \qquad (7.54)$$

It turns out to be a good idea to redefine the charge including all vacuum polarization loops as given by (7.54).

We showed how the infinities can be removed by working in terms of the physical (renormalized) charge e given by (7.53) at $Q^2 = \mu^2$. In fact, we could have used any value of μ^2. However, different choices $Q^2 = \mu_1^2, \mu_2^2, \ldots$ correspond to perturbation expansions in terms of numerically different values of the physical charge $e(\mu_i^2)$. Indeed, using the notation of (7.43), we have, from (7.54),

$$e^2(Q^2) = e_0^2 \left(\frac{1}{1 + I(q^2)} \right). \qquad (7.55)$$

Equation (7.55) explicitly displays the fact that the charge the experimentalist measures depends on the Q^2 of the experiment; $\alpha(Q^2) \equiv e^2(Q^2)/4\pi$ is referred to as the "running coupling constant."

In the large $Q^2 \equiv -q^2$ limit, $I(q^2)$ is given by (7.24), and (7.55) becomes

$$\alpha(Q^2) = \frac{\alpha_0}{1 - \dfrac{\alpha_0}{3\pi} \log\left(\dfrac{Q^2}{M^2} \right)}. \qquad (7.56)$$

To eliminate the explicit dependence of $\alpha(Q^2)$ on the cutoff M, we choose a renormalization or reference momentum μ. The renormalization procedure is then to subtract $\alpha(\mu^2)$ from $\alpha(Q^2)$. We find

$$\boxed{\alpha(Q^2) = \frac{\alpha(\mu^2)}{1 - \dfrac{\alpha(\mu^2)}{3\pi} \log\left(\dfrac{Q^2}{\mu^2} \right)}} \qquad (7.57)$$

for large Q^2. Equation (7.57) now contains only finite, physically measurable quantities.

The running coupling constant, $\alpha(Q^2)$, describes how the effective charge depends on the separation of the two charged particles. By summing part of all orders of perturbation theory, we have obtained the charge screening of electrodynamics, see Fig. 1.5. As Q^2 increases, the photon sees more and more charge until, at some astronomically large but finite Q^2, the coupling $\alpha(Q^2)$ is infinite. However, by inserting numerical values, we find that for all practically attainable Q^2, the variation of α with Q^2 is extremely small; α increases from $1/137$ very slowly as Q^2 increases. Of course, as Q^2 increases, other loops (formed, for example, by a $\mu^+\mu^-$-pair or a $\bar{u}u$-quark pair) will also contribute to the variation.

7.9 Running Coupling Constant for QCD

The Q^2 behavior of the QCD coupling, $\alpha_s(Q^2)$, turns out to be very different to that for $\alpha(Q^2)$. The manipulations of the QCD graphs needed for the calculation of $\alpha_s(Q^2)$ carries over from the discussion of $\alpha(Q^2)$. The final answer, (7.57), is therefore also true for $\alpha_s(Q^2)$, but there is a crucial difference: the coefficient of $\log(Q^2/\mu^2)$ is not the same. To determine the coefficient, we must calculate $I(q^2)$ in QCD. The equivalent of Fig. 7.4 is

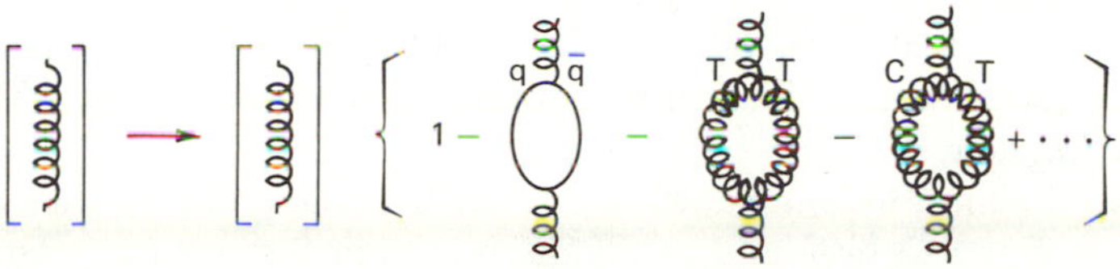

$$\text{(7.58)}$$

where the extra terms arise from the color self-coupling of the gluons, and where C and T stand for "Coulomb" and "transverse" gluons, respectively, see Section 6.13. In the covariant gauge, it can be shown that these QCD diagrams yield the following coefficient of $\log(Q^2/\mu^2)$:

$$\frac{\alpha_s(\mu^2)}{4\pi}\left(-\frac{2}{3}n_f - 5 + 16\right), \tag{7.59}$$

in contrast to the QED coefficient of (7.57),

$$\frac{\alpha(\mu^2)}{4\pi}\left(-\frac{4}{3}\right). \tag{7.60}$$

The consecutive terms in (7.59) represent the contribution of the consecutive loops in (7.58). The first loop is familiar; the gluon can fluctuate into a virtual $q\bar{q}$ pair, just as a photon can fluctuate into an e^+e^- pair. There is, however, one loop for each quark flavor; hence, $-\frac{2}{3}n_f$, where n_f is the number of flavors. The QED result, (7.57), was written for one flavor, "n_f" $\to 1$ (i.e., just the e^+e^- loop). However, (7.60) is consistent with the first term of (7.59) because there is a factor of 2 mismatch in the definitions of α and α_s, see (2.95). The relation between α and α_s is discussed in more detail in Sections 10.4 and 10.7. We see that the fermion loops contribute a negative coefficient. So does the loop with two transverse gluons with a coefficient -5. One can in fact prove a theorem that all these loops have to lead to the same (negative) sign because they are all related to physical cross sections for producing lepton, quark, or gluon pairs. Symbolically,

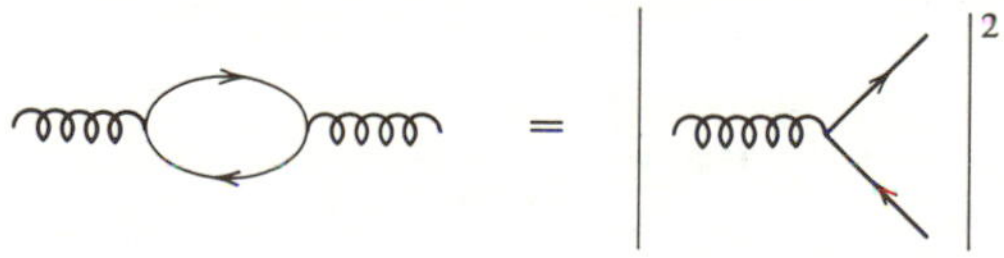

$$\text{(7.61)}$$

for leptons or quarks, and

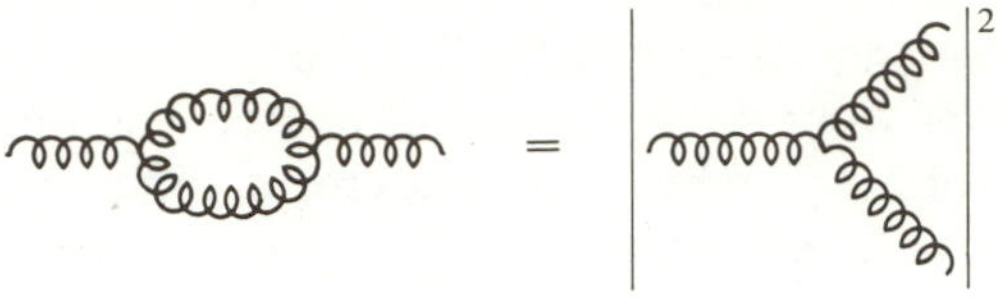

(7.62)

for the production of two transverse gluons in, for example, $q\bar{q} \to g \to gg$. The theorem implies that any state which can be physically produced when the propagator becomes time-like will lead to screening of the charge and hence to a negative coefficient.

How can the third loop in (7.58) violate the theorem and give a coefficient $+16$ in (7.59)? A clue can be found in the discussion in Section 6.13. There, we saw that the instantaneous Coulomb interaction was associated with the exchange of virtual longitudinal and scalar photons. Such photons are never produced as real physical states since the probability for producing scalar photons cancels that for producing longitudinal photons. In QCD the cancellation is more complicated. It requires the introduction of "ghost" particles to cancel the unphysical polarizations. They can contribute to loops without leading to the production of physical particles. This is how the theorem is sidestepped. The sign of the third loop is not restricted and it is found to be positive. It is not only positive, it is sufficiently large to reverse the overall sign of the coefficient of $\log(Q^2/\mu^2)$ relative to that in QED. This is related to the fact that there are eight gluons but only three colors of quarks.

Combining (7.59) with (7.57), we obtain the QCD "running coupling constant"

$$\alpha_s(Q^2) = \frac{\alpha_s(\mu^2)}{1 + \dfrac{\alpha_s(\mu^2)}{12\pi}(33 - 2n_f)\log(Q^2/\mu^2)}. \tag{7.63}$$

Only in a world with more than 16 quark flavors (we are safely below this number at present energies) is the sign of the coefficient the same as in QED, see (7.57). As anticipated in Chapter 1, $\alpha_s(Q^2)$ decreases with increasing Q^2 and therefore becomes small for short-distance interactions. We say that the theory is "asymptotically free."

One parameter, μ, with the dimensions of mass, remains as a relic of the renormalization. From (7.63) we see that at sufficiently low Q^2, the effective coupling will become large. It is customary to denote the Q^2 scale at which this happens by Λ^2, where

$$\Lambda^2 = \mu^2 \exp\left[\frac{-12\pi}{(33 - 2n_f)\alpha_s(\mu^2)}\right]. \tag{7.64}$$

It then follows that (7.63) may be written

$$\boxed{\alpha_s(Q^2) = \frac{12\pi}{(33 - 2n_f)\log(Q^2/\Lambda^2).}} \tag{7.65}$$

For Q^2 values much larger than Λ^2, the effective coupling is small and a perturbative description in terms of quarks and gluons interacting weakly makes sense. For Q^2 of order Λ^2, we cannot make such a picture, since quarks and gluons will arrange themselves into strongly bound clusters, namely, hadrons. Thus, we can think of Λ as marking the boundary between a world of quasi-free quarks and gluons, and the world of pions, protons, and so on. The value of Λ is not predicted by the theory; it is a free parameter to be determined from experiment. We should expect that it is of the order of a typical hadronic mass.

In Chapters 10 and 11, we find indeed that Λ has a value somewhere in the range 0.1 to 0.5 GeV. Thus, for experiments with $Q^2 = (30 \text{ GeV})^2$, it follows from (7.65) that α_s is of order 0.1. We may therefore apply QCD perturbation theory, just as we have done for QED. In the large Q^2 limit, all the quark masses can be neglected, and they contribute no mass scale to QCD. Nevertheless, there is a mass scale, Λ, inherent in the theory which enters through renormalization.

7.10 Summary and Comments

We should mention that the story of infinities in field theory does not stop here. We have in particular omitted a discussion of infrared divergences which are connected with the $Q^2 \to 0$ limit, namely, the limit of very soft photons. This discussion is postponed until Chapter 11, where it plays a crucial role in the discussion of perturbative QCD.

In this chapter, we have shown how renormalization allows us to compute the physical effects due to the presence of loops in the perturbative expansion of QED (and QCD) amplitudes. The infinities appearing in loop diagrams were a consequence of a naive definition of the electric (or color) charge in the previous chapters. After a proper reparametrization, which takes us from the bare to the physical charge, loops lead to finite and measurable effects. The calculations based on renormalization agree with experiment. The Lamb shift and the anomalous magnetic moment are dramatic illustrations. The loops "dress" the bare leptons so that they no longer appear to be simple point particles. They acquire, for instance, an anomalous magnetic moment like the truly composite neutron and proton, see Chapter 2. A complete calculation of all loops will require a reparametrization of the mass and wavefunction of the particles as well as their charge. The procedure is completely analogous to the one discussed for the charge.

8

The Structure of Hadrons

In Chapters 3 through 7, we have learned how to perform quantitative calculations for the electromagnetic interactions of leptons and quarks. The same techniques will allow us to compute the color interactions of quarks and gluons. There is, however, an immediate problem: experiments to study color (strong) interactions are performed with hadrons (e.g., proton beams or secondary π-beams interacting with nuclear targets), not with the quarks and gluons that are described by quantum field theory. This situation is similar to that encountered in atomic physics where experiments involving complex atoms have to be interpreted through the electromagnetic interactions of the constituent electrons. This analogy reveals the problem: we need to find the "wavefunctions" that describe, for example, a proton in terms of its constituent quarks and gluons. In this chapter, we present an experimental technique that allows us to determine the quark and gluon structure of hadrons; namely the deep inelastic scattering of leptons off hadronic targets. The structure functions so obtained will be presented in Chapter 9. Finally, in Chapters 10 and 11 we reach our goal and translate these quark–gluon QCD calculations into predictions for the results of experiments involving leptons and hadrons.

The discussion will soon reveal that these structure functions are not the static quark wavefunctions introduced in Chapter 2, although they are indirectly related to them.

8.1 Probing a Charge Distribution with Electrons. Form Factors

"Photographing" an object by scattering an electron beam off it is a well-proved technique in physics. Suppose we want to determine the charge distribution shown in Fig. 8.1, which could, for example, be the electron cloud of an atom. The procedure is to measure the angular distribution of the scattered electrons and compare it to the (known) cross section for scattering electrons from a point charge, in the form

$$\frac{d\sigma}{d\Omega} = \left(\frac{d\sigma}{d\Omega}\right)_{\text{point}} |F(q)|^2, \tag{8.1}$$

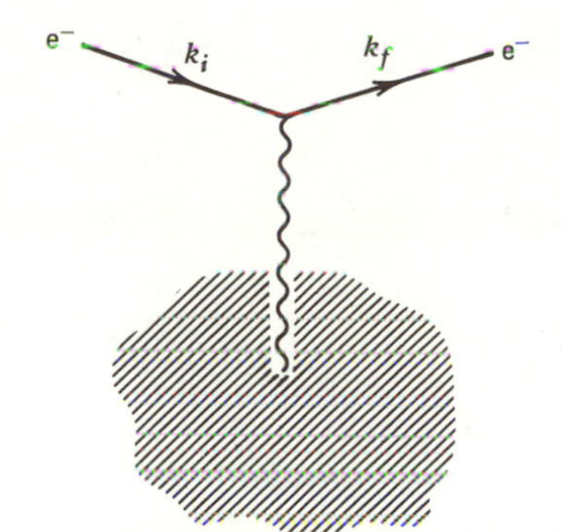

Fig. 8.1 Lowest-order electron scattering by a charge cloud.

where q is the momentum transfer between the incident electron and the target, $q = k_i - k_f$, see Fig. 8.1. We then attempt to deduce the structure of the target from the form factor $F(q)$ so determined.

We can gain insight into this technique by first looking at the scattering of unpolarized electrons of energy E from a static, spinless charge distribution $Ze\,\rho(\mathbf{x})$, normalized so that

$$\int \rho(\mathbf{x})\, d^3x = 1. \tag{8.2}$$

For a static target, it is found that the form factor in (8.1) is just the Fourier transform of the charge distribution

$$F(\mathbf{q}) = \int \rho(\mathbf{x}) e^{i\mathbf{q}\cdot\mathbf{x}}\, d^3x, \tag{8.3}$$

while the reference cross section for a structureless target is

$$\left(\frac{d\sigma}{d\Omega}\right)_{\text{point}} \equiv \left(\frac{d\sigma}{d\Omega}\right)_{\text{Mott}} = \frac{(Z\alpha)^2 E^2}{4k^4 \sin^4 \frac{\theta}{2}} \left(1 - v^2 \sin^2 \frac{\theta}{2}\right), \tag{8.4}$$

where $k = |\mathbf{k}_i| = |\mathbf{k}_f|$, $v = k/E$, and θ is the angle through which the electron is scattered.

EXERCISE 8.1 It is useful practice of the techniques developed in the previous chapters to derive (8.3) and (8.4). We outline the various steps below. The electromagnetic field due to $Ze\,\rho(\mathbf{x})$ is $A^\mu = (\phi, \mathbf{0})$ where, using (6.59),

$$\nabla^2\phi = -Ze\,\rho(\mathbf{x}).$$

Use (6.4) and (6.6) to show that the scattering amplitude is (see also Section 7.1)

$$T_{fi} = -i2\pi\,\delta(E_f - E_i)(-e\bar{u}_f\gamma_0 u_i)\int e^{i\mathbf{q}\cdot\mathbf{x}}\phi(\mathbf{x})\,d^3x. \tag{8.5}$$

Justify

$$\int e^{i\mathbf{q}\cdot\mathbf{x}}\nabla^2\phi\,d^3x = -|\mathbf{q}|^2\int e^{i\mathbf{q}\cdot\mathbf{x}}\phi\,d^3x$$

and hence show that the integral in (8.5) is $Ze\,F(\mathbf{q})/|\mathbf{q}|^2$, see (7.9). Following the arguments of Section 4.3, verify that the differential cross section from a fixed target is

$$d\sigma = \frac{|T_{fi}|^2}{T}\,\frac{d^3k_f}{(2\pi)^3 2E_f}\left(\frac{1}{v2E_i}\right), \tag{8.6}$$

with

$$d^3k_f\,\delta(E_f - E_i) = kE\,d\Omega.$$

Summing final, and averaging initial, electron spins give

$$\frac{1}{2}\sum_{s_f, s_i}|\bar{u}_f\gamma_0 u_i|^2 = 4E^2\left(1 - v^2\sin^2\frac{\theta}{2}\right), \tag{8.7}$$

where θ is the angle introduced in Section 7.1. Check this answer with (6.25). Putting all this together yields the advertised result

$$\frac{d\sigma}{d\Omega} = \left(\frac{d\sigma}{d\Omega}\right)_{\text{Mott}}|F(\mathbf{q})|^2,$$

with the form factor given by (8.3).

EXERCISE 8.2 Show that if the electron beam is replaced by a beam of "point" spinless particles, the only change is that factor (8.7) is replaced by $4E^2$. This raises a question: why does the electron spin make no difference in the nonrelativistic limit, $v \to 0$? The remarks following (6.13) are the clue.

EXERCISE 8.3 By considering the electron helicity, explain why you would anticipate the $\cos^2(\theta/2)$ behavior of factor (8.7) in the extreme relativistic limit, see Section 6.6.

By virtue of the normalization condition, (8.2),

$$F(0) \equiv 1. \tag{8.8}$$

If $|\mathbf{q}|$ is not too large, we can expand the exponential in (8.3), giving

$$F(\mathbf{q}) = \int \left(1 + i\mathbf{q}\cdot\mathbf{x} - \frac{(\mathbf{q}\cdot\mathbf{x})^2}{2} + \cdots \right) \rho(\mathbf{x})\, d^3x$$

$$= 1 - \frac{1}{6}|\mathbf{q}|^2 \langle r^2 \rangle + \cdots, \tag{8.9}$$

where we have assumed that ρ is spherically symmetric, that is, a function of $r \equiv |\mathbf{x}|$ alone. The small-angle scattering therefore just measures the mean square radius $\langle r^2 \rangle$ of the charge cloud. This is because in the small $|\mathbf{q}|$ limit the photon in Fig. 8.1 is soft and with its large wavelength can resolve only the size of the charge distribution $\rho(r)$ and is not sensitive to its detailed structure.

EXERCISE 8.4 If the charge distribution $\rho(r)$ has an exponential form, e^{-mr}, show, using (8.3), that the form factor

$$F(|\mathbf{q}|) = \left(1 - \frac{q^2}{m^2} \right)^{-2}$$

with $q^2 = -|\mathbf{q}|^2$.

8.2 Electron–Proton Scattering. Proton Form Factors

The above discussion cannot be applied directly to yield the structure of the proton. First, the proton's magnetic moment is involved in the scattering of the electron, not just its charge. Second, the proton is not static, but will recoil under the electron's bombardment. If, however, the proton were a point charge e with a Dirac magnetic moment $e/2M$, then we already know the answer. We can take over the result for electron–muon scattering, (6.50), and simply replace the mass of the muon by that of the proton:

$$\left.\frac{d\sigma}{d\Omega}\right|_{\text{lab}} = \left(\frac{\alpha^2}{4E^2 \sin^4 \frac{\theta}{2}} \right) \frac{E'}{E} \left\{ \cos^2 \frac{\theta}{2} - \frac{q^2}{2M^2} \sin^2 \frac{\theta}{2} \right\}, \tag{8.10}$$

where the factor

$$\frac{E'}{E} = \frac{1}{1 + \frac{2E}{M} \sin^2 \frac{\theta}{2}},$$

given by (6.48), arises from the recoil of the target.

Copying the calculation of the electron–muon cross section, the lowest-order amplitude for electron–proton elastic scattering, Fig. 8.2, is given by [see (6.8)]

$$T_{fi} = -i \int j_\mu \left(-\frac{1}{q^2} \right) J^\mu \, d^4x,$$

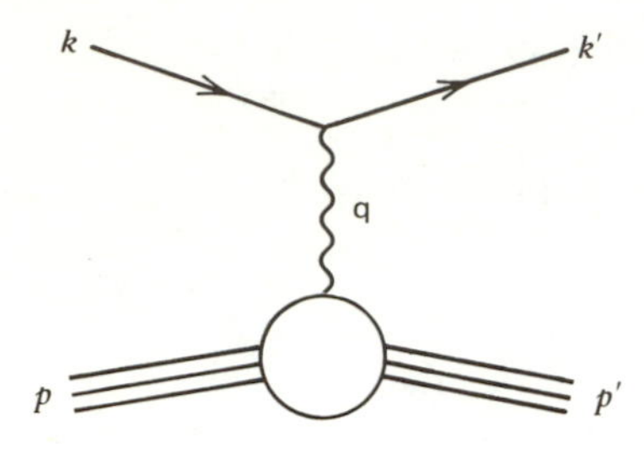

Fig. 8.2 Lowest-order electron–proton elastic scattering.

where $q = p' - p$ and the electron and proton transition currents are, respectively,

$$j^\mu = -e\,\bar{u}(k')\gamma^\mu u(k)\, e^{i(k'-k)\cdot x} \tag{8.11}$$

$$J^\mu = e\,\bar{u}(p')[\quad]u(p)\, e^{i(p'-p)\cdot x}, \tag{8.12}$$

see (6.6). Since the proton is an extended structure, we cannot replace the square brackets in (8.12) by γ^μ, as for point spin-$\frac{1}{2}$ particles in (8.11). But we know that J^μ must be a Lorentz four-vector, and so we must use the most general four-vector form that can be constructed from p, p', q and the Dirac γ-matrices sandwiched between $\bar{u}$ and u. There are only two independent terms, γ^μ and $i\sigma^{\mu\nu}q_\nu$, and their coefficients are functions of q^2 (q^2 is the only independent scalar variable at the proton vertex). Terms involving γ^5 are ruled out by the conservation of parity. Therefore, quite generally, we may write the square bracket of (8.12) in the form

$$[\quad] = \left[F_1(q^2)\gamma^\mu + \frac{\kappa}{2M}F_2(q^2)\, i\sigma^{\mu\nu}q_\nu\right] \tag{8.13}$$

where F_1 and F_2 are two independent form factors and κ is the anomalous magnetic moment (see Exercise 6.2).

EXERCISE 8.5 Show that current conservation, $\partial_\mu J^\mu = 0$, rules out $(p - p')^\mu$ as a possible four-vector. Why do we not show a term involving $(p + p')^\mu$ in (8.13)?

EXERCISE 8.6 Show that $p \cdot q$ is not an independent scalar variable by expressing it in terms of the variable q^2.

For $q^2 \to 0$, that is, when we probe with long-wavelength photons, it does not make any difference that the proton has structure at the order of 1 fermi. We effectively see a particle of charge e and magnetic moment $(1 + \kappa)e/2M$, where κ, the anomalous moment, is measured to be 1.79. The factors in (8.13) must therefore be chosen so that in this limit

$$F_1(0) = 1, \qquad F_2(0) = 1. \tag{8.14}$$

The corresponding values for the neutron are $F_1(0) = 0$, $F_2(0) = 1$, and experimentally $\kappa_n = -1.91$.

If we use (8.13) to calculate the differential cross section for electron–proton elastic scattering, we find an expression similar to (8.10):

$$\left.\frac{d\sigma}{d\Omega}\right|_{\text{lab}} = \left(\frac{\alpha^2}{4E^2 \sin^4 \frac{\theta}{2}}\right)\frac{E'}{E}\left\{\left(F_1^2 - \frac{\kappa^2 q^2}{4M^2}F_2^2\right)\cos^2\frac{\theta}{2} - \frac{q^2}{2M^2}(F_1 + \kappa F_2)^2 \sin^2\frac{\theta}{2}\right\}, \qquad (8.15)$$

see (6.50). This is known as the Rosenbluth formula. The two form factors, $F_{1,2}(q^2)$, parametrize our ignorance of the detailed structure of the proton represented by the blob in Fig. 8.2. These form factors can be determined experimentally by measuring $d\sigma/d\Omega$ as a function of θ and q^2. Note that if the proton were a point particle like the muon, then $\kappa = 0$ and $F_1(q^2) = 1$ for all q^2, and (8.15) would revert to (8.10).

In practice, it is better to use linear combinations of $F_{1,2}$,

$$G_E \equiv F_1 + \frac{\kappa q^2}{4M^2}F_2$$

$$G_M \equiv F_1 + \kappa F_2, \qquad (8.16)$$

defined so that no interference terms, $G_E G_M$, occur in the cross section. Equation (8.15) becomes

$$\left.\frac{d\sigma}{d\Omega}\right|_{\text{lab}} = \frac{\alpha^2}{4E^2 \sin^4 \frac{\theta}{2}}\frac{E'}{E}\left(\frac{G_E^2 + \tau G_M^2}{1+\tau}\cos^2\frac{\theta}{2} + 2\tau G_M^2 \sin^2\frac{\theta}{2}\right), \qquad (8.17)$$

with $\tau \equiv -q^2/4M^2$.

Now that interference terms have disappeared, these proton form factors may be regarded as generalizations of the nonrelativistic form factor introduced in Section 8.1, and so it would be nice if we could interpret their Fourier transforms as the charge and magnetic moment distributions of the proton. Unfortunately, the recoil of the proton makes this impossible. However, it is possible to show that the form factors $G_E(q^2)$ and $G_M(q^2)$ are closely related to the proton charge and magnetic moment distributions, respectively, in a particular Lorentz frame, called the Breit (or brick wall) frame, defined by $\mathbf{p}' = -\mathbf{p}$.

EXERCISE 8.7 Show that the proton transition current, $J^\mu(x)$ of (8.12), can be rewritten in the form

$$J^\mu(0) = e\,\bar{u}(p')\left[\gamma^\mu(F_1 + \kappa F_2) - \frac{(p^\mu + p'^\mu)}{2M}\kappa F_2\right]u(p). \qquad (8.18)$$

Evaluate $J^\mu(0) \equiv (\rho, \mathbf{J})$ in the Breit frame ($\mathbf{p}' = -\mathbf{p}$). There is no energy

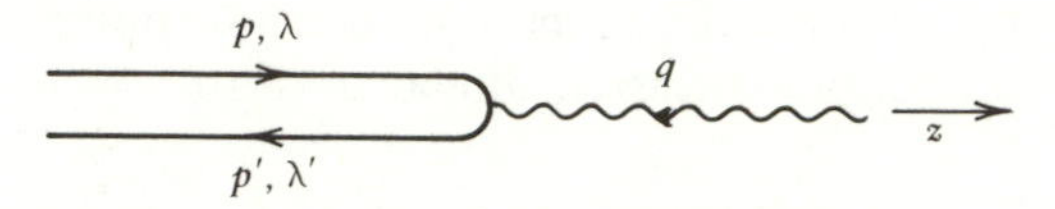

Fig. 8.3 The Breit or brick-wall frame, $\mathbf{p}' = -\mathbf{p}$.

transferred to the proton in this frame, and it behaves as if it had bounced off a brick wall, see Fig. 8.3. If the z axis is chosen along $\mathbf{p}$ and helicity spinors are used, show that

$$\rho = 2Me\, G_E(q^2) \qquad \text{for } \lambda = -\lambda',$$
$$J_1 \pm iJ_2 = \mp 2|\mathbf{q}|e\, G_M(q^2) \qquad \text{for } \lambda = \lambda' = \mp \tfrac{1}{2}, \tag{8.19}$$

and that all other matrix elements are zero; λ and λ' denote the initial and final proton helicities, respectively. Determine the corresponding values of the helicity of the virtual photon.

In generalizing the form factor of Section 8.1, we have replaced $F(|\mathbf{q}|)$ by $F(q^2)$. However, as long as $|\mathbf{q}|^2 \ll M^2$, we can take over the Fourier transform interpretation of Section 8.1.

EXERCISE 8.8 Show that for $|\mathbf{q}|^2 \ll M^2$, the form factors G_E and G_M are the Fourier transforms of the proton's charge and magnetic moment distributions, respectively.

G_E and G_M are referred to as the electric and magnetic form factors, respectively. The data on the angular dependence of ep → ep scattering can be used to separate G_E, G_M at different values of q^2, see (8.17). The result for $G_E(q^2)$ is

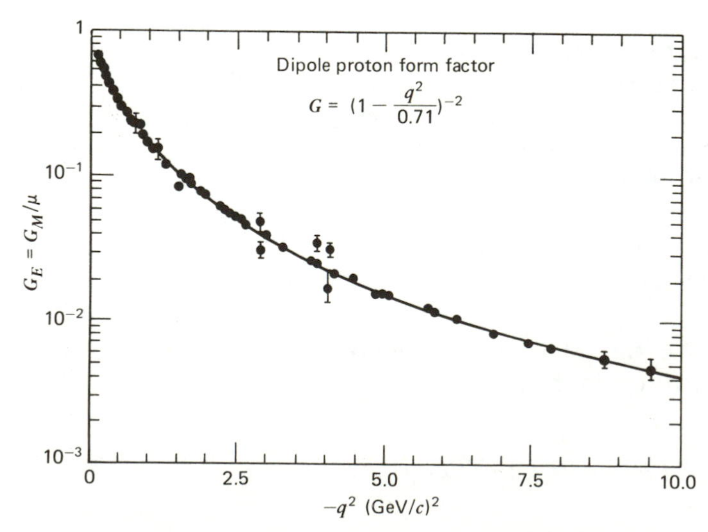

Fig. 8.4 The proton form factors as a function of q^2.

shown in Fig. 8.4. $G_M(q^2)$ has the same q^2 dependence. A closer look at Fig. 8.4 reveals that

$$G_E(q^2) \approx \left(1 - \frac{q^2}{0.71}\right)^{-2} \qquad \text{(in units of GeV}^2\text{)}. \tag{8.20}$$

The behavior for small $-q^2$ can be used to determine the residual terms in the expansion of (8.9). In particular, the mean square proton charge radius is

$$\langle r^2 \rangle = 6\left(\frac{dG_E(q^2)}{dq^2}\right)_{q^2=0} = (0.81 \times 10^{-13}\ \text{cm})^2. \tag{8.21}$$

The same radius of about 0.8 fm is obtained for the magnetic moment distribution. Using the result of Exercise 8.4, we conclude that the charge distribution of the nucleon has an exponential shape in configuration space.

8.3 Inelastic Electron–Proton Scattering ep → eX

Having measured the size of the proton, one might like to take a more detailed look at its structure by increasing the $-q^2$ of the photon to give better spatial resolution. This can be done simply by requiring a large energy loss of the bombarding electron. There is, however, a catch: because of the large transfer of energy, the proton will often break up, and the picture of Fig. 8.2 has to be generalized to Fig. 8.5. For modest $-q^2$, one might just excite the proton into a Δ-state and hence produce an extra π-meson, that is, ep → eΔ^+→ epπ^0. In these events, the invariant mass (see Fig. 8.5) is $W^2 \simeq M_\Delta^2$. When $-q^2$ is very large, however, the debris becomes so messy that the initial state proton loses its identity completely and a new formalism must be devised to extract information from the measurements. Figure 8.6 shows the invariant mass distribution. One notices the peak when the proton does not break up ($W \simeq M$) and broader peaks when the target is excited to resonant baryon states. Beyond the resonances, the complicated multiparticle states with large invariant mass result in a smooth distribution in missing mass W.

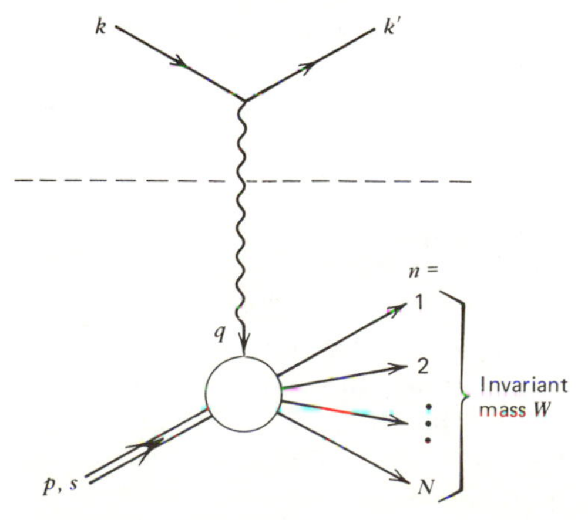

Fig. 8.5 Lowest-order diagram for ep → eX.

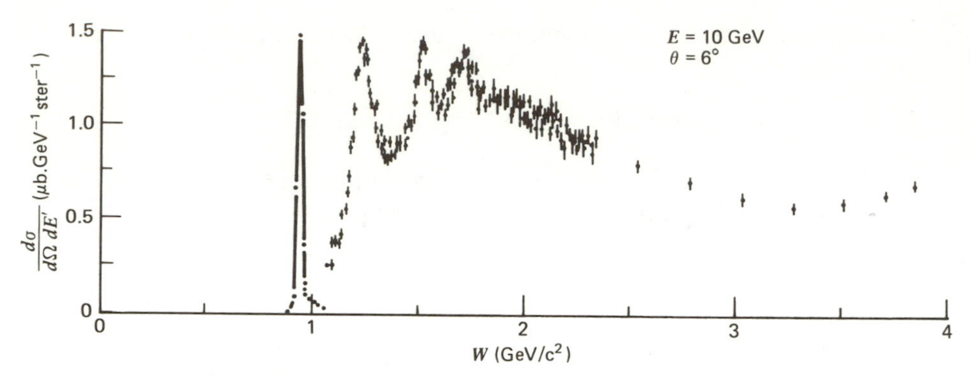

Fig. 8.6 The ep → eX cross section as a function of the missing mass W. Data are from the Stanford Linear Accelerator. The elastic peak at $W = M$ has been reduced by a factor of 8.5.

The problem now facing us is illustrated by recalling (8.11), (8.12), and Fig. 8.2. The switch from a muon to a proton target was made by replacing the lepton current j^μ ($\sim \bar{u}\gamma^\mu u$) by a proton current J^μ ($\sim \bar{u}\Gamma^\mu u$), and the most general form of Γ^μ was constructed. This is inadequate to describe the inelastic events of Fig. 8.5. Although everything above the dashed line in Fig. 8.5 remains unchanged (a fact which we shall exploit), the final state below it is not a single fermion described by a Dirac "$\bar{u}$" entry in the matrix element or current. Therefore, J^μ must have a more complex structure than (8.12). Instead, the expression for the cross section [see (6.18)]

$$d\sigma \sim L^e_{\mu\nu}(L^p)^{\mu\nu} \tag{8.22}$$

is directly generalized to

$$d\sigma \sim L^e_{\mu\nu}W^{\mu\nu}, \tag{8.23}$$

where $L^e_{\mu\nu}$ represents the lepton tensor of (6.20), since everything in the leptonic part of the diagram above the photon propagator in Fig. 8.5 is left unchanged. The hadronic tensor $W^{\mu\nu}$ serves to parametrize our total ignorance of the form of the current at the other end of the propagator. The most general form of the tensor $W^{\mu\nu}$ must now be constructed out of $g^{\mu\nu}$ and the independent momenta p and q ($p' = p + q$). γ^μ is not included, as we are parametrizing the cross section which is already summed and averaged over spins. We write

$$W^{\mu\nu} = -W_1 g^{\mu\nu} + \frac{W_2}{M^2}p^\mu p^\nu + \frac{W_4}{M^2}q^\mu q^\nu + \frac{W_5}{M^2}(p^\mu q^\nu + q^\mu p^\nu). \tag{8.24}$$

We have omitted antisymmetric contributions to $W^{\mu\nu}$, since their contribution to the cross section vanishes after insertion into (8.23) because the tensor $L^e_{\mu\nu}$ is symmetric. Note the omission of W_3 in our notation; this spot is reserved for a parity-violating structure function when a neutrino beam is substituted for the electron beam, so that the virtual photon probe is replaced by a weak boson; see Perl (1974), Close (1979), or Llewellyn Smith (1972).

EXERCISE 8.9 Show that indeed $L^e_{\mu\nu} = L^e_{\nu\mu}$ and that

$$q^\mu L^e_{\mu\nu} = q^\nu L^e_{\mu\nu} = 0. \tag{8.25}$$

EXERCISE 8.10 Show that current conservation at the hadronic vertex requires

$$q_\mu W^{\mu\nu} = q_\nu W^{\mu\nu} = 0. \tag{8.26}$$

The proof may be left until after (8.39); it follows from $\partial_\mu \tilde{J}^\mu = 0$. As a result of (8.26), verify that

$$W_5 = -\frac{p \cdot q}{q^2} W_2,$$

$$W_4 = \left(\frac{p \cdot q}{q^2}\right)^2 W_2 + \frac{M^2}{q^2} W_1.$$

Thus, only two of the four inelastic structure functions of (8.24) are independent; so we may write

$$W^{\mu\nu} = W_1\left(-g^{\mu\nu} + \frac{q^\mu q^\nu}{q^2}\right) + W_2 \frac{1}{M^2}\left(p^\mu - \frac{p \cdot q}{q^2} q^\mu\right)\left(p^\nu - \frac{p \cdot q}{q^2} q^\nu\right), \tag{8.27}$$

where the W_i's are functions of the Lorentz scalar variables that can be constructed from the four-momenta at the hadronic vertex. Unlike elastic scattering, there are two independent variables, and we choose

$$q^2 \quad \text{and} \quad \nu \equiv \frac{p \cdot q}{M}. \tag{8.28}$$

The invariant mass W of the final hadronic system is related to ν and q^2 by

$$W^2 = (p + q)^2 = M^2 + 2M\nu + q^2. \tag{8.29}$$

EXERCISE 8.11 It is common to replace ν and q^2 by the dimensionless variables

$$x = \frac{-q^2}{2p \cdot q} = \frac{-q^2}{2M\nu}, \qquad y = \frac{p \cdot q}{p \cdot k}, \tag{8.30}$$

where the four-momenta are shown on Fig. 8.5. Show that the allowed kinematic region for ep → eX is $0 \le x \le 1$ and $0 \le y \le 1$. Sketch this physical region in the ν, q^2 plane and check your answer with Fig. 9.3.

EXERCISE 8.12 Show that in the rest frame of the target proton,

$$\nu = E - E', \qquad y = \frac{E - E'}{E},$$

where E and E' are the initial and final electron energies, respectively.

Evaluation of the cross section for ep $\to$ eX is a straightforward repetition of the same calculation for $e^-\mu^- \to e^-\mu^-$ (or ep $\to$ ep) scattering with the substitution of $W_{\mu\nu}$, given by (8.27), for $L_{\mu\nu}^{\text{muon}}$ (or $L_{\mu\nu}^{p}$). Using the expression (6.25) for $(L^e)^{\mu\nu}$ and noting (8.25), we find

$$(L^e)^{\mu\nu} W_{\mu\nu} = 4W_1(k \cdot k') + \frac{2W_2}{M^2}\left[2(p \cdot k)(p \cdot k') - M^2 k \cdot k'\right]. \tag{8.31}$$

In the laboratory frame, this becomes

$$(L^e)^{\mu\nu} W_{\mu\nu} = 4EE'\left\{\cos^2\frac{\theta}{2} W_2(\nu, q^2) + \sin^2\frac{\theta}{2} 2W_1(\nu, q^2)\right\}, \tag{8.32}$$

see (6.44). By including the flux factor, (4.32), and the phase space factor for the outgoing electron, (4.24), we can obtain the inclusive differential cross section for inelastic electron–proton scattering, ep $\to$ eX,

$$d\sigma = \frac{1}{4\left((k \cdot p)^2 - m^2 M^2\right)^{1/2}} \left\{\frac{e^4}{q^4}(L^e)^{\mu\nu} W_{\mu\nu} 4\pi M\right\} \frac{d^3k'}{2E'(2\pi)^3}, \tag{8.33}$$

where $\overline{|\mathcal{M}|^2}$ is given by the expression in the braces [recall (6.18)]. The extra factor of $4\pi M$ arises because we have adopted the standard convention for the normalization of $W^{\mu\nu}$. Inserting (8.32) in (8.33) yields

$$\boxed{\left.\frac{d\sigma}{dE'\,d\Omega}\right|_{\text{lab}} = \frac{\alpha^2}{4E^2 \sin^4\frac{\theta}{2}}\left\{W_2(\nu, q^2)\cos^2\frac{\theta}{2} + 2W_1(\nu, q^2)\sin^2\frac{\theta}{2}\right\},} \tag{8.34}$$

where, as usual, we neglect the mass of the electron.

8.4 Summary of the Formalism for Analyzing ep Scattering

Although we have achieved our objective, it is informative to take a second look at the formalism. The final result, (8.34), may be reexpressed in the form

$$\frac{d\sigma}{dE'\,d\Omega} = \frac{\alpha^2}{q^4}\frac{E'}{E}(L^e)^{\mu\nu}W_{\mu\nu}, \tag{8.35}$$

see (8.32) and (6.44). For comparison, recall the cross section for $e\mu^- \to e\mu^-$ scattering, (6.46),

$$d\sigma = \frac{1}{4ME}\frac{d^3k'}{(2\pi)^3 2E'}\frac{d^3p'}{(2\pi)^3 2p_0'}\left\{\frac{e^4}{q^4}(L^e)^{\mu\nu}L_{\mu\nu}^{\text{muon}}\right\}(2\pi)^4\delta^4(p+q-p'), \tag{8.36}$$

which can also be written in the form (8.35) with

$$W_{\mu\nu} = \frac{1}{4\pi M}\left(\frac{1}{2}\sum_s\sum_{s'}\right)\int\frac{d^3p'}{(2\pi)^3 2p_0'}\langle p,s|\tilde{J}_\mu^\dagger|p',s'\rangle$$
$$\times\langle p',s'|\tilde{J}_\nu|p,s\rangle(2\pi)^4\delta^4(p+q-p') \tag{8.37}$$

where

$$\langle p',s'|\tilde{J}_\nu|p,s\rangle \equiv \bar{u}^{(s')}(p')\gamma_\nu u^{(s)}(p). \tag{8.38}$$

Insertion of (8.37) into (8.35) reproduces (6.49). All we have done is to use a particular regrouping of the factors in our earlier derivation. If we replace (8.38) by $\bar{u}[\quad]u$ with $[\quad]$ given by (8.13), then we can recover the result for ep → ep scattering. The reason for all this is to note that $W_{\mu\nu}$ for ep → eX is nothing but a generalization of (8.37) to the case where a proton breaks up into many particles in the final hadronic state X. It can be formally written as

$$W_{\mu\nu} = \frac{1}{4\pi M}\sum_N\left(\frac{1}{2}\sum_s\right)\int\prod_{n=1}^{N}\left(\frac{d^3p_n'}{2E_n'(2\pi)^3}\right)\sum_{s_n}\langle p,s|\tilde{J}_\mu^\dagger|X\rangle$$
$$\times\langle X|\tilde{J}_\nu|p,s\rangle(2\pi)^4\delta^4\left(p+q-\sum_n p_n'\right), \tag{8.39}$$

where a sum over all possible many-particle (i.e., N particle) states X is included, see Fig. 8.5.

For future reference, it is useful to make a compendium of our results on form factors. We keep to the laboratory kinematics of Fig. 6.10 and neglect the mass of the electron. For all the reactions, the differential cross section in the energy (E') and angle (θ) of the scattered electron can be written in the form

$$\frac{d\sigma}{dE'\,d\Omega} = \frac{4\alpha^2 E'^2}{q^4}\{\quad\}. \tag{8.40}$$

First, for a muon target of mass m (or a quark target of mass m after substitution $\alpha^2 \to \alpha^2 e_q^2$ where e_q is the quark's fractional charge),

$$\{\quad\}_{e\mu\to e\mu} = \left(\cos^2\frac{\theta}{2} - \frac{q^2}{2m^2}\sin^2\frac{\theta}{2}\right)\delta\left(\nu + \frac{q^2}{2m}\right). \tag{8.41}$$

For elastic scattering from a proton target,

$$\{\quad\}_{ep\to ep} = \left(\frac{G_E^2 + \tau G_M^2}{1+\tau}\cos^2\frac{\theta}{2} + 2\tau G_M^2 \sin^2\frac{\theta}{2}\right)\delta\left(\nu + \frac{q^2}{2M}\right), \tag{8.42}$$

where $\tau = -q^2/4M^2$ and M is the mass of the proton. Finally, for the case when the proton target is broken up by the bombarding electron,

$$\{\quad\}_{ep\to eX} = W_2(\nu, q^2)\cos^2\frac{\theta}{2} + 2W_1(\nu, q^2)\sin^2\frac{\theta}{2}. \tag{8.43}$$

Making use of the delta function, (8.41) and (8.42) can be integrated over E' with the result [see (6.50)]

$$\frac{d\sigma}{d\Omega} = \frac{\alpha^2}{4E^2\sin^4\frac{\theta}{2}}\frac{E'}{E}(\quad). \tag{8.44}$$

EXERCISE 8.13 The above results assume (lowest-order) single photon exchange is dominant. If two-photon exchange were significant, convince yourself that the e^-p and e^+p cross sections would not be equal.

8.5 Inelastic Electron Scattering as a (Virtual) Photon–Proton Total Cross Section

It is clear from the above discussion that the important issue is what happens below the dashed line in Fig. 8.5, where a (virtual) photon interacts with a proton. The role of the electron beam is simply that it is responsible for the presence of the virtual photon. It is useful to display these facts in our formalism. We start by writing the total cross section for scattering a *real* photon, with energy $q^0 = \nu \equiv K$ and (transverse) polarization ε, off the same unpolarized proton target producing two or more final-state particles. Using the Feynman rules and cross section kinematics which we have developed, we obtain

$$\sigma^{\text{tot}}(\gamma p \to X) = \frac{1}{(2K)(2M)}\sum_N\left(\frac{1}{2}\sum_s\right)\int\prod_{n=1}^{N}\left(\frac{d^3p_n'}{2E_n'(2\pi)^3}\right)$$
$$\times\sum_{s_n}(2\pi)^4\delta^4\left(p + q - \sum_n p_n'\right)\varepsilon^{\mu*}\varepsilon^{\nu}e^2\langle p, s|\tilde{J}_\mu^\dagger|X\rangle\langle X|\tilde{J}_\nu|p, s\rangle, \tag{8.45}$$

where, as before, a sum over all final states X is included. If W is the invariant mass of the final state, then

$$W^2 = (p + q)^2 = M^2 + 2MK. \tag{8.46}$$

We immediately note the striking similarity between (8.45) and the formal expression for $W_{\mu\nu}$ given by (8.39). Indeed, (8.45) appears to be simply

$$\sigma^{\text{tot}}(\gamma \text{p} \to \text{X}) = \frac{4\pi^2\alpha}{K} \varepsilon^{\mu *} \varepsilon^{\nu} W_{\mu\nu}. \tag{8.47}$$

However, there is a crucial proviso: for real photons, we must sum only over the two transverse polarizations of the incident photon. On the other hand, to interpret the ep → eX hadronic tensor $W_{\mu\nu}$ as a photon–proton total cross section, it is vital to remember that the photon is *virtual* and not limited to two polarization states. In fact, the cross section for virtual photons is not a well-defined concept. When $q^2 = 0$, the flux factor is the $4MK$ with $K = \nu$; but for virtual photons ($q^2 \neq 0$), the flux is arbitrary. The conventional choice is to require K to continue to satisfy (8.46), that is,

$$K = \frac{W^2 - M^2}{2M} = \nu + \frac{q^2}{2M}, \tag{8.48}$$

in the laboratory frame. This is known as the Hand convention. Another possible choice is $K = |\mathbf{q}|$.

To complete the interpretation of the structure functions, we must specify the polarization vectors ε_λ^μ of virtual photons (helicity λ). We take the z axis along $\mathbf{q}$ and use [see (6.92)]

$$\lambda = \pm 1: \varepsilon_{\pm} = \mp \sqrt{\tfrac{1}{2}}\,(0; 1, \pm i, 0), \tag{8.49}$$

$$\lambda = 0: \quad \varepsilon_0 = \frac{1}{\sqrt{-q^2}} \left(\sqrt{\nu^2 - q^2}\,; 0, 0, \nu \right). \tag{8.50}$$

EXERCISE 8.14 Verify that $q \cdot \varepsilon = 0$ for each λ, and show that

$$\sum \varepsilon^{\mu *} \varepsilon^{\nu} = -g^{\mu\nu} + \frac{q^\mu q^\nu}{q^2}, \tag{8.51}$$

where the sum runs over the three polarization states of (8.49) and (8.50). For the above choice of polarization vectors, the sum (8.51) with q^2 replaced by M^2 is identical to that over the spin states of a massive vector particle, see Section 6.12.

We can now evaluate the total cross sections for polarized photons (helicity λ) interacting with unpolarized protons:

$$\sigma_\lambda^{\text{tot}} = \frac{4\pi^2\alpha}{K} \varepsilon_\lambda^{\mu *} \varepsilon_\lambda^{\nu} W_{\mu\nu}. \tag{8.52}$$

Using (8.24) for $W_{\mu\nu}$, together with the above polarization vectors, we find that the transverse and longitudinal cross sections are, respectively,

$$\sigma_T \equiv \frac{1}{2}(\sigma_+^{\text{tot}} + \sigma_-^{\text{tot}}) = \frac{4\pi^2\alpha}{K} W_1(\nu, q^2) \tag{8.53}$$

$$\sigma_L \equiv \sigma_0^{\text{tot}} = \frac{4\pi^2\alpha}{K}\left[\left(1 - \frac{\nu^2}{q^2}\right)W_2(\nu, q^2) - W_1(\nu, q^2)\right] \tag{8.54}$$

EXERCISE 8.15 Verify eqs. (8.53) and (8.54). The calculation can be greatly simplified by writing the tensor decomposition for $W_{\mu\nu}$, (8.27), in the laboratory frame, where

$$p = (M; 0, 0, 0),$$

$$q = \left(\nu; 0, 0, \sqrt{\nu^2 - q^2}\right).$$

EXERCISE 8.16 Express the ep → eX differential cross section (8.34) in terms of $\sigma_{T,L}$. That is, show that

$$\left.\frac{d\sigma}{dE'\,d\Omega}\right|_{\text{lab}} = \Gamma(\sigma_T + \varepsilon\sigma_L), \tag{8.55}$$

where

$$\Gamma = \frac{\alpha K}{2\pi^2|q^2|}\frac{E'}{E}\frac{1}{1-\varepsilon}, \tag{8.56}$$

$$\varepsilon = \left(1 - 2\frac{\nu^2 - q^2}{q^2}\tan^2\frac{\theta}{2}\right)^{-1}. \tag{8.57}$$

EXERCISE 8.17 The formalism has been set up in such a way that, when $q^2 \to 0$,

$$\sigma_T \to \sigma^{\text{tot}}(\gamma\text{p})$$

$$\sigma_L \to 0, \tag{8.58}$$

where γ is a real photon and $\sigma^{\text{tot}}(\gamma\text{p})$ is given by (8.45).

Despite its appearance, convince yourself that $W_{\mu\nu}$ must not be singular at $q^2 = 0$. Hence, show that

$$W_2 \to 0 \qquad \text{and} \qquad \left(W_1 + \frac{\nu^2}{q^2} W_2\right) \to 0 \tag{8.59}$$

as $q^2 \to 0$, and so establish that σ_L vanishes.

Can we extract additional information about the structure of the proton from these complex events where the proton breaks up? Both the structure of the events and their phenomenological interpretation look quite forbidding. The answer to this question is of great importance and is the subject of the next two chapters.

9

Partons

Equipped with the formalism of Chapter 8, we can now turn to the experimental information and ask the question, "Can small-wavelength photons resolve the quarks inside the proton target?"

9.1 Bjorken Scaling

If simple, point-like, spin-$\frac{1}{2}$ quarks reside inside the proton, we should be able to illuminate them with a small-wavelength (large $-q^2$) virtual photon beam (Fig. 9.1). The fact that such photons break up the proton target can be handled by using the inelastic form factors discussed in the previous chapter. The sign that there are structureless particles inside a complex system like a proton is that for small wavelengths, the proton described by (8.43) suddenly starts behaving like a free Dirac particle (a quark) and (8.43) turns into (8.41). The proton structure functions thus become simply

$$2W_1^{\text{point}} = \frac{Q^2}{2m^2}\delta\left(\nu - \frac{Q^2}{2m}\right),$$

$$W_2^{\text{point}} = \delta\left(\nu - \frac{Q^2}{2m}\right). \tag{9.1}$$

For convenience, we have introduced the positive variable

$$Q^2 \equiv -q^2.$$

Here, m is the quark mass; the "point" notation reminds us the quark is a structureless Dirac particle. Equation (9.1) can be pictured as

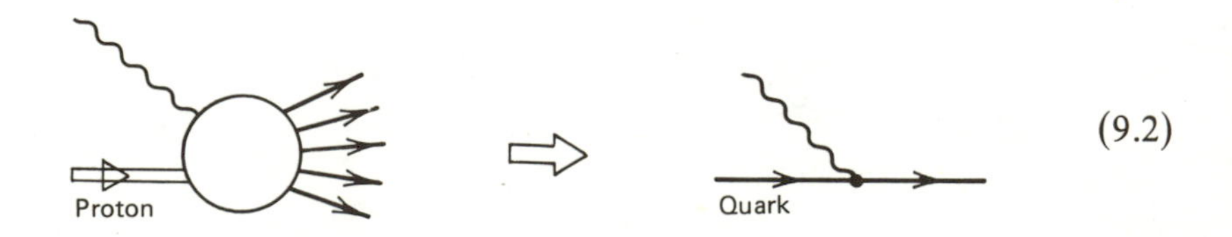

(9.2)

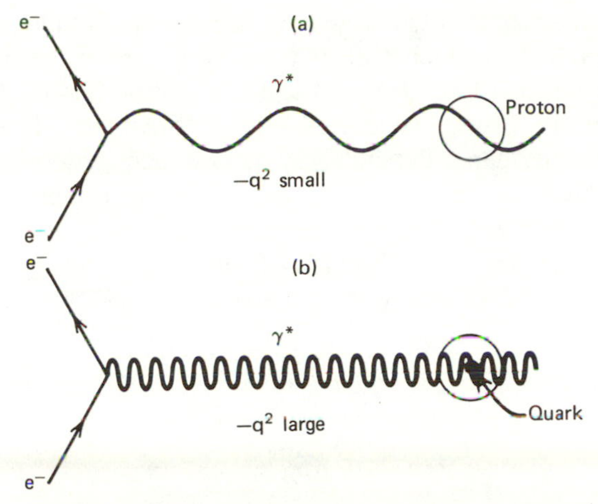

Fig. 9.1 (a) Elastic ep → ep scattering in which a large-wavelength "photon beam" measures the size of the proton through the elastic form factor analysis. (b) In deep inelastic scattering a short-wavelength "photon beam" resolves the quarks within the proton provided $\lambda\left(\approx 1/\sqrt{-q^2}\right) \ll 1\text{F}$.

that is, at large Q^2, *inelastic* electron–proton scattering is viewed simply as *elastic* scattering of the electron on a "free" quark within the proton. Using the identity $\delta(x/a) = a\,\delta(x)$, (9.1) may be rearranged to introduce dimensionless structure functions

$$\begin{aligned} 2mW_1^{\text{point}}(\nu, Q^2) &= \frac{Q^2}{2m\nu}\delta\left(1 - \frac{Q^2}{2m\nu}\right), \\ \nu W_2^{\text{point}}(\nu, Q^2) &= \delta\left(1 - \frac{Q^2}{2m\nu}\right). \end{aligned} \tag{9.3}$$

These "point" functions now display the intriguing property that they are only functions of the ratio $Q^2/2m\nu$ and *not* of Q^2 and ν independently. This behavior can be contrasted with that for ep elastic scattering. For simplicity, set $\kappa = 0$, so that $G_E = G_M \equiv G$; then, comparing (8.42) and (8.43), we have

$$\begin{aligned} W_1^{\text{elastic}} &= \frac{Q^2}{4M^2}G^2(Q^2)\,\delta\left(\nu - \frac{Q^2}{2M}\right), \\ W_2^{\text{elastic}} &= G^2(Q^2)\,\delta\left(\nu - \frac{Q^2}{2M}\right). \end{aligned} \tag{9.4}$$

In contrast to (9.1), the structure functions of (9.4) contain a form factor $G(Q^2)$, and so cannot be rearranged to be functions of a single dimensionless variable. A mass scale is explicitly present; it is set by the empirical value 0.71 GeV in the dipole formula for $G(Q^2)$ which reflects the inverse size of the proton, see (8.20). As Q^2 increases above $(0.71\text{ GeV})^2$, the form factor depresses the chance of elastic scattering; the proton is more likely to break up. The point structure functions, on the other hand, depend only on a dimensionless variable $Q^2/2m\nu$, and no scale of mass is present. The mass m merely serves as a scale for the momenta Q^2, ν.

The discussion can be summarized as follows: if large Q^2 virtual photons resolve "point" constituents inside the proton, then

$$MW_1(\nu, Q^2) \underset{\text{large } Q^2}{\longrightarrow} F_1(\omega),$$
$$\nu W_2(\nu, Q^2) \underset{\text{large } Q^2}{\longrightarrow} F_2(\omega), \tag{9.5}$$

where

$$\omega = \frac{2q \cdot p}{Q^2} = \frac{2M\nu}{Q^2}. \tag{9.6}$$

Note that in (9.5) we have changed the scale from what it was in (9.3). We have introduced the proton mass instead of the quark mass to define the dimensionless variable ω. The presence of free quarks is signaled by the fact that the inelastic structure functions are independent of Q^2 at a given value of ω [see (9.5)]. This is equivalent to the onset of $\sin^{-4}(\theta/2)$ behavior for large momentum transfers in the Rutherford experiment, which reveals the "point" charge of the nucleus in the atom. A sample of data is shown in Fig. 9.2. νW_2 at $\omega = 4$ is independent of Q^2; the photon is indeed interacting with point-like particles. No form factors, leading to additional Q^2 dependence as in (9.4), are present. Are these particles (called partons by Bjorken) the same as the quarks discovered in the spectroscopy of hadrons (Chapter 2)?

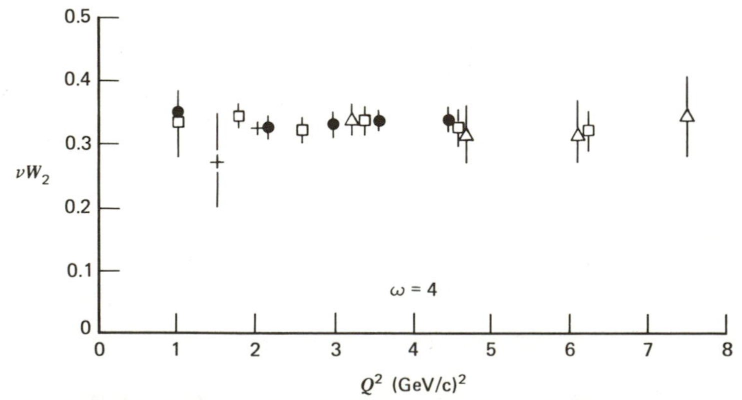

Fig. 9.2 The structure function νW_2 determined by electron–proton scattering as a function of Q^2 for $\omega = 4$. Data are from the Stanford Linear Accelerator.

9.2 Partons and Bjorken Scaling

Now that scaling is an approximate experimental fact, we attempt to make the identification of (9.2) explicit:

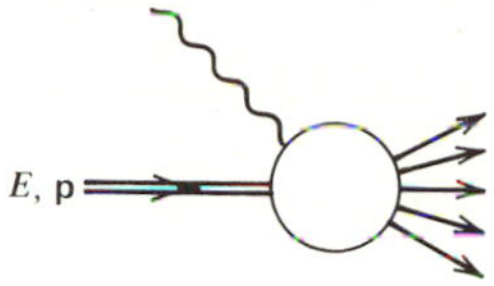

$$= \sum_i \int dx \, e_i^2 \quad (9.7)$$

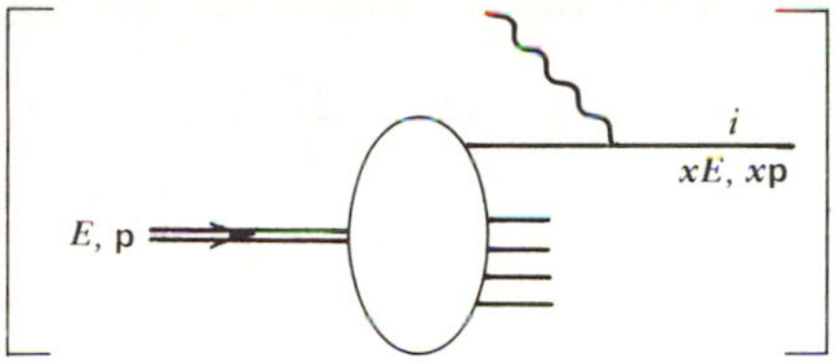

Equation (9.7) recognizes the fact that various types of "point" partons make up the proton ($i = \text{u}, \text{d}, \ldots$, quarks, with various charges e_i, as well as gluons; the latter do not interact with the photon, of course). They can each carry a different fraction x of the parent proton's momentum and energy. We introduce the parton momentum distribution

$$f_i(x) = \frac{dP_i}{dx} = \quad (9.8)$$

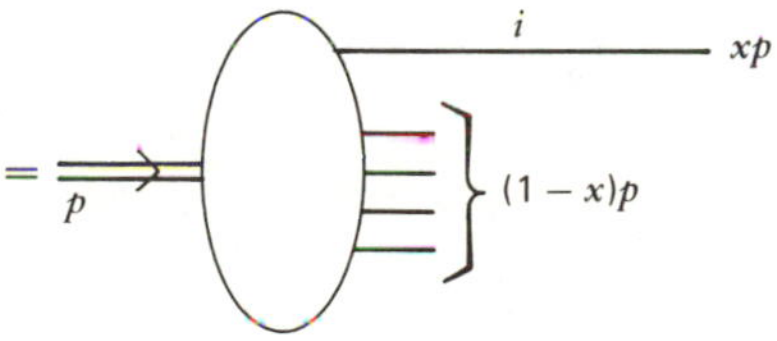

which describes the probability that the struck parton i carries a fraction x of the proton's momentum p. All the fractions x have to add up to 1; therefore,

$$\sum_{i'} \int dx \, x f_{i'}(x) = 1. \quad (9.9)$$

Here, i' sums over all the partons, not just the charged ones i which interact with the photon. The kinematics can be summarized as follows:

	Proton ↓	Parton ↓	
Energy	E	xE	(9.10)
Momentum	p_L	xp_L	
	$p_T = 0$	$p_T = 0$	
Mass	M	$m = (x^2E^2 - x^2p_L^2)^{1/2} = xM.$	

Both the proton and its parton progeny move along the z axis (i.e., $p_T = 0$) with longitudinal momenta p_L and xp_L. The definition of the reference frame has to be made with more care; we shall return to it later on.

For an electron hitting a parton with momentum fraction x and unit charge, we see from (9.3) and (9.5) that the dimensionless structure functions are

$$F_1(\omega) = \frac{Q^2}{4m\nu x}\delta\left(1 - \frac{Q^2}{2m\nu}\right) = \frac{1}{2x^2\omega}\delta\left(1 - \frac{1}{x\omega}\right),$$

$$F_2(\omega) = \delta\left(1 - \frac{Q^2}{2m\nu}\right) = \delta\left(1 - \frac{1}{x\omega}\right). \tag{9.11}$$

We have used the kinematics of (9.10); ω is the dimensionless variable defined in (9.6). Summing our results for $F_{1,2}$ for one parton, (9.11), over the partons making up a proton, (9.7) and (9.8), we obtain

$$F_2(\omega) = \sum_i \int dx\, e_i^2 f_i(x) x\, \delta\left(x - \frac{1}{\omega}\right),$$

$$F_1(\omega) = \frac{\omega}{2} F_2(\omega). \tag{9.12}$$

It is conventional to redefine $F_{1,2}(\omega)$ as $F_{1,2}(x)$ and to express the results in terms of x. Recalling the identification (9.5), we see that (9.12) become, at large Q^2,

$$\boxed{\nu W_2(\nu, Q^2) \to F_2(x) = \sum_i e_i^2 x f_i(x),} \tag{9.13}$$

$$\boxed{MW_1(\nu, Q^2) \to F_1(x) = \frac{1}{2x} F_2(x),} \tag{9.14}$$

with

$$x = \frac{1}{\omega} = \frac{Q^2}{2M\nu}. \tag{9.15}$$

That is, the momentum fraction is found to be identical to the (dimensionless) kinematic variable x of the virtual photon that we introduced in Chapter 8; see (8.30). In other words, the virtual photon must have just the right value of the variable x to be absorbed by a parton with momentum fraction x. It is the delta function in (9.12) that equates these two distinct physical variables.

The inelastic structure functions $F_{1,2}$ of (9.13) and (9.14) are functions of only one variable, namely, x. They are independent of Q^2 at fixed x. We say they satisfy Bjorken scaling.

> ***EXERCISE 9.1*** Prove that $0 \leq x \leq 1$, as it must be if x represents a momentum fraction; recall Exercise 8.11.

Note that the kinematics, (9.10), are a bit funny. Assigning a variable mass xM to the parton is of course out of the question. Clearly, if the parton's momentum is xp, its energy can only be xE if we put $m = M = 0$. Equivalently, a proton can only emit a parton moving parallel to it ($p_T = 0$ for both) if they both have zero mass. A corollary to this statement is that if a massive particle decays, there is a nonzero angle between its decay products. We justify our previous calculation by working in a Lorentz frame where

$$|\mathbf{p}| \gg m, M, \tag{9.16}$$

so that all masses can be neglected. In this frame, where the proton is moving with infinite momentum, the kinematics of (9.10) and the structure of $F_{1,2}(x)$ given by (9.13) and (9.14) become exact. In this frame, relativistic time dilation slows down the rate at which partons interact with one another; that is, during the short time the virtual photon interacts with the quark [see (9.7)], it is essentially a free particle, not interacting with its friends in the proton. We implicitly used this *incoherence* assumption in the derivation of $F_{1,2}(x)$; (9.7) represents an addition of probabilities (not amplitudes) of scattering from *single* free partons. It is the analogue of the impulse approximation in nuclear physics. However, there is a difference. A struck nucleon can escape from the nucleus as a completely free particle, but the struck colored parton has to recombine with the noninteracting spectator partons to form the colorless hadrons into which the proton breaks up. This has to happen with probability 1, because of color confinement (see Chapter 1); and, due to the size of the proton, it requires a much longer time scale than the quick punch the parton receives from the virtual photon. In summary, we argue that in a hard collision, the parton recoils as if it were free enabling the ep → eX cross section to be calculated [see (8.43), (9.13), and (9.14)] and that the subsequent confining final state interactions do not affect the result. This picture is valid when both the Q^2 of the virtual photon and the invariant mass of the final-state hadronic system, W, are large.

EXERCISE 9.2 Convince yourself that the interaction time is much shorter than the time scale over which the partons inside the target interact with one another. Read the explicit derivation in J. D. Bjorken and E. A. Paschos, *Phys. Rev. 185*, 1975 (1969); see also Perl (1974).

EXERCISE 9.3 It is helpful to work through an alternative derivation of the parton model result, (9.13)–(9.15), in terms of the invariant variables of (6.29):

$$s \simeq 2k \cdot p, \qquad u \simeq -2k' \cdot p, \qquad t \equiv -Q^2 = -2k \cdot k',$$

with particle masses neglected. Set the parton momentum $\hat{p} = xp$, that is, neglect its component transverse to the proton momentum p. We outline the steps below. Using the basic idea of (9.7), we can write

$$\left(\frac{d\sigma}{dt\,du}\right)_{\mathrm{ep}\to\mathrm{eX}} = \sum_i \int dx\, f_i(x) \left(\frac{d\sigma}{dt\,du}\right)_{\mathrm{eq}_i\to\mathrm{eq}_i}, \tag{9.17}$$

that is, the ep → eX rate is simply the incoherent sum over all the contributing partons. Show that the invariant variables for the parton subprocess are given by

$$\hat{s} = xs, \qquad \hat{u} = xu, \qquad \hat{t} = t.$$

Use these relations, together with the eμ scattering amplitude of (6.30), to show that

$$\left(\frac{d\sigma}{dt\,du}\right)_{\mathrm{eq}_i \to \mathrm{eq}_i} = x\frac{d\sigma}{d\hat{t}\,d\hat{u}} = x\frac{2\pi\alpha^2 e_i^2}{t^2}\left(\frac{s^2+u^2}{s^2}\right)\delta(t + x(s+u)). \tag{9.18}$$

Now, also express the left-hand side of (9.17) in terms of s, t, and u. It is simplest to use (8.31). Verify that

$$\left(\frac{d\sigma}{dt\,du}\right)_{\mathrm{ep}\to\mathrm{eX}} = \frac{4\pi\alpha^2}{t^2 s^2}\frac{1}{s+u}\left[(s+u)^2 xF_1 - usF_2\right], \tag{9.19}$$

where $F_1 \equiv MW_1$ and $F_2 \equiv \nu W_2$. Insert (9.18) and (9.19) into (9.17). Compare coefficients of us and $s^2 + u^2$ and so obtain the master formula of the parton model:

$$\boxed{2x\,F_1(x) = F_2(x) = \sum_i e_i^2 x\, f_i(x).}$$

As before, we see that $F_{1,2}$ are functions only of the scaling variable x, here fixed by the delta function in (9.18):

$$x = \frac{-t}{s+u} = \frac{Q^2}{2M\nu}. \tag{9.20}$$

EXERCISE 9.4 In Chapter 8, we evaluated the ep → eX cross section for the electron to be scattered into the $dE'\,d\Omega$ element in the target proton rest frame (the laboratory frame). Show that

$$dE'\,d\Omega = \frac{\pi}{EE'}dQ^2\,d\nu = \frac{2ME}{E'}\pi y\,dx\,dy, \tag{9.21}$$

where x and y are the dimensionless variables

$$x = \frac{Q^2}{2M\nu}, \qquad y = \frac{p\cdot q}{p\cdot k} \underset{\text{(lab.)}}{=} \frac{\nu}{E}, \tag{9.22}$$

see (8.30). The allowed kinematic region ($0 \le x, y \le 1$) is shown in Fig. 9.3.

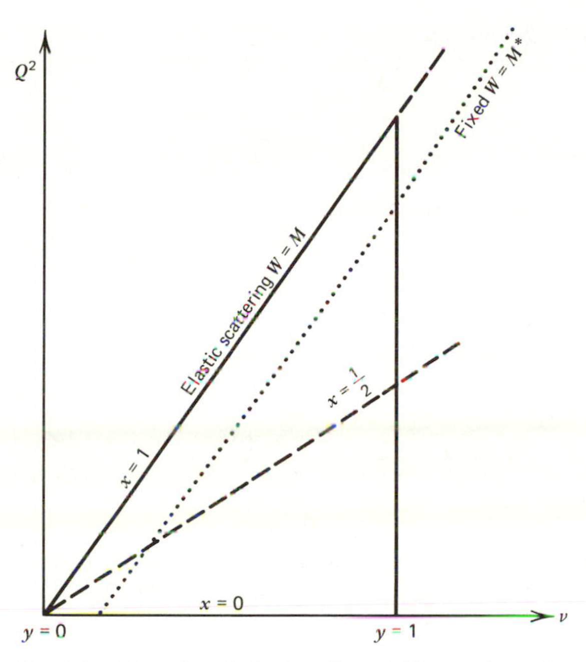

Fig. 9.3 The triangle is the allowed kinematic region for ep → eX. $\nu_{max} = E$ in the laboratory frame. W is the invariant mass of the hadronic state X, see (8.29).

Verify that the ep → eX cross section may be written in the invariant form

$$M\nu_{max}\frac{d\sigma}{dx\,dy} = \frac{2\pi\alpha^2}{x^2y^2}\left\{xy^2F_1 + \left[(1-y) - \frac{Mxy}{2\nu_{max}}\right]F_2\right\}, \tag{9.23}$$

where $\nu_{max} = E$ in the laboratory frame.

EXERCISE 9.5 Use (9.23) to show that the parton model predicts

$$\left(\frac{d\sigma}{dx\,dy}\right)_{ep\to eX} = \frac{2\pi\alpha^2}{Q^4}s\left[1 + (1-y)^2\right]\sum_i e_i^2 x\, f_i(x) \tag{9.24}$$

if particle masses are neglected.

EXERCISE 9.6 Show that

$$1 - y = \frac{p\cdot k'}{p\cdot k} \simeq \frac{1}{2}(1 + \cos\theta), \tag{9.25}$$

where θ is the scattering angle in the electron–quark center-of-mass frame.

Hence, identify the $(1 - y)^2$ and 1 terms in (9.24) with scattering between an electron and quark with opposite helicities and with the same helicity, respectively.

The parton model result $2xF_1 = F_2$ is known as the Callan–Gross relation. It is a consequence of the quarks having spin $\frac{1}{2}$ and is well borne out by the data.

EXERCISE 9.7 In the limit ν, $Q^2 \to \infty$, with x fixed, show that the Callan–Gross relation implies that the virtual photon–quark cross sections of (8.53), (8.54) satisfy

$$\frac{\sigma_L}{\sigma_T} \to 0. \tag{9.26}$$

EXERCISE 9.8 Starting from (6.51), show that if quarks had spin 0, $F_2(x)$ would still be given by (9.13) but that $F_1(x) = 0$ and hence $\sigma_T = 0$.

Thus, in contrast to (9.26), spin-0 quarks would yield $\sigma_T/\sigma_L = 0$. We can understand this difference by glancing at Fig. 9.4, which shows the head-on collision between the quark and the virtual photon. By conservation of J_z (with z along $\mathbf{p}$), we see that a spin-0 quark cannot absorb a photon of helicity $\lambda = \pm 1$, so $\sigma_T = 0$. Suppose now the quark has spin $\frac{1}{2}$. We recall that its helicity is conserved in a high-energy interaction (see Section 6.6). This can only be achieved by a $\lambda = \pm 1$ photon; hence, $\sigma_L \to 0$ in this case.

9.3 The Quarks Within the Proton

Does the photon see the proton structure described in Chapter 1, that is, does virtual photon–proton scattering look like Fig. 9.5? Just as measurements of elastic form factors provided us with information on the size of the proton, measurements on the inelastic structure function at large Q^2 reveal the quark structure of the proton. After establishing Bjorken scaling, telling us that the constituents are there, eqs. (9.13) and (9.14) become the tools for extracting further information. The sum in (9.13) runs over the charged partons in the

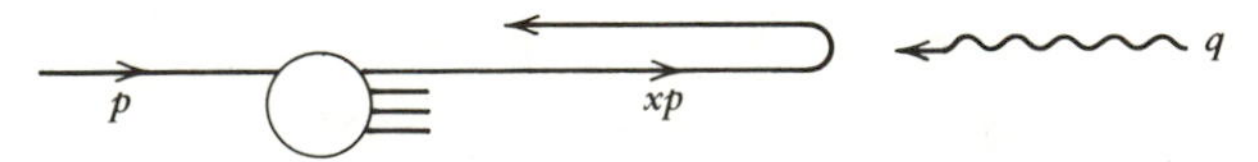

Fig. 9.4 Head-on collision between a constituent quark and the virtual photon.

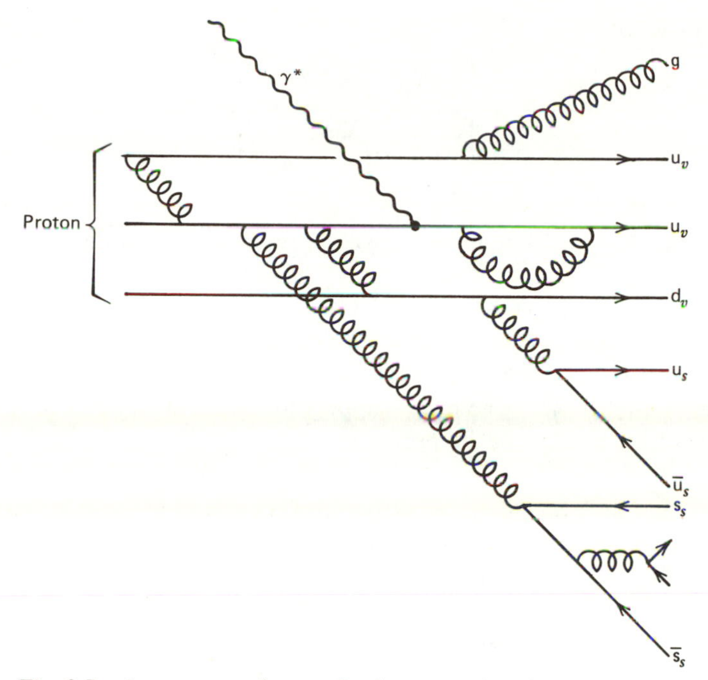

Fig. 9.5 A proton made up of valence quarks, gluons, and slow debris consisting of quark–antiquark pairs.

proton:

$$\frac{1}{x}F_2^{ep}(x) = \left(\frac{2}{3}\right)^2[u^p(x) + \bar{u}^p(x)] + \left(\frac{1}{3}\right)^2[d^p(x) + \bar{d}^p(x)]$$

$$+\left(\frac{1}{3}\right)^2[s^p(x) + \bar{s}^p(x)], \tag{9.27}$$

where $u^p(x)$ and $\bar{u}^p(x)$ are the probability distributions of u quarks and antiquarks within the proton. We have neglected the possibility of a sizable presence of charm and heavier quarks inside the proton.

We have six unknown quark structure functions, $f_i(x)$. However, the inelastic structure functions for neutrons are experimentally accessible by scattering electrons from a deuterium target. The counterpart to (9.27) is

$$\frac{1}{x}F_2^{en} = \left(\frac{2}{3}\right)^2[u^n + \bar{u}^n] + \left(\frac{1}{3}\right)^2[d^n + \bar{d}^n] + \left(\frac{1}{3}\right)^2[s^n + \bar{s}^n]; \tag{9.28}$$

and as the proton and neutron are members of an isospin doublet, their quark content is related. There are as many u quarks in a proton as d quarks in a neutron, and so on. So

$$\begin{aligned} u^p(x) &= d^n(x) \equiv u(x), \\ d^p(x) &= u^n(x) \equiv d(x), \\ s^p(x) &= s^n(x) \equiv s(x). \end{aligned} \tag{9.29}$$

EXERCISE 9.9 Show that the above expressions lead to the bounds

$$\frac{1}{4} \leq \frac{F_2^{en}(x)}{F_2^{ep}(x)} \leq 4$$

whatever the value of x. The lower (upper) limit would be realized if only u (d) quarks were present in the proton.

Further constraints on the quark structure functions $f_i(x)$ result from the fact that the quantum numbers of the proton must be exactly those of the uud combination of "valence" quarks of Chapter 2. As shown in Fig. 9.5, we describe the proton as three-constituent or three-valence quarks $u_v u_v d_v$ accompanied by many quark–antiquark pairs $u_s\bar{u}_s$, $d_s\bar{d}_s$, $s_s\bar{s}_s$, and so on. These are known as "sea" quarks. If we picture them as being radiated by the valence quarks, as in Fig. 9.5, then as a first approximation we may assume that the three lightest flavor quarks (u, d, s) occur in the "sea" with roughly the same frequency and momentum distribution, and neglect the heavier flavor quark pairs $c_s\bar{c}_s$, and so on. This picture of the proton can then be summarized as follows:

$$u_s(x) = \bar{u}_s(x) = d_s(x) = \bar{d}_s(x) = s_s(x) = \bar{s}_s(x) = S(x), \tag{9.30a}$$

$$u(x) = u_v(x) + u_s(x), \tag{9.30b}$$

$$d(x) = d_v(x) + d_s(x), \tag{9.30c}$$

where $S(x)$ is the sea quark distribution common to all quark flavors. Clearly, the heavier strange quarks are penalized in the radiation process by some threshold suppression so that (9.30a) is only approximately true.

By summing over all contributing partons, we must recover the quantum numbers of the proton: charge 1, baryon number 1, strangeness 0. It follows that

$$\begin{aligned} \int_0^1 [u(x) - \bar{u}(x)]\, dx &= 2, \\ \int_0^1 [d(x) - \bar{d}(x)]\, dx &= 1, \\ \int_0^1 [s(x) - \bar{s}(x)]\, dx &= 0. \end{aligned} \tag{9.31}$$

These sum rules express the requirement that the net number of each kind of valence quark corresponds to the uud combination of constituents discussed in

Chapter 2. Equation (9.31) clearly follows from (9.30) with

$$u - \bar{u} = u - \bar{u}_s = u - u_s = u_v,$$
$$d - \bar{d} = d - \bar{d}_s = d - d_s = d_v,$$
$$s - \bar{s} = s_s - \bar{s}_s = 0.$$

Note however that the sum rules are true in any picture where the sea is taken to be made of quark–antiquark pairs, and so does not affect the quantum numbers of the proton, which are exclusively determined by the valence quarks, as expressed by (9.31).

Combining (9.30) with (9.27) and (9.28), we obtain

$$\frac{1}{x}F_2^{ep} = \frac{1}{9}[4u_v + d_v] + \frac{4}{3}S,$$
$$\frac{1}{x}F_2^{en} = \frac{1}{9}[u_v + 4d_v] + \frac{4}{3}S, \tag{9.32}$$

where $\frac{4}{3}$ is the sum of e_i^2 over the six sea quark distributions. Since gluons create the $q\bar{q}$ pairs in the sea, we expect $S(x)$ to have a bremsstrahlung-like spectrum at small x, so that the number of sea quarks grows logarithmically as $x \to 0$ (see Exercise 9.10).

EXERCISE 9.10 Assume that the virtual photon–proton total cross section of Section 8.5 behaves like a constant as $x \to 0$, $\nu \to \infty$ for fixed Q^2, and hence show that

$$f_i(x) \xrightarrow[x \to 0]{} \frac{1}{x}. \tag{9.33}$$

Thus, we have a logarithmic growth of partons at small x.

When probing the small-momentum ($x \approx 0$) debris of the proton, we therefore anticipate that the presence of the three valence quarks will be overshadowed by these multiple, low-momentum $q\bar{q}$ pairs that make up the sea $S(x)$. According to (9.32), this means that

$$\frac{F_2^{en}(x)}{F_2^{ep}(x)} \xrightarrow[x \to 0]{} 1. \tag{9.34}$$

This is indeed the case experimentally, as can be checked from the data in Fig. 9.6. On the other hand, when probing the large-momentum part of the proton structure ($x \approx 1$), the fast-valence quarks u_v, d_v leave little momentum unoccupied for sea pairs. In this limit, the valence quarks dominate in (9.32), and therefore

$$\frac{F_2^{en}(x)}{F_2^{ep}(x)} \xrightarrow[x \to 1]{} \frac{u_v + 4d_v}{4u_v + d_v}. \tag{9.35}$$

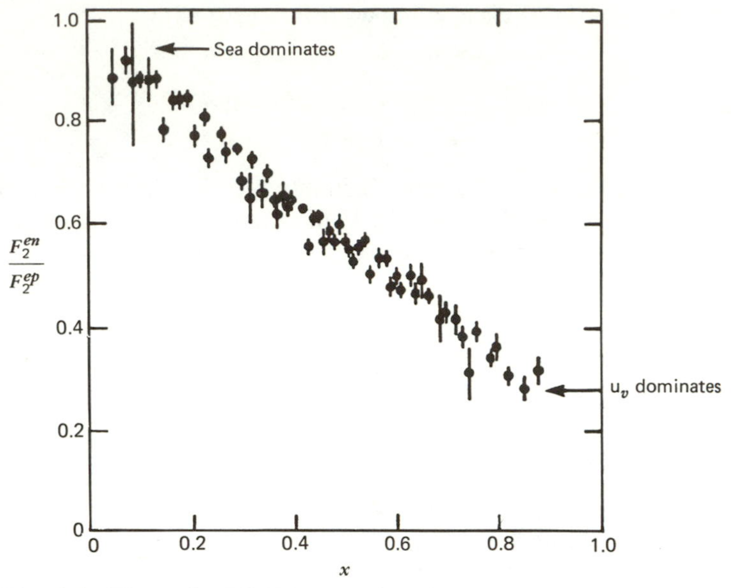

Fig. 9.6 The ratio F_2^{en}/F_2^{ep} as a function of x, measured in deep inelastic scattering. Data are from the Stanford Linear Accelerator.

For the proton, there is evidence that $u_v \gg d_v$ at large x, and the ratio (9.35) tends to $\frac{1}{4}$, as hinted at in Fig. 9.6.

EXERCISE 9.11 Discuss, on physical grounds, the behavior of $f_i(x)$ in the limit as $x \to 1$ when parton i carries all the momentum of the proton. Counting rules have been proposed which argue that

$$f_i(x) \xrightarrow[x \to 1]{} (1 - x)^{2n_s - 1},$$

where n_s is the number of spectator valence quarks which share between them the residual, vanishingly small momentum of the proton. Contrast the $x \to 1$ behavior of $u^p(x)$ with that of $u^\pi(x)$, the u-quark structure function of a π^+-meson.

What should the complete $F_2(x)$ look like according to our picture of the proton (Fig. 9.5)? Its shape can be guessed by successive approximation; see Fig. 9.7. Figure 9.7 is self-explanatory. The transition from scenario 2 to 3 is of course due to the fact that once the quarks interact, they can redistribute the momenta among themselves, and the sharply defined momentum $x = \frac{1}{3}$ is washed out, becoming a distribution of momenta peaked around $x = \frac{1}{3}$. Data on $F_2^{ep}(x)$ at large Q^2 do indeed have the general shape of scenario 4 corresponding to Fig. 9.5.

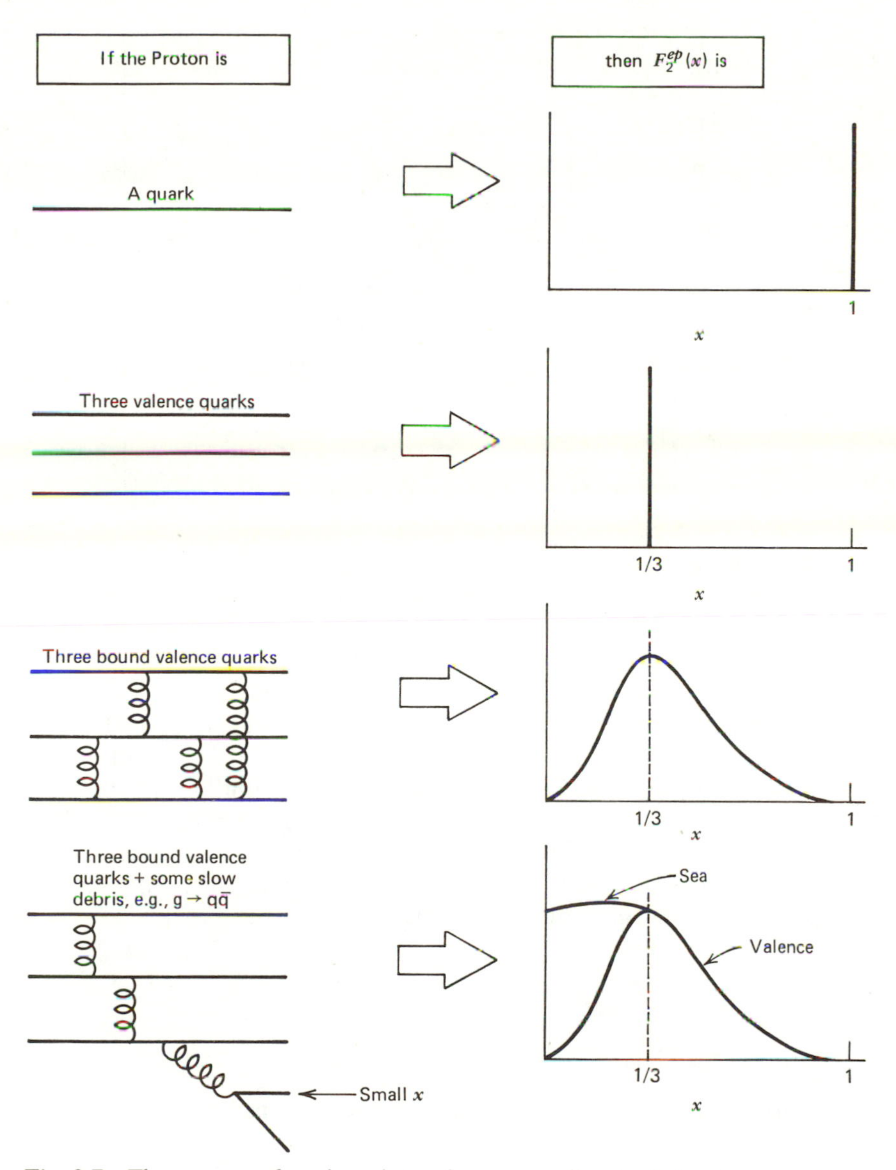

Fig. 9.7 The structure function pictured corresponding to different compositions assumed for the proton.

However, by subtracting eqs. (9.32),

$$\frac{1}{x}\left[F_2^{ep}(x) - F_2^{en}(x)\right] = \frac{1}{3}\left[u_v(x) - d_v(x)\right], \tag{9.36}$$

we can observe the valence quarks without their sea quark partners. The result should look like scenario 3 of Fig. 9.7 and should peak around $\frac{1}{3}$. It does, as can be seen in Fig. 9.8.

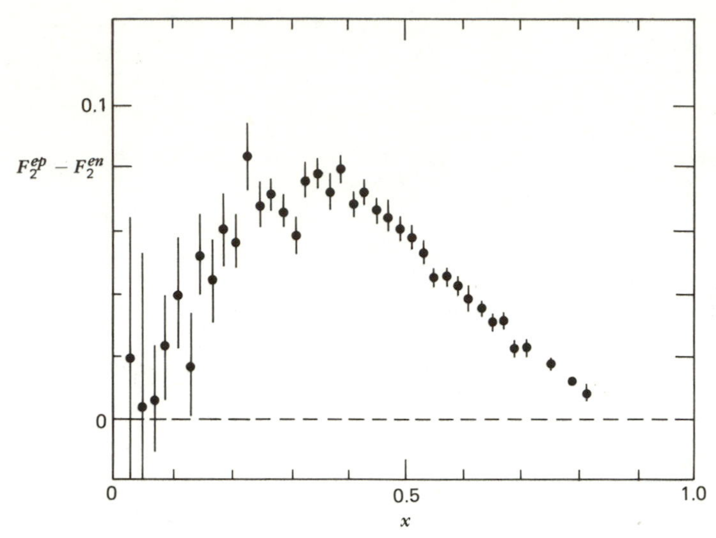

Fig. 9.8 The difference $F_2^{ep} - F_2^{en}$ as a function of x, as measured in deep inelastic scattering. Data are from the Stanford Linear Accelerator.

An alternative phenomenological approach is to parametrize all large Q^2 data on $F_2^{ep,\,en}(x)$ in terms of the valence and sea distributions and extract the quark structure functions at each x subject to the sum rules (9.31). The result of such an analysis is shown in Fig. 9.9. Fig. 9.9a shows the $\bar{q}$ distribution, which has been subjected also to assumption (9.30). We see how $u(x) = u_v(x) + u_s(x)$ approaches $\bar{u}(x)$ at small x as $xu_v(x) \to 0$. Figure 9.9b displays the general shape of the total valence and sea quark components, corresponding to scenarios 3 and 4 in Fig. 9.7. Note how slow the sea quarks are as compared to their valence partners.

9.4 Where Are the Gluons?

If we sum over the momenta of all the partons, we must reconstruct the total momentum p of the proton [see (9.8)],

$$\int_0^1 dx\,(xp)[u + \bar{u} + d + \bar{d} + s + \bar{s}] = p - p_g,$$

or, dividing by p, (9.37)

$$\int_0^1 dx\, x(u + \bar{u} + d + \bar{d} + s + \bar{s}) = 1 - \varepsilon_g.$$

The momentum fraction $\varepsilon_g \equiv p_g/p$ carried by the gluons is not directly exposed

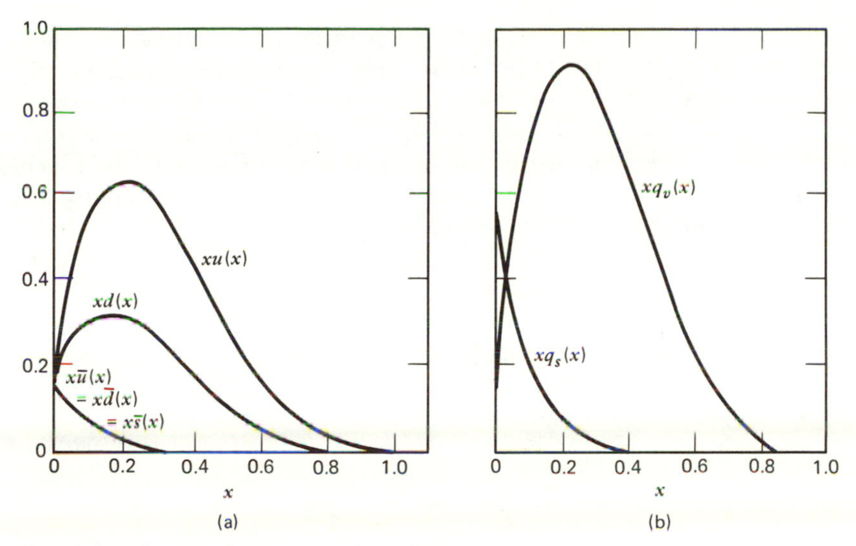

Fig. 9.9 The quark structure functions extracted from an analysis of deep inelastic scattering data. Figure (b) shows the total valence and sea quark contributions to the structure of the proton.

by the photon probe (since gluons carry no electric charge) and is therefore subtracted from the right-hand side. Integrating over the experimental data on $F_2^{ep,\,en}(x)$ gives us the following information:

$$\int dx\, F_2^{ep}(x) = \tfrac{4}{9}\varepsilon_u + \tfrac{1}{9}\varepsilon_d = 0.18,$$

$$\int dx\, F_2^{en}(x) = \tfrac{1}{9}\varepsilon_u + \tfrac{4}{9}\varepsilon_d = 0.12, \tag{9.38}$$

where

$$\varepsilon_u \equiv \int_0^1 dx\, x(u + \bar{u})$$

is the momentum carried by u quarks and antiquarks, and similarly for ε_d. Equation (9.38) follows from (9.27) and (9.28) after neglecting the strange quarks which carry a small fraction of the nucleon's momentum. From (9.37), we have

$$\varepsilon_g \simeq 1 - \varepsilon_u - \varepsilon_d,$$

and on solving (9.38), we obtain

$$\varepsilon_u = 0.36, \qquad \varepsilon_d = 0.18, \qquad \varepsilon_g = 0.46. \tag{9.39}$$

Hence, the gluons carry about 50% of the momentum, which was unaccounted for by the charged quarks.

In summary, an analysis of data on deep inelastic scattering of leptons by nucleons reveals the presence of point-like Dirac particles inside hadrons through Bjorken scaling. A study of the quantum number of these partons allows us to identify them with the quarks introduced in the study of the hadron spectrum in Chapter 2. The momentum distribution of the quarks forces us to the conclusion that a substantial fraction of the proton's momentum is carried by neutral partons, not by quarks. These are the gluons of QCD.

10

Quantum Chromodynamics

10.1 The Dual Role of Gluons

We have seen in the previous chapter that deep inelastic scattering measurements actually require the existence of electrically neutral as well as charged constituents of the proton. We concluded that the charged partons could be identified with the colored quarks postulated in Chapter 2 to explain the observed systematics of the hadron spectrum. It is tempting to identify the neutral partons with gluons. Is this identification justified? That is, does experiment actually require the existence of gluons independent of the formal arguments which were the original motivation for postulating their existence?

Recall that the color charge of quarks was originally introduced to remedy a statistics problem in constructing the Δ^{++} wavefunction (see Section 2.11). It is interesting to reflect on the fact that although it is economical to associate the same color charge with the charge of the strong interaction and although it is helpful that such a force, mediated by the exchange of gluons, is asymptotically free (see Sections 1.3 and 7.9) so that we can apply perturbation theory, none of these arguments constitutes direct evidence that quantum chromodynamics (QCD) is the correct physical theory.

It is clearly crucial to check that the gluons, introduced in Chapters 1 and 2, can indeed be identified with the neutral partons discovered as "missing momentum" in deep inelastic scattering. Subsequently, we can ask more probing questions and verify that their dynamical properties correspond to those of the carriers of the color force.

In order to proceed, it is not necessary at this stage to formally develop QCD as a color gauge theory. It is sufficient to recall the essential properties of the theory. We have already discussed them in Chapters 1 and 2.

- Quarks carry color as well as electric charge; there are three colors, R, G, and B.
- Color is exchanged by eight bicolored gluons (see Fig. 1.4).
- Color interactions are assumed to be "a copy of electromagnetic interactions." To be precise, quark–gluon interactions are computed by the rules of QED with the substitution $\sqrt{\alpha} \rightarrow \sqrt{\alpha_s}$ at each vertex (see Fig. 1.9) and

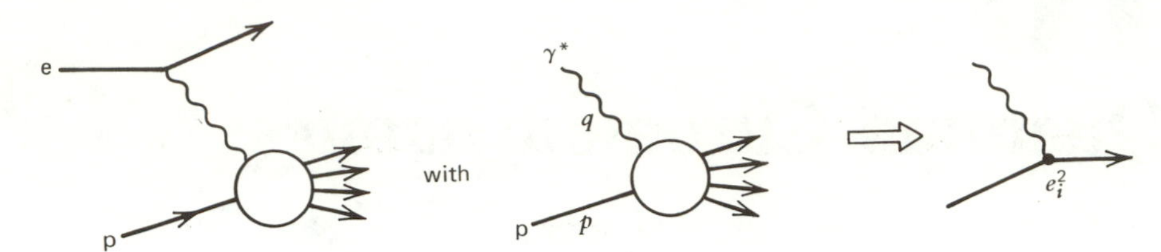

Fig. 10.1 Pictorial representation of the parton model for ep → eX.

the introduction of a color factor, which may be computed by the methods of Section 2.15. That is, the qqg vertex has the same structure as the eeγ vertex. The (eight) gluons are massless and have spin 1.

- Gluons themselves carry color charge, and so they can interact with other gluons. That is, there is a ggg as well as a qqg vertex in the theory (see Fig. 1.4d).
- At short distances, α_s is sufficiently small so that we can compute color interactions using the perturbative techniques familiar from QED.

The crucial statement that "color interactions are a copy of electromagnetic interactions" will be given a formal meaning in Chapter 14, where both QED and QCD are shown to be a consequence of local gauge symmetries.

How does the color dynamics of the partons (quarks *and* gluons) affect our discussion of deep inelastic scattering in the two previous chapters? The parton model of Chapter 9, symbolically represented by Fig. 10.1, completely ignores the dynamical role of gluons as the carriers of the strong force associated with colored quarks. We have, for instance, neglected the fact that quarks can radiate gluons. We must therefore allow for the possibility that the quark in Fig. 10.1 may radiate a gluon before or after being struck by the virtual photon, γ^*. These possibilities are shown in Fig. 10.2. Moreover, a gluon constituent in the target can contribute to deep inelastic scattering via $\gamma^* g \to q\bar{q}$ pair production as shown in Fig. 10.3. In a computational sense, the processes in Figs. 10.2 and 10.3 are $O(\alpha\alpha_s)$ contributions to the cross section, whereas the leading contribution in Fig. 10.1 is $O(\alpha)$.

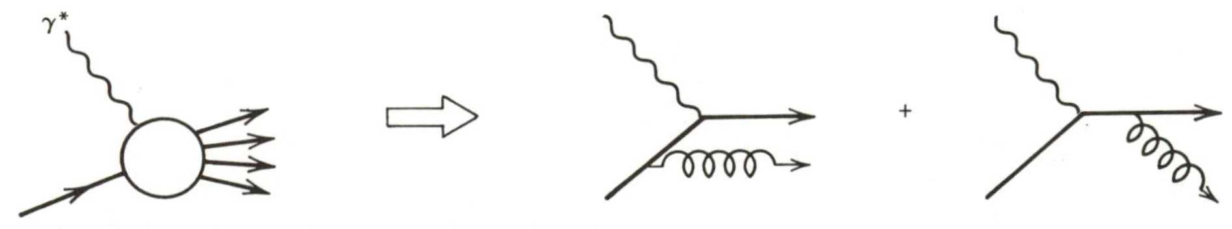

Fig. 10.2 $O(\alpha\alpha_s)$ contributions ($\gamma^* q \to qg$) to ep → eX.

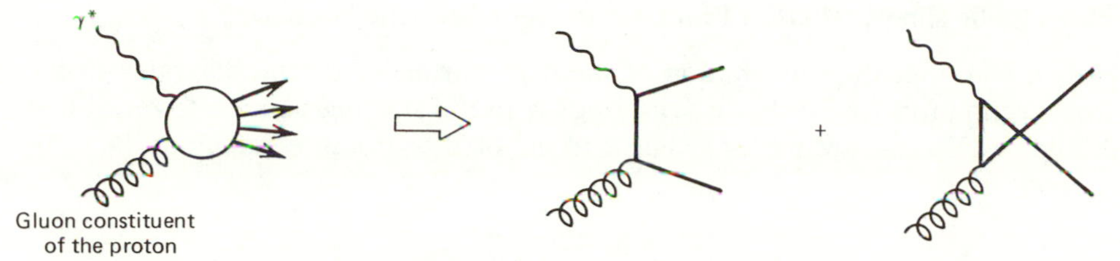

Fig. 10.3 $O(\alpha\alpha_s)$ gluon-initiated "hard" scattering contributions ($\gamma^* g \to q\bar{q}$) to ep $\to$ eX.

The inclusion of QCD diagrams of the type shown in Figs. 10.2 and 10.3 has two experimentally observable consequences: (1) the scaling property of the structure functions will no longer be true, and (2) the outgoing quark (and therefore the direction of its hadron jet) will no longer be collinear with the virtual photon. Point (2) is visualized in Fig. 10.4. In the parton model of Chapter 9, final-state hadrons in the jet of the struck quark are produced in the direction of the virtual photon (which is experimentally measured) as sketched in Fig. 10.4a. But if gluons are emitted (Fig. 10.4b), the quark can recoil against a radiated gluon and two jets are produced (indicated by arrows) each of which has a transverse momentum p_T relative to the virtual photon.

Quark–gluon color theory (QCD) allows us to compute the contribution from the diagrams of Figs. 10.2 and 10.3. Using QCD, we can therefore predict the scaling violations as well as the jet angular distributions relative to the virtual photon. They can then be compared with experiment and so the quark–gluon dynamics can be confronted with data in a quantitative and completely dynamical sense.

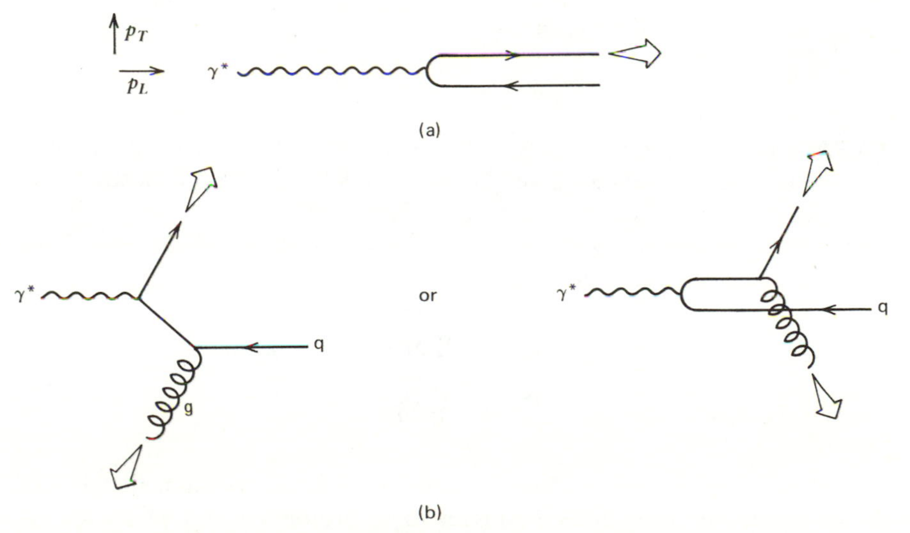

Fig. 10.4 (a) Parton model diagram for $\gamma^* q \to q$, producing a jet with $p_T = 0$. (b) Gluon emission diagrams which produce jets with $p_T \neq 0$.

10.2 Embedding γ^*-Parton Processes in Deep Inelastic Scattering

How do we find the contribution of these γ^*-parton scattering diagrams to the cross section for deep inelastic scattering? A useful starting point is to recall that the ep $\to$ eX cross section is given in terms of the structure functions $W_{1,2}$ or, equivalently,

$$F_1 = MW_1(\nu, Q^2),$$
$$F_2 = \nu W_2(\nu, Q^2), \tag{10.1}$$

see (8.34). Since we are going beyond the parton model, we shall find $F_{1,2}$ no longer scale; that is, they are functions of both

$$\nu \equiv \frac{p \cdot q}{M} \quad \text{and} \quad Q^2 = -q^2, \tag{10.2}$$

rather than simply the ratio $x = Q^2/2M\nu$. In Chapter 8, we interpreted the structure functions in terms of the virtual photon–proton cross sections; see (8.53) and (8.54). In the deep inelastic limit, these relations simplify to

$$2F_1 = \frac{\sigma_T}{\sigma_0} \tag{10.3}$$

$$\frac{F_2}{x} = \frac{\sigma_T + \sigma_L}{\sigma_0} \tag{10.4}$$

where σ_T and σ_L are the γ^*p total cross sections for transverse and longitudinal virtual photons, respectively, and

$$\sigma_0 \equiv \frac{4\pi^2\alpha}{2MK} \simeq \frac{4\pi^2\alpha}{s}. \tag{10.5}$$

EXERCISE 10.1 Derive (10.3)–(10.5). It is easiest to work in the laboratory frame; the variables are given in Section 6.8. In the deep inelastic limit, the electron beam energy $E \gg E'$. Then, as $\nu^2 \sim E^2$ and $Q^2 \simeq 4EE'\sin^2(\theta/2)$, we have $\nu^2 \gg Q^2$. We also have introduced the γ^*-proton center-of-mass energy squared:

$$s = (q + p)^2 = M^2 + 2M\nu - Q^2 \simeq 2MK, \tag{10.6}$$

where K is associated with the γ^*-flux factor of (8.48).

In calculating the γ^*-parton contributions to σ_T and σ_L, we immediately face a problem. Equations (10.3) and (10.4) refer to γ^*-proton and not γ^*-parton cross sections. For example, consider single gluon emission (γ^*q $\to$ qg), which is shown

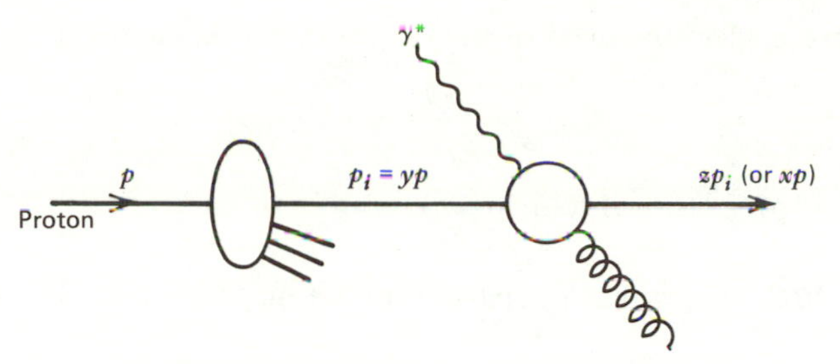

Fig. 10.5 Embedding $\gamma^*q \to qg$ in $\gamma^*p \to X$.

in Fig. 10.5. The relation between the two frames is given by

γ*-Proton Frame		**γ*-Parton Frame**
p	$\longrightarrow$	$p_i = yp$
$x = \dfrac{Q^2}{2p \cdot q}$	$\longrightarrow$	$z = \dfrac{Q^2}{2p_i \cdot q} = \dfrac{x}{y}$

Here, we have relied extensively on the fact that both collinear frames move with infinite momentum.

We are now ready to relate cross section ratios in the two systems. For instance, for the ratio (10.3), we have

$$\left(\frac{\sigma_T(x, Q^2)}{\sigma_0}\right)_{\gamma^* p} = \sum_i \int_0^1 dz \int_0^1 dy\, f_i(y)\, \delta(x - zy) \left(\frac{\hat{\sigma}_T(z, Q^2)}{\hat{\sigma}_0}\right)_{\gamma^* i} \tag{10.7}$$

where $f_i(y)$ are the parton structure functions (which give the probability that there is a parton i carrying a fraction y of the proton's momentum p) and $\hat{\sigma}_T$ is the cross section for the absorption of a transverse photon of momentum q by a parton of momentum p_i; x is fixed, and we have to integrate over all z, y subject to the constraint $x = zy$. Using the delta function to perform the z integration, we obtain

$$\boxed{\left(\frac{\sigma_T}{\sigma_0}(x, Q^2)\right) = \sum_i \int_x^1 \frac{dy}{y} f_i(y) \left(\frac{\hat{\sigma}_T}{\hat{\sigma}_0}(x/y, Q^2)\right).} \tag{10.8}$$

Throughout this chapter, we use $\hat{\sigma}$, $\hat{s}$, $\hat{t}$, and so on, to distinguish γ^*-parton quantities from those of the parent process, σ, s, t, and so on.

10.3 The Parton Model Revisited

If all gluon effects were absent, we should be able to recover the parton model result from (10.8). In this case, $\hat{\sigma}_T$ and $\hat{\sigma}_L$ are given by the diagram of Fig. 10.1,

γ^*q $\to$ q. If we neglect the mass of the outgoing quark, $(q + p_i)^2 = 0$. Hence,

$$z = \frac{Q^2}{2p_i \cdot q} = 1, \tag{10.9}$$

and $\hat{\sigma}(z, Q^2)$ is proportional to $\delta(1 - z)$.

EXERCISE 10.2 Show that the parton model diagram, Fig. 10.1, gives

$$\frac{\hat{\sigma}_T(z, Q^2)}{\hat{\sigma}_0} = e_i^2 \delta(1 - z) \tag{10.10}$$

$$\hat{\sigma}_L(z, Q^2) = 0. \tag{10.11}$$

We outline the various stages of the calculation. First, show that for $\gamma^*(q)\,\text{q}(p) \to \text{q}(p')$,

$$\overline{|\mathcal{M}|^2} = 2e_i^2 e^2 p \cdot q, \tag{10.12}$$

where we have averaged over transverse polarization states of the incoming γ^*. From Section 4.3, we have

$$F\,d\hat{\sigma}_T = \overline{|\mathcal{M}|^2}(2\pi)^4\,\delta^{(4)}(p' - p - q)\frac{d^3p'}{2p_0'(2\pi)^3}. \tag{10.13}$$

where F is the γ^*q flux factor. Calculate $F\hat{\sigma}_T$ by making use of (6.47). Use $F\hat{\sigma}_0 = 8\pi^2\alpha$, see (10.5).

To determine the parton model prediction for F_2/x of (10.4), we input in (10.8) the following cross section ratio for γ^*q $\to$ q:

$$\frac{1}{\hat{\sigma}_0}(\hat{\sigma}_T + \hat{\sigma}_L) = e_i^2\,\delta(1 - z), \tag{10.14}$$

see (10.10) and (10.11). After substitution, we obtain

$$\frac{F_2(x, Q^2)}{x} = \sum_i e_i^2 \int_x^1 \frac{dy}{y} f_i(y)\,\delta\left(1 - \frac{x}{y}\right) = \sum_i e_i^2 f_i(x). \tag{10.15}$$

An identical expression is found for $2F_1$. The parton model results of (9.13) and (9.14) are indeed reproduced.

10.4 The Gluon Emission Cross Section

We are now ready to include the gluon emission diagrams of Fig. 10.2. To calculate γ^*q $\to$ qg, we take over our results for Compton scattering. In Section

6.14, we evaluated the closely related QED process $\gamma^* e \to \gamma e$ and found

$$\overline{|\mathfrak{M}|^2} = 32\pi^2\alpha^2\left(-\frac{u}{s} - \frac{s}{u} + \frac{2tQ^2}{su}\right), \tag{10.16}$$

using for the γ^*-polarization sum $\Sigma\varepsilon_\mu^*\varepsilon_\nu = -g_{\mu\nu}$, see (6.98). The result for $\gamma^* q \to qg$ follows directly on substitution of $\alpha^2 \to e_i^2\alpha\alpha_s$ and insertion of the color factor $\frac{4}{3}$, and $u \leftrightarrow t$ on account of the different ordering of the outgoing particles; so,

$$\overline{|\mathfrak{M}|^2} = 32\pi^2(e_i^2\alpha\alpha_s)\frac{4}{3}\left(-\frac{\hat{t}}{\hat{s}} - \frac{\hat{s}}{\hat{t}} + \frac{2\hat{u}Q^2}{\hat{s}\hat{t}}\right). \tag{10.17}$$

The invariant variables are denoted $\hat{s}$, $\hat{t}$, $\hat{u}$ to indicate that we are considering a parton subprocess.

The factor $\frac{4}{3}$ takes into account the summation/averaging over final/initial colors. It can readily be deduced from counting the color lines shown in Fig. 10.6. Each color line can take one of three possible colors, so there are $(3 \times 3 - 1)$ different configurations, if we subtract the color singlet configuration as explained in Section 2.15. We now average over the three initial quark colors and obtain $\frac{8}{3}$. However, this result has to be divided by a factor 2 because of the unfortunate historical definition of α_s (see Section 2.15).

Most simple QCD diagrams are exactly analogous to QED diagrams, and the QCD cross section is obtained by the replacement

$$\alpha^n \to C_F\alpha_s^n, \tag{10.18}$$

where n is the number of quark–gluon (or gluon–gluon) vertices in the diagram. The color factor C_F is obtained by simply summing and averaging in much the same way as we do for spin. However, without changing the cross section, the color factor can be altered by a redefinition of α_s. For example,

$$\alpha_s^n C_F \to (2\alpha_s)^n(C_F/2^n) = (\alpha_s')^n C_F'. \tag{10.19}$$

The conventional definition of α_s corresponds to the second convention in (10.19). The final rule to obtain the conventional color factor is simple: "count color on your fingers" and divide the result by 2^n. This is why we divided the color factor $\frac{8}{3}$ of Fig. 10.6 by 2^1.

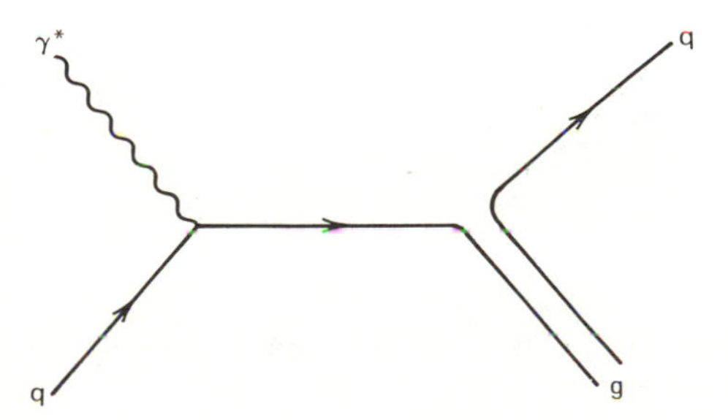

Fig. 10.6 Color lines for $\gamma^* q \to qg$.

Now back to the main problem. In preparation for the calculation of the $\gamma^* q \to qg$ cross section from (10.17), we gather together the relevant kinematics.

EXERCISE 10.3 In the center-of-mass frame of the parton process $\gamma^* q_1 \to q_2 g$ of Fig. 10.7, show that

$$\hat{s} = 2k^2 + 2kq_0 - Q^2 = 4k'^2, \tag{10.20}$$

$$\hat{t} = -Q^2 - 2k'q_0 + 2kk'\cos\theta = -2kk'(1 - \cos\theta), \tag{10.21}$$

$$\hat{u} = -2kk'(1 + \cos\theta), \tag{10.22}$$

where k, k' are the magnitudes of the center-of-mass momenta $\mathbf{k}, \mathbf{k}'$. Note that for the virtual photon, $q_0^2 = k^2 - Q^2$. A useful result is

$$4kk' = -\hat{t} - \hat{u} = \hat{s} + Q^2. \tag{10.23}$$

The interesting quantity is the transverse momentum of the outgoing quark, $p_T = k'\sin\theta$. Show that

$$p_T^2 = \frac{\hat{s}\hat{t}\hat{u}}{(\hat{s} + Q^2)^2} \tag{10.24}$$

or, in the limit of small-angle scattering, $-\hat{t} \ll \hat{s}$, that

$$p_T^2 = \frac{\hat{s}(-\hat{t})}{\hat{s} + Q^2}. \tag{10.25}$$

Further, show that for small scattering angles ($\cos\theta \simeq 1$),

$$d\Omega = \frac{4\pi}{\hat{s}} dp_T^2. \tag{10.26}$$

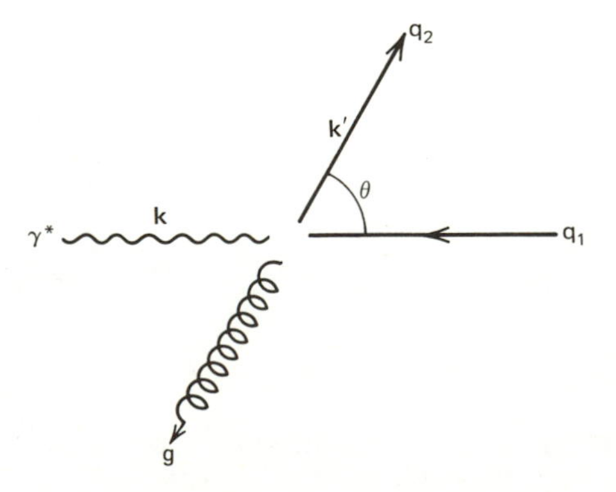

Fig. 10.7 Center-of-mass frame for $\gamma^* q_1 \to q_2 g$.

It is clear from (10.17) that at high energy ($\hat{s}$ large), the $\gamma^*q \to qg$ cross section peaks as $-\hat{t} \to 0$. Referring back to Fig. 4.8 and the related discussion, we see that this is due to quark exchange in the t channel. We can therefore approximate the cross section by its forward peak. For forward scattering, we find, from (10.26) and (4.35), that

$$\frac{d\hat{\sigma}}{dp_T^2} \simeq \frac{1}{16\pi\hat{s}^2}\overline{|\mathfrak{M}|^2}. \tag{10.27}$$

EXERCISE 10.4 Derive (10.27). Make use of (10.26) and (10.20), together with (4.34). Show that the γ^*q flux factor is given by $2\hat{s}$ using the convention of (8.48).

On substituting (10.17) into (10.27), the $\gamma^*q \to qg$ cross section becomes

$$\frac{d\hat{\sigma}}{dp_T^2} \simeq \frac{8\pi e_i^2\alpha\alpha_s}{3\hat{s}^2}\left(\frac{1}{-\hat{t}}\right)\left[\hat{s} + \frac{2(\hat{s}+Q^2)Q^2}{\hat{s}}\right], \tag{10.28}$$

where we have used $-\hat{t} \ll \hat{s}$. Using (10.25) and

$$z \equiv \frac{Q^2}{2p_i\cdot q} = \frac{Q^2}{(p_i+q)^2 - q^2} = \frac{Q^2}{\hat{s}+Q^2}, \tag{10.29}$$

equation (10.28) can finally be rewritten as

$$\boxed{\frac{d\hat{\sigma}}{dp_T^2} \simeq e_i^2\hat{\sigma}_0\frac{1}{p_T^2}\frac{\alpha_s}{2\pi}P_{qq}(z),} \tag{10.30}$$

where $\hat{\sigma}_0 = 4\pi^2\alpha/\hat{s}$, see (10.5), and where

$$P_{qq}(z) = \frac{4}{3}\left(\frac{1+z^2}{1-z}\right) \tag{10.31}$$

represents the probability of a quark emitting a gluon and so becoming a quark with momentum reduced by a fraction z. The $z \to 1$ singularity is associated with the emission of a "soft" massless gluon. It is an example of an infrared divergence (mentioned in Chapter 7). We explain how it is canceled by virtual gluon diagrams in Section 10.8.

The cross section (10.30) is also singular as $p_T^2 \to 0$. Now the diagrams of Figs. 10.1 and 10.3 either do not contribute (remember that for the parton diagram the p_T of the outgoing parton relative to the virtual photon always vanishes) or are negligible in comparison to $\gamma^*q \to qg$ in the limit $-\hat{t} \ll \hat{s}$. This is the case for the pair production diagrams of Fig. 10.3. Therefore, in the region $-\hat{t} \ll \hat{s}$, (10.30) represents the full p_T^2 distribution of the final-state parton jets.

What is the experimental signature of this result? We refer to Fig. 10.4. The presence of gluon emission is signaled by a quark jet and gluon jet in the final state, neither of which is moving along the direction of the virtual photon. The transverse momentum p_T of the jet or of the bremsstrahlung hadrons contained in this jet (see Chapter 1) is nonzero; in fact, we expect the p_T distribution to be given by (10.30). It has to be embedded into the electron–proton system, using (10.8), exactly as was done for the parton cross section in Section 10.3. Figure 10.8 shows the result of comparing such a calculation with data.

Several remarks about this comparison are in order. The calculation shown in Fig. 10.8 actually includes all diagrams in Figs. 10.1, 10.2, and 10.3, and the limit $(-\hat{t}) \ll \hat{s}$ was not made. Moreover, one has to model the way a quark or gluon fragments into hadrons or, alternatively, one can infer this information from a different experiment. How this is done is explained in the next chapter. What is important is the qualitative observation that hadrons emerge with $p_T \neq 0$ signaling the presence of gluon emission. In a parton model without gluons, all final-state jets would be collinear with the virtual photon. Their hadron fragments will therefore be nearly collinear with the photon, too, that is, with a spread of p_T of about 300 MeV as required by the uncertainty principle for confined quarks. This prediction is represented by the dashed line in Fig. 10.8. The data clearly establish an excess of large p_T hadrons which are the fragments of the quark and gluon jets recoiling against one another.

This "Rutherford experiment of QCD" can be repeated in many disguises such as $e^+e^- \to$ hadrons or pp $\to$ large-p_T hadrons (see Chapter 1). The ideas and computational techniques are very similar to the example discussed here. Finally, the large Q^2 of the photon guarantees that we are dealing with a short-distance

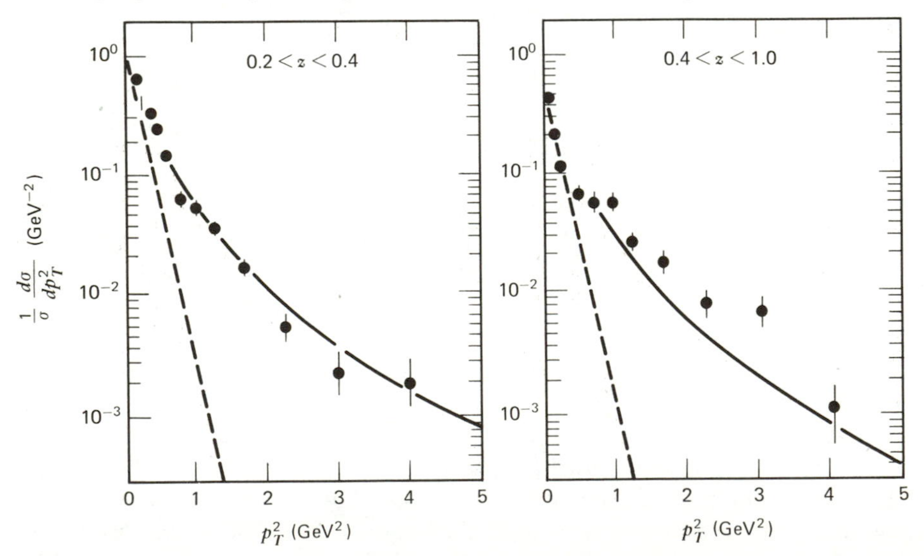

Fig. 10.8 The p_T^2 distribution of hadrons produced in μN interactions relative to the direction of the virtual photon. The dashed line is the expectation in the absence of gluon emission. Data are from the EMC collaboration at CERN. (μN and eN interactions give the same curves.)

interaction where α_s is small (see Chapters 1 and 7). The number of large-p_T hadrons in Fig. 10.8 measures the normalization of the cross section of (10.30), that is, it measures $\alpha_s(Q^2)$. At these values of Q^2 the data imply that $\alpha_s \simeq 0.2$.

10.5 Scaling Violations. The Altarelli–Parisi Equation

How does the gluon bremsstrahlung diagram of (10.30) contribute to the structure functions? Recall that the structure functions are related to parton cross sections via (10.8) [and (10.3), (10.4)]. This involves the integrated γ^*-parton cross section. We therefore have to compute

$$\begin{aligned}\hat{\sigma}(\gamma^* q \to qg) &= \int_{\mu^2}^{\hat{s}/4} dp_T^2 \frac{d\hat{\sigma}}{dp_T^2}\\ &\simeq e_i^2 \hat{\sigma}_0 \int_{\mu^2}^{\hat{s}/4} \frac{dp_T^2}{p_T^2} \frac{\alpha_s}{2\pi} P_{qq}(z)\\ &\simeq e_i^2 \hat{\sigma}_0 \left(\frac{\alpha_s}{2\pi} P_{qq}(z) \log \frac{Q^2}{\mu^2} \right). \end{aligned} \tag{10.32}$$

EXERCISE 10.5 Show that the maximal transverse momentum for the two-body interaction $\gamma^* q \to qg$ is given by

$$(p_T^2)_{\max} = \frac{\hat{s}}{4} = Q^2 \frac{1-z}{4z}. \tag{10.33}$$

In deriving (10.32), we integrated up to the maximum p_T of the gluon and then used (10.33) to write $\log(\hat{s}/4) \simeq \log Q^2$ in the large Q^2 limit. The lower limit μ on the transverse momentum is introduced as a cutoff to regularize the divergence when $p_T^2 \to 0$.

Adding $\hat{\sigma}(\gamma^* q \to qg)$ to the parton model cross section, (10.14), we find QCD modifies (10.15) to

$$\frac{F_2(x, Q^2)}{x} = \left| \text{[diagram]} \right|^2 + \left| \text{[diagram]} + \text{[diagram]} \right|^2$$

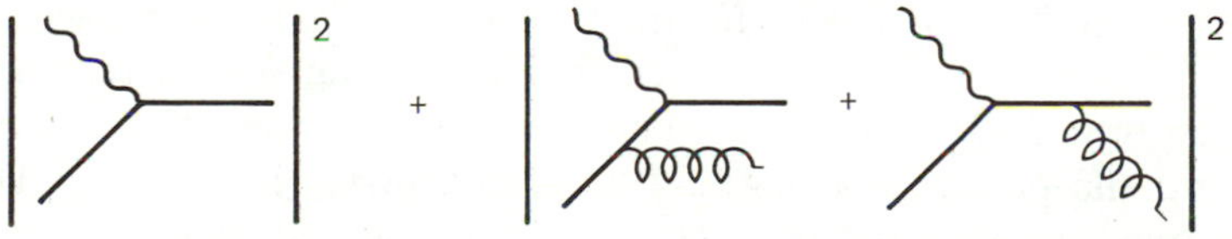

$$= \sum_q e_q^2 \int_x^1 \frac{dy}{y} q(y) \left(\delta\left(1 - \frac{x}{y}\right) + \frac{\alpha_s}{2\pi} P_{qq}\left(\frac{x}{y}\right) \log \frac{Q^2}{\mu^2} \right), \tag{10.34}$$

where we have introduced the notation that the quark structure function $q(y) \equiv$

$f_q(y)$. The presence of the $\log Q^2$ factor means that the parton model scaling prediction for the structure functions should be violated. That is, in QCD, F_2 is a function of Q^2 as well as of x, but the variation with Q^2 is only logarithmic. The violation of Bjorken scaling is a signature of gluon emission.

EXERCISE 10.6 Study the origin of the $\log Q^2$ term. Recall that the $\gamma^* q \to qg$ cross section, $d\hat{\sigma}/dp_T^2$, is dominated by the forward peak. The $\hat{t}$-channel quark propagator leads to a factor $1/p_T^4$. Show that helicity conservation at the gluon vertex weakens this singularity by introducing a factor p_T^2 in the numerator.

Equation (10.34) may be regarded as the first two terms in a power series in α_s; α_s is a useful expansion parameter at large Q^2 since $\alpha_s \sim (\log Q^2)^{-1}$. But comparing the leading and next-order terms in (10.34), we find that the expansion parameter α_s is multiplied by $\log Q^2$. From (7.65), we know that $\alpha_s(Q^2)$ $\log(Q^2/\mu^2)$ does not vanish at large Q^2, and so (10.34) does not seem very useful as it stands. How should we proceed? Can we absorb the $\log Q^2$ term into a modified quark probability distribution? To this end, we rewrite (10.34) in the "parton-like" form

$$\begin{aligned}\frac{F_2(x, Q^2)}{x} &\equiv \sum_q e_q^2 \int_x^1 \frac{dy}{y}\left(q(y) + \Delta q(y, Q^2)\right)\delta\left(1 - \frac{x}{y}\right)\\ &= \sum_q e_q^2\left(q(x) + \Delta q(x, Q^2)\right)\end{aligned} \tag{10.35}$$

where

$$\Delta q(x, Q^2) \equiv \frac{\alpha_s}{2\pi}\log\left(\frac{Q^2}{\mu^2}\right)\int_x^1 \frac{dy}{y} q(y)\, P_{qq}\left(\frac{x}{y}\right). \tag{10.36}$$

The quark densities $q(x, Q^2)$ now depend on Q^2. We interpret this as arising from a photon with larger Q^2 probing a wider range of p_T^2 within the proton.

We can picture this as follows. As Q^2 is increased to $Q^2 \sim Q_0^2$, say, the photon starts to "see" evidence for the point-like valence quarks within the proton; see Fig. 10.9a. If the quarks were noninteracting, no further structure would be resolved as Q^2 was increased and exact scaling [described by $q(x)$] would set in, and the parton model would be satisfactory. However, QCD predicts that on increasing the resolution ($Q^2 \gg Q_0^2$), we should "see" that each quark is itself surrounded by a cloud of partons. We have calculated one particular diagram, shown in Fig. 10.9b, but there are of course other diagrams with a greater number of partons. The number of resolved partons which share the proton's momentum increases with Q^2. There is an increased probability of finding a quark at small x and a decreased chance of finding one at high x, because high-momentum quarks lose momentum by radiating gluons.

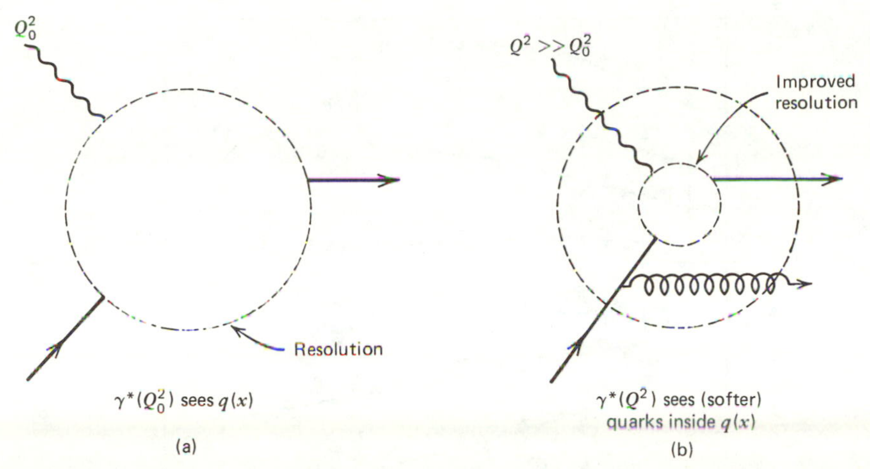

Fig. 10.9 The quark structure of the proton as seen by a virtual photon as Q^2 increases.

The Q^2 evolution of the quark densities is determined by QCD through (10.36). By considering the change in the quark density, $\Delta q(x, Q^2)$, when one probes a further interval of $\Delta \log Q^2$, (10.36) can be rewritten as an integro-differential equation for $q(x, Q^2)$:

$$\boxed{\frac{d}{d\log Q^2} q(x, Q^2) = \frac{\alpha_s}{2\pi} \int_x^1 \frac{dy}{y} q(y, Q^2)\, P_{qq}\left(\frac{x}{y}\right).} \tag{10.37}$$

This is an "Altarelli–Parisi evolution equation." The equation mathematically expresses the fact that a quark with momentum fraction x [$q(x, Q^2)$ on the left-hand side] could have come from a parent quark with a larger momentum fraction y [$q(y, Q^2)$ on the right-hand side] which has radiated a gluon. The probability that this happens is proportional to $\alpha_s P_{qq}(x/y)$. The integral in (10.37) is the sum over all possible momentum fractions $y(> x)$ of the parent.

To summarize: QCD predicts the breakdown of scaling and allows us to compute explicitly the dependence of the structure function on Q^2. Given the quark structure function at some reference point $q(x, Q_0^2)$, we can compute it for any value of Q^2 using the Altarelli–Parisi equation (10.37). The experimental results for $q(x, Q^2)$, or, to be precise, $F_2(x, Q^2)$, are displayed in Fig. 10.10. Moment analysis is often used to show that the Q^2 variation of the structure function is described by the differential equation (10.37). This procedure is purely technical and of no interest to us (see, however, Exercise 10.16). The systematics of the Q^2 dependence should be noted, however. Around $x = 0.25$ ($\omega = 4$), the structure function is found to scale, and Fig. 9.2 displays the absence of Q^2 dependence at this particular x value. But for $x \lesssim 0.25$, the structure function increases with Q^2, while for $x \gtrsim 0.25$, it decreases. Another way to state this result is that we resolve increasing numbers of "soft" quarks with increasing Q^2.

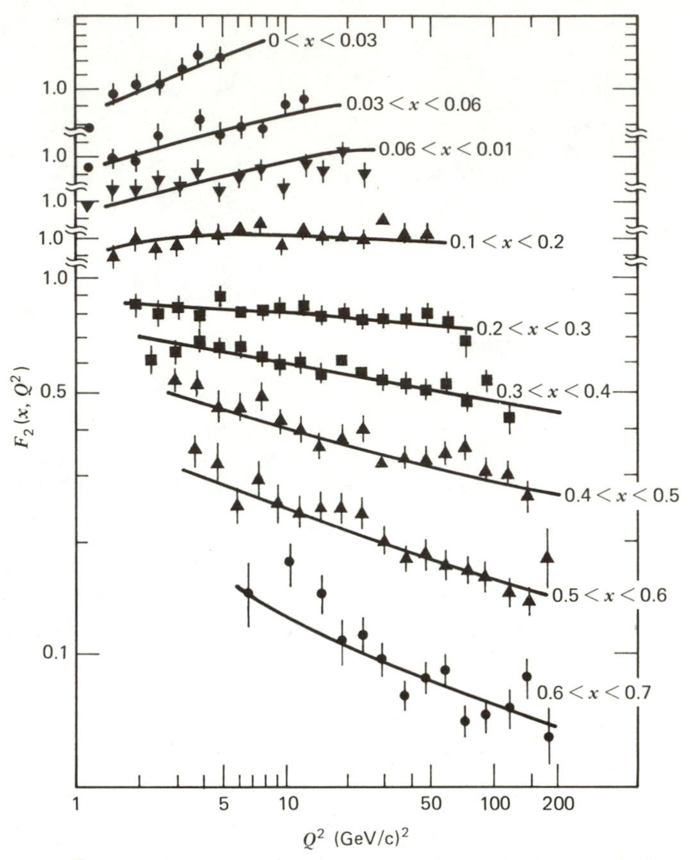

Fig. 10.10 Deviations from scaling. With increasing Q^2, the structure function $F_2(x, Q^2)$ increases at small x and decreases at large x. The data are from the CDHS counter experiment at CERN.

The large-momentum quark component ($x \simeq 1$) is depleted and shifted toward low momentum ($x \simeq 0$). This agrees with our earlier discussion that large-Q^2, high-resolution photons have more chance of seeing "softer" quarks whose momentum has been degraded by gluon emission.

EXERCISE 10.7 The origin of the scaling violation of $q(x, Q^2)$ given by (10.37) can be traced back to (10.32). There, we assumed that α_s is a constant. Show that (10.37) is also obtained for a running coupling constant. Assume that α_s in (10.32) is $\alpha_s(p_T^2)$ as given by (7.65).

Comment The reason for choosing p_T^2 as the argument of α_s cannot be exhibited without a discussion of the higher-order diagrams. However, note that we are working in a kinematic regime with two large quantities, p_T^2 and

Q^2, and the dominant region is $p_T^2 \ll Q^2$. In this limit, a discussion of higher orders introduces p_T^2 as the argument of α_s [see, for example, Reya (1981) or Dokshitzer et al. (1980)].

10.6 Including Gluon Pair Production

So far, we have only incorporated contributions to deep inelastic scattering, ep → eX, from the quark-initiated processes $\gamma^* q \to q$ and $\gamma^* q \to qg$. To order α_s, we should also include contributions where a gluon in the initial proton produces a quark–antiquark pair to which the virtual photon then couples, that is, the process $\gamma^* g \to q\bar{q}$ of Fig. 10.3. This is similar to the Compton diagrams of Section 10.4.

EXERCISE 10.8 Show that the color factor for $\gamma^* g \to q\bar{q}$ is $\frac{1}{2}$.

EXERCISE 10.9 Verify that for $\gamma^* g \to q\bar{q}$,

$$\overline{|\mathcal{M}|^2} = 32\pi^2 \left(e_q^2 \alpha \alpha_s\right) \frac{1}{2} \left(\frac{\hat{u}}{\hat{t}} + \frac{\hat{t}}{\hat{u}} - \frac{2\hat{s}Q^2}{\hat{t}\hat{u}} \right), \tag{10.38}$$

using $\sum \varepsilon_\mu^* \varepsilon_\nu = -g_{\mu\nu}$ for the γ^*-polarization sum. Hence, show that (10.34) for the proton structure function contains the additional contribution

$$\left. \frac{F_2(x, Q^2)}{x} \right|_{\gamma^* g \to q\bar{q}} = \tag{10.39}$$

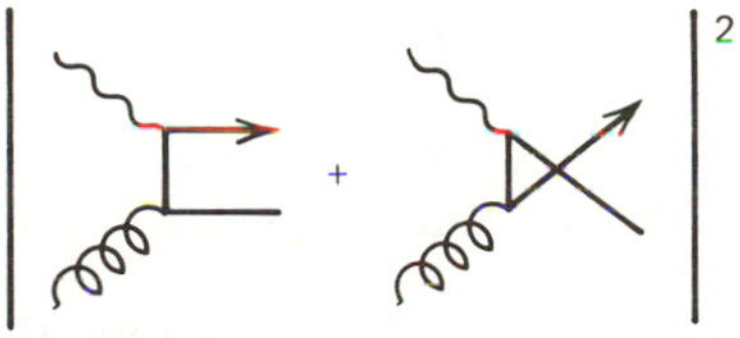

+

2

$$= \sum_q e_q^2 \int_x^1 \frac{dy}{y} g(y) \frac{\alpha_s}{2\pi} P_{qg}\left(\frac{x}{y}\right) \log \frac{Q^2}{\mu^2}, \tag{10.40}$$

where $g(y)$ is the gluon density in the proton and where

$$P_{qg}(z) = \tfrac{1}{2}\left(z^2 + (1-z)^2\right) \tag{10.41}$$

represents the probability that a gluon annihilates into a $q\bar{q}$ pair such that the quark has a fraction z of its momentum. Detailed measurements of the scaling violations of $F_2(x, Q^2)$ probe the gluon distribution inside the proton through (10.40).

10.7 Complete Evolution Equations for the Parton Densities

If we include the pair production contribution to the evolution of the quark density, (10.37) becomes

$$\frac{dq_i(x,Q^2)}{d\log Q^2} = \frac{\alpha_s}{2\pi}\int_x^1 \frac{dy}{y}\left(q_i(y,Q^2)P_{qq}\left(\frac{x}{y}\right) + g(y,Q^2)P_{qg}\left(\frac{x}{y}\right)\right) \tag{10.42}$$

for each quark flavor i. The second term, previously omitted, considers the possibility that a quark with momentum fraction x is the result of $q\bar{q}$ pair creation by a parent gluon with momentum fraction $y(> x)$. The probability is $\alpha_s P_{qg}(x/y)$. The framework is obviously still incomplete since we require an equation for the evolution of the gluon density in the proton. Repeating our previous arguments, we can give a symbolic representation of the gluon evolution equation,(10.43)

$\dfrac{d}{d\log Q^2}$ ($g(x, Q^2)$) $= \sum_i$ [$q_i(y, Q^2)$ → $g(x, Q^2)$, $P_{gq}(\frac{x}{y})$] + [$g(y, Q^2)$ → $g(x, Q^2)$, $P_{gg}(\frac{x}{y})$] (10.43)

which tells us that

$$\frac{dg(x,Q^2)}{d\log Q^2} = \frac{\alpha_s}{2\pi}\int_x^1 \frac{dy}{y}\left(\sum_i q_i(y,Q^2)P_{gq}\left(\frac{x}{y}\right) + g(y,Q^2)P_{gg}\left(\frac{x}{y}\right)\right), \tag{10.44}$$

where the sum $i = 1,\ldots,2n_f$ runs over quarks and antiquarks of all flavors. P_{gq} does not depend on the index i if the quark masses can be neglected.

EXERCISE 10.10 How would you set about verifying that

$$P_{gq}(z) = \frac{4}{3}\frac{1+(1-z)^2}{z}, \tag{10.45}$$

$$P_{gg}(z) = 6\left(\frac{1-z}{z} + \frac{z}{1-z} + z(1-z)\right)? \tag{10.46}$$

EXERCISE 10.11 Express (10.42) in symbolic form.

EXERCISE 10.12 Obtain the evolution equations for the combinations

$$q_{NS} \equiv q_i - q_j \tag{10.47}$$

$$q_S \equiv \sum_i q_i. \tag{10.48}$$

The subscripts are conventional and, in fact, are used to indicate that the combinations refer to nonsinglet and singlet combinations of the quark flavor group.

10.8 Physical Interpretation of the *P* Functions

Let us take stock of where QCD has taken us. At large Q^2, the structure functions for ep → eX are given by a "parton model-like" formula with parton densities which do not scale but evolve with Q^2 in a way which is calculable from QCD. Although we have arrived at these Q^2-dependent parton densities by studying deep inelastic scattering, these densities should simply characterize the target proton and should not depend on the nature of the probe. In other words, the parton densities are universal in the sense that they can relate the structure functions found in different processes.

The evolution is governed by the P_{qq}, P_{gq}, P_{qg}, and P_{gg} functions. The physical meaning of the P functions becomes transparent if we go back to (10.35) and rewrite it in the form of (10.7):

$$q(x, Q^2) + \Delta q(x, Q^2) = \int_0^1 dy \int_0^1 dz\, q(y, Q^2) \mathcal{P}_{qq}(z, Q^2)\, \delta(x - zy) \tag{10.49}$$

where

$$\mathcal{P}_{qq}(z, Q^2) \equiv \delta(1 - z) + \frac{\alpha_s}{2\pi} P_{qq}(z) \log\frac{Q^2}{\mu^2}. \tag{10.50}$$

It is natural to interpret $\mathcal{P}_{qq}(z)$ as the *probability density* of finding a quark inside a quark with fraction z of the parent quark momentum to first order in α_s. The $\delta(1 - z)$ term corresponds to there being no change in $q(x, Q^2)$. Clearly, this probability for a quark to remain unchanged will be reduced when the $O(\alpha_s)$ contributions are included.

At this point, it is crucial to include some virtual gluon diagrams which we have neglected until now. When these diagrams are included, the first term on the right-hand side of (10.34) is enlarged to

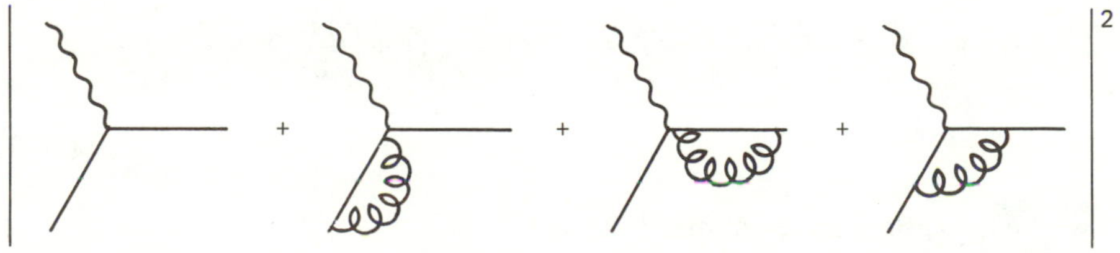

There is thus an $O(\alpha\alpha_s)$ contribution from the interference of the parton diagram with the (three) diagrams containing virtual gluons. These additional interference contributions are also singular at $z = 1$. It turns out that their singularity will exactly cancel the $z = 1$ singularity present in the incomplete $O(\alpha_s)$ calculation (10.50).

Rather than calculate these contributions explicitly, we can easily see what they must give. The additional contribution is entirely concentrated at $z = 1$ and is of the form $\delta(1 - z)$. It must be such that the total probability $P_{qq}(z)$ satisfies the constraint

$$\int_0^1 P_{qq}(z)\, dz = 0. \tag{10.51}$$

This constraint expresses the fact that the total number of quarks minus antiquarks is conserved; the probability, $\mathcal{P}_{qq}$, of finding a quark in a quark integrated over all z must add up to 1. From (10.50), we see that if the integral of $\mathcal{P}_{qq}(z)$ is to be unity, we need the condition (10.51).

The virtual diagrams regularize the $1/(1 - z)$ singularity in $P_{qq}(z)$ of (10.31) so that (10.51) holds. This modification to $P_{qq}(z)$ can be conveniently expressed in terms of the so-called "+ prescription" for regularization in which $1/(1 - z)$ is replaced by $1/(1 - z)_+$ defined so that

$$\int_0^1 dz\, \frac{f(z)}{(1-z)_+} = \int_0^1 dz\, \frac{f(z) - f(1)}{1 - z} \tag{10.52}$$

where $(1 - z)_+ = (1 - z)$ for $z < 1$ but is infinite at $z = 1$.

EXERCISE 10.13 Use (10.51) and (10.52) to show that

$$P_{qq}(z) = \frac{4}{3}\frac{1 + z^2}{(1 - z)_+} + 2\delta(1 - z). \tag{10.53}$$

EXERCISE 10.14 Use momentum conservation to justify

$$\int_0^1 dz\, z\big(q_S(z, Q^2) + g(z, Q^2)\big) = 1. \tag{10.54}$$

Hence, determine the $\delta(1 - z)$ term in P_{gg} and verify

$$P_{gg}(z) = 6\left(\frac{1 - z}{z} + \frac{z}{(1 - z)_+} + z(1 - z)\right) + \left(\frac{11}{2} - \frac{n_f}{3}\right)\delta(1 - z), \tag{10.55}$$

where n_f is the number of quark flavors.

EXERCISE 10.15 Show that momentum conservation at the QCD vertex requires (for $z < 1$)

$$
\begin{aligned}
P_{qq}(z) &= P_{gq}(1-z), \\
P_{qg}(z) &= P_{qg}(1-z), \qquad (10.56)\\
P_{gg}(z) &= P_{gg}(1-z).
\end{aligned}
$$

Check that the explicit formulas for the "splitting" functions satisfy these relations.

EXERCISE 10.16 If $\alpha_s(Q^2) = c/\log Q^2$, show that (10.37) leads to

$$\frac{\int x^{n-1} q(x, Q^2)\, dx}{\int x^{n-1} q(x, Q_0^2)\, dx} = \left(\frac{\log Q^2}{\log Q_0^2}\right)^{A_n},$$

where

$$
\begin{aligned}
A_n &= \frac{c}{2\pi}\int_0^1 x^{n-1} P_{qq}(x)\, dx \\
&= \frac{c}{2\pi}\frac{4}{3}\left(-\frac{1}{2} + \frac{1}{n(n+1)} - 2\sum_{j=2}^{n}\frac{1}{j}\right).
\end{aligned}
$$

That is, in QCD, the moments ($n \geq 1$) of the quark structure functions decrease as calculable powers of $\log Q^2$. c is given by (7.65).

The observant reader may have noticed what appears to be a contradiction in our interpretation of the P functions. For example, we regard the P_{qq} term as the correction factor to the quark density that arises from allowing for gluon emission. However, there are two diagrams: one with the gluon emitted from the initial quark line and the other with the gluon radiated from the final quark line; see Fig. 10.2. Our picture is only valid if the first diagram dominates. Then, the emitted gluon can be considered as part of the proton structure. It is a "parton-like" diagram. It turns out that both diagrams are required to ensure gauge invariance of the amplitude, but that the second only plays the role of canceling the contributions from the unphysical polarization states of the gluon. Adopting a physical gauge, in which we sum only over transverse gluons, only the first diagram remains.

10.9 The Altarelli–Parisi Techniques Also Apply to Leptons and Photons: The Weizsäcker–Williams Formula

We conclude this chapter by emphasizing the astonishing simplicity of the Altarelli–Parisi formalism. Let us return to (10.30), which computes the probability for producing a gluon with momentum fraction $1 - z$ and transverse momentum p_T in the process $\gamma^* q \to qg$. We should really have written it as a double-differential cross section

$$\frac{d\hat{\sigma}}{dz\, dp_T^2} = \left(e_i^2 \hat{\sigma}_0\right) \gamma_{qq}\left(z, p_T^2\right), \tag{10.57}$$

with

$$\gamma_{qq}\left(z, p_T^2\right) = \frac{\alpha_s}{2\pi} \frac{1}{p_T^2} P_{qq}(z). \tag{10.58}$$

This cross section for $\gamma^* q \to qg$ can be pictured symbolically as

$$\frac{d\hat{\sigma}}{dz\, dp_T^2} = \tag{10.57$'$}$$

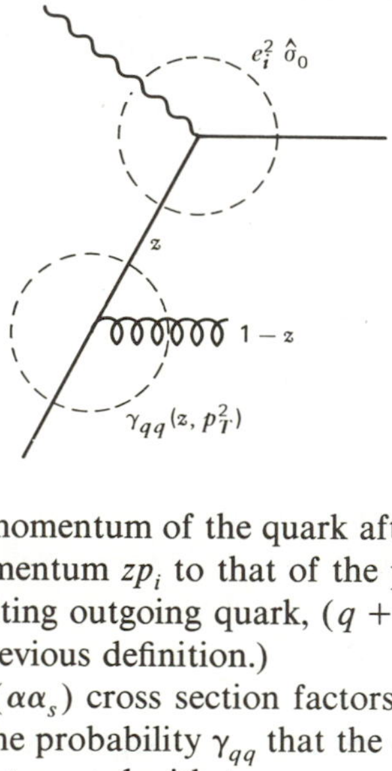

(That z is indeed the fractional momentum of the quark after gluon radiation can be easily seen by adding its momentum zp_i to that of the photon and noting the mass-shell condition of the resulting outgoing quark, $(q + zp_i)^2 = 0$. This shows that z is given by (10.29), our previous definition.)

In (10.57$'$) we see that the $O(\alpha\alpha_s)$ cross section factors into the $O(\alpha)$ parton model cross section $(e_i^2 \hat{\sigma}_0)$ and the probability γ_{qq} that the quark radiates a gluon with fraction $1 - z$ of its momentum and with transverse momentum p_T. How is this possible? Cross sections are calculated from probability *amplitudes*, where the different amplitudes for the process are added and then the square modulus of the sum is taken. What we have found is that, in the limit that the gluons have a p_T that is not too large, the calculation of (10.57$'$) can be viewed as two sequential events, that is, provided we retain only the singular part, $1/p_T^2$, of the full p_T distribution as we did in deriving (10.30). The probabilities for the interaction $e_i^2 \hat{\sigma}_0$ and the gluon emission γ_{qq} can then be calculated separately and multiplied. That is, the probabilistic parton picture thus applies to gluon emission.

This technique is not special to quarks and gluons (QCD). In fact, it was known, since the work of Weizsäcker and Williams in 1934, that it applies equally well to leptons and photons (QED). Consider, for example, the process ep → eX. The cross section may be written

$$\frac{d\sigma}{dz\,dp_T^2} =$$

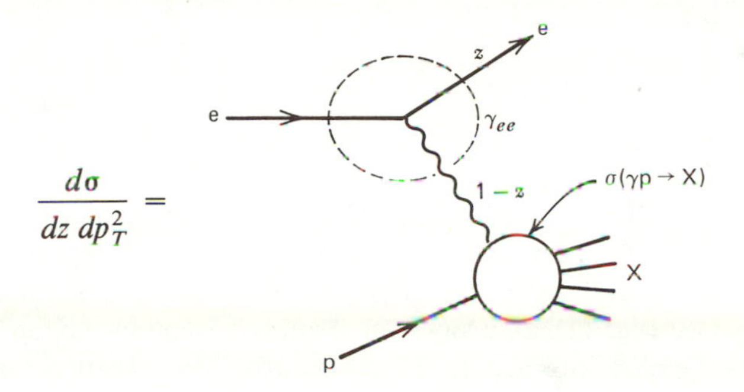

$$= \gamma_{ee}(z, p_T^2)[\sigma(\gamma p \to X)]_{E_\gamma=(1-z)E} \tag{10.59}$$

with

$$\gamma_{ee}(z, p_T^2) = \frac{\alpha}{\pi}\frac{1}{p_T^2}P_{ee}(z), \tag{10.60}$$

where z and p_T are, respectively, the momentum fraction and transverse momentum of the outgoing electron, and

$$P_{ee}(z) = \frac{1+z^2}{1-z}; \tag{10.61}$$

see, for example, Chen and Zerwas (1975) *Phys. Rev.* D12, 187. Equation (10.60) follows from (10.58) after taking account of the factor 2 mismatch in the definitions of α and α_s, namely, $\alpha_s \to 2\alpha$, see (10.19). $P_{ee}(z)$ is just $P_{qq}(z)$ as given by (10.31) but without the color factor. Equation (10.60) is known in QED as the equivalent photon distribution. Similarly, one can define the QED equivalent of P_{qg} given by (10.41):

$$P_{e\gamma}(z) = z^2 + (1-z)^2. \tag{10.62}$$

In the next chapter, we further illustrate the power of this technique by calculating the cross section for the process $e^+e^- \to q\bar{q}g$, which we view as $e^+e^- \to q\bar{q}$ followed by the emission of a gluon from the quark (or antiquark). $\gamma_{qq}(z, p_T^2)$ has been computed once and for all and can just be substituted into other diagrams.

11

e^+e^- Annihilation and QCD

In the previous chapter, we studied QCD in the framework of a truly historic type of experiment, namely, the deep inelastic scattering of leptons by hadrons. The large-Q^2, short-wavelength, virtual photon, prepared by the inelastic scattering of the lepton (see Fig. 11.1a), probes the proton, revealing its constituents (Chapter 9) and their color interactions (Chapter 10). The resulting picture is easy to interpret, because the short-distance (small α_s) nature of the quark–gluon interactions allows us to confront the experimental results with quantitative perturbative calculations.

High-resolution photons can also be prepared by colliding high-energy electron and positron beams head-on (see Fig. 11.1b). The exceptional power of this experimental technique is illustrated by the gallery of diagrams in Fig. 11.2: e^+e^- colliders can be used to study QED, weak interactions, quarks, and gluons and also to study or search for heavy quarks and leptons. Moreover, e^+e^- annihilation is a "clean" process in the sense that leptons (rather than hadrons, which are complex structures made of partons) appear in the initial state. For these reasons, we choose e^+e^- processes as our main working example to illustrate how the ideas and techniques of Chapters 9 and 10 carry over to other experimental situations.

11.1 e^-e^+ Annihilation into Hadrons: $e^-e^+ \rightarrow q\bar{q}$

The bulk of hadrons produced in e^-e^+ annihilations are fragments of a quark and antiquark produced by the process $e^-e^+ \rightarrow q\bar{q}$. (We shall justify this statement by computing the higher-order process $e^-e^+ \rightarrow q\bar{q}g$ in the next section.) The cross section for the (QED) process $e^-e^+ \rightarrow q\bar{q}$ of Fig. 11.2c is readily obtained from that for the process of Fig. 11.2a,

$$\sigma(e^-e^+ \rightarrow \mu^-\mu^+) = \frac{4\pi\alpha^2}{3Q^2}, \tag{11.1}$$

a result we obtained in (6.33). Here, the center-of-mass energy squared is

$$s = Q^2 = 4E_b^2, \tag{11.2}$$

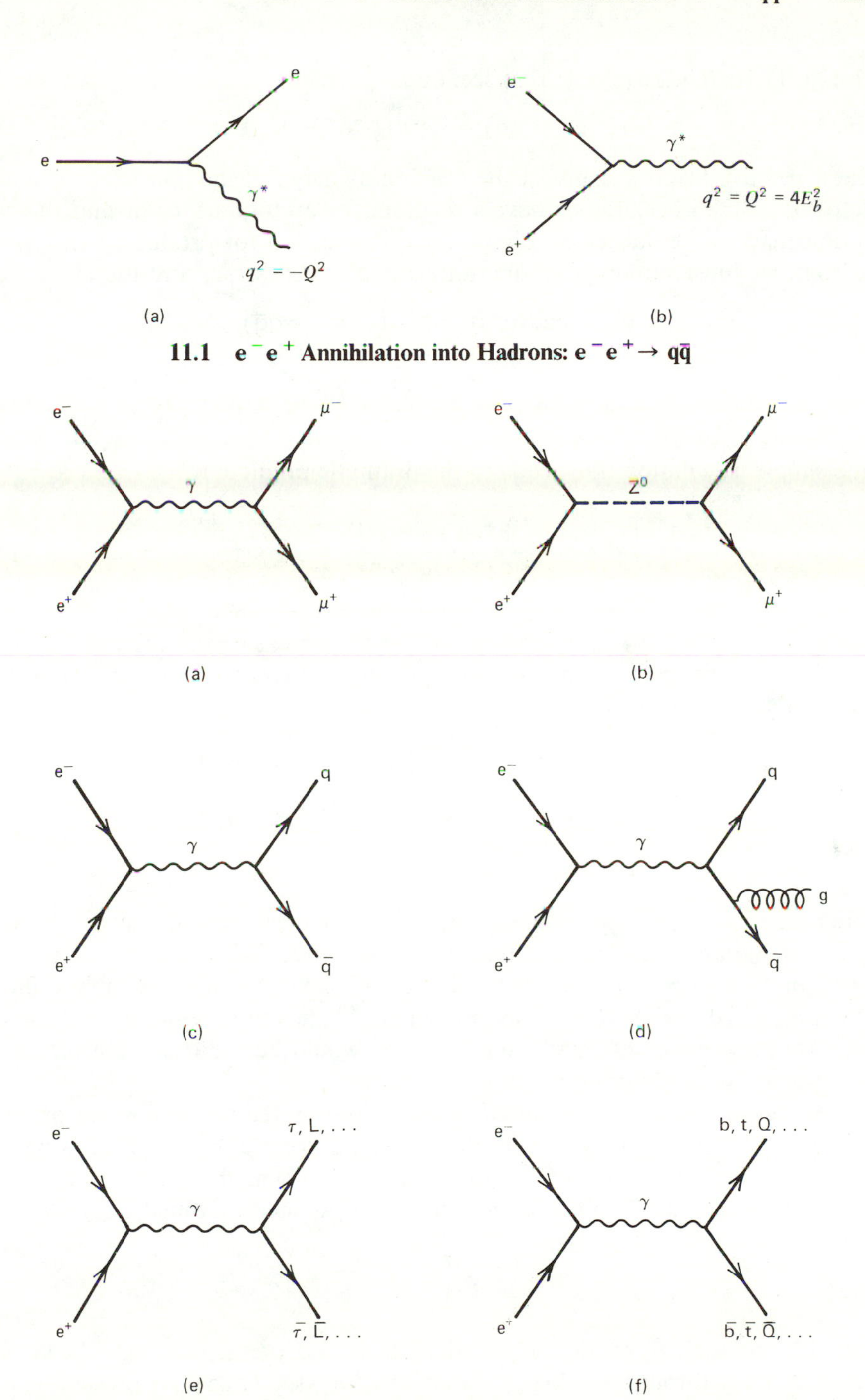

11.1 e^-e^+ Annihilation into Hadrons: $e^-e^+ \to q\bar{q}$

Fig. 11.2 Some experimental possibilities resulting from e^+e^- annihilation.

see Fig. 11.1b. The required cross section is

$$\sigma(e^-e^+ \to q\bar{q}) = 3e_q^2 \sigma(e^-e^+ \to \mu^-\mu^+), \tag{11.3}$$

where we have taken account of the fractional charge of the quark, e_q. The extra factor of 3 arises because we have a diagram for each quark color and the cross sections have to be added. To obtain the cross section for producing all types of hadrons, we must sum over all quark flavors q = u, d, s, ..., and therefore

$$\sigma(e^-e^+ \to \text{hadrons}) = \sum_q \sigma(e^-e^+ \to q\bar{q}) \tag{11.4}$$

$$= 3\sum_q e_q^2 \sigma(e^-e^+ \to \mu^-\mu^+). \tag{11.5}$$

This simple calculation thus leads to the dramatic prediction

$$\boxed{R \equiv \frac{\sigma(e^-e^+ \to \text{hadrons})}{\sigma(e^-e^+ \to \mu^-\mu^+)} = 3\sum_q e_q^2.} \tag{11.6}$$

As $\sigma(e^-e^+ \to \mu^-\mu^+)$ is well known (see Fig. 6.6), a measurement of the total e^-e^+ annihilation cross section into hadrons therefore directly counts the number of quarks, their flavors, as well as their colors. We have

$$\begin{aligned} R &= 3\left[\left(\tfrac{2}{3}\right)^2 + \left(\tfrac{1}{3}\right)^2 + \left(\tfrac{1}{3}\right)^2\right] = 2 && \text{for u, d, s,} \\ &= 2 + 3\left(\tfrac{2}{3}\right)^2 = \tfrac{10}{3} && \text{for u, d, s, c,} \\ &= \tfrac{10}{3} + 3\left(\tfrac{1}{3}\right)^2 = \tfrac{11}{3} && \text{for u, d, s, c, b.} \end{aligned} \tag{11.7}$$

In Fig. 11.3, these predictions are compared to the measurements of R. The value $R \simeq 2$ is apparent below the threshold for producing charmed particles at $Q = 2(m_c + m_u) \simeq 3.7$ GeV. Above the threshold for all five quark flavors ($Q > 2m_b \simeq 10$ GeV), $R \simeq \frac{11}{3}$ as predicted. These measurements confirm that there are three colors of quark, since $R = \frac{11}{3}$ would be reduced by a factor 3 if there was only one color, see (11.3).

These results for R will be modified when interpreted in the context of QCD. Equation (11.4) is based on the (leading order) process $e^-e^+ \to q\bar{q}$. However, we should also include contributions from diagrams where the quark and/or antiquark radiate gluons. To $O(\alpha_s)$, the result (11.6) is then modified to

$$R = 3\sum_q e_q^2 \left(1 + \frac{\alpha_s(Q^2)}{\pi}\right).$$

That is, the scaling result, (11.6), that R is independent of Q^2 is violated logarithmically through the $\log Q^2$ behavior of α_s, see (7.65). At present, experiments are unable to detect this additional contribution to R, and our previous comparison with the data remains a good approximation. We postpone the

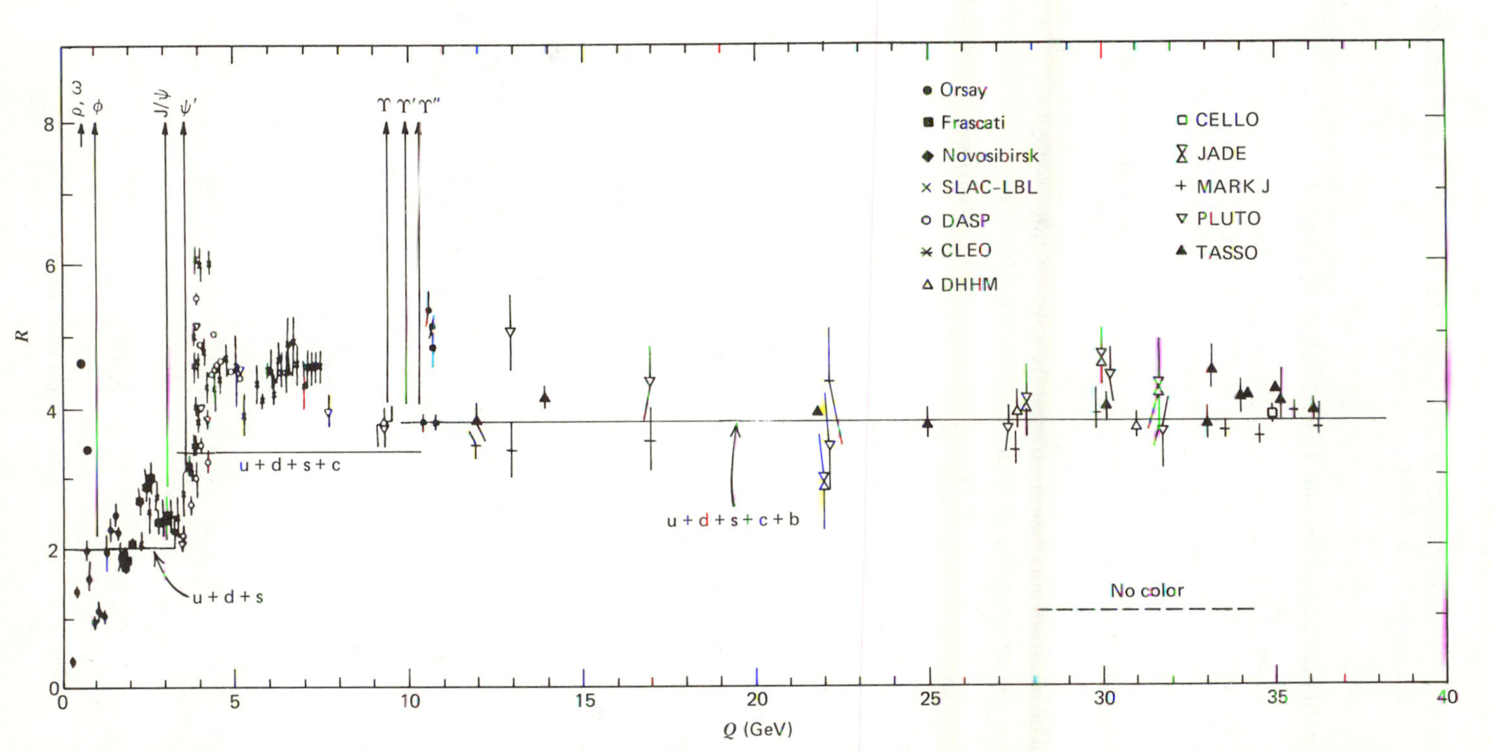

Fig. 11.3 Ratio R of (11.6) as a function of the total e^-e^+ center-of-mass energy. (The sharp peaks correspond to the production of narrow 1^- resonances just below or near the flavor thresholds.)

derivation of the α_s/π correction until Section 11.7, where it will emerge as a by-product of the study of the three-jet events $e^-e^+ \to q\bar{q}g$.

11.2 Fragmentation Functions and Their Scaling Properties

So far, we have not faced the problem of how the quarks turn into the hadrons that hit the detector. It was sufficient to state that the quarks must fragment into hadrons with unit probability. This gives (11.4). For more detailed calculations, this problem cannot be sidestepped.

In the center-of-mass frame, the produced quark and antiquark separate with equal and opposite momentum and materialize into two back-to-back jets of hadrons which have momenta roughly collinear with the original q and $\bar{q}$ directions. The hadrons may be misaligned by a momentum transverse to the q or $\bar{q}$ direction by an amount not exceeding about 300 MeV.

In Chapter 1, we visualized jet formation as hadron bremsstrahlung once the q and $\bar{q}$ separate by a distance of around 1 fm. Then, α_s becomes large, and strong color forces pull on the separating q and $\bar{q}$. The potential energy becomes so large that one or more $q\bar{q}$ pairs are created (see Fig. 1.14). Eventually, all the energy is degraded into two jets of hadrons moving more or less in the direction of the q and $\bar{q}$.

To describe the fragmentation of quarks into hadrons, we use an analogous formalism to that introduced in Chapter 9 to describe the quarks inside hadrons. Figure 11.4 shows the observation of a hadron h, whose energy is measured to be E_h. The corresponding differential cross section can be written as

$$\frac{d\sigma}{dz}(e^-e^+ \to hX) = \sum_q \sigma(e^-e^+ \to q\bar{q})\left[D_q^h(z) + D_{\bar{q}}^h(z)\right]. \tag{11.8}$$

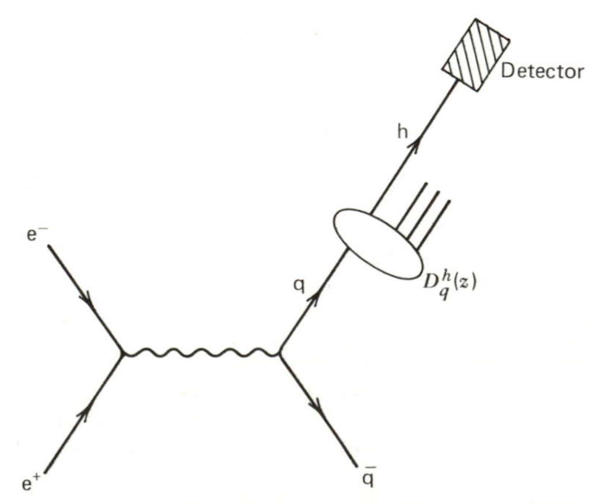

Fig. 11.14 The Drell–Yan process, $pp \to l^-l^+X$.

It describes Fig. 11.4 as two sequential events: the production of a $q\bar{q}$ pair, followed by the fragmentation of either the q or the $\bar{q}$ producing the detected hadron h. The D functions therefore represent the probability that the hadron h is found in the debris of a quark (or antiquark) carrying a fraction z of its energy. That is,

$$z \equiv \frac{E_h}{E_q} = \frac{E_h}{E_b} = \frac{2E_h}{Q}. \tag{11.9}$$

The summation in (11.8) is over all quark flavors and recognizes the fact that the detector is unaware of the quantum numbers of the parent of the observed hadron.

The fragmentation function, $D(z)$, describes the transition (parton → hadron) in the same way that the structure function, $f(x)$, of Chapter 9 describes the embedding (hadron → parton). Like the f functions, the D functions are subject to constraints imposed by momentum and probability conservation:

$$\sum_h \int_0^1 z\, D_q^h(z)\, dz = 1 \tag{11.10}$$

$$\sum_q \int_{z_{\min}}^1 \left[D_q^h(z) + D_{\bar{q}}^h(z) \right] dz = n_h, \tag{11.11}$$

where $z_{\min}$ is the threshold energy $2m_h/Q$ for producing a hadron of mass m_h, and n_h is the average multiplicity of hadrons of type h. Equation (11.10) simply states that the sum of the energies of all hadrons is the energy of the parent quark. Clearly, the same relation holds for $D_{\bar{q}}^h(z)$. Equation (11.11) says that the number n_h of hadrons of type h is given by the sum of probabilities of obtaining h from all possible parents, namely, from q or $\bar{q}$ of any flavor.

EXERCISE 11.1 Fragmentation functions are often parametrized by the form

$$D_q^h(z) = N \frac{(1-z)^n}{z}$$

where n and N are constants. Show that

$$N = (n+1)\langle z \rangle,$$

where $\langle z \rangle$ is the average fraction of the quark energy carried by hadrons of type h after fragmentation. Further, show that

$$n_h \sim \log\left(\frac{Q}{2m_h}\right)$$

for the two-jet process of Fig. 11.4. That is, the multiplicity of hadrons h grows logarithmically with the annihilation energy.

Taking the ratio of (11.8) and (11.4), and using (11.3), we find

$$\frac{1}{\sigma}\frac{d\sigma}{dz}(e^-e^+ \to hX) = \frac{\sum_q e_q^2\left[D_q^h(z) + D_{\bar{q}}^h(z)\right]}{\sum_q e_q^2} \tag{11.12}$$

$$= \mathcal{F}(z) \tag{11.13}$$

That is, the inclusive cross section $d\sigma/dz$ divided by the total annihilation cross section into hadrons, σ, is predicted to scale. The cross sections σ and $d\sigma/dz$ depend on the annihilation energy Q, but (11.12) predicts that the ratio is

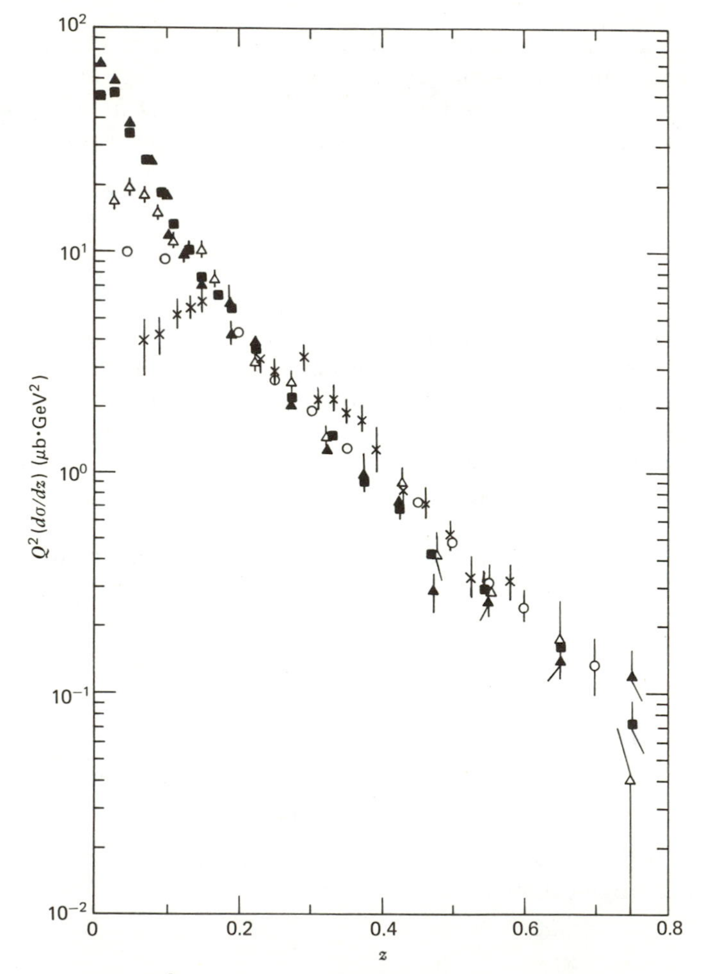

Fig. 11.5 $Q^2(d\sigma/dz) \sim 1/\sigma(d\sigma/dz)$ for $e^-e^+ \to hX$ as measured as a function z at different center-of-mass energies Q: (×) 5 GeV; (○) 7.4 GeV; (Δ) 12 GeV; (■) 27.4–31.6 GeV; (▲) 35.0–36.6 GeV. Data are from the Stanford Linear Accelerator and PETRA.

independent of Q. Such a scaling result is not a complete surprise, because we have relied on the scaling parton model to derive (11.12), see Fig. 11.4.

Figure 11.5 shows $(1/\sigma)(d\sigma/dz)$ as a function of z for different values of Q^2. The scaling is not perfect. Gluon emission from the q or $\bar{\text{q}}$ will introduce $\log Q^2$ scaling violations in (11.13). Their qualitative trend is the same as in electroproduction, that is, $\mathcal{F}(z, Q^2)$ will increase at small z with increasing values of Q^2 but decrease for z near 1. The large violations of scaling for $z \lesssim 0.2$, seen in Fig. 11.5, are not exclusively due to gluon emission, however, and are the subject of the next section.

EXERCISE 11.2 The fragmentation functions $D(z)$ describe properties of partons and are therefore the same, no matter how the partons are produced. Consider the inclusive leptoproduction cross section $\sigma(\text{ep} \to \text{hX})$ and show that

$$\frac{1}{\sigma}\frac{d\sigma}{dz}(\text{ep} \to \text{hX}) = \frac{\sum_q e_q^2 f_q(z)\, D_q^h(z)}{\sum_q e_q^2 f_q(x)},$$

where $f_q(x)$ are the proton structure functions of Chapter 9, see Fig. 11.6. The sum runs over the quarks and antiquarks that can be a parent of h.

EXERCISE 11.3 Using charge conjugation and isospin invariance, show that

$$D_u^{\pi^+} = D_{\bar{u}}^{\pi^-} = D_d^{\pi^-} = D_{\bar{d}}^{\pi^+},$$

$$D_u^{\pi^-} = D_{\bar{u}}^{\pi^+} = D_d^{\pi^+} = D_{\bar{d}}^{\pi^-},$$

$$D_s^{\pi^+} = D_s^{\pi^-}.$$

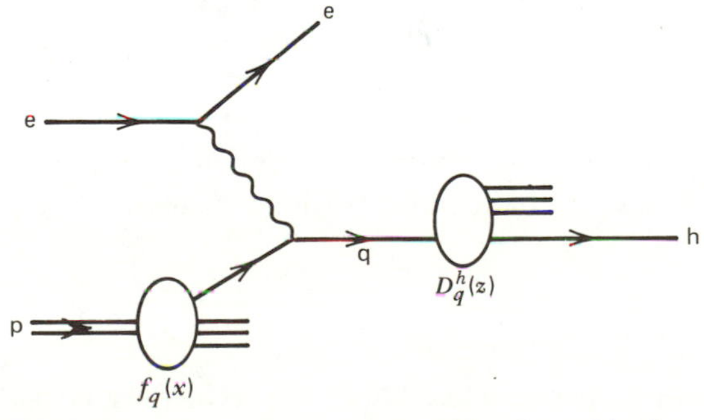

Fig. 11.6 Deep inelastic leptoproduction of hadron h; ep → hX.

EXERCISE 11.4 Using the notation

$$N_p^{\pi}(z) \equiv \frac{1}{\sigma}\frac{d\sigma}{dz}(\text{ep} \to \pi\text{X}),$$

show that, in the valence quark approximation for p, n,

$$\frac{\int dz\left[N_n^{\pi^+} - N_n^{\pi^-}\right]}{\int dz\left[N_p^{\pi^+} - N_p^{\pi^-}\right]} = \frac{2}{7}.$$

11.3 A Comment on Heavy Quark Production

Although the similarities between e^-e^+ annihilation and leptoproduction are becoming more and more evident, we must not forget one major difference. In leptoproduction, u, d, and s quarks play a dominant role because they are plentiful inside the nucleon target. Charm quarks occur in roughly one in ten events, and were indeed ignored in the phenomenological discussion of Chapter 9. In e^-e^+ annihilation, the situation is quite different. Beyond threshold ($Q^2 > 4m_c^2$), the cross section rises steeply and readily attains a sizable fraction of its asymptotic value, see Fig. 11.3. It is clear from (11.3) that c and u quarks are produced with the same cross section, since both have $e_q = +\frac{2}{3}$. However, close to threshold, the final-state hadron structure of these relatively frequent charm–quark events is very different from the two-jet structure of a typical event involving light quarks. Then, the c and $\bar{c}$ are produced almost at rest, and subsequently decay weakly into a rather large number of soft hadrons with low z. Therefore, an increase of low-z events is associated with the crossing of the charm threshold. This leads to a violation of scaling which will confuse, indeed simulate, the violations resulting from gluon emission.

The drastic violations of scaling, seen for $z \lesssim 0.2$ in Fig. 11.5, are associated with the production of c and b quarks, resulting in events with a large multiplicity and low z values when the c and b thresholds are crossed. This mechanism has a positive aspect. The characteristic features of heavy quark events can serve as an experimental signature in the search for yet heavier quarks, such as the t quark in Table 1.5, as the energy of e^-e^+ colliders is increased. A "step" in R also signals a new quark, for example, charm in Fig. 11.3. But for $e_q = -\frac{1}{3}$, this step is four times smaller and not easy to observe experimentally. Looking for deviations from the two-jet structure, characteristic of light quarks, is often a more sensitive test.

11.4 Three-Jet Events: $e^-e^+ \to q\bar{q}g$

From the viewpoint of perturbative QCD, we have only considered the leading $O(\alpha^2)$ contribution to $\sigma(e^-e^+ \to \text{hadrons})$. In order $\alpha^2\alpha_s$, the q or $\bar{q}$ can emit a gluon, see Fig. 11.2d; and these $e^-e^+ \to q\bar{q}g$ events have three fragmenting jets in the final state. The additional jet has a gluon as its parent parton. For larger values of the annihilation energy Q, such that $\alpha_s(Q^2) \simeq 0.1$–0.2, we expect that three-jet events will account for roughly 10% of the final states. Our first task is to introduce kinematical variables to describe such events.

The momentum vectors of the q, $\bar{q}$, and g, which are produced by a virtual photon (γ^*) at rest, are displayed in Fig. 11.7. As in (11.9), we work with the energies, and with the longitudinal and transverse momenta of the partons, scaled to the e^- (and e^+) beam energy. That is, we introduce

$$x_q \equiv \frac{2E_q}{Q}, \qquad x_{\bar{q}} \equiv \frac{2E_{\bar{q}}}{Q}, \qquad x_g \equiv \frac{2E_g}{Q} \tag{11.14}$$

and

$$x_T \equiv \frac{2p_T}{Q}. \tag{11.15}$$

Like z in (11.9), all these ratios are bounded by 0 and 1. The four-momentum fractions in Fig. 11.7 are

$$\begin{aligned} &(x_q; 0, 0, x_q) && \text{for q},\\ &(x_{\bar{q}}; x_T, 0, x_L) && \text{for } \bar{\text{q}},\\ &(x_g; -x_T, 0, x_L - x_q) && \text{for g}. \end{aligned} \tag{11.16}$$

The variables are defined relative to the most energetic jet, for example, the q jet in Fig. 11.7. Its direction is referred to as the "thrust axis." The q, $\bar{q}$, and g are coplanar in the plane $y = 0$. Longitudinal and transverse momentum conservation are embodied in (11.16), but energy conservation introduces the additional requirement that

$$x_q + x_{\bar{q}} + x_g = 2. \tag{11.17}$$

The zero mass of the $\bar{q}$ and g leads to the further constraints [see (11.16)]

$$\begin{aligned} x_{\bar{q}}^2 - x_T^2 - x_L^2 &= 0,\\ x_g^2 - x_T^2 - (x_L - x_q)^2 &= 0. \end{aligned} \tag{11.18}$$

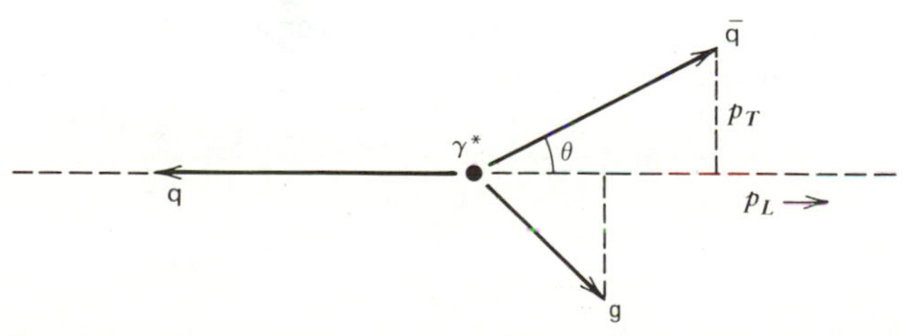

Fig. 11.7 The process $e^-e^+ \to \gamma^* \to q\bar{q}g$ in the center-of-mass frame.

From (11.18) and (11.17), it follows that

$$x_T^2 = \frac{4}{x_q^2}(1 - x_q)(1 - x_{\bar{q}})(1 - x_g). \tag{11.19}$$

EXERCISE 11.5 Derive (11.19). Also show that the angle θ between the q and $\bar{\text{q}}$ directions in Fig. 11.7 is determined by the relation

$$x_{\bar{q}} = \frac{2(1 - x_q)}{2 - x_q - x_q \cos\theta}. \tag{11.20}$$

Let us now compute the cross section corresponding to Fig. 11.7. For this particular graph, the $\bar{\text{q}}$ emits a softer gluon, so that

$$x_q \geq x_{\bar{q}} \geq x_g. \tag{11.21}$$

The most obvious experimental signature of gluon emission is that the q and $\bar{\text{q}}$ are no longer produced back to back. The $\bar{\text{q}}$ is produced with a transverse momentum fraction x_T relative to the direction of the quark. The relevant observable quantity is therefore $d\sigma/dx_T^2$. This cross section can readily be obtained using the Altarelli–Parisi–Weizsäcker–Williams technique of Section 10.9. Referring to Fig. 11.8, we obtain

$$\frac{d\sigma}{dx_{\bar{q}}\, dp_T^2} = \sigma(\text{e}^-\text{e}^+ \to \text{q}\bar{\text{q}})\,\gamma_{\bar{q}\bar{q}}\left(x_{\bar{q}}, p_T^2\right), \tag{11.22}$$

see (10.57), where σ gives the probability for producing a q$\bar{\text{q}}$ pair and $\gamma_{\bar{q}\bar{q}}$ is the probability that the $\bar{\text{q}}$ subsequently emits a gluon with a fraction $(1 - x_{\bar{q}})$ of its momentum and a transverse momentum $|p_T|$. From (11.3) and (10.58), we have

$$\sigma(\text{e}^-\text{e}^+ \to \text{q}\bar{\text{q}}) = \frac{4\pi\alpha^2}{Q^2} e_q^2,$$

$$\gamma_{\bar{q}\bar{q}}\left(x_{\bar{q}}, p_T^2\right) = \gamma_{qq}\left(x_{\bar{q}}, p_T^2\right) = \frac{\alpha_s}{2\pi}\frac{1}{p_T^2} P_{qq}(x_{\bar{q}}). \tag{11.23}$$

On substitution of (11.23) into (11.22), we find

$$\frac{1}{\sigma}\frac{d\sigma}{dx_{\bar{q}}\, dx_T^2} = \frac{\alpha_s}{2\pi}\frac{1}{x_T^2} P_{qq}(x_{\bar{q}}). \tag{11.24}$$

To calculate $d\sigma/dx_T^2$, it remains to integrate over all possible $\bar{\text{q}}$ energy fractions $x_{\bar{q}}$. Using (10.31) for P_{qq}, we obtain

$$\frac{1}{\sigma}\frac{d\sigma}{dx_T^2} = 2\frac{\alpha_s}{2\pi}\frac{1}{x_T^2}\int_{(x_{\bar{q}})_{\min}}^{(x_{\bar{q}})_{\max}} dx\, \frac{4}{3}\left(\frac{1 + x^2}{1 - x}\right). \tag{11.25}$$

An extra factor of 2 is included to allow for the equally probable diagram with q $\leftrightarrow$ $\bar{\text{q}}$.

The integrand in (11.25) diverges when $x_{\bar{q}} \to 1$. The kinematic situation where $x_{\bar{q}}$ reaches its maximum value is therefore of special interest. From (11.21), we see

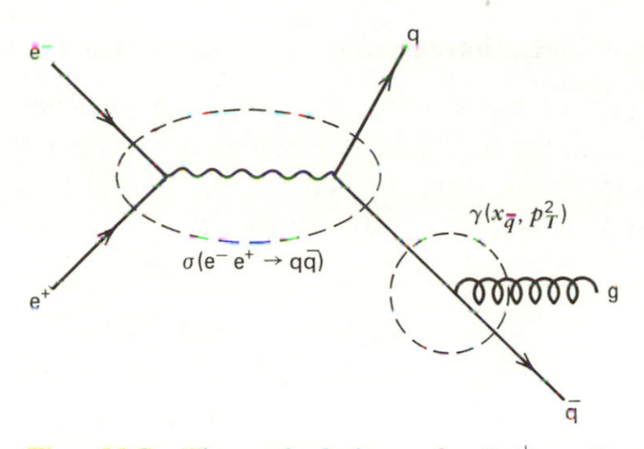

Fig. 11.8 The calculation of $e^-e^+ \to q\bar{q}g$ using the probabilistic techniques of §10.9.

that the largest value allowed for $x_{\bar{q}}$ is

$$x_{\bar{q}} = x_q. \tag{11.26}$$

This value can be approached if we make the emitted gluon as soft as possible, which, remembering that x_T is fixed, occurs when

$$x_g = x_T. \tag{11.27}$$

This kinematic configuration is shown in Fig. 11.9. Thus, from (11.17), we have

$$(x_q)_{\min} = (x_{\bar{q}})_{\max} \simeq 1 - \frac{x_T}{2}. \tag{11.28}$$

Momentum is conserved for θ or x_T not too large, but x_g is only exactly equal to x_T when $\theta \to 0$. Such approximations are implicit in the Altarelli–Parisi calculation of (11.25), see Chapter 10. Using (11.28), (11.25) becomes

$$\frac{1}{\sigma}\frac{d\sigma}{dx_T^2} \simeq \frac{8\alpha_s}{3\pi}\frac{1}{x_T^2}\int_{(x_{\bar{q}})_{\min}}^{1-\frac{1}{2}x_T}\frac{dx}{1-x}, \tag{11.29}$$

where we have approximated $1 + x^2$ by 2. Finally, omitting all but the leading logarithmic term, we obtain

$$\boxed{\frac{1}{\sigma}\frac{d\sigma}{dx_T^2} \simeq \frac{4\alpha_s}{3\pi}\frac{1}{x_T^2}\log\left(\frac{1}{x_T^2}\right).} \tag{11.30}$$

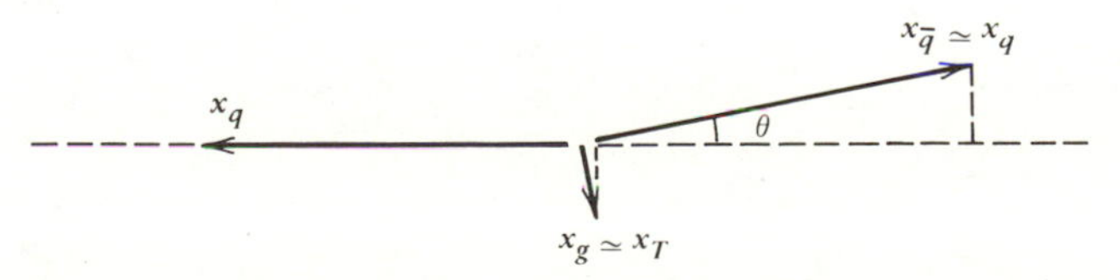

Fig. 11.9 The kinematic configuration giving the maximum value of $x_{\bar{q}}$ for a given x_T.

11.5 An Alternative Derivation of the $e^-e^+ \to q\bar{q}g$ Cross Section

The cross section (11.30) for $e^-e^+ \to \gamma^* \to q\bar{q}g$ can also be obtained using the Feynman rules of Chapter 6. We have calculated similar processes before. For instance, we found that $\gamma^*q \to qg$ is given by (10.31) and that $\gamma^*g \to q\bar{q}$ is given by (10.38). Proceeding in the same way, we find that the square of the amplitude for the sum of the $\gamma^* \to q\bar{q}g$ diagrams of Fig. 11.10 is

$$|\mathcal{M}|^2 = N\left(\frac{t}{s} + \frac{s}{t} + \frac{2uQ^2}{st}\right) \tag{11.31}$$

where N represents the normalization factors and coupling constants and where

$$\begin{aligned} s &= (p_\gamma - p_q)^2, \\ t &= (p_\gamma - p_{\bar{q}})^2, \\ u &= (p_\gamma - p_g)^2, \end{aligned} \tag{11.32}$$

where $Q^2 \equiv p_\gamma^2$. Strictly speaking, for a parton process we should have denoted the Mandelstam variables by $\hat{s}$, $\hat{t}$, $\hat{u}$.

In order to rewrite (11.31) in terms of the energy fraction variables x_i of (11.14), we use

$$\begin{aligned} s &= Q^2(1 - x_q), \\ t &= Q^2(1 - x_{\bar{q}}), \\ u &= Q^2(1 - x_g). \end{aligned} \tag{11.33}$$

These relations follow from four-momentum conservation, which gives

$$p_\gamma^2 = (p_q + p_{\bar{q}} + p_g)^2 = 2p_q \cdot p_{\bar{q}} + 2p_q \cdot p_g + 2p_{\bar{q}} \cdot p_g. \tag{11.34}$$

For instance,

$$\begin{aligned} s &= (p_{\bar{q}} + p_g)^2 = 2p_{\bar{q}} \cdot p_g = p_\gamma^2 - 2p_q \cdot (p_{\bar{q}} + p_g) \\ &= p_\gamma^2 - 2p_q \cdot p_\gamma = Q^2\left(1 - \frac{2E_q}{Q}\right) = Q^2(1 - x_q). \end{aligned}$$

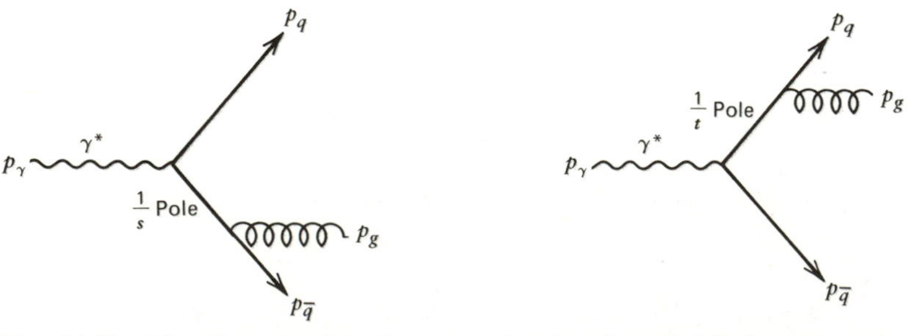

Fig. 11.10 The diagrams for $\gamma^* \to q\bar{q}g$ showing the particle four-momenta.

Substituting (11.33) into (11.31), we obtain for $\gamma^* \to q\bar{q}g$

$$|\mathfrak{M}|^2 = N \frac{x_q^2 + x_{\bar{q}}^2}{(1 - x_q)(1 - x_{\bar{q}})}, \tag{11.35}$$

where we have used (11.17) to eliminate x_g. If we "attach" the e^-e^+ pair, then, up to a factor (which changes $N \to N'$), (11.35) gives the cross section for $e^-e^+ \to q\bar{q}g$

$$\boxed{\frac{d\sigma}{dx_q\, dx_{\bar{q}}} = N' \frac{x_q^2 + x_{\bar{q}}^2}{(1 - x_q)(1 - x_{\bar{q}})}.} \tag{11.36}$$

To verify that (11.36) is equivalent to our previous result (11.24), we must change the variable x_q to x_T^2. We have

$$\frac{d\sigma}{dx_{\bar{q}}\, dx_T^2} = \frac{d\sigma}{dx_{\bar{q}}\, dx_q} \frac{dx_q}{dx_T^2}. \tag{11.37}$$

It is sufficient to use the small p_T approximation inherent in the Altarelli–Parisi result (11.24). Using (11.19), we find

$$\left| \frac{dx_T^2}{dx_q} \right| \simeq 4x_{\bar{q}}(1 - x_{\bar{q}}) \qquad \text{for } x_q \simeq 1. \tag{11.38}$$

Thus, (11.36) may be written

$$\frac{d\sigma}{dx_{\bar{q}}\, dx_T^2} \simeq N' \left(\frac{1 + x_{\bar{q}}^2}{1 - x_{\bar{q}}} \right) \left[\frac{1}{4(1 - x_q)(1 - x_{\bar{q}}) x_{\bar{q}}} \right], \tag{11.39}$$

where here again we have assumed $x_q \simeq 1$. In this limit, $x_{\bar{q}} \simeq (1 - x_g)$, and so the factor in square brackets is just x_T^{-2}, as can be seen from (11.19). Thus, (11.39) becomes

$$\frac{d\sigma}{dx_{\bar{q}}\, dx_T^2} \simeq N' \frac{3}{4} P_{qq}(x_{\bar{q}}) \frac{1}{x_T^2}, \tag{11.40}$$

where P_{qq} is given by (10.31). This is the same as the Altarelli–Parisi result of (11.24), and, indeed, we can identify the normalization coefficient to be

$$N' = \frac{2\alpha_s}{3\pi} \sigma.$$

The exact $O(\alpha_s)$ result is therefore

$$\boxed{\frac{1}{\sigma} \frac{d\sigma}{dx_q\, dx_{\bar{q}}} = \frac{2\alpha_s}{3\pi} \frac{x_q^2 + x_{\bar{q}}^2}{(1 - x_q)(1 - x_{\bar{q}})},} \tag{11.41}$$

whereas (11.24) is the leading logarithmic approximation.

11.6 A Discussion of Three-Jet Events

The $e^-e^+ \to q\bar{q}g$ events led to three hadronic jets in the final state. The distribution (11.30) can be written as

$$\frac{1}{\sigma}\frac{d\sigma}{dp_T^2} \sim \alpha_s \frac{1}{p_T^2} \log\left(\frac{Q^2}{4p_T^2}\right), \tag{11.42}$$

where p_T is the transverse momentum between the q and $\bar{q}$ as a result of the emission of the gluon, recall Fig. 11.7. Only when the $\bar{q}$ (or q) recoils against g can its p_T relative to the q (or $\bar{q}$) be nonzero. For two-jet events, $e^-e^+ \to q\bar{q}$, we have $p_T = 0$. Equation (11.42) shows that, for a fixed p_T, the cross section increases with increasing Q^2. That is, the number of $\bar{q}$ jets with a transverse momentum p_T relative to the q jet increases with Q^2. This is a result of the increased probability of emitting a gluon with a given p_T value when the annihilation energy increases. The physics is identical to that in electroproduction. There, also, the cross section

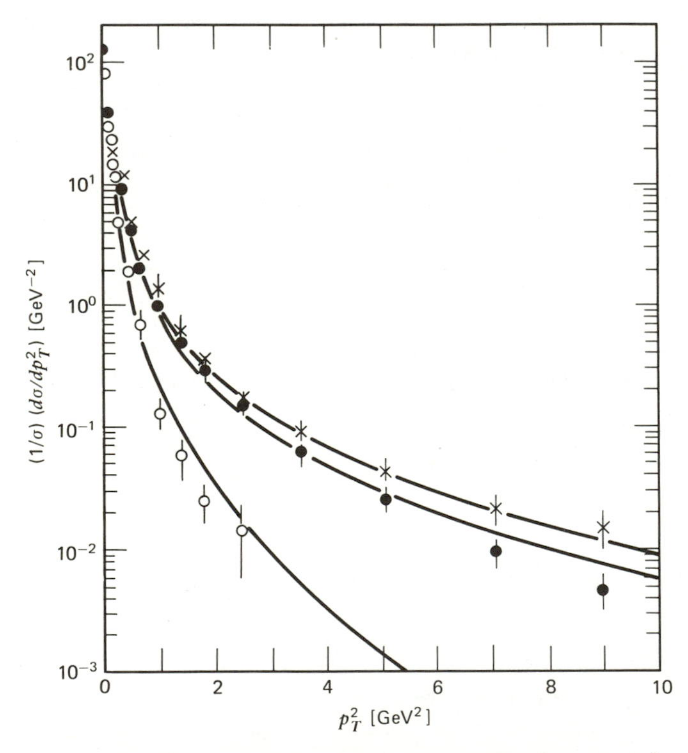

Fig. 11.11 The transverse momentum distribution $d\sigma/dp_T^2$ of hadrons relative to the thrust axis for different e^-e^+ center-of-mass energies Q: (○) $Q = 12$ GeV; (●) $27.4 \leq Q \leq 31.6$ GeV; (×) $35.0 \leq Q \leq 36.6$ GeV. The curves are a QCD calculation. Data are from PETRA.

for producing jets with transverse momentum p_T relative to the γ^*-direction increased as $\log Q^2/p_T^2$ [compare (10.30) and (11.24)].

The hadron fragments of the $\bar{q}$ jet will also have large p_T relative to the quark direction since the p_T distribution of these hadrons should follow the trend of the p_T distribution of jets. The two distributions can be explicitly related using the D functions introduced in Section 11.2. The resulting p_T and Q^2 dependence of hadrons relative to the thrust axis (whichever parton it refers to) is shown in Fig. 11.11. In some events, all three jets will be well separated despite the $k_T \simeq 300$ MeV of the daughter hadrons relative to their parent jet. One such event is shown in Fig. 11.12.

The assumption that α_s is constant in (11.42) requires explanation. If we had taken $\alpha_s(Q^2) \sim 1/\log Q^2$, then the above discussion would be meaningless. The

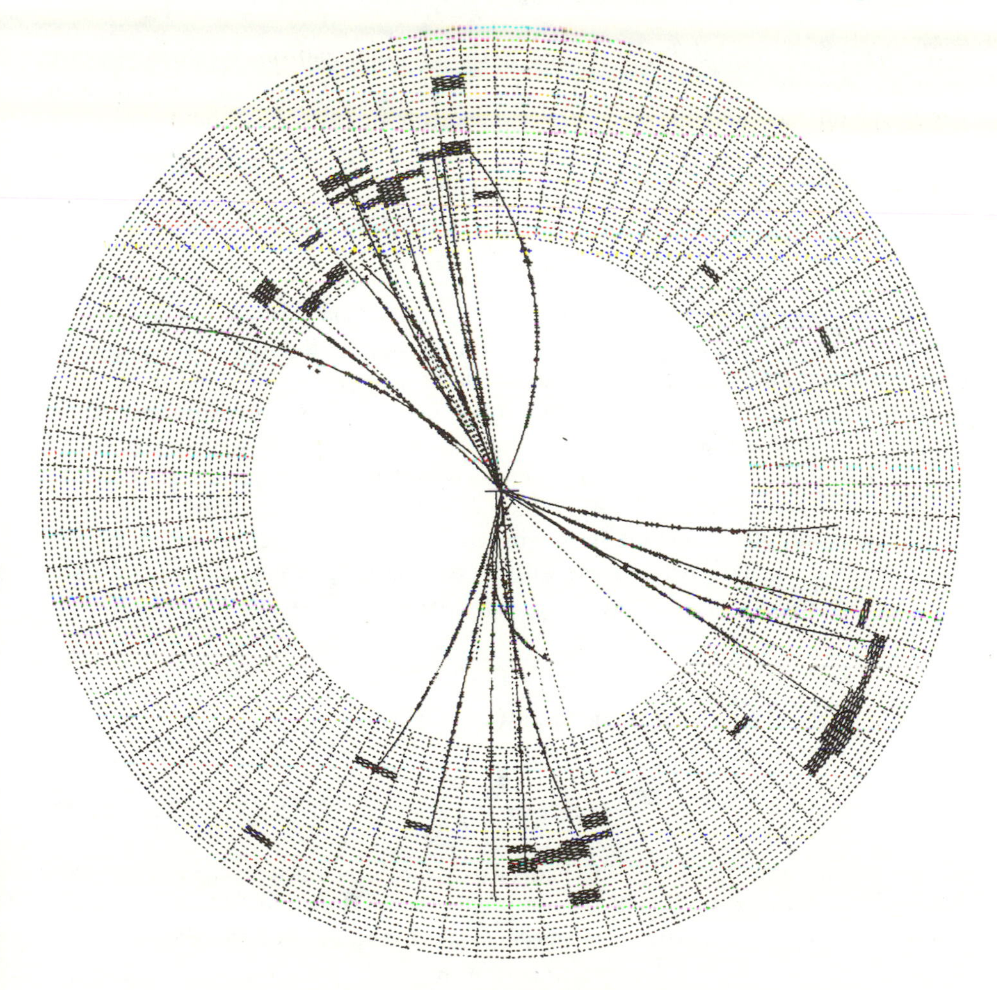

Fig. 11.12 A three-jet event observed by the JADE detector at PETRA.

crucial point is that we are discussing a process with two momentum scales, p_T^2 and Q^2 with $p_T^2 \ll Q^2$, a situation already encountered in Exercise 10.7. We argued there that α_s is in fact $\alpha_s(p_T^2)$, but noted that keeping α_s constant gave the correct result to leading order.

EXERCISE 11.6 The x_T distribution (11.30) can be translated into an "acollinearity" distribution $d\sigma/d\theta$, where θ is the angle between the q and $\bar{\text{q}}$ jet directions defined in (11.20) and Fig. 11.7. Show that for θ not too large,

$$\frac{1}{\sigma}\frac{d\sigma}{d\theta} \simeq \frac{8\alpha_s}{3\pi}\frac{1}{\theta}\log\left(\frac{1}{\theta^2}\right).$$

An exact result can be obtained using (11.41) instead of (11.30).

We have repeatedly drawn attention to the fact that hadrons fragmenting from a quark, or any other parton, form a cone around the direction $\mathbf{p}_q$. It is an experimental fact that the $\langle k_T \rangle$ for the hadron fragments is about 300 MeV, where T refers to the direction transverse to $\mathbf{p}_q$. We might anticipate that at higher energies the fragmentation cone would narrow,

$$\langle \theta \rangle \simeq \frac{\langle k_T \rangle}{p_q} \simeq \frac{0.3 \text{ GeV}}{Q/2}, \tag{11.43}$$

and the jets become narrow bundles of energetic particles when Q increases. This is not the case. Gluon radiation results in an increase of the $\langle k_T \rangle$ of the hadrons which we associate with the original quark or antiquark jet. It is increasingly probable that in a very high-energy two-jet event, one of the observed jets is actually the fragmentation product of a qg or $\bar{\text{q}}$g state as a result of gluon emission by the q or $\bar{\text{q}}$. The experimentalist will recognize such a jet as being "fatter" or having increased $\langle k_T \rangle$. This dynamical broadening of $\langle \theta \rangle$ almost compensates for the kinematic narrowing given by (11.43). As a consequence, it turns out that the narrowing of jets with increasing Q^2 is a logarithmic and not a linear effect, with

$$\langle \theta \rangle \sim \frac{1}{\log Q^2}. \tag{11.44}$$

This result has to be derived with care. We shall only briefly sketch how it comes about. By definition,

$$\langle k_T \rangle \sim \int_0^Q k_T \frac{d\sigma}{dk_T}. \tag{11.45}$$

Let us assume that the k_T of the hadrons in the broadened jet just reflects the relative k_T of the q or $\bar{\text{q}}$ and the emitted gluon. Then, $d\sigma/dk_T$ in (11.45) is nothing but the familiar transverse momentum distribution of gluon emission

$$\frac{d\sigma}{dk_T} \sim \frac{\alpha_s}{k_T}.$$

We here neglect all logarithms and therefore also take α_s to be constant. In this approximation, $d\sigma/dk_T$ is the same as (11.42). Substituting into (11.45) gives

$$\langle k_T \rangle \sim \alpha_s \int_0^Q dk_T \sim \alpha_s Q,$$

and from (11.43) we see that $\langle \theta \rangle \sim \alpha_s$. The logarithmic narrowing of the jets (11.44) comes about when we incorporate the running of $\alpha_s \sim 1/\log Q^2$. A proper justification of this result is complicated (e.g., Dokshitzer et al., 1980). However, the phenomenological message is clear: jet identification which at low energies is only possible through clever statistical analysis of $e^+e^- \to$ hadron events, should be simpler for high-energy experiments due to the increased collimation predicted by (11.44).

11.7 QCD Corrections to $e^-e^+ \to$ Hadrons

The parton model result (11.6),

$$R \equiv \frac{\sigma(e^-e^+ \to \text{hadrons})}{\sigma(e^-e^+ \to \mu^-\mu^+)} = 3\sum_q e_q^2, \tag{11.46}$$

which we derived from $\sigma(e^-e^+ \to q\bar{q})$, will be modified by the possibility of gluon emission from the q or $\bar{q}$. The order α_s diagrams are shown in Fig. 11.10, and the cross section $d\sigma/dx_q\,dx_{\bar{q}}$ is given in (11.41). To calculate the order α_s correction to R, we must integrate over both x_q and $x_{\bar{q}}$ from 0 to 1. In doing this, we encounter a problem which is common in perturbative QCD calculations. The integrand, (11.41), diverges as x_q or $x_{\bar{q}}$ goes to 1. To trace the origin of the problem, consider, for instance, the factor $1 - x_q$ in the denominator. Using (11.33),

$$\begin{aligned} 1 - x_q &= \frac{s}{Q^2} = \frac{2p_{\bar{q}} \cdot p_g}{Q^2} \\ &= \frac{2}{Q^2} E_{\bar{q}} E_g (1 - \cos\theta_{\bar{q}g}) \end{aligned} \tag{11.47}$$

and so $1 - x_q$ vanishes when the gluon becomes soft ($E_g \to 0$) or when $\bar{q}$ and g become collinear ($\cos\theta_{\bar{q}g} \to 1$). The first type of divergence is called an *infrared divergence* (see Chapter 7), and the second is called a *collinear divergence* (or mass singularity since, if the quark *or* the gluon had nonzero mass, $\cos\theta_{\bar{q}g} = 1$ would be kinematically impossible). In QED, where leptons do have a mass, such mass singularities do not occur.

To proceed, we must regularize these infrared and mass singularities. One way to accomplish this is to give the gluon a fictitious mass m_g and to repeat the calculation of the Feynman diagrams of Fig. 11.10 which led to (11.41). A straightforward but lengthy calculation yields

$$\sigma_{\text{real}} =$$

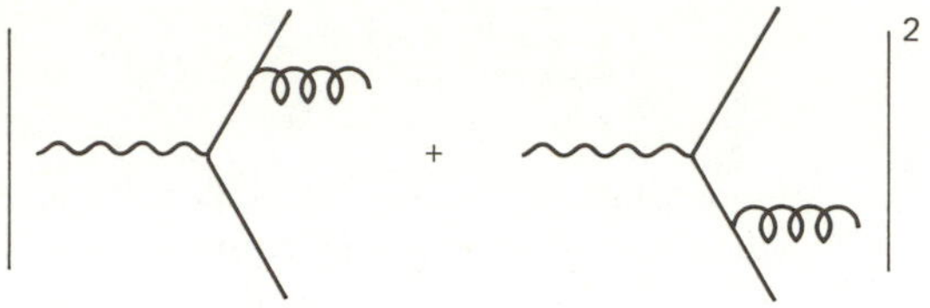

$$= \int dx_q \, dx_{\bar{q}} \frac{d\sigma}{dx_q \, dx_{\bar{q}}}$$

$$= \sigma_q \frac{\alpha_s}{2\pi} \frac{4}{3} \left\{ \log^2\left(\frac{m_g}{Q}\right) + 3 \log\left(\frac{m_g}{Q}\right) - \frac{\pi^2}{3} + 5 \right\}, \tag{11.48}$$

where σ_q denotes $\sigma(e^- e^+ \to q\bar{q})$. As expected, (11.48) is divergent when $m_g \to 0$.
Obviously, (11.48) cannot be the final answer since it depends on the fictitious mass m_g, whereas the correct result should not. However, there is another $O(\alpha_s)$ contribution. This additional $O(\alpha_s)$ term occurs in

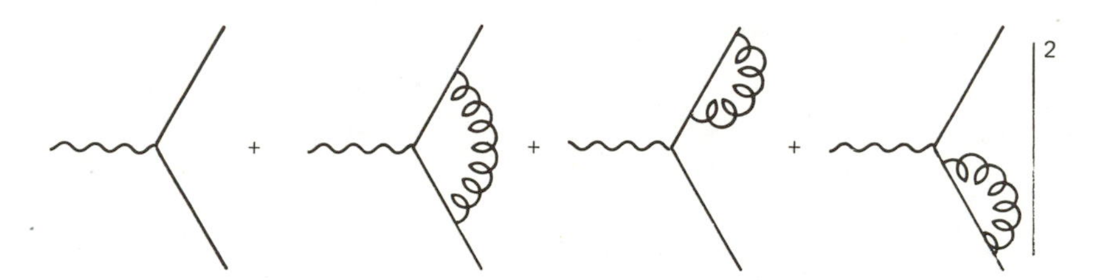

and corresponds to the interference of the $\gamma^* \to q\bar{q}$ diagram, with the sum of the three diagrams containing virtual gluon loops (see Section 10.8). This interference term gives a contribution

$$\sigma_{\text{virtual}} = \sigma_q \frac{\alpha_s}{2\pi} \frac{4}{3} \left\{ -\log^2\left(\frac{m_g}{Q}\right) - 3 \log\left(\frac{m_g}{Q}\right) + \frac{\pi^2}{3} - \frac{7}{2} \right\}. \tag{11.49}$$

The total order α_s contribution is

$$\sigma = \sigma_{\text{real}} + \sigma_{\text{virtual}} = \sigma_q \frac{\alpha_s}{\pi}, \tag{11.50}$$

which is finite and independent of m_g, as indeed it must be.
The cancellation of the singularities between the contributions with the emission of real and virtual gluons is not just a property of this particular process. It occurs over and over again. For example, in Section 10.8, we already discussed the analogous cancellation in deep inelastic scattering. These particular cancellations are examples of a general theorem due to Kinoshita, Lee, and Nauenberg. The reason that we did not meet this mechanism earlier in the chapter is because we choose to compute the cross section for the production of gluons with a fixed,

nonzero p_T. Then, x_q and $x_{\bar{q}}$ are unable to reach 1, and the denominator in (11.41) cannot vanish. This choice also excludes the diagrams with the virtual gluon loops for which $p_T = 0$.

Including the order α_s correction to R of (11.46), we obtain

$$R = 3\sum_q e_q^2\left[1 + \frac{\alpha_s(Q^2)}{\pi}\right]. \tag{11.51}$$

For a typical Q^2 where $\alpha_s \sim 0.2$, the correction is small and cannot easily be distinguished within the errors of the present measurements. Note, however, that R is no longer exactly constant since α_s is a function of Q^2, see (7.65). This is another example of a scaling parton model result being modified by $\log Q^2$ corrections arising from the emission of gluons.

11.8 Perturbative QCD

We have seen that hadrons with unexpectedly large transverse momenta in both deep inelastic scattering (ep → hX) and e^-e^+ annihilation ($e^-e^+ \to$ h) have a common origin. This is an example of parton diagrams for one process being used, after crossing, in another process. In this way, the same underlying QCD process can be tested in totally different experimental situations. Some examples are shown in Fig. 11.13. The ingredients are q, $\bar{q}$, g, γ, and γ^*; a lepton pair (i.e., e^-e^+, $\mu^-\mu^+$); the structure functions $f(x)$, giving the probability of finding partons in the parent hadron; and, finally, the fragmentation functions $D(z)$, denoting the probability of final-state hadrons emerging from the partons. Figure 11.13 shows just some of the possible ways of combining these ingredients.

To be a useful test of QCD, it is important that the process is a short-distance interaction, so that $\alpha_s(Q^2)$ is small enough for perturbation theory to be valid. The processes of Fig. 11.13 satisfy this requirement. We must also recall from Chapter 9 the criteria for the validity of the parton model (which is the lowest-order approximation to QCD). A large energy and a large momentum transfer are required to justify the impulse approximation used in the parton model calculation. Through the uncertainty principle, the large energy requirement guarantees that the time scale of the parton interaction is short, so that we can ignore the interactions with "spectator" partons (that is, with constituents which are not directly involved in the QCD subprocess). A large momentum (e.g., Q^2, p_T^2, a heavy quark mass) guarantees that the process occurs at a short range so that α_s is small.

As an example, consider processes (d) and (e) of Fig. 11.13. Even when the initial protons collide with very high energy, the momentum transferred in the process also has to be large for perturbative QCD to be valid. This can be ensured by requiring that the photon be produced with a large transverse momentum,

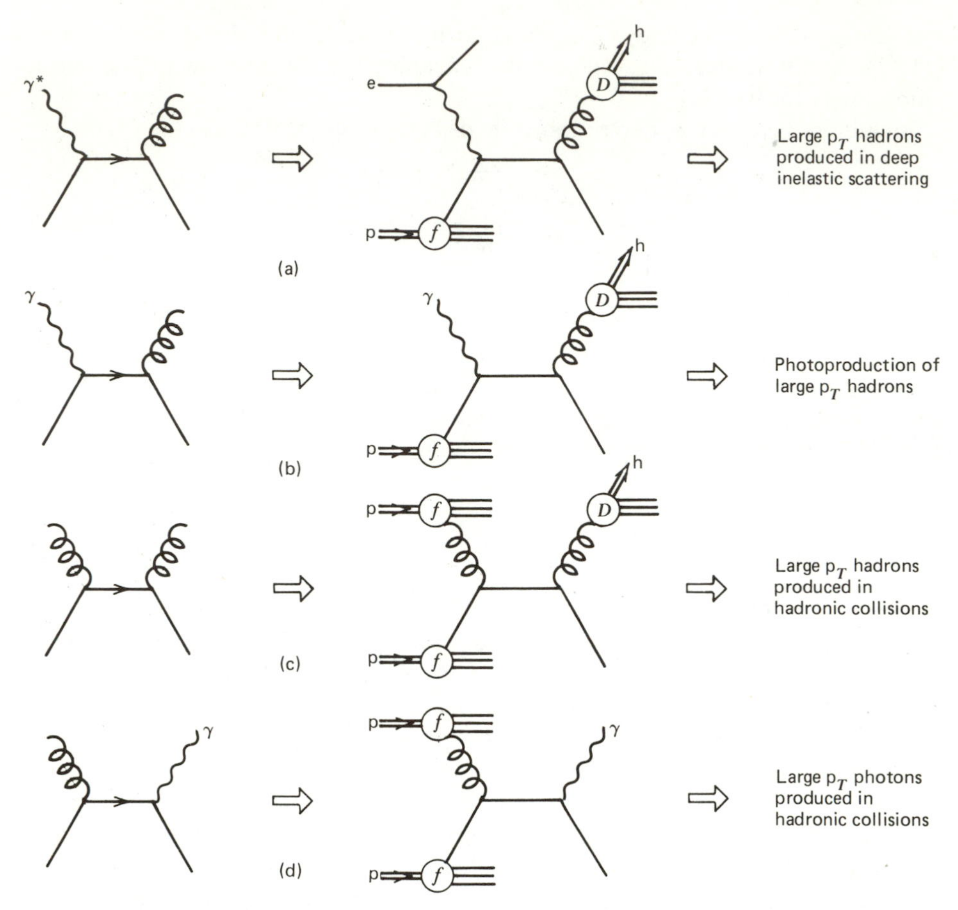

Fig. 11.13 Various observable processes which contain γ (or gluon)–quark Compton scattering, or the crossed reaction, as a parton subprocess; f and D denote structure and fragmentation functions, respectively.

process (d), or by giving it a large Q^2 by detecting a lepton pair in the final state with a large invariant mass, process (e).

Figure 11.13 shows only one parton diagram for each process. To the same order of α_s, other diagrams have to be computed to obtain a result which can be compared with experiment. For instance, in process (c), another possibility is quark–quark scattering via gluon exchange, whereby quarks in the beam scatter at large angles off target quarks producing two quark jets in the final state with large transverse momentum (see also Exercise 11.7).

There are many applications of perturbative QCD. Reviews are given, for example, by Field (1979), Ellis and Sachrajda (1979), Reya (1981), Altarelli (1982), Collins and Martin (1982), and Pennington (1983). The phenomenological evidence is that the theory can successfully confront experiment in all such

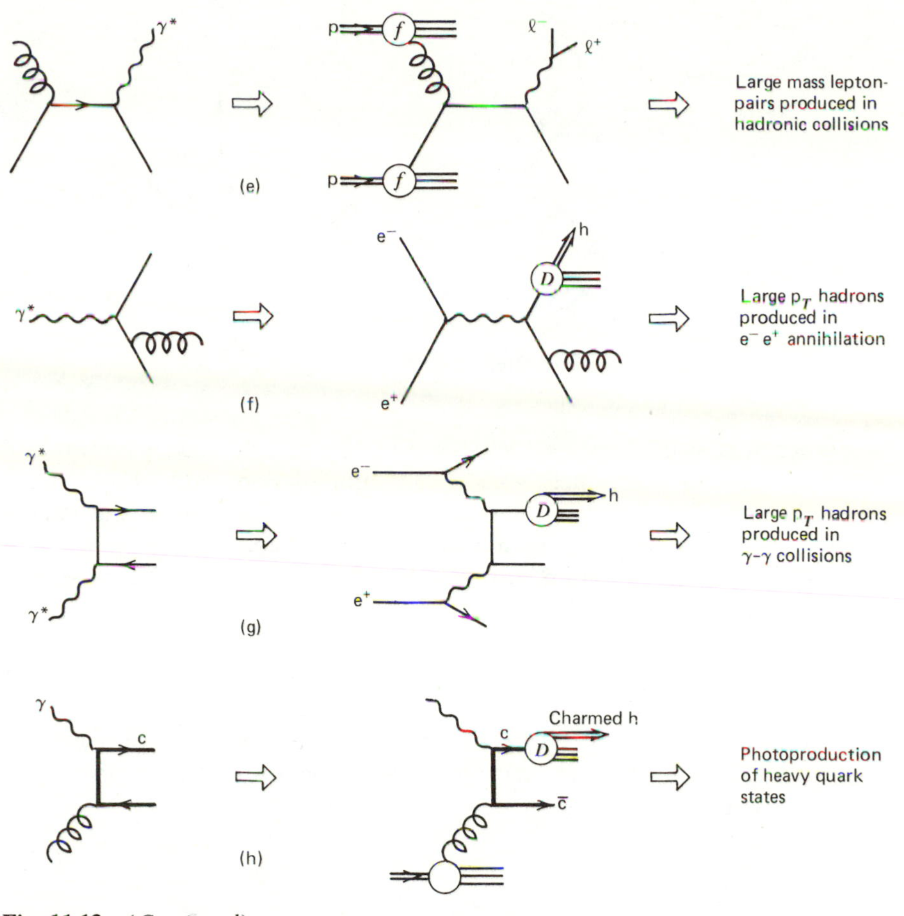

Fig. 11.13 (*Continued*)

situations. Unfortunately, however, due primarily to color confinement, no QCD tests exist (or can even be envisaged) that are comparable to the accuracy of the lepton magnetic moment calculation in QED.

> ***EXERCISE 11.7*** List all parton processes, in addition to those shown in Fig. 11.13, that can contribute to reaction (c) to the same order of α_s. Comment on the relative strengths of the subprocesses, comparing, in particular, pp- and p$\bar{\text{p}}$-initiated reactions.

11.9 A Final Example: The Drell–Yan Process

Diagram (e) of Fig. 11.13 gives an $O(\alpha^2\alpha_s)$ contribution to the cross section for the hadronic production of lepton pairs. However, this is a correction to the

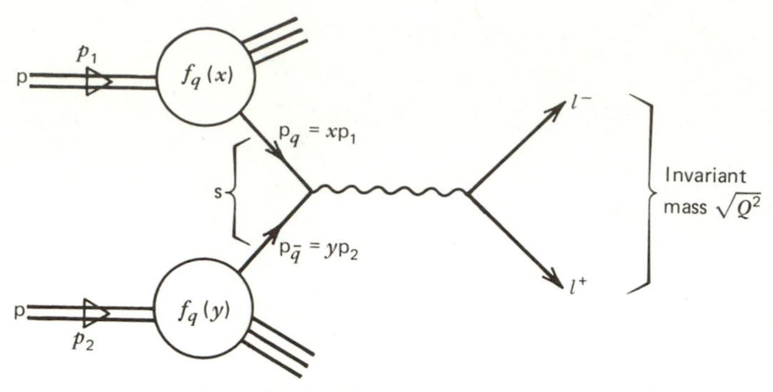

Fig. 11.14 The Drell–Yan process, $pp \to l^- l^+ X$.

lower-order process $q\bar{q} \to \gamma^* \to l^- l^+$ shown in Fig. 11.14, which is known as the Drell–Yan process. Together with e^-e^+ annihilation and deep inelastic leptoproduction, it has played an important role in determining the structure functions and in testing the parton model and its QCD corrections.

To calculate the cross section corresponding to Fig. 11.14, we begin with that for the parton subprocess,

$$\hat{\sigma}(q\bar{q} \to l^- l^+) = \frac{4\pi\alpha^2}{3Q^2} e_q^2, \tag{11.52}$$

see (11.3). In order to embed (11.52) in the hadronic process, we first rewrite it as a differential cross section, $d\sigma/dQ^2$, for the production of lepton pairs having invariant mass $\sqrt{Q^2}$, where

$$Q^2 = \hat{s} = (p_q + p_{\bar{q}})^2. \tag{11.53}$$

That is,

$$\frac{d\hat{\sigma}}{dQ^2} = \frac{4\pi\alpha^2}{3Q^2} e_q^2 \, \delta(Q^2 - \hat{s}). \tag{11.54}$$

The hadronic cross section can now be obtained using the structure functions $f_i(x)$, familiar from deep inelastic scattering (Chapter 9). We obtain

$$\frac{d\sigma}{dQ^2}(pp \to l\bar{l}X) = 3\sum_q \int dx \int dy\, f_q(x)\, f_{\bar{q}}(y) \frac{d\hat{\sigma}}{dQ^2}, \tag{11.55}$$

where the sum is over all the possible $q\bar{q}$ pairs that can be formed from the constituents of the colliding protons. There is only a factor 3 from color, as the q and $\bar{q}$ must annihilate to form a colorless virtual photon. The q and $\bar{q}$ carry fractions x and y of the momenta of the protons, respectively, and so (11.53) becomes

$$\hat{s} = (xp_1 + yp_2)^2 \simeq xys, \tag{11.56}$$

where $s \simeq 2p_1 \cdot p_2$ is the center-of-mass energy squared of the colliding protons. From (11.54)–(11.56), we obtain

$$\boxed{\frac{d\sigma}{dQ^2}(\text{pp} \to l\bar{l}\text{X}) = \frac{4\pi\alpha^2}{Q^4}\sum_q e_q^2 \int dx \int dy\, f_q(x)\, f_{\bar{q}}(y)\, \delta\left(1 - xy\frac{s}{Q^2}\right).} \tag{11.57}$$

In lowest order, with no gluon emission, we expect a scaling result. This is hidden in (11.57). Although the cross section is a function of both the collision energy $\sqrt{s}$ and the lepton pair mass $\sqrt{Q^2}$, the quantity

$$Q^4 \frac{d\sigma}{dQ^2} = \mathcal{F}\left(\frac{s}{Q^2}\right) \tag{11.58}$$

is a function only of the ratio s/Q^2. Figure 11.15 shows that the data satisfy this scaling law rather well.

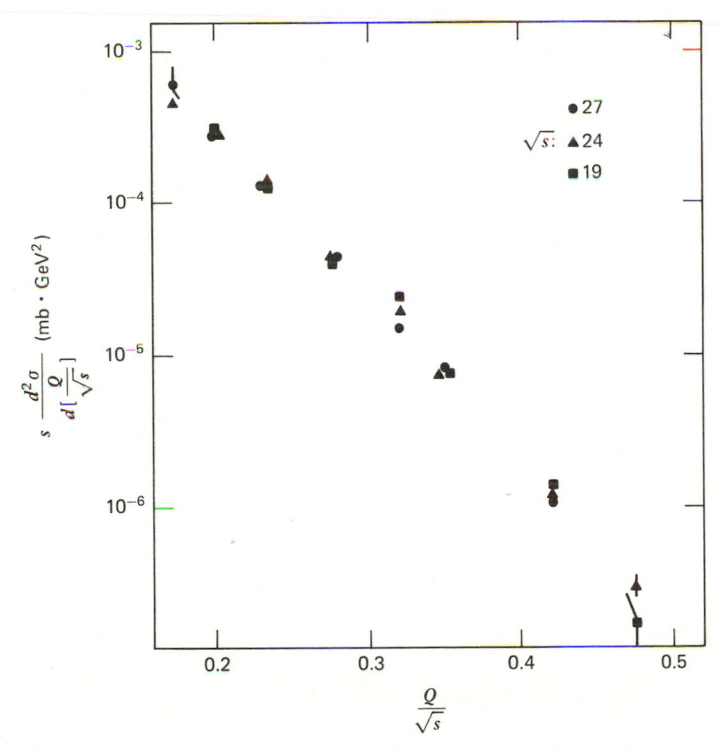

Fig. 11.15 Scaling in the Drell–Yan process. The scaling test shown is equivalent to (11.58); $\sqrt{s}$ in GeV. Data are from the Fermi laboratory.

EXERCISE 11.8 Express the sum in (11.57) in terms of the valence and sea quark structure functions introduced in Chapter 9. Do the same for lepton pair production in $\bar{p}p$ and $\pi^{\pm}p$ interactions.

EXERCISE 11.9 The data for an (isoscalar) carbon target indicate that the ratio

$$\frac{\sigma(\pi^+ C \rightarrow \mu^- \mu^+ X)}{\sigma(\pi^- C \rightarrow \mu^- \mu^+ X)}$$

is approximately unity for small values of Q^2/s but decreases toward $\frac{1}{4}$ as Q^2/s approaches 1. Explain why this is in agreement with the Drell–Yan prediction. Note that $xy = Q^2/s$.

EXERCISE 11.10 Including diagrams with gluons introduces logarithmic scaling violations in (11.58). Draw the diagrams which give an $O(\alpha^2 \alpha_s)$ contribution to the lepton pair cross section. Compute the cross section for the diagram shown in Fig. 11.13e.

In explicit calculations it is important to realize that f_q, $f_{\bar{q}}$ in (11.55) describe quarks having specific colors and are therefore 1/3 of the corresponding functions measured in deep inelastic scattering.

12
Weak Interactions

The observed lifetimes of the pion and muon are considerably longer than those of particles which decay either through color (i.e., strong) or electromagnetic interactions. It is found that

$$\begin{aligned} \pi^- &\to \mu^- \bar{\nu}_\mu \qquad \text{with } \tau = 2.6 \times 10^{-8} \text{ sec}, \\ \mu^- &\to e^- \bar{\nu}_e \nu_\mu \qquad \text{with } \tau = 2.2 \times 10^{-6} \text{ sec}, \end{aligned} \tag{12.1}$$

whereas particles decay by color interactions in about 10^{-23} sec and through electromagnetic interactions in about 10^{-16} sec (for example, $\pi^0 \to \gamma\gamma$). The lifetimes are inversely related to the coupling strength of these interactions, with the longer lifetime of the π^0 reflecting the fact that $\alpha \ll \alpha_s$. The pion and muon decays are evidence for another type of interaction with an even weaker coupling than electromagnetism.

Though all hadrons and leptons experience this weak interaction, and hence can undergo weak decays, they are often hidden by the much more rapid color or electromagnetic decays. However, the $\pi^\pm$ and μ are special. They cannot decay via the latter two interactions. The π is the lightest hadron. Whereas the neutral π can decay into photons, the charged pions cannot. As a result, the weak decay given in (12.1) is the dominant one. The reason why (12.1) is the dominant decay of the μ is interesting. In principle, the μ could decay electromagnetically via $\mu \to e\gamma$. The fact that the decay mode $\mu \to e\gamma$ is not seen and that the particular decay modes (12.1) occur are evidence for additive conserved lepton numbers: the electron number (L_e) and the muon number (L_μ). For example, the electron number assignments are

$$\begin{aligned} L_e &= +1: \qquad e^- \text{ and } \nu_e, \\ L_e &= -1: \qquad e^+ \text{ and } \bar{\nu}_e, \\ L_e &= 0: \qquad \text{all other particles.} \end{aligned} \tag{12.2}$$

Similar assignments are made for L_μ and L_τ. Clearly, $L_\mu = 1$ and $L_e = 0$ for both the initial and final states of $\mu^- \to e^- \bar{\nu}_e \nu_\mu$, so this decay is consistent with the conservation of these quantum numbers; but $\mu^- \to e^- \gamma$ is not. In fact, known reactions conserve these three lepton numbers separately (see Section 12.12 for a further discussion).

EXERCISE 12.1 Give the π^+ and μ^+ decay processes. List the possible decay modes of the τ^- lepton (the τ is the third lepton in the sequence e, μ, τ with a mass $m_\tau = 1.8$ GeV).

The two examples of weak decays given in (12.1) involve neutrinos. Neutrinos are unique in that they can only interact by weak interactions. They are colorless and electrically neutral and, within experimental limits, also massless. Neutrinos are frequently found among the products of a weak decay, but not always. For example, a K^+ meson has the following weak decay modes:

$$\left.\begin{aligned} &K^+ \to \mu^+\nu_\mu, e^+\nu_e \\ &K^+ \to \pi^0\mu^+\nu_\mu, \pi^0 e^+\nu_e \end{aligned}\right\} \quad \text{semileptonic decays,}$$
$$K^+ \to \pi^+\pi^0, \pi^+\pi^+\pi^-, \pi^+\pi^0\pi^0 \quad \text{nonleptonic decays.} \tag{12.3}$$

The customary terminology is given on the right.

The weak interaction is also responsible for the β-decay of atomic nuclei, which involves the transformation of a proton to a neutron (or vice versa). Examples involving the emission of an $e^+\nu_e$ lepton pair are

$$\begin{aligned} {}^{10}\mathrm{C} &\to {}^{10}\mathrm{B}^* + e^+ + \nu_e, \\ {}^{14}\mathrm{O} &\to {}^{14}\mathrm{N}^* + e^+ + \nu_e. \end{aligned} \tag{12.4}$$

Here, one of the protons in the nucleus transforms into a neutron via

$$p \to n e^+ \nu_e. \tag{12.5}$$

For free protons, this is energetically impossible (check the particle masses), but the crossed reaction, the β-decay process

$$n \to p e^- \bar{\nu}_e, \tag{12.6}$$

is allowed and is the reason for the neutron's instability (mean life 920 sec). Without the weak interaction, the neutron would be as stable as the proton, which has a lifetime in excess of 10^{30} years.

12.1 Parity Violation and the *V–A* Form of the Weak Current

Fermi's explanation of β-decay (1932) was inspired by the structure of the electromagnetic interaction. Recall that the invariant amplitude for electromagnetic electron–proton scattering (Fig. 12.1) is

$$\mathfrak{M} = \left(e\bar{u}_p\gamma^\mu u_p\right)\left(\frac{-1}{q^2}\right)\left(-e\bar{u}_e\gamma_\mu u_e\right), \tag{12.7}$$

see (6.8), where we have treated the proton as a structureless Dirac particle. $\mathfrak{M}$ is the product of the electron and proton electromagnetic currents, together with the propagator of the exchanged photon, see Section 6.2. To facilitate the comparison

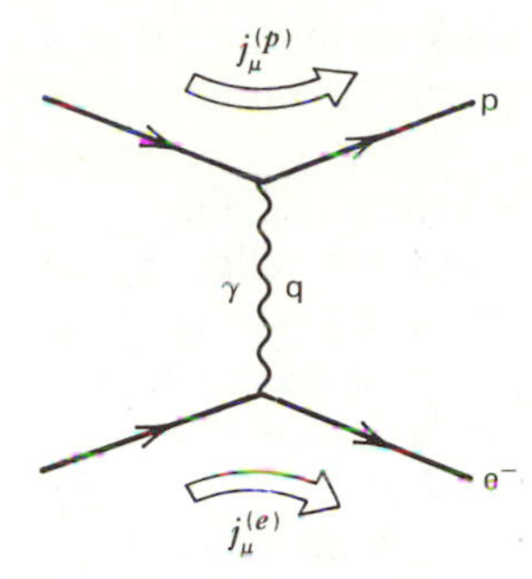

Fig. 12.1 Electron–proton (electromagnetic) scattering.

with weak interactions, we define, for example, an electron electromagnetic current of the form

$$e j_\mu^{em} \equiv j_\mu^{fi}(0) = -e\bar{u}_f \gamma_\mu u_i,$$

where $j_\mu^{fi}(x)$ is given by (6.6). Thus, the invariant amplitude, (12.7), becomes

$$\mathfrak{M} = -\frac{e^2}{q^2}\left(j_\mu^{em}\right)_p \left(j^{em\mu}\right)_e.$$

The β-decay process (12.5), or its crossed form

$$\mathrm{pe}^- \to \mathrm{n}\nu_e,$$

is shown in Fig. 12.2. By analogy with the current–current form of (12.7), Fermi proposed that the invariant amplitude for β-decay be given by

$$\mathfrak{M} = G\left(\bar{u}_n \gamma^\mu u_p\right)\left(\bar{u}_{\nu_e} \gamma_\mu u_e\right), \tag{12.8}$$

where G is the weak coupling constant which remains to be determined by experiment; G is called the Fermi constant. Note the charge-raising or charge-lowering structure of the weak current. We speak of these as the "charged weak currents." (The existence of a weak current that is electrically neutral, like the electromagnetic current, was not revealed until much later in 1973, see Section 12.9). Also note the absence of a propagator in (12.8). We return to this point in Section 12.2.

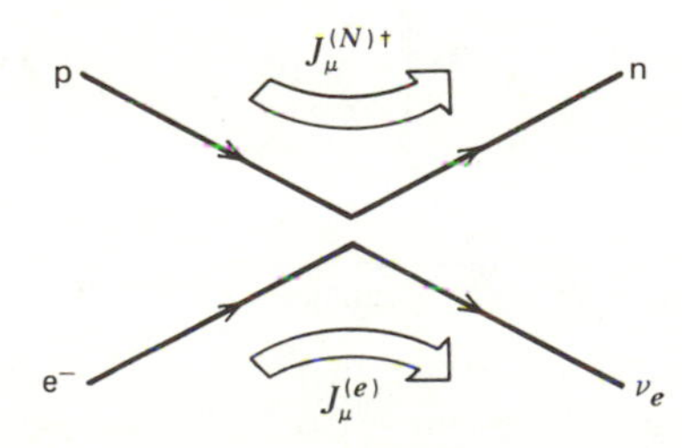

Fig. 12.2 The diagram for β-decay, $\mathrm{p} \to \mathrm{ne}^+\nu_e$, showing the weak currents.

Fermi's inspired guess of a vector–vector form of the weak amplitude $\mathcal{M}$ is a very specific choice from among the various Lorentz invariant amplitudes that can in general be constructed using the bilinear covariants of (5.52). There is *a priori* no reason to use only vectors. The amplitude (12.8) explained the properties of some features of β-decay, but not others. Over the following 25 years or so, attempts to unravel the true form of the weak interaction led to a whole series of ingenious β-decay experiments, reaching a climax with the discovery of parity violation in 1956. Amazingly, the only essential change required in Fermi's original proposal was the replacement of γ^μ by $\gamma^\mu(1 - \gamma^5)$. Fermi had not foreseen parity violation and had no reason to include a $\gamma^5\gamma^\mu$ contribution; a mixture of γ^μ and $\gamma^5\gamma^\mu$ terms automatically violates parity conservation, see (5.67).

In 1956, Lee and Yang made a critical survey of all the weak interaction data. A particular concern at the time was the observed nonleptonic decay modes of the kaon, $K^+ \to 2\pi$ and 3π, in which the two final states have opposite parities. (People, in fact, believed that two different particles were needed to explain the two final states.) Lee and Yang argued persuasively that parity was not conserved in weak interactions. Experiments to check their assertion followed immediately. The first of these historic experiments serves as a good illustration of the effects of parity violation. The experiment studied β-transitions of polarized cobalt nuclei:

$$^{60}\text{Co} \to {}^{60}\text{Ni}^* + e^- + \bar{\nu}_e.$$

The nuclear spins in a sample of ^{60}Co were aligned by an external magnetic field, and an asymmetry in the direction of the emitted electrons was observed. The asymmetry was found to change sign upon reversal of the magnetic field such that electrons prefer to be emitted in a direction opposite to that of the nuclear spin. The essence of the argument is sketched in Fig. 12.3. The observed correlation between the nuclear spin and the electron momentum is explained if the required $J_z = 1$ is formed by a right-handed antineutrino, $\bar{\nu}_R$, and a left-handed electron, e_L.

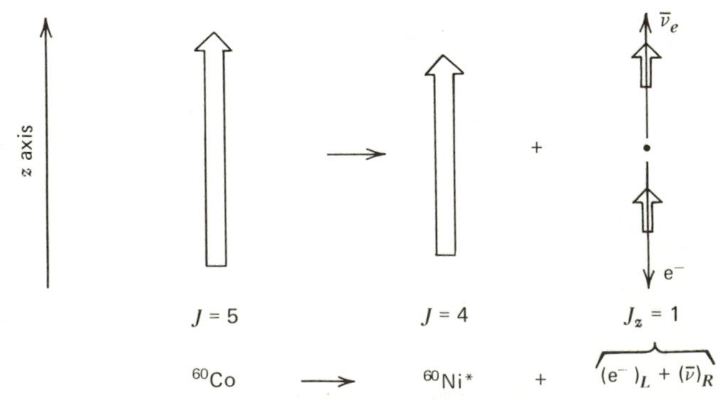

Fig. 12.3 The ^{60}Co experiment: the electron is emitted preferentially opposite the direction of the spin of the ^{60}Co nucleus.

The cumulative evidence of many experiments is that indeed only $\bar{\nu}_R$ (and ν_L) are involved in weak interactions. The absence of the "mirror image" states, $\bar{\nu}_L$ and ν_R, is a clear violation of parity invariance (see Section 5.7). Also, charge conjugation, C, invariance is violated, since C transforms a ν_L state into a $\bar{\nu}_L$ state. However, the $\gamma^\mu(1-\gamma^5)$ form leaves the weak interaction invariant under the combined CP operation. For instance,

$$\Gamma(\pi^+ \to \mu^+ \nu_L) \neq \Gamma(\pi^+ \to \mu^+ \nu_R) = 0 \qquad \text{P violation,}$$

$$\Gamma(\pi^+ \to \mu^+ \nu_L) \neq \Gamma(\pi^- \to \mu^- \bar{\nu}_L) = 0 \qquad \text{C violation,}$$

but

$$\Gamma(\pi^+ \to \mu^+ \nu_L) = \Gamma(\pi^- \to \mu^- \bar{\nu}_R) \qquad \text{CP invariance.}$$

In this example, ν denotes a muon neutrino. We discuss CP invariance in Section 12.13.

EXERCISE 12.2 Show that a (charge-lowering) weak current of the form

$$\bar{u}_e \gamma^\mu \tfrac{1}{2}(1-\gamma^5) u_\nu \tag{12.9}$$

involves only left-handed electrons (or right-handed positrons). In the relativistic limit ($v \approx c$), show that the electrons have negative helicity.

The $\frac{1}{2}(1-\gamma^5)$ in (12.9) automatically selects a left-handed neutrino (or a right-handed antineutrino). This V–A (vector–axial vector) structure of the weak current can be directly exposed by scattering ν_e's off electrons (see Section 12.7), just as the γ^μ structure of electromagnetism was verified by measurements of the angular distribution of e^+e^- scattering.

It is natural to hope that all weak interaction phenomena are described by a V–A current–current interaction with a universal coupling G. For example, β-decay of Fig. 12.2 and μ-decay of Fig. 12.4 can be described by the amplitudes

$$\mathcal{M}(\mathrm{p} \to \mathrm{n} e^+ \nu_e) = \frac{G}{\sqrt{2}} \left[\bar{u}_n \gamma^\mu (1-\gamma^5) u_p\right]\left[\bar{u}_{\nu_e} \gamma_\mu (1-\gamma^5) u_e\right] \tag{12.10}$$

and

$$\mathcal{M}(\mu^- \to e^- \bar{\nu}_e \nu_\mu) = \frac{G}{\sqrt{2}} \left[\bar{u}_{\nu_\mu} \gamma^\sigma (1-\gamma^5) u_\mu\right]\left[\bar{u}_e \gamma_\sigma (1-\gamma^5) u_{\nu_e}\right], \tag{12.11}$$

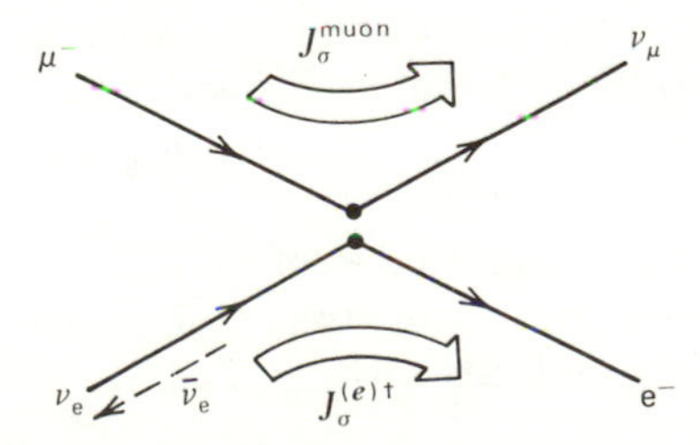

Fig. 12.4 The diagram for μ^- decay: $\mu^- \to e^- \bar{\nu}_e \nu_\mu$.

respectively. The $1/\sqrt{2}$ is pure convention (to keep the original definition of G which did not include γ^5). We then proceed in analogy with the Feynman rules for QED. The calculations only involve particles, and the diagrams show only particle lines. Antiparticles do not appear. Thus, the outgoing $\bar{\nu}_e$ (of momentum k) in μ-decay is shown in Fig. 12.4 as an ingoing ν_e (of momentum $-k$). As before, the spinor $u_{\nu_e}(-k)$ of (12.11) will be denoted $v_{\nu_e}(k)$, see (5.33). The same remarks apply to the outgoing e^+ of (12.10).

EXERCISE 12.3 Show that the *charge-raising* weak current

$$J^\mu = \bar{u}_\nu \gamma^\mu \tfrac{1}{2}(1 - \gamma^5) u_e \tag{12.12}$$

couples an ingoing negative helicity electron to an outgoing negative helicity neutrino. Neglect the mass of the electron. Besides the configuration (e_L^-, ν_L), show that J^μ also couples the following (ingoing, outgoing) lepton pair configurations: $(\bar{\nu}_R, e_R^+)$, $(0, \nu_L e_R^+)$, and $(e_L^- \bar{\nu}_R, 0)$.

Further, show that the *charge-lowering* weak current, (12.9), is the hermitian conjugate of (12.12):

$$J_\mu^\dagger = \bar{u}_e \gamma_\mu \tfrac{1}{2}(1 - \gamma^5) u_\nu .$$

List the lepton pair configurations coupled by $J_\mu^\dagger$.

Weak interaction amplitudes are of the form

$$\boxed{\mathcal{M} = \frac{4G}{\sqrt{2}} J^\mu J_\mu^\dagger .} \tag{12.13}$$

Charge conservation requires that $\mathcal{M}$ is the product of a charge-raising and a charge-lowering current; see, for example, (12.10) and (12.11). The factor 4 arises because the currents, (12.13), are defined with the normalized projection operator $\frac{1}{2}(1 - \gamma^5)$ rather than the old-fashioned $(1 - \gamma^5)$.

12.2 Interpretation of the Coupling G

We can use the observed rates for nuclear β-decay and for μ-decay to obtain a numerical value for G. It is also crucial to check the universality of the strength of the weak coupling constant G of (12.10) and (12.11). We do not want to introduce a new interaction for every weak process! However, because we do this, let us cast G in a form that can be directly compared to the couplings of the color and electromagnetic interactions.

Examination of the electromagnetic and the weak amplitudes of (12.7) and (12.10) shows that in Fermi's model the analogy between the two interactions has not been fully developed. We see that G essentially replaces e^2/q^2. Thus, G, in

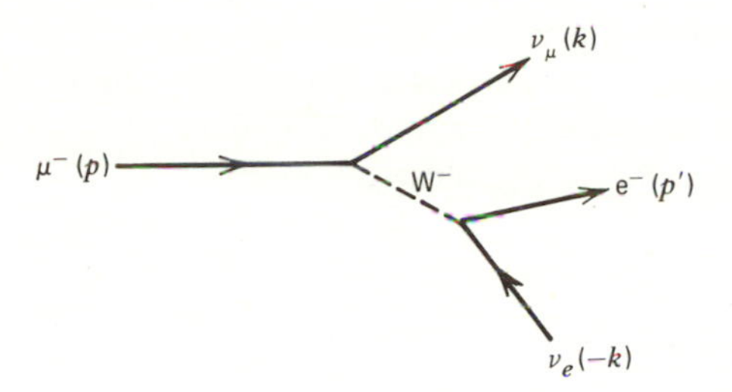

Fig. 12.5 Muon decay.

contrast to the dimensionless coupling e, has dimensions GeV^{-2}. It is tempting to try and extend the analogy by postulating that the weak interactions are generated by the emission and absorption of charged vector bosons, which we call weak bosons, $W^{\pm}$. The weak bosons are the analogues of the photon for the electromagnetic force and the gluons for the color force. For example, μ^- decay is mediated by a W^- boson (see Fig. 12.5) and the amplitude is of the form [see (12.7)]

$$\mathfrak{M} = \left(\frac{g}{\sqrt{2}}\bar{u}_{\nu_\mu}\gamma^\sigma \tfrac{1}{2}(1-\gamma^5)u_\mu\right)\frac{1}{M_W^2 - q^2}\left(\frac{g}{\sqrt{2}}\bar{u}_e\gamma_\sigma \tfrac{1}{2}(1-\gamma^5)u_{\nu_e}\right), \tag{12.14}$$

where $g/\sqrt{2}$ is a dimensionless weak coupling and q is the momentum carried by the weak boson (the factors $1/\sqrt{2}$ and $\frac{1}{2}$ are inserted so that we have the conventional definition of g). In contrast to the photon, the weak boson must be massive, otherwise it would have been directly produced in weak decays. Indeed, it turns out that $M_W \sim 80$ GeV (see Chapter 15).

In (12.14), we have been cavalier about the spin sum in the boson propagator, see (6.87). However, at the moment, we are interested in situations where $q^2 \ll M_W^2$ (e.g., β-decay and μ-decay). Then, (12.14) reverts to (12.11) with

$$\frac{G}{\sqrt{2}} = \frac{g^2}{8M_W^2} \tag{12.15}$$

and the weak currents interact essentially at a point. That is, in the limit (12.15), the propagator between the currents disappears. Equation (12.15) prompts the idea that weak interactions are weak not because $g \ll e$, but because M_W^2 is large. If indeed $g \approx e$, then at energies $O(M_W)$ and above, the weak interaction would become of comparable strength to the electromagnetic interaction.

We may think of $g \approx e$ as a unification of weak and electromagnetic interactions in much the same way as the unification of the electric and magnetic forces in Maxwell's theory of electromagnetism, where

$$\mathbf{F} = e\mathbf{E} + e_M \mathbf{v} \times \mathbf{B}$$

with $e_M = e$. At low velocities, the magnetic forces are very weak, whereas for high-velocity particles, the electric and magnetic forces play a comparable role.

The velocity of light c is the scale which governs the relative strength. The analogue for the electroweak force is M_W on the energy scale. The unification of the electromagnetic and weak forces is the subject of Chapters 13 and 15.

12.3 Nuclear β-Decay

Let us use the observed transition rate of the process

$$^{14}\mathrm{O} \to {}^{14}\mathrm{N}^* + e^+ + \nu_e$$

to estimate G. By analogy with the QED calculations of Section 6.2, we write the transition amplitude for this process (Fig. 12.2) in the form

$$T_{fi} = -i\frac{4G}{\sqrt{2}}\int J_\mu^{(N)\dagger}(x) J^{(e)\mu}(x)\, d^4x \tag{12.16}$$

$$= -i\frac{4G}{\sqrt{2}}\int \left[\bar{\psi}_n(x)\gamma_\mu \tfrac{1}{2}(1-\gamma^5)\psi_p(x)\right]\left[\bar{\psi}_\nu(x)\gamma^\mu \tfrac{1}{2}(1-\gamma^5)\psi_e(x)\right] d^4x, \tag{12.17}$$

see (12.10). For this problem, it is easier not to perform the x integration at this stage. Remember that usually we carry out the x integration and obtain $(2\pi)^4$ times the "energy–momentum conserving" delta function (see Section 6.2). We then define

$$T_{fi} = -i(2\pi)^4\,\delta^{(4)}(p_p - p_n - p_e - p_\nu)\mathfrak{M}.$$

Thus, (12.16) reduces to the form (12.13).

In writing down (12.16), we have assumed that the other nucleons in ^{14}O are simply spectators to the decaying proton. However, *a priori*, we cannot ignore the fact the nucleons participating in β-decay are bound inside nuclei. We also have no reason to believe the idealized form of weak nucleon current, $J^{(N)}$, shown in (12.17), since the nucleons themselves are composite objects and not structureless Dirac particles. Despite these problems, it turns out to be quite easy to get an accurate estimate of G. There are several reasons for this. First, the low-energy weak interaction is essentially a point interaction, and we can ignore the longer-range strong interaction effects we just mentioned. Actually, there is a beautiful and more precise justification for this vague argument. The weak current $(\bar{\psi}_p\gamma^\mu\psi_n)$ and its conjugate $(\bar{\psi}_n\gamma^\mu\psi_p)$, together with the electromagnetic current $(\bar{\psi}_p\gamma^\mu\psi_p)$, are believed to form an isospin triplet of conserved vector currents. This is referred to as the *conserved vector current hypothesis*. The intimate connection with the electromagnetic current "protects" the vector part of the weak current from any strong interaction corrections, just as the electromagnetic charge is protected. The axial vector part, $\bar{\psi}_n\gamma^\mu\gamma^5\gamma_p$, will not contribute to the process as we are considering a transition between two $J^P = 0^+$ nuclear states, which precludes a change of parity. Moreover, since the process occurs between two $J = 0$ nuclei, we can safely assume that the nuclear wavefunction is essentially unchanged by the transition.

Another simplifying feature is that the energy released in the decay (about 2 MeV) is small compared to the rest energy of the nuclei. We can therefore use nonrelativistic spinors for the nucleons [see (6.11)], and then only γ^μ with $\mu = 0$ contributes [see (6.13)]. Thus,

$$T_{fi} \simeq -i\frac{G}{\sqrt{2}}\left[\bar{u}(p_\nu)\gamma^0(1-\gamma^5)v(p_e)\right]\int \psi_n^\dagger(x)\psi_p(x)e^{-i(p_\nu+p_e)\cdot x}\,d^4x, \tag{12.18}$$

where, as remarked after (12.11), the v spinor $v(p_e)$ describes an outgoing positron of momentum p_e. The e^+ and ν are emitted with an energy of the order of 1 MeV, and so their de Broglie wavelengths are about 10^{-11} cm, which is much larger than the nuclear diameter. We can therefore set

$$e^{i(\mathbf{p}_\nu+\mathbf{p}_e)\cdot\mathbf{x}} \simeq 1$$

and perform the spatial integration of (12.18). Noting the relation between T_{fi} and the invariant amplitude $\mathcal{M}$ (see Section 6.2), we obtain

$$\mathcal{M} = \frac{G}{\sqrt{2}}\left(\bar{u}(p_\nu)\gamma^0(1-\gamma^5)v(p_e)\right)(2m_N)\left(2\sqrt{\tfrac{1}{2}}\right), \tag{12.19}$$

where $2m_N$ arises from the normalization of the nucleon spinors [see (6.13)] and $2/\sqrt{2}$ is the hadronic isospin factor for the $^{14}O \rightarrow {}^{14}N^*$ transition (see Exercise 12.4).

EXERCISE 12.4 Verify the isospin factor $\sqrt{2}$ in (12.19). Note that ^{14}C, $^{14}N^*$, ^{14}O form an isospin triplet, which can be viewed as nn, np, pp, together with an isospin zero ^{12}C core (see Fig. 2.2). Keep in mind that for indistinguishable proton decays, we must add amplitudes, not probabilities.

The rate $d\Gamma$ for the "p" → "n"$e^+\nu$ transition is related to $|\mathcal{M}|^2$ by (4.36). We obtain

$$d\Gamma = G^2 \sum_{\text{spins}} \left|\bar{u}(p_\nu)\gamma^0(1-\gamma^5)v(p_e)\right|^2 \frac{d^3p_e}{(2\pi)^3 2E_e}$$
$$\times \frac{d^3p_\nu}{(2\pi)^3 2E_\nu} 2\pi\,\delta(E_0 - E_e - E_\nu), \tag{12.20}$$

where E_0 is the energy released to the lepton pair. The normalization factor $(2m_N)^2$ cancels with the equivalent $2E_p 2E_n$ factor in (4.36), as indeed it must. The summation over spins can be performed using the techniques we introduced in Chapter 6. Neglecting the mass of the electron, we have

$$\begin{aligned}
\sum_{\text{spins}} \left|\bar{u}\gamma^0(1-\gamma^5)v\right|^2 &= \sum\left(\bar{u}\gamma^0(1-\gamma^5)v\right)\left(\bar{v}(1+\gamma^5)\gamma^0 u\right) \\
&= \mathrm{Tr}\left(\not{p}_\nu\gamma^0(1-\gamma^5)\not{p}_e(1+\gamma^5)\gamma^0\right) \\
&= 2\mathrm{Tr}\left(\not{p}_\nu\gamma^0\not{p}_e(1+\gamma^5)\gamma^0\right) \\
&= 8(E_e E_\nu + \mathbf{p}_e\cdot\mathbf{p}_\nu) \\
&= 8E_e E_\nu(1+v_e\cos\theta),
\end{aligned} \tag{12.21}$$

where θ is the opening angle between the two leptons and where the electron velocity $v_e \simeq 1$ in our approximation. Here, we have used the trace theorems of Section 6.4; see also (12.25) and (12.26). Substituting (12.21) into (12.20), the transition rate becomes

$$d\Gamma = \frac{2G^2}{(2\pi)^5}(1+\cos\theta)\left[(2\pi d\cos\theta\, p_e^2\, dp_e)(4\pi E_\nu^2 dE_\nu)\right]\delta(E_0 - E_e - E_\nu), \tag{12.22}$$

where $d^3p_e\, d^3p_\nu$ has been replaced by the expression in the square brackets.

Many experiments focus attention on the energy spectrum of the emitted positron. From (12.22), we obtain

$$\begin{aligned}
\frac{d\Gamma}{dp_e} &= \frac{4G^2}{(2\pi)^3}p_e^2(E_0 - E_e)^2\int d\cos\theta(1+\cos\theta) \\
&= \frac{G^2}{\pi^3}p_e^2(E_0-E_e)^2.
\end{aligned} \tag{12.23}$$

Thus, if from the observed positron spectrum we plot $p_e^{-1}(d\Gamma/dp_e)^{1/2}$ as a function of E_e, we should obtain a linear plot with end point E_0. This is called the Kurie plot. It can be used to check whether the neutrino mass is indeed zero. A nonvanishing neutrino mass destroys the linear behavior, particularly for E_e near E_0. (In practice, of course, we must examine the approximations we have made, correct E_e for the energy gained from the nuclear Coulomb field, and allow for the experimental energy resolution.)

Our immediate interest here is of a different nature. We wish to determine G from the observed value E_0 and the measured lifetime τ of the nuclear state to β-decay. We therefore carry out the dp_e integration of (12.23) over the interval $0, E_0$. Making the relativistic approximation $p_e \approx E_e$, we find

$$\boxed{\Gamma = \frac{1}{\tau} = \frac{G^2E_0^5}{30\pi^3}.}$$

Now, for $^{14}O \rightarrow {}^{14}N^*e^+\nu$, the nuclear energy difference E_0 is 1.81 MeV, and the measured half-life is $\tau \log 2 = 71$ sec. Using this information, we find

$$G \simeq 10^{-5}/m_N^2. \tag{12.24}$$

Recall that G has dimension $(\text{mass})^{-2}$. We have chosen to quote the value with respect to the nucleon mass.

EXERCISE 12.5 Calculate G from the data for the β-transition $^{10}\text{C} \to {}^{10}\text{B}^* e^+ \nu$. The measured half-life is $\tau \log 2 = 20$ sec, and $E_0 = 2$ MeV. (^{10}C and ^{10}B* are both isospin 1, $J^P = 0^+$ states.)

EXERCISE 12.6 Accepting the vector boson exchange picture of weak interactions with coupling, $g = e$, estimate the mass M_W of the weak boson. (In the standard model of weak interactions, introduced in Chapter 13, $g \sin\theta_W = e$, with $\sin^2\theta_W \approx \frac{1}{4}$.)

12.4 Further Trace Theorems

We collect together some results that are useful for the computation of weak interaction processes and that follow directly from the trace theorems of Section 6.4:

$$\text{Tr}(\gamma^\mu \not{p}_1 \gamma^\nu \not{p}_2) = 4\left[p_1^\mu p_2^\nu + p_1^\nu p_2^\mu - (p_1 \cdot p_2) g^{\mu\nu}\right], \tag{12.25}$$

$$\text{Tr}\left[\gamma^\mu (1-\gamma^5) \not{p}_1 \gamma^\nu (1-\gamma^5) \not{p}_2\right] = 2\text{Tr}(\gamma^\mu \not{p}_1 \gamma^\nu \not{p}_2) + 8i\varepsilon^{\mu\alpha\nu\beta} p_{1\alpha} p_{2\beta}, \tag{12.26}$$

$$\begin{aligned}&\text{Tr}(\gamma^\mu \not{p}_1 \gamma^\nu \not{p}_2)\, \text{Tr}(\gamma_\mu \not{p}_3 \gamma_\nu \not{p}_4)\\ &\quad = 32\left[(p_1 \cdot p_3)(p_2 \cdot p_4) + (p_1 \cdot p_4)(p_2 \cdot p_3)\right],\end{aligned} \tag{12.27}$$

$$\begin{aligned}&\text{Tr}(\gamma^\mu \not{p}_1 \gamma^\nu \gamma^5 \not{p}_2)\, \text{Tr}(\gamma_\mu \not{p}_3 \gamma_\nu \gamma^5 \not{p}_4)\\ &\quad = 32\left[(p_1 \cdot p_3)(p_2 \cdot p_4) - (p_1 \cdot p_4)(p_2 \cdot p_3)\right],\end{aligned} \tag{12.28}$$

$$\begin{aligned}&\text{Tr}\left[\gamma^\mu (1-\gamma^5) \not{p}_1 \gamma^\nu (1-\gamma^5) \not{p}_2\right] \text{Tr}\left[\gamma_\mu (1-\gamma^5) \not{p}_3 \gamma_\nu (1-\gamma^5) \not{p}_4\right]\\ &\quad = 256 (p_1 \cdot p_3)(p_2 \cdot p_4).\end{aligned} \tag{12.29}$$

EXERCISE 12.7 Verify these results.

12.5 Muon Decay

Muon decay,

$$\mu^-(p) \to e^-(p') + \bar{\nu}_e(k') + \nu_\mu(k), \tag{12.30}$$

is the model reaction for weak decays. The particle four-momenta are defined in (12.30), and the Feynman diagram is shown in Fig. 12.5. According to the Feynman rules, it must be drawn using only particle lines; and so the outgoing $\bar{\nu}_e$ is shown as an incoming ν_e. The invariant amplitude for muon decay is

$$\mathfrak{M} = \frac{G}{\sqrt{2}}\left[\bar{u}(k)\gamma^\mu(1-\gamma^5)u(p)\right]\left[\bar{u}(p')\gamma_\mu(1-\gamma^5)v(k')\right], \tag{12.31}$$

see (12.11), where the spinors are labeled by the particle momenta. Recall that the outgoing $\bar{\nu}_e$ is described by $v(k')$. The muon decay rate can now be obtained using (4.36),

$$d\Gamma = \frac{1}{2E}\overline{|\mathfrak{M}|^2}dQ \tag{12.32}$$

where the invariant phase space is

$$\begin{aligned} dQ &= \frac{d^3p'}{(2\pi)^3 2E'}\frac{d^3k}{(2\pi)^3 2\omega}\frac{d^3k'}{(2\pi)^3 2\omega'}(2\pi)^4\delta^{(4)}(p-p'-k-k') \\ &= \frac{1}{(2\pi)^5}\frac{d^3p'}{2E'}\frac{d^3k'}{2\omega'}\theta(E-E'-\omega')\delta\big((p-p'-k')^2\big), \end{aligned} \tag{12.33}$$

with $p^0 \equiv E$, $k^0 \equiv \omega$, and so on, and where in reaching the last line we have performed the d^3k integration using

$$\int \frac{d^3k}{2\omega} = \int d^4k\, \theta(\omega)\delta(k^2). \tag{12.34}$$

EXERCISE 12.8 Derive (12.34) by performing the $d\omega$ integration on the right-hand side (see Exercise 6.7).

Using (12.31) and (12.29), we find the spin-averaged probability is

$$\overline{|\mathfrak{M}|^2} \equiv \frac{1}{2}\sum_{\text{spins}} |\mathfrak{M}|^2 = 64G^2(k\cdot p')(k'\cdot p), \tag{12.35}$$

where $p = p' + k + k'$ on account of the d^4k integration performed in (12.33). Since $m_\mu > 200m_e$, we can safely neglect the mass of the electron.

EXERCISE 12.9 Verify (12.35). Neglect the mass of the electron, but not that of the muon.

EXERCISE 12.10 Show that

$$2(k\cdot p')(k'\cdot p) = (p-k')^2(k'\cdot p) = (m^2 - 2m\omega')m\omega' \tag{12.36}$$

in the muon rest frame, where $p = (m, 0, 0, 0)$.

Gathering these results together, the decay rate in the muon rest frame is

$$\begin{aligned} d\Gamma &= \frac{G^2}{2m\pi^5}\frac{d^3p'}{2E'}\frac{d^3k'}{2\omega'}m\omega'(m^2 - 2m\omega') \\ &\quad \times\delta\big(m^2 - 2mE' - 2m\omega' + 2E'\omega'(1-\cos\theta)\big), \end{aligned} \tag{12.37}$$

and, as for β-decay, we can replace $d^3p'\,d^3k'$ by

$$4\pi E'^2\,dE'\,2\pi\omega'^2\,d\omega' d\cos\theta.$$

We now use the fact that

$$\delta(\cdots + 2E'\omega'\cos\theta) = \frac{1}{2E'\omega'}\delta(\cdots - \cos\theta)$$

to perform the integration over the opening angle θ between the emitted e^- and $\bar{\nu}_e$ and obtain

$$d\Gamma = \frac{G^2}{2\pi^3}dE'\,d\omega'\,m\omega'(m - 2\omega'). \tag{12.38}$$

The δ-function integration introduces the following restrictions on the energies E', ω', stemming from the fact that $-1 \leq \cos\theta \leq 1$:

$$\tfrac{1}{2}m - E' \leq \omega' \leq \tfrac{1}{2}m, \tag{12.39}$$

$$0 \leq E' \leq \tfrac{1}{2}m. \tag{12.40}$$

These limits are easily understood in terms of the various limits in which the three-body decay $\mu \to e\bar{\nu}_e\nu_\mu$ becomes effectively a two-body decay. For example, when the electron energy E' vanishes, (12.39) yields $\omega' = m/2$, which is expected because then the two neutrinos share equally the muon's rest energy.

To obtain the energy spectrum of the emitted electron, we perform the ω' integration of (12.38):

$$\frac{d\Gamma}{dE'} = \frac{mG^2}{2\pi^3}\int_{\frac{1}{2}m - E'}^{\frac{1}{2}m} d\omega'\,\omega'(m - 2\omega')$$

$$= \frac{G^2}{12\pi^3}m^2E'^2\left(3 - \frac{4E'}{m}\right). \tag{12.41}$$

This prediction is in excellent agreement with the observed electron spectrum. Finally, we calculate the muon decay rate

$$\boxed{\Gamma \equiv \frac{1}{\tau} = \int_0^{m/2} dE'\,\frac{d\Gamma}{dE'} = \frac{G^2m^5}{192\pi^3}.} \tag{12.42}$$

Inserting the measured muon lifetime $\tau = 2.2 \times 10^{-6}$ sec, we can calculate the Fermi coupling G. We find

$$G \sim 10^{-5}/m_N^2. \tag{12.43}$$

Comparison of the values of G obtained in (12.24) and (12.43) supports the assertion that the weak coupling constant is the same for leptons and nucleons, and hence universal. It means that nuclear β-decay and the decay of the muon have the same physical origin. Indeed, when all corrections are taken into account, G_β and G_μ are found to be equal to within a few percent:

$$G_\mu = (1.16632 \pm 0.00002) \times 10^{-5}\ \mathrm{GeV}^{-2},$$
$$G_\beta = (1.136 \pm 0.003) \times 10^{-5}\ \mathrm{GeV}^{-2}. \tag{12.44}$$

The reason for the small difference is important and is discussed in Section 12.11 [see (12.107)].

EXERCISE 12.11 Draw a diagram showing the particle helicities in the μ^- rest frame in the case where the emitted electron has its maximum permissible energy. In this limit, explain why the electron angular distribution has the form $1 - P\cos\alpha$, where $\mathbf{P}$ is the polarization of the muon and α is the angle between the polarization direction and the direction of the emitted electron:

$$P \equiv \frac{N_+ - N_-}{N_+ + N_-},$$

where $N_\pm$ are the numbers of spin-up, spin-down muons.

EXERCISE 12.12 "Predict" the rate for the decay $\tau^- \to e^- \bar{\nu}_e \nu_\tau$, where the τ-lepton has mass 1.8 GeV. The observed branching ratio of this decay mode is approximately 20%. Calculate the lifetime of the τ-lepton. Can you explain this branching ratio?

12.6 Pion Decay

Can we now also understand the lifetime of the $\pi^\pm$-mesons? To be specific, we take the decay

$$\pi^-(q) \to \mu^-(p) + \bar{\nu}_\mu(k), \tag{12.45}$$

which is shown in Fig. 12.6. The amplitude is of the form

$$\mathfrak{M} = \frac{G}{\sqrt{2}}(\ldots)^\mu \bar{u}(p)\gamma_\mu(1-\gamma^5)v(k) \tag{12.46}$$

where $(\ldots)$ represents the weak quark current of Fig. 12.6. It is tempting to write it as $\bar{u}_d\gamma^\mu(1-\gamma^5)v_{\bar{u}}$, but this is incorrect since the ū, d quarks in Fig. 12.6 are not free quark states but are quarks bound into a π^--meson. We know, however, that

- $\mathfrak{M}$ is Lorentz invariant, so that $(\ldots)^\mu$ must be a vector or axial-vector, as indicated.
- The π^- is spinless, so that q is the only four-vector available to construct $(\ldots)^\mu$.

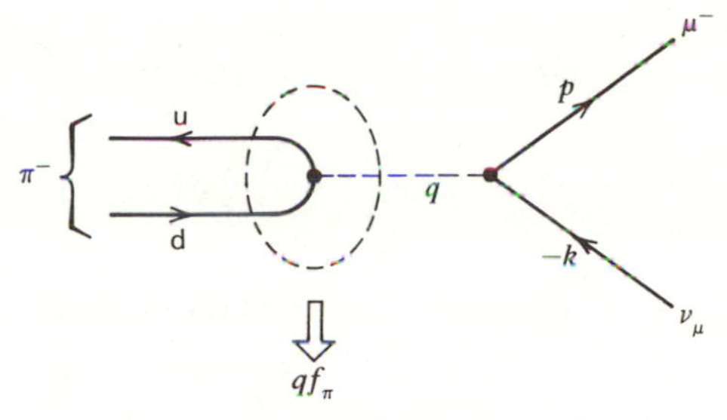

Fig. 12.6 Feynman diagram for the decay $\pi^-(\bar{u}d) \to \mu^- \bar{\nu}_\mu$ with four-momentum $q = p + k$.

We therefore have

$$(\ldots)^\mu = q^\mu f(q^2) \equiv q^\mu f_\pi, \tag{12.47}$$

where f is a function of q^2 since it is the only Lorentz scalar that can be formed from q, but $q^2 = m_\pi^2$ and $f(m_\pi^2) \equiv f_\pi$ is a constant. Inserting (12.47) into (12.46), the $\pi^- \to \mu^- \bar{\nu}$ decay amplitude is

$$\begin{aligned}\mathcal{M} &= \frac{G}{\sqrt{2}}(p^\mu + k^\mu) f_\pi \left[\bar{u}(p)\gamma_\mu(1-\gamma^5)v(k)\right] \\ &= \frac{G}{\sqrt{2}} f_\pi m_\mu \bar{u}(p)(1-\gamma^5)v(k).\end{aligned} \tag{12.48}$$

Here, we have used $\not{k}v(k) = 0$ and $\bar{u}(p)(\not{p} - m_\mu) = 0$, the Dirac equations for the neutrino and muon, respectively. In its rest frame, the π-decay rate is

$$d\Gamma = \frac{1}{2m_\pi}\overline{|\mathcal{M}|^2}\frac{d^3p}{(2\pi)^3 2E}\frac{d^3k}{(2\pi)^3 2\omega}(2\pi)^4\,\delta(q-p-k), \tag{12.49}$$

where the sum over the spins of the outgoing lepton pair can be performed by familiar traceology [see (6.22) and (6.23)]:

$$\begin{aligned}\overline{|\mathcal{M}|^2} &= \frac{G^2}{2} f_\pi^2 m_\mu^2 \operatorname{Tr}\left((\not{p} + m_\mu)(1-\gamma^5)\not{k}(1+\gamma^5)\right) \\ &= 4G^2 f_\pi^2 m_\mu^2 (p \cdot k).\end{aligned} \tag{12.50}$$

In the π rest frame ($\mathbf{k} = -\mathbf{p}$),

$$p \cdot k = E\omega - \mathbf{k}\cdot\mathbf{p} = E\omega + \mathbf{k}^2 = \omega(E+\omega). \tag{12.51}$$

Gathering these results together, we have

$$\Gamma = \frac{G^2 f_\pi^2 m_\mu^2}{(2\pi)^2 2m_\pi}\int \frac{d^3p\, d^3k}{E\omega}\delta(m_\pi - E - \omega)\,\delta^{(3)}(\mathbf{k}+\mathbf{p})\,\omega(E+\omega).$$

The d^3p integration is taken care of by the $\delta^{(3)}$ function and, since there is no angular dependence, we are left with only the integration over $d\omega$:

$$\Gamma = \frac{G^2 f_\pi^2 m_\mu^2}{(2\pi)^2 2m_\pi} 4\pi \int d\omega\, \omega^2 \left(1 + \frac{\omega}{E}\right) \delta(m_\pi - E - \omega), \tag{12.52}$$

where $E = (m_\mu^2 + \omega^2)^{1/2}$. The result of the integration is ω_0^2, where

$$\omega_0 \equiv \frac{m_\pi^2 - m_\mu^2}{2m_\pi}. \tag{12.53}$$

This can be seen by rewriting the δ-function in (12.52) as

$$\delta[f(\omega)] = \delta(\omega - \omega_0) \Big/ \left|\frac{\partial f}{\partial \omega}\right|_{\omega=\omega_0} = \delta(\omega - \omega_0) \Big/ \left(1 + \frac{\omega_0}{E}\right).$$

Therefore, finally we obtain

$$\boxed{\Gamma = \frac{1}{\tau} = \frac{G^2}{8\pi} f_\pi^2 m_\pi m_\mu^2 \left(1 - \frac{m_\mu^2}{m_\pi^2}\right)^2.} \tag{12.54}$$

Taking the universal value of $G = 10^{-5}\, m_N^{-2}$ obtained from β- or μ-decay and assuming that $f_\pi = m_\pi$ (a guess which at least guarantees the correct dimension), we indeed obtain the π^- lifetime announced at the beginning of the chapter. Although the theory can clearly accommodate the long lifetime of the charged π, the decay does not provide a quantitative test, as $f_\pi = m_\pi$ is a pure guess.

A quantitative test, however, is possible. If we repeat the calculation for the decay mode $\pi^- \to e^- \bar{\nu}_e$, we obtain (12.54) with m_μ replaced by m_e. Therefore,

$$\frac{\Gamma(\pi^- \to e^- \bar{\nu}_e)}{\Gamma(\pi^- \to \mu^- \bar{\nu}_\mu)} = \left(\frac{m_e}{m_\mu}\right)^2 \left(\frac{m_\pi^2 - m_e^2}{m_\pi^2 - m_\mu^2}\right)^2 = 1.2 \times 10^{-4}, \tag{12.55}$$

where the numerical value comes from inserting the particle masses. The charged π prefers (by a factor of 10^4) to decay into a muon, which has a similar mass, rather than into the much lighter electron. This is quite contrary to what one would expect from phase–space considerations, so some dynamical mechanism must be at work.

The pion is spinless, and so, by the conservation of angular momentum, the outgoing lepton pair ($e^- \bar{\nu}_e$) must have $J = 0$. As the $\bar{\nu}_e$ has positive helicity, the e^- is also forced into a positive helicity state, see Fig. 12.7. But recall that this is

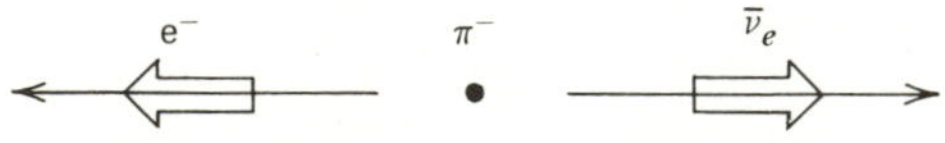

e^- π^- $\bar{\nu}_e$

Fig. 12.7 The decay $\pi^- \to e^- \bar{\nu}_e$ showing the right-handed helicity of the outgoing leptons.

the "wrong" helicity state for the electron. In the limit $m_e = 0$, the weak current only couples negative helicity electrons, and hence the positive helicity coupling is highly suppressed. Thus, in the π^--decay, the e^- (or μ^-) is forced by angular momentum conservation into its "wrong" helicity state. This is much more likely to happen for the μ^- than for the relatively light e^-, in fact, 10^4 times more likely. Experiment confirms this result, which is a direct consequence of the $1 - \gamma^5$ or left-handed structure of the weak current, (12.12). It is however interesting to note that prior to the discovery of parity violation, an argument for (12.55) based on helicity conservation was proposed by Ruderman and Finkelstein (1949) (*Phys. Rev.* **76**, 1458).

EXERCISE 12.13 Predict the ratio of the $K^- \to e^- \bar{\nu}_e$ and $K^- \to \mu^- \bar{\nu}_\mu$ decay rates. Given that the lifetime of the K^- is $\tau = 1.2 \times 10^{-8}$ sec and the $K \to \mu\nu$ branching ratio is 64%, estimate the decay constant f_K. Comment on your assumptions and on your result.

12.7 Charged Current Neutrino–Electron Scattering

Although the experiments exposing the violation of parity in weak interactions (polarized ^{60}Co decay, K decay, π-decay, etc.) are some of the highlights in the development of particle physics, parity violation and its V–A structure can now be demonstrated experimentally much more directly. In fact, these days neutrinos, particularly muon neutrinos, can be prepared in intense beams which are scattered off hadronic, or even leptonic, targets to probe the structure of the weak interaction. A common method is to allow a high-energy monoenergetic beam of pions (or kaons) to decay (e.g., $\pi^+ \to \mu^+ \nu_\mu$) in a long decay tunnel and then to absorb the muons by passing the, approximately collinear, decay products through a thick shield which absorbs the charged particles, letting only the neutrinos through. This technological achievement opens up the possibility of exhibiting the $\gamma^\mu(1 - \gamma^5)$ structure of the weak coupling by measuring the angular distribution of ν_ee or $\bar{\nu}_e$e scattering. This is the analogue of the confirmation of the γ^μ structure of the electromagnetic vertex by studying ee or eμ scattering, which we discussed in Chapter 6.

The relevant diagrams are shown in Fig. 12.8, where the particle four-momenta are defined. The invariant amplitude for $\nu_e e^- \to \nu_e e^-$ is computed from diagram (a):

$$\mathcal{M} = \frac{G}{\sqrt{2}} \left(\bar{u}(k') \gamma^\mu (1 - \gamma^5) u(p) \right) \left(\bar{u}(p') \gamma_\mu (1 - \gamma^5) u(k) \right). \quad (12.56)$$

The calculation now proceeds along the lines of that for $e^-\mu^-$ scattering in Section 6.3, except for the replacement of γ^μ by $\gamma^\mu(1 - \gamma^5)$. Squaring (12.56), summing over the final-state spins, and averaging over the two spin states of the

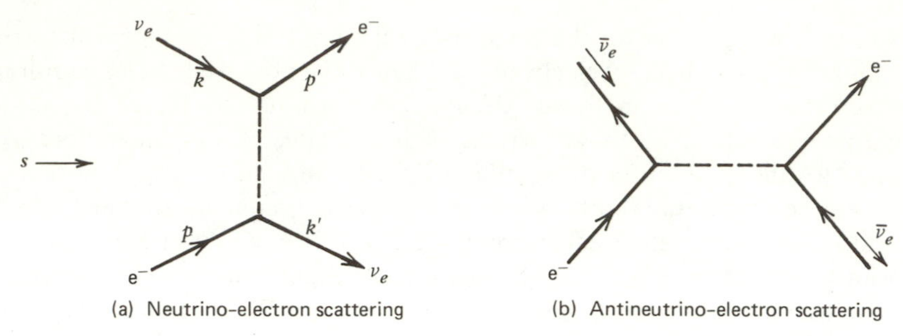

Fig. 12.8 Charged current contributions to elastic $\nu_e e^-$ and $\bar{\nu}_e e^-$ scattering.

initial e^- gives

$$\begin{aligned}\frac{1}{2}\sum_{\text{spins}}|\mathfrak{M}|^2 &= \frac{G^2}{4}\operatorname{Tr}\left(\gamma^\mu(1-\gamma^5)\not{p}\gamma^\nu(1-\gamma^5)\not{k}'\right)\operatorname{Tr}\left(\gamma_\mu(1-\gamma^5)\not{k}\gamma_\nu(1-\gamma^5)\not{p}'\right)\\ &= 64G^2(k\cdot p)(k'\cdot p')\\ &= 16G^2s^2\end{aligned} \tag{12.57}$$

where we have used (12.29). Also we are working in the relativistic limit $m_e = 0$ and have made use of

$$s = (k+p)^2 = 2k\cdot p = 2k'\cdot p'. \tag{12.58}$$

The angular distribution in the center of mass follows from (4.35) (with $p_f = p_i$ in the limit $m_e = 0$):

$$\frac{d\sigma(\nu_e e^-)}{d\Omega} = \frac{1}{64\pi^2 s}\overline{|\mathfrak{M}|^2} = \frac{G^2 s}{4\pi^2}. \tag{12.59}$$

Integration over this isotropic angular distribution gives

$$\sigma(\nu_e e^-) = \frac{G^2 s}{\pi}. \tag{12.60}$$

EXERCISE 12.14 On purely dimensional grounds, show that the cross section (for a point interaction) must behave as $\sigma(\nu_e e^-) \sim G^2 s$ at high energies. Comment on the significance of this result.

EXERCISE 12.15 Show that

$$\sigma(\nu_e e^-) \approx (E_\nu \text{ in GeV}) \times 10^{-41}\ \text{cm}^2,$$

where E_ν is the laboratory energy of the neutrino.

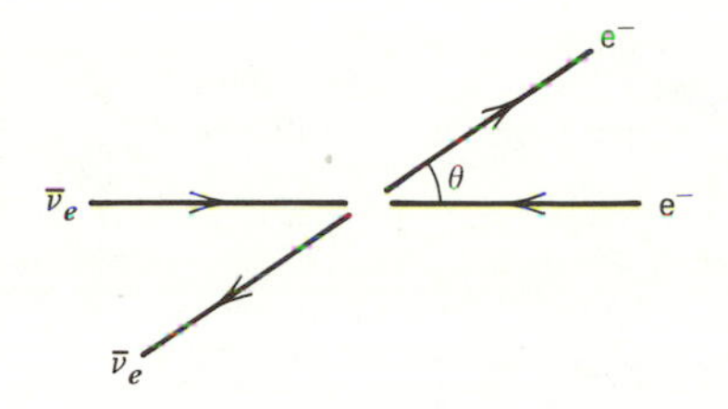

Fig. 12.9 Definition of θ for $\bar{\nu}_e e^-$ scattering.

The Feynman diagram for $\bar{\nu}_e e^- \to e^- \bar{\nu}_e$ is shown in Fig. 12.8b. We see that it can be obtained by crossing the neutrinos in $\nu_e e^- \to e^- \nu_e$ of diagram (a); see Sections 4.6 and 4.7. We therefore simply replace s by t in (12.57):

$$\frac{1}{2} \sum_{\text{spins}} |\mathfrak{M}|^2 = 16 G^2 t^2$$

$$= 4G^2 s^2 (1 - \cos\theta)^2, \tag{12.61}$$

where θ is the angle between the incoming $\bar{\nu}_e$ and the outgoing e^- (see Fig. 12.9), and

$$t \simeq -\frac{s}{2}(1 - \cos\theta),$$

see (4.45). From (12.61), we obtain

$$\frac{d\sigma(\bar{\nu}_e e^-)}{d\Omega} = \frac{G^2 s}{16\pi^2}(1 - \cos\theta)^2, \tag{12.62}$$

and integrating over angles yields

$$\sigma(\bar{\nu}_e e^-) = \frac{G^2 s}{3\pi}. \tag{12.63}$$

Comparing with (12.60) gives

$$\sigma(\bar{\nu}_e e^-) = \tfrac{1}{3}\sigma(\nu_e e^-). \tag{12.64}$$

Results (12.59), (12.62), and (12.64) expose the $\gamma^\mu(1 - \gamma^5)$ structure of the weak current in a way that can be experimentally checked. We can convince ourselves of this important statement by comparing the results with those obtained for the γ^μ vertex in the electromagnetic process $e\mu \to e\mu$ or by performing the following exercise.

EXERCISE 12.16 If the weak current had had a $V + A$ structure, $\gamma^\mu(1 + \gamma^5)$, show that

$$\frac{d\sigma}{d\Omega} = \frac{G^2 s}{4\pi^2}(1 + \cos\theta)^2$$

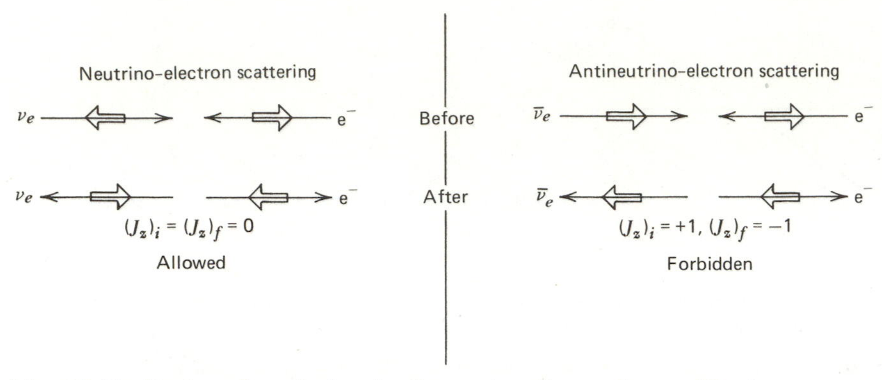

Fig. 12.10 Backward scattering in the center-of-mass frame. The long arrows represent the particle momenta and the short arrows represent their helicities in the limit in which the masses are negligible. The z axis is along the incident neutrino direction.

for both ν_ee and $\bar{\nu}_e$e elastic scattering. If this were the case, then, in contrast to (12.64), we would have

$$\sigma(\bar{\nu}_e \mathrm{e}) = \sigma(\nu_e \mathrm{e}).$$

The most striking difference between the two angular distributions, (12.59) and (12.62), is that $\bar{\nu}_e$e scattering vanishes for $\cos\theta = 1$, whereas ν_ee scattering does not. With our definition of θ, see Fig. 12.9, this corresponds to backward scattering of the beam particle. We could have anticipated these results from the helicity arguments we used to interpret previous calculations. The by now familiar pictures are shown in Fig. 12.10. Backward $\bar{\nu}_e$e scattering is forbidden by angular momentum conservation. In fact, the process $\bar{\nu}_e \mathrm{e} \to \bar{\nu}_e \mathrm{e}$ proceeds entirely in a $J = 1$ state with net helicity $+1$; that is, only one of the three helicity states is allowed. This is the origin of the factor $\frac{1}{3}$ in (12.64). With our definition of θ, the allowed amplitude is proportional to $d^1_{-11}(\theta) = \frac{1}{2}(1 - \cos\theta)$, see (6.39), in agreement with result (12.61).

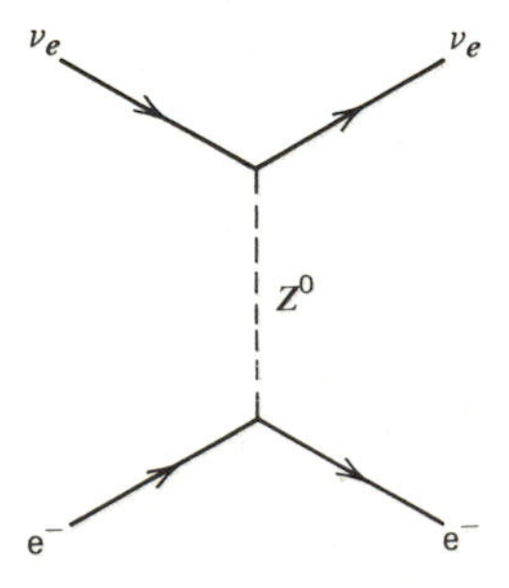

Fig. 12.11 Neutral current contributions to neutrino–electron elastic scattering.

Elastic $\nu_e e^-$ and $\bar{\nu}_e e^-$ scattering can also proceed via a weak neutral current interaction (see Fig. 12.11) which interferes with the charged current interaction (Fig. 12.8a). This is discussed in Section 13.5. However, high-energy neutrino beams are predominantly ν_μ (or $\bar{\nu}_\mu$), and so the most accessible (charged current) purely leptonic scattering process is

$$\nu_\mu + e^- \to \mu^- + \nu_e \tag{12.65}$$

(i.e., inverse muon decay). Here, there is no neutral current contribution, and so the cross section is given just by (12.59).

12.8 Neutrino–Quark Scattering

A study of the scattering of ν_μ's (or $\bar{\nu}_\mu$'s) from quarks is experimentally feasible by impinging high-energy neutrino beams on proton or nuclear targets. This is analogous to the study of the electromagnetic lepton–quark interaction by scattering high-energy electron or muon beams off hadronic targets, which we described in Chapters 8 and 9.

To predict the neutrino–quark cross sections, we clearly need to know the form of the quark weak currents. Quarks interact electromagnetically just like leptons, apart from their fractional charge. Our inclination therefore is to construct the quark weak current just as we did for leptons. For instance, we model the charge-raising quark current

$$J_q^\mu = \bar{u}_u \gamma^\mu \tfrac{1}{2}(1 - \gamma^5) u_d \quad \Longrightarrow \tag{12.66}$$

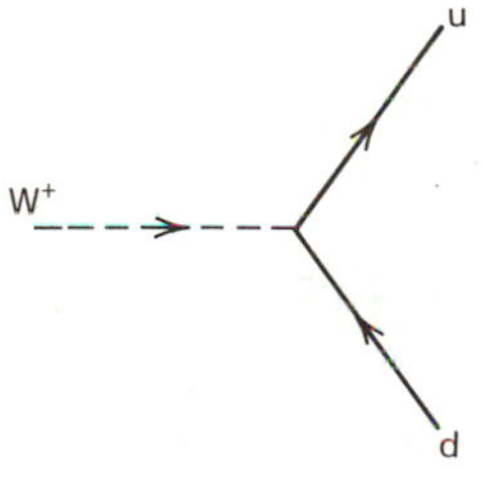

W⁺ u d

on the electron weak current (12.12)

$$J_e^\mu = \bar{u}_\nu \gamma^\mu \tfrac{1}{2}(1 - \gamma^5) u_e \quad \Longrightarrow \tag{12.67}$$

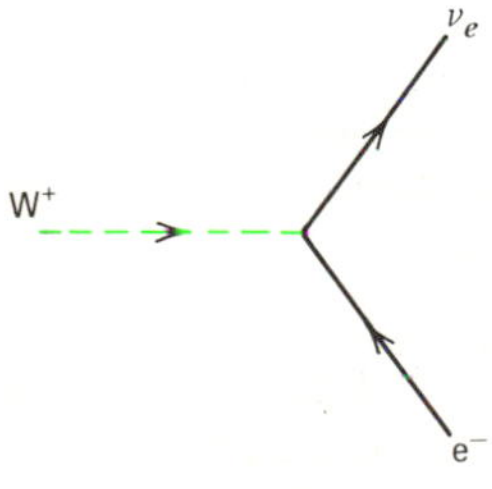

W⁺ ν_e e^-

The hermitian conjugates of (12.66) and (12.67) give the charge-lowering weak currents

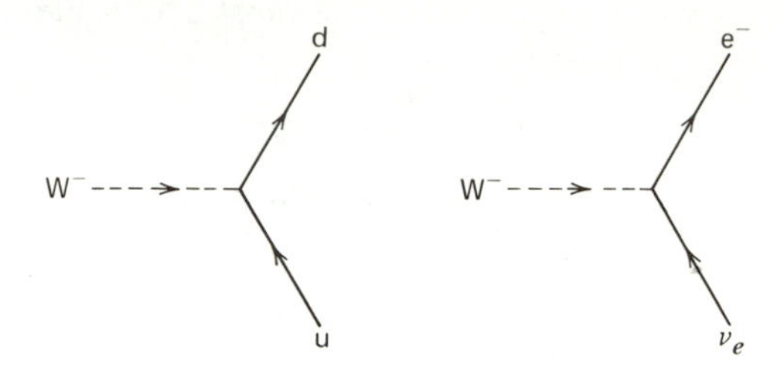

The $V–A$ structure means that the weak current couples only left-handed u and d quarks (or right-handed $\bar{u}$ and $\bar{d}$ quarks), see Exercise 12.3. At high energies, this means only negative helicity u and d quarks are coupled, or positive helicity $\bar{u}$ and $\bar{d}$ quarks.

Using the above currents, we can evaluate diagrams such as Fig. 12.12. That is, we can calculate the amplitude for

$$\mathrm{d} \rightarrow \mathrm{u} e^- \bar{\nu}_e$$

which is responsible for a constituent description of neutron β-decay. The "spectator" u and d quarks, shown in Fig. 12.12, can be treated just like the spectator nucleons in the nuclear β-transitions of Section 12.3. The same d → u transition is responsible for the $\pi^- \rightarrow \pi^0 e^- \nu_e$ decay mode, the spectator quark now being a $\bar{u}$; alternatively, we may have a $\bar{u} \rightarrow \bar{d}$ transition with a spectator d quark.

EXERCISE 12.17 Using the above approach, show that

$$\Gamma(\pi^- \rightarrow \pi^0 e^- \bar{\nu}_e) = \frac{G^2}{30\pi^3}(\Delta m)^5, \tag{12.68}$$

where $\Delta m = m(\pi^-) - m(\pi^0) = 4.6$ MeV. Evaluate the decay rate and compare with $\Gamma(\pi^- \rightarrow \mu^- \bar{\nu}_\mu)$.

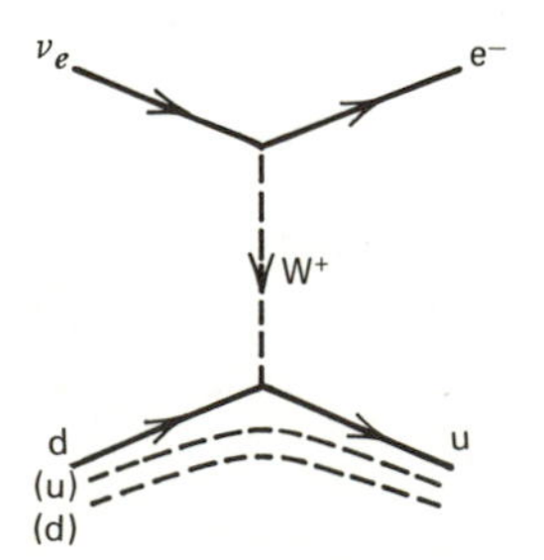

Fig. 12.12 The quark diagram responsible for neutron β-decay, $\mathrm{n} \rightarrow \mathrm{p} e^- \bar{\nu}_e$. The two spectator quarks which do not take part in the weak interaction are shown by dashed lines.

We are now ready to tackle neutrino–quark scattering. As the quarks and lepton weak currents have identical forms, we can carry over the results for νe scattering that we obtained in Section 12.7. From (12.59) and (12.62), we obtain in the center-of-mass frame

$$\frac{d\sigma(\nu_\mu \mathrm{d} \to \mu^- \mathrm{u})}{d\Omega} = \frac{G^2 s}{4\pi^2} \tag{12.69}$$

$$\frac{d\sigma(\bar{\nu}_\mu \mathrm{u} \to \mu^+ \mathrm{d})}{d\Omega} = \frac{G^2 s}{16\pi^2}(1 + \cos\theta)^2, \tag{12.70}$$

where θ is defined as in Fig. 12.13. From the figure, it is immediately apparent that the backward $\bar{\nu}_\mu \mathrm{u} \to \mu^+ \mathrm{d}$ scattering ($\theta = \pi$) is forbidden by helicity considerations. The cross sections for scattering from antiquarks, $\bar{\nu}_\mu \bar{\mathrm{d}} \to \mu^+ \bar{\mathrm{u}}$ and $\nu_\mu \bar{\mathrm{u}} \to \mu^- \bar{\mathrm{d}}$, are given by (12.69) and (12.70), respectively. We see that, for instance, ν_μ does not interact with either u or $\bar{\mathrm{d}}$ quarks.

To compare these results with experiment, we have to embed the constituent cross sections, (12.69) and (12.70), in the overall νN inclusive cross section. The procedure is familiar from Chapter 9. We obtain

$$\frac{d\sigma(\nu N \to \mu X)}{dx\, dy} = \sum_i \quad \text{[diagram]} \tag{12.71}$$

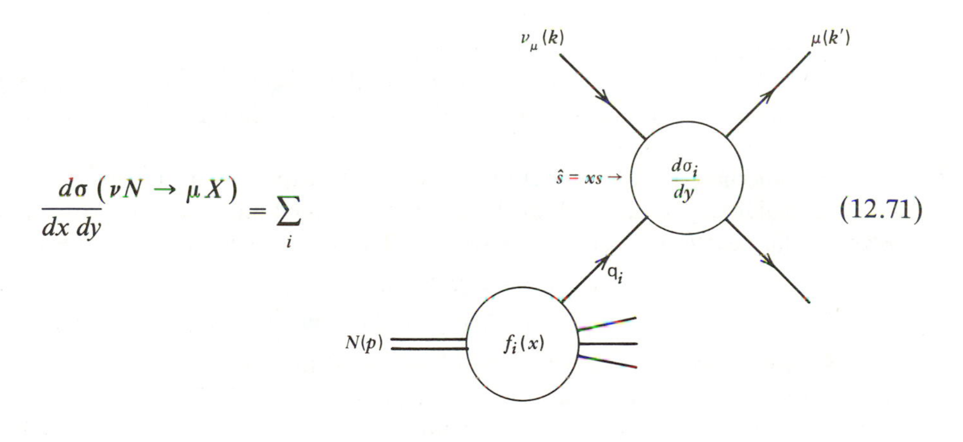

$$= \sum_i f_i(x) \left(\frac{d\sigma_i}{dy} \right)_{\hat{s} = xs}. \tag{12.72}$$

First, note that the angular distributions of the constituent process have been expressed in terms of the dimensionless variable y. It is related to $\cos\theta$ by

$$1 - y \equiv \frac{p \cdot k'}{p \cdot k} \simeq \tfrac{1}{2}(1 + \cos\theta),$$

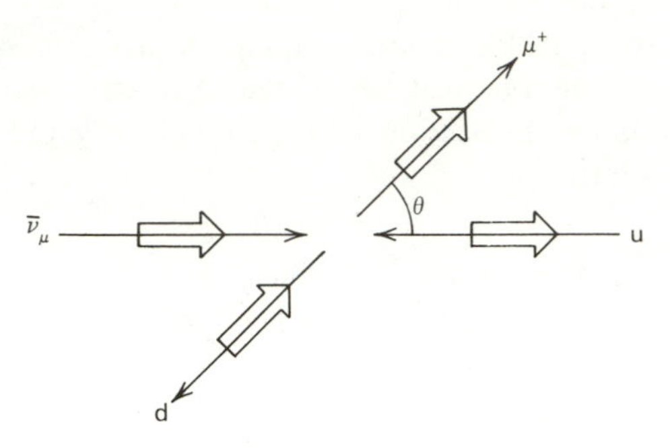

Fig. 12.13 The helicity configuration for high-energy $\bar{\nu}_\mu u \to \mu^+ d$ scattering.

see (9.25). The four-momenta are given in (12.71). The constituent cross sections, (12.69) and (12.70), are therefore

$$\frac{d\sigma}{dy}\left(\nu_\mu d \to \mu^- u\right) = \frac{G^2 xs}{\pi}, \tag{12.73}$$

$$\frac{d\sigma}{dy}\left(\bar{\nu}_\mu u \to \mu^+ d\right) = \frac{G^2 xs}{\pi}(1-y)^2. \tag{12.74}$$

The appropriate $\nu q \to \mu q'$ center-of-mass energy is xs, where now s refers to $\nu N \to \mu X$ (see Exercise 9.3). Using these results, together with the nucleon structure functions $f_i(x)$ introduced in Chapter 9, we can calculate the deep inelastic $\nu_\mu N \to \mu X$ cross section.

To confront these parton model predictions with experiment, it is simplest to take an isoscalar target, in which the nuclei contain equal numbers of protons and neutrons. The neutrinos interact only with d or $\bar{u}$ quarks. They therefore measure

$$\begin{aligned} d^p(x) + d^n(x) &= d(x) + u(x) \equiv Q(x) \\ \bar{u}^p(x) + \bar{u}^n(x) &= \bar{u}(x) + \bar{d}(x) \equiv \bar{Q}(x), \end{aligned} \tag{12.75}$$

see (9.29), where we have denoted the distribution functions, $f_i(x)$, of up- and down-quarks in a proton by $u(x)$ and $d(x)$. Inserting (12.73) and the cross section for $\nu_\mu \bar{u} \to \mu^- \bar{d}$ into (12.72), we find the $\nu_\mu N \to \mu^- X$ cross section per nucleon to be

$$\frac{d\sigma}{dx\,dy}\left(\nu_\mu N \to \mu^- X\right) = \frac{G^2 xs}{2\pi}\left(Q(x) + (1-y)^2 \bar{Q}(x)\right). \tag{12.76}$$

On the other hand, antineutrinos interact with $\bar{d}$ and u constituents; and going through the same steps, we obtain

$$\frac{d\sigma}{dx\,dy}\left(\bar{\nu}_\mu N \to \mu^+ X\right) = \frac{G^2 xs}{2\pi}\left[\bar{Q}(x) + (1-y)^2 Q(x)\right]. \tag{12.77}$$

EXERCISE 12.18 Show that deep inelastic electron electromagnetic scattering on an isoscalar target gives

$$\frac{d\sigma\,(\mathrm{eN} \to \mathrm{eX})}{dx\,dy} = \frac{2\pi\alpha^2}{Q^4} xs\left[1 + (1-y)^2\right] \frac{5}{18}\left[Q(x) + \bar{Q}(x)\right] \qquad (12.78)$$

per nucleon, see Exercise 9.5. Note that, in contrast to $\nu \mathrm{N} \to \mu \mathrm{X}$, (12.78) embodies parity conservation so Q and $\bar{Q}$ appear symmetrically.

If there were just three valence quarks in a nucleon, $\bar{Q} = 0$, the $\nu \mathrm{N} \to \mu^- \mathrm{X}$ and $\bar{\nu} \mathrm{N} \to \mu^+ \mathrm{X}$ data would exhibit the dramatic V–A properties of the weak interaction exactly. That is,

$$\frac{d\sigma(\nu)}{dy} = c, \qquad \frac{d\sigma(\bar{\nu})}{dy} = c(1-y)^2, \qquad (12.79)$$

where c can be found from (12.76); and for the integrated cross sections,

$$\frac{\sigma(\bar{\nu})}{\sigma(\nu)} = \frac{1}{3}.$$

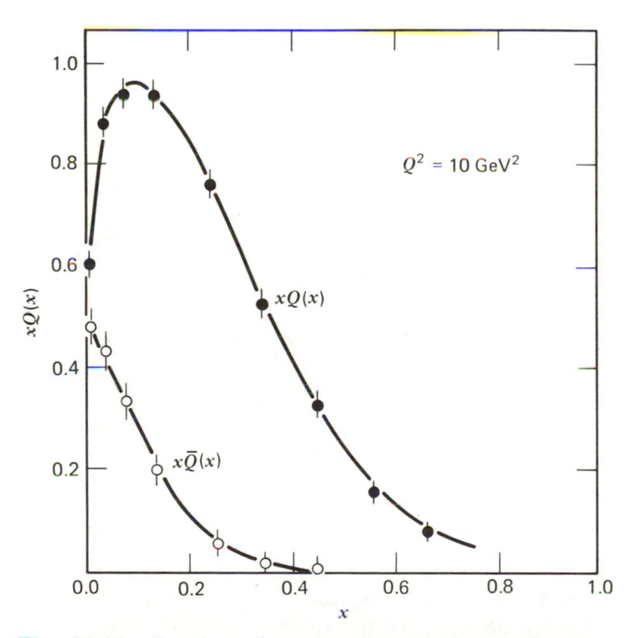

Fig. 12.14 Quark and antiquark momentum distributions in a nucleon as measured at CERN and the Fermi laboratory. The experiments reveal that only about half the proton's momentum is carried by quarks. We have associated the remainder with the gluon constituents (see Section 9.4).

The data approximately reproduce these expectations. In fact, (12.76) and (12.77) allow a determination of $Q(x)$ and $\overline{Q}(x)$. An example is shown in Fig. 12.14. There is about a 5% $\overline{Q}$ component in a proton.

EXERCISE 12.19 If $\sigma(\bar{\nu})/\sigma(\nu) = R$, show that

$$\frac{\int x\,\overline{Q}(x)\,dx}{\int x\,Q(x)\,dx} = \frac{3R-1}{3-R}.$$

Detailed analyses show that the functions $u(x)$, $d(x), \ldots$, are indeed the same whether one extracts them from electroproduction or neutrino experiments. This is a definitive success of the parton model: the $u(x)$, $d(x)$, describe the intrinsic structure of the hadronic target and are the same whatever experimental probe is used to determine them.

12.9 First Observation of Weak Neutral Currents

The detection in 1973 of neutrino events of the type

$$\bar{\nu}_\mu e^- \rightarrow \bar{\nu}_\mu e^-, \tag{12.80}$$

$$\left.\begin{aligned} \nu_\mu N &\rightarrow \nu_\mu X \\ \bar{\nu}_\mu N &\rightarrow \bar{\nu}_\mu X \end{aligned}\right\} \tag{12.81}$$

heralded a new chapter in particle physics. These events are evidence of a weak neutral current. Until then, no weak neutral current effects had been observed, and indeed very stringent limits had been set on the (strangeness changing) neutral current by the absence of decay modes such as

$$\begin{aligned} &K^0 \rightarrow \mu^+\mu^-, \\ &K^+ \rightarrow \pi^+ e^+ e^-, \\ &K^+ \rightarrow \pi^+ \nu\bar{\nu}. \end{aligned}$$

Induced weak neutral current effects are expected to occur by the combined action of the (neutral) electromagnetic and the (charged) weak current (for example, $K^+ \rightarrow \pi^+ e^+ e^-$ can proceed via a virtual photon: $K^+ \rightarrow \pi^+\gamma$ with $\gamma \rightarrow e^+e^-$), but these effects are very small. The rate, compared to the corresponding allowed weak decay, is of the order

$$\frac{\Gamma(K^+ \rightarrow \pi^+ e^+ e^-)}{\Gamma(K^+ \rightarrow \pi^0 e^+ \nu_e)} \sim \left(\frac{\alpha G}{G}\right)^2 \sim 10^{-5}, \tag{12.82}$$

in agreement with the data. (Here, the $1/q^2$ behavior of the propagator of the virtual photon is canceled by the helicity suppression of the $0^- \rightarrow 0^-\gamma$ coupling.)

However, reactions (12.80) and (12.81) were found to occur at rates very similar to those of other weak scattering processes.

12.10 Neutral Current Neutrino–Quark Scattering

A quantitative comparison of the strength of neutral current (NC) to charged current (CC) weak processes has been obtained, for example, by scattering neutrinos off an iron target. The present experimental values are

$$R_\nu \equiv \frac{\sigma^{NC}(\nu)}{\sigma^{CC}(\nu)} \equiv \frac{\sigma(\nu_\mu \mathrm{N} \to \nu_\mu \mathrm{X})}{\sigma(\nu_\mu \mathrm{N} \to \mu^- \mathrm{X})} = 0.31 \pm 0.01,$$
$$R_{\bar\nu} \equiv \frac{\sigma^{NC}(\bar\nu)}{\sigma^{CC}(\bar\nu)} \equiv \frac{\sigma(\bar\nu_\mu \mathrm{N} \to \bar\nu_\mu \mathrm{X})}{\sigma(\bar\nu_\mu \mathrm{N} \to \mu^+ \mathrm{X})} = 0.38 \pm 0.02. \tag{12.83}$$

The $\nu\mathrm{N} \to \nu\mathrm{X}$ data can be explained in terms of neutral current–current $\nu\mathrm{q} \to \nu\mathrm{q}$ interactions, see Fig. 12.15, with amplitudes

$$\mathcal{M} = \frac{G_N}{\sqrt{2}}\left[\bar u_\nu \gamma^\mu (1-\gamma^5) u_\nu\right]\left[\bar u_q \gamma_\mu (c_V^q - c_A^q \gamma^5) u_q\right] \tag{12.84}$$

where $q = \mathrm{u}, \mathrm{d}, \ldots$ are the quarks in the target. *A priori*, there is no reason why the neutral weak interaction should have the four-vector current–current form of (12.84). It is decided by experiment, for instance, by the observed y distribution (see Exercise 12.20).

It is appropriate at this stage to introduce the conventional normalization of the weak neutral currents, J_μ^{NC}. The invariant amplitude for an arbitrary neutral current process is written

$$\boxed{\mathcal{M} = \frac{4G}{\sqrt{2}} 2\rho J_\mu^{NC} J^{NC\mu},} \tag{12.85}$$

compare (12.13) for a charged current process. The $\nu\mathrm{q} \to \nu\mathrm{q}$ amplitude of (12.84)

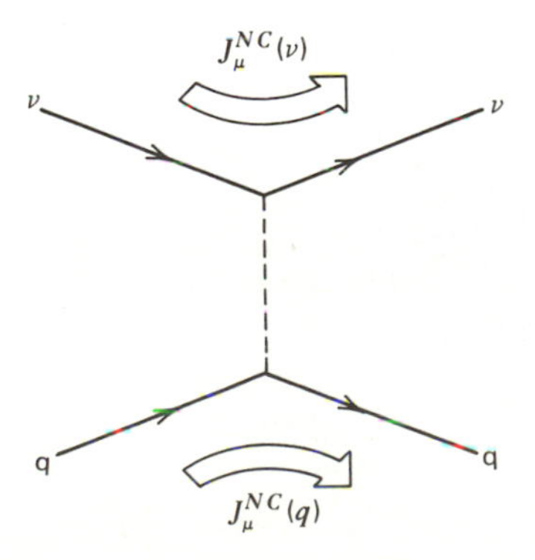

Fig. 12.15 Neutral current $\nu\mathrm{q} \to \nu\mathrm{q}$ scattering.

is of this form; the customary definition of the neutral currents is

$$J_\mu^{NC}(\nu) = \tfrac{1}{2}\left(\bar{u}_\nu \gamma_\mu \tfrac{1}{2}(1-\gamma^5)u_\nu\right), \tag{12.86}$$

$$J_\mu^{NC}(q) = \left(\bar{u}_q \gamma_\mu \tfrac{1}{2}(c_V^q - c_A^q\gamma^5)u_q\right). \tag{12.87}$$

In general, the J_μ^{NC}, unlike the charged current J_μ, are not pure V–A currents ($c_V \neq c_A$); they have right-handed components. However, the neutrino is left-handed; and so, $c_V^\nu = c_A^\nu \equiv \frac{1}{2}$ in (12.86). The parameter ρ in (12.85) determines the relative strength of the neutral and charged current processes. In the standard theoretical model all the c_V^i, c_A^i (with $i = \nu, \mathrm{e}, \mathrm{u}\ldots$) are given in terms of one parameter, and $\rho = 1$ (see Chapters 13 and 15). In other words, if the model is successful, all neutral current phenomena will be described by a common parameter. In fact, the present experiments give $\rho = 1$ to within small errors. However, for the moment let us leave c_V^i, c_A^i and ρ as free parameters to be determined by experiment. Upon inserting the currents (12.86) and (12.87) into (12.85), we obtain the $\nu\mathrm{q} \to \nu\mathrm{q}$ amplitude of (12.84) with

$$G_N = \rho G (\simeq G). \tag{12.88}$$

We now return to our interpretation of the $\nu\mathrm{N} \to \nu\mathrm{X}$ data. The calculation of the $\nu\mathrm{q} \to \nu\mathrm{q}$ cross sections proceeds exactly as that for the charged current processes $\nu\mathrm{q} \to \mu\mathrm{q}'$. For example, using the results (12.73) and (12.74),

$$\frac{d\sigma(\nu_L \mathrm{d}_L \to \mu\mathrm{u})}{dy} = \frac{G^2 xs}{\pi},$$

$$\frac{d\sigma(\nu_L \bar{\mathrm{u}}_R \to \mu\bar{\mathrm{d}})}{dy} = \frac{G^2 xs}{\pi}(1-y)^2, \tag{12.89}$$

we obtain directly

$$\frac{d\sigma(\nu\mathrm{q} \to \nu\mathrm{q})}{dy} = \frac{G_N^2 xs}{\pi}\left((g_L^q)^2 + (g_R^q)^2(1-y)^2\right), \tag{12.90}$$

where we have introduced

$$g_L^q \equiv \tfrac{1}{2}(c_V^q + c_A^q), \qquad g_R^q \equiv \tfrac{1}{2}(c_V^q - c_A^q). \tag{12.91}$$

As compared to (12.89), the new feature of (12.90) is the possibility of a right-handed component g_R^q of $J_\mu^{NC}(q)$.

EXERCISE 12.20 Show that $|\mathfrak{M}(\nu\mathrm{q} \to \nu\mathrm{q})|^2$ behaves like s^2, $s^2(1-y)^2$, s^2y^2 for pure $V - A$, pure $V + A$, and S, P neutral couplings of the quark, respectively. Pure $V \pm A$ denote $\gamma^\mu(1 \pm \gamma^5)$ couplings, and S, P stands for the scalar, pseudoscalar interaction amplitude

$$\mathfrak{M} = \frac{G_N}{\sqrt{2}}\left(\bar{u}_\nu(1-\gamma^5)u_\nu\right)\left(\bar{u}_q(g_S - g_P\gamma^5)u_q\right).$$

The parton model predictions for the neutral current (NC) processes $\nu N \to \nu X$ and $\bar{\nu} N \to \bar{\nu} X$ are obtained by following the calculation of the CC processes $\nu N \to \mu^- X$ and $\bar{\nu} N \to \mu^+ X$ of Section 12.8. For an isoscalar target, we find that the cross section per nucleon is

$$\frac{d\sigma(\nu N \to \nu X)}{dx\,dy} = \frac{G_N^2 xs}{2\pi}\Big[g_L^2\big(Q(x) + (1-y)^2\overline{Q}(x)\big) + g_R^2\big(\overline{Q}(x) + (1-y)^2 Q(x)\big)\Big], \tag{12.92}$$

where, if we assume only u, d, $\bar{\mathrm{u}}$, $\bar{\mathrm{d}}$ quarks within the nucleon,

$$g_L^2 \equiv (g_L^u)^2 + (g_L^d)^2, \tag{12.93}$$

and similarly for g_R^2. We may integrate over x and define

$$Q \equiv \int x\, Q(x)\, dx = \int x[u(x) + d(x)]\, dx, \tag{12.94}$$

see (12.75). Cross section (12.92) and that for $\bar{\nu} N \to \bar{\nu} X$ become

$$\frac{d\sigma^{NC}(\nu)}{dy} = \frac{G_N^2 s}{2\pi}\left\{g_L^2\big(Q + (1-y)^2\overline{Q}\big) + g_R^2\big(\overline{Q} + (1-y^2)Q\big)\right\},$$

$$\frac{d\sigma^{NC}(\bar{\nu})}{dy} = \frac{G_N^2 s}{2\pi}\left\{g_L^2\big(\overline{Q} + (1-y)^2 Q\big) + g_R^2\big(Q + (1-y^2)\overline{Q}\big)\right\}, \tag{12.95}$$

which are to be contrasted with the charged current expressions (12.76) and (12.77)

$$\frac{d\sigma^{CC}(\nu)}{dy} = \frac{G^2 s}{2\pi}\big(Q + (1-y)^2\overline{Q}\big),$$

$$\frac{d\sigma^{CC}(\bar{\nu})}{dy} = \frac{G^2 s}{2\pi}\big(\overline{Q} + (1-y)^2 Q\big). \tag{12.96}$$

Correcting (12.95) and (12.96) for the neutron excess in an iron target and for an s quark contribution, the present data give

$$g_L^2 = 0.300 \pm 0.015, \qquad g_R^2 = 0.024 \pm 0.008. \tag{12.97}$$

The experimental verdict is that the weak neutral current is predominantly V–A (i.e., left-handed) but, since $g_R \neq 0$, not pure V–A. The NC and the CC have a tantalizingly similar structure, but the CC is believed to have a pure V–A form. Chapter 13 takes up this point, but first we must look more carefully at the quark sector.

12.11 The Cabibbo Angle

So far, we have seen that leptons and quarks participate in weak interactions through charged V–A currents constructed from the following pairs of (left-handed) fermion states:

$$\begin{pmatrix} \nu_e \\ e^- \end{pmatrix}, \quad \begin{pmatrix} \nu_\mu \\ \mu^- \end{pmatrix}, \quad \text{and} \quad \begin{pmatrix} u \\ d \end{pmatrix}. \tag{12.98}$$

All these charged currents couple with a universal coupling constant G. It is natural to attempt to extend this universality to embrace the doublet

$$\begin{pmatrix} c \\ s \end{pmatrix} \tag{12.99}$$

formed from the heavier quark states. However, we already know that this cannot be quite correct. For instance, the decay $K^+ \to \mu^+ \nu_\mu$ occurs. The K^+ is made of u and $\bar{s}$ quarks. There must thus be a weak current which couples a u to an $\bar{s}$ quark (see Fig. 12.16). This contradicts the above scheme, which only allows weak transitions between u $\leftrightarrow$ d and c $\leftrightarrow$ s.

Instead of introducing new couplings to accommodate observations like $K^+ \to \mu^+ \nu_\mu$, let us try to keep universality but modify the quark doublets. We assume that the charged current couples "rotated" quark states

$$\begin{pmatrix} u \\ d' \end{pmatrix}, \quad \begin{pmatrix} c \\ s' \end{pmatrix}, \ldots, \tag{12.100}$$

where

$$\begin{aligned} d' &= d \cos \theta_c + s \sin \theta_c \\ s' &= -d \sin \theta_c + s \cos \theta_c . \end{aligned} \tag{12.101}$$

This introduces an arbitrary parameter θ_c, the quark mixing angle, known as the Cabibbo angle. In 1963, Cabibbo first introduced the doublet u, d′ to account for the weak decays of strange particles. Indeed, the mixing of the d and s quark can be determined by comparing $\Delta S = 1$ and $\Delta S = 0$ decays. For example,

$$\frac{\Gamma(K^+ \to \mu^+ \nu_\mu)}{\Gamma(\pi^+ \to \mu^+ \nu_\mu)} \sim \sin^2 \theta_c ,$$

$$\frac{\Gamma(K^+ \to \pi^0 e^+ \nu_e)}{\Gamma(\pi^+ \to \pi^0 e^+ \nu_e)} \sim \sin^2 \theta_c .$$

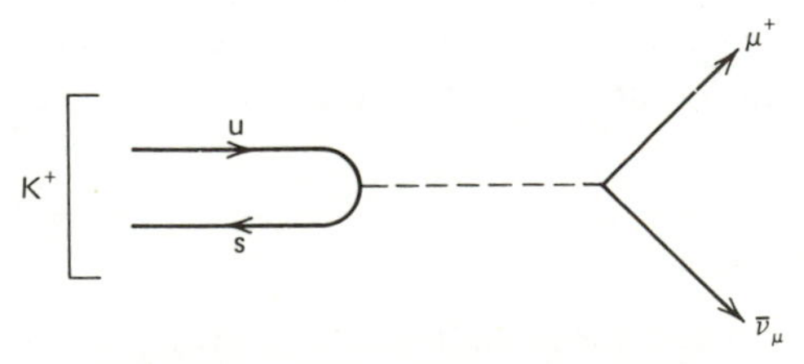

Fig. 12.16 The decay $K^+ \to \mu^+ \bar{\nu}_\mu$.

After allowing for the kinematic factors arising from the different particle masses, the data show that the $\Delta S = 1$ transitions are suppressed by a factor of about 20 as compared to the $\Delta S = 0$ transitions. This corresponds to a Cabibbo angle $\theta_c \approx 13°$.

What we have done is to change our mind about the charged current (12.66). We now have "Cabibbo favored" transitions (proportional to $\cos\theta_c$)

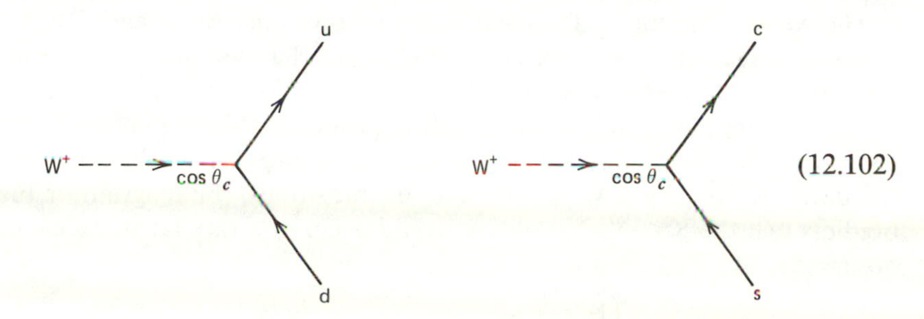

(12.102)

and "Cabibbo suppressed" transitions

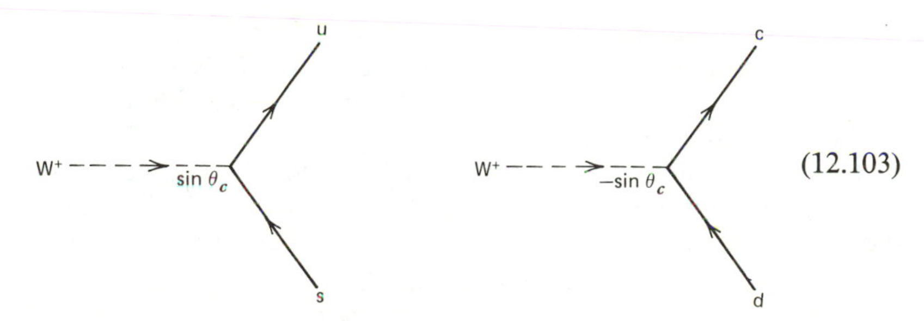

(12.103)

[see (12.101)], and similar diagrams for the charge-lowering transitions. We can summarize this by writing down the explicit form of the matrix element describing the charged current weak interactions of the quarks. From (12.13),

$$\mathcal{M} = \frac{4G}{\sqrt{2}} J^{\mu} J_{\mu}^{\dagger} \tag{12.104}$$

with

$$J^{\mu} = (\bar{u} \quad \bar{c}) \frac{\gamma^{\mu}(1-\gamma^5)}{2} U \begin{pmatrix} d \\ s \end{pmatrix}. \tag{12.105}$$

The unitary matrix U performs the rotation (12.101) of the d and s quark states:

$$U = \begin{pmatrix} \cos\theta_c & \sin\theta_c \\ -\sin\theta_c & \cos\theta_c \end{pmatrix}. \tag{12.106}$$

Of course, there will also be amplitudes describing semileptonic decays constructed from the product of a quark with a lepton current, $J^{\mu}(\text{quark})\, J_{\mu}^{\dagger}(\text{lepton})$.

All this has implications for our previous calculations. For instance, we must replace G in the formula for the nuclear β-decay rate by

$$G_\beta = G\cos\theta_c, \tag{12.107}$$

whereas the purely leptonic μ-decay rate, which involves no mixing, is unchanged: $G_\mu = G$. The detailed comparison of these rates, (12.44), supports Cabibbo's hypothesis.

The weak interactions discussed so far involve only the u and d′ quark states. However, in (12.100), we have coupled s′ to the charmed quark c, so that we have weak transitions c $\leftrightarrow$ s′ as well as d′ $\leftrightarrow$ u. In fact, following this line of argument, Glashow, Iliopoulos, and Maiani (GIM) proposed the existence of the c quark some years before its discovery. A reason for doing this can be seen by studying the decay $K^0 \to \mu^+\mu^-$. With only u $\leftrightarrow$ d′ transitions, the diagram of Fig. 12.17a predicts that the $K_L^0 \to \mu^+\mu^-$ decay would occur at a rate far in excess of what is observed:

$$\frac{\Gamma(K_L^0 \to \mu^+\mu^-)}{\Gamma(K_L^0 \to \text{all modes})} = (9.1 \pm 1.9) \times 10^{-9}.$$

However, with the introduction of the c quark, a second diagram, Fig. 12.17b, occurs which would exactly cancel with diagram 12.17a if it were not for the mass difference of the u and c quarks. We take up this discussion in the next section.

The c, s′ weak current is responsible for the weak decays of charmed particles. A delightful example is the decay of a D^+ meson. The D^+ meson consists of a c and $\bar{d}$ quark. Since $\cos^2\theta_c \gg \sin^2\theta_c$, it follows from (12.100) and (12.101) that the favored quark decay pattern is that shown in Fig. 12.18, with amplitude

$$\mathcal{M}(c \to su\bar{d}) \sim \cos^2\theta_c. \tag{12.108}$$

That is, a $\overline{K}$ meson should preferentially feature among the decay products of a D^+. On the other hand, the decay $D^+ \to K\ \ldots$ is highly (Cabibbo) suppressed since

$$\mathcal{M}(c \to \bar{s}ud) \sim \sin^2\theta_c. \tag{12.109}$$

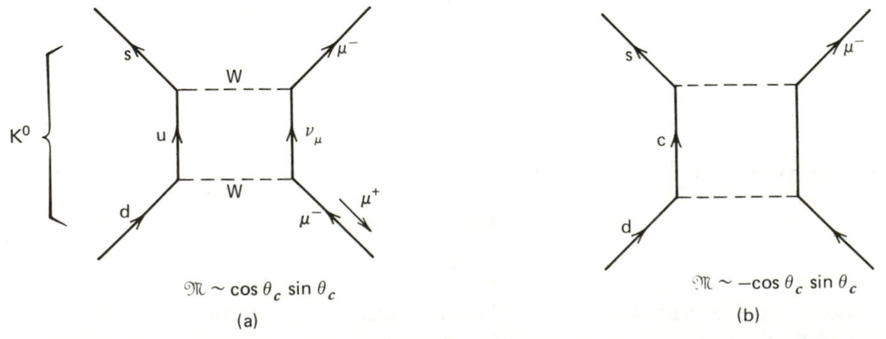

Fig. 12.17 Two contributions to $K^0 \to \mu^+\mu^-$. Diagram (b) is simply (a) with u $\to$ c.

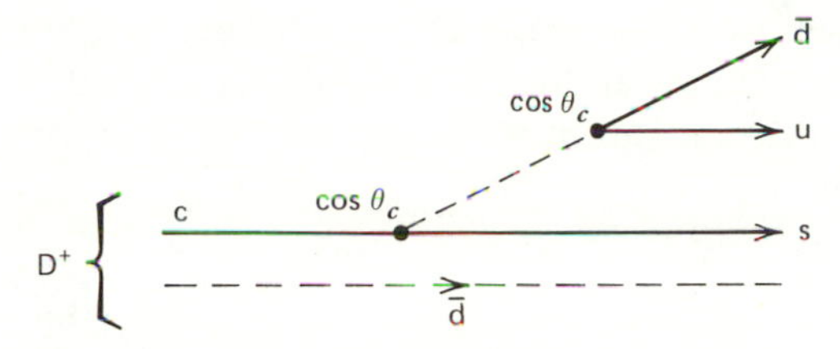

Fig. 12.18 A quark description of D^+ decay: the Cabibbo-favored process is $c \to su\bar{d}$ with a spectator $\bar{d}$ quark.

We thus have a very characteristic signature for D^+ decay, that for instance the decay mode $K^-\pi^+\pi^+$ is highly favored as compared to the $K^+\pi^+\pi^-$ mode.

EXERCISE 12.21 Estimate the relative rates for the following three decay modes of the $D^0(c\bar{u})$ meson: $D^0 \to K^-\pi^+, \pi^-\pi^+, K^+\pi^-$.

EXERCISE 12.22 Given that the partial rate

$$\Gamma(K^+ \to \pi^0 e^+ \nu) = 4 \times 10^6 \text{ sec}^{-1},$$

calculate the rate for $D^0 \to K^- e^+ \nu$. Hence, estimate the lifetime of the D^0 meson.

EXERCISE 12.23 Show, in the "spectator" quark model approach, that the charmed meson lifetimes satisfy

$$\tau(D^0) = \tau(D^+) = \tau(F^+),$$

where F^+ is made of c and $\bar{s}$ quarks, see Chapter 2.

12.12 Weak Mixing Angles

We can summarize the above Cabibbo-GIM scheme as follows. The charged (or flavor-changing) current couples $u \leftrightarrow d'$ or $c \leftrightarrow s'$ (left-handed) quark states, where d' and s' are orthogonal combinations of the physical (i.e., mass) eigenstates of quarks of definite flavor d, s:

$$\begin{pmatrix} d' \\ s' \end{pmatrix} = \begin{pmatrix} \cos\theta_c & \sin\theta_c \\ -\sin\theta_c & \cos\theta_c \end{pmatrix} \begin{pmatrix} d \\ s \end{pmatrix}. \tag{12.110}$$

The quark mixing is described by a single parameter, the Cabibbo angle θ_c.

The original motivation for the GIM proposal was to ensure that there are no $s \leftrightarrow d$ transitions, which change flavor but not charge. The experimental evidence for the absence of strangeness-changing neutral currents is compelling. For instance, decays such as $K^0 \to \mu^+\mu^-$, $K^+ \to \pi^+ e^+ e^-$, $K^+ \to \pi^+ \nu\bar{\nu}$, which would be otherwise allowed, are either absent or highly suppressed.

How does the GIM mechanism work? To see this, it is convenient to rewrite (12.110) in the form

$$d_i' = \sum_j U_{ij} d_j, \tag{12.111}$$

with $d_1 \equiv d_L$ and $d_2 \equiv s_L$ where L denotes a left-handed quark state. Now the matrix U, introduced in (12.105), is unitary (provided we adopt the universal weak coupling hypothesis), and therefore we have

$$\begin{aligned}\sum_i \bar{d}_i' d_i' &= \sum_{i,j,k} \bar{d}_j U_{ji}^\dagger U_{ik} d_k \\ &= \sum_j \bar{d}_j d_j.\end{aligned} \tag{12.112}$$

That is, only transitions $d \to d$ and $s \to s$ are allowed; flavor-changing transitions, $s \leftrightarrow d$, are forbidden.

Before we extend the GIM mechanism to incorporate additional quark flavors, we must answer two questions that may have come to mind. First, why is the mixing taken in the d, s sector? In fact, the mixing could equally well have been formulated in the u, c sector; no observable difference would result since the absolute phases of the quark wavefunctions are not observable. Indeed, a more involved mixing in both the u, c and d, s sectors can be used, but it can always be simplified (by appropriately choosing the phases of the quark states) to the one-parameter form given in (12.110). This will become clearer in a moment.

A second question is, "Why is there no Cabibbo-like angle in the leptonic sector?"

$$\begin{pmatrix} \nu_e \\ e^- \end{pmatrix}, \quad \begin{pmatrix} \nu_\mu \\ \mu^- \end{pmatrix}. \tag{12.113}$$

The reason is that if ν_e and ν_μ are massless, then lepton mixing is unobservable. Any Cabibbo-like rotation still leaves us with neutrino mass eigenstates. By definition, we take ν_e to be the partner of the electron. This guarantees conserved lepton numbers L_e and L_μ. By contrast, the weak interaction eigenstates d′, s′ are not the same as the mass eigenstates but are related by (12.110).

Now consider the generalization of the Cabibbo-GIM ideas to more than four quark flavors. Imagine for a moment that weak interactions operate on N doublets of left-handed quarks,

$$\begin{pmatrix} u_i \\ d_i' \end{pmatrix} \qquad \text{with } i = 1, 2, \ldots, N \tag{12.114}$$

where d_i' are mixtures of the mass eigenstates d_i:

$$d_i' = \sum_{j=1}^{N} U_{ij} d_j. \tag{12.115}$$

U is a unitary $N \times N$ matrix to be determined by the flavor-changing weak processes. How many observable parameters does U contain? We can change the phase of each of the $2N$ quark states independently without altering the physics. Therefore, U contains

$$N^2 - (2N - 1)$$

real parameters. One phase is omitted as an overall phase change still leaves U invariant. On the other hand, an orthogonal $N \times N$ matrix has only $\frac{1}{2}N(N-1)$ real parameters [e.g., (12.110)]. Therefore, by redefining the quark phases, it is not possible, in general, to make U real. U must contain

$$N^2 - (2N - 1) - \tfrac{1}{2}N(N-1) = \tfrac{1}{2}(N-1)(N-2) \qquad (12.116)$$

residual phase factors. Thus, for two doublets ($N = 2$), there is one real parameter (θ_c), whereas for three doublets, there are three real parameters and one phase factor, $e^{i\delta}$.

We have in fact conclusive evidence for a fifth flavor of quark, the bottom quark b with charge $Q = -\frac{1}{3}$ (see Chapter 2), and it is widely believed that its partner, the top quark t with $Q = +\frac{2}{3}$, exists. Weak interactions would then operate on three doublets of left-handed quarks,

$$\begin{pmatrix} u \\ d' \end{pmatrix}, \quad \begin{pmatrix} c \\ s' \end{pmatrix}, \quad \begin{pmatrix} t \\ b' \end{pmatrix}. \qquad (12.117)$$

Why should we expect quarks to come in pairs? There are two reasons for this. First, it provides a natural way to suppress the flavor-changing neutral current; the argument leading to (12.112) applies just as well for three as for two doublets. The second reason is concerned with the desire to obtain a renormalizable gauge theory of weak interactions (see Chapters 14 and 15). This requires a delicate cancellation between different diagrams, relations which can easily be upset by "anomalies" due to fermion loops such as Fig. 12.19. These anomalies must be canceled for a renormalizable theory. Each triangle is proportional to $c_A^f Q_f^2$, where Q_f is the charge and c_A^f is the axial coupling of the weak neutral current. Thus, for an equal number N of lepton and quark doublets, the total anomaly is proportional to

$$\sum_{i=1}^{N} \left(\tfrac{1}{2}(0)^2 - \tfrac{1}{2}(-1)^2 + \tfrac{1}{2}N_c\left(+\tfrac{2}{3}\right)^2 - \tfrac{1}{2}N_c\left(-\tfrac{1}{3}\right)^2 \right) = 0. \qquad (12.118)$$

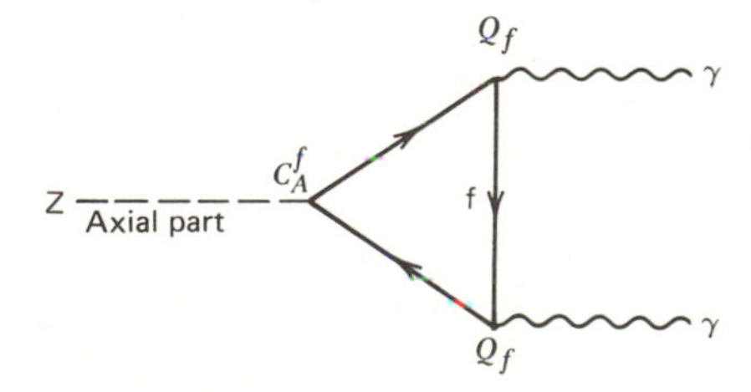

Fig. 12.19 A fermion (quark or lepton) triangle diagram which potentially could cause an anomaly.

The values used for c_A^f are determined in the next chapter (see Table 13.2). Thus, taking account of the three colors of each quark ($N_c = 3$), the anomalies are canceled. Since we have three lepton doublets (electron, muon, and tau), it is therefore natural to anticipate the three quark doublets of (12.117).

It is straightforward to extend the weak current, (12.105), to embrace the new doublet of quarks:

$$J^{\mu} = (\bar{u}\,\bar{c}\,\bar{t})\frac{\gamma_{\mu}(1-\gamma^5)}{2}U\begin{pmatrix} d \\ s \\ b \end{pmatrix}. \tag{12.119}$$

The 3×3 mixing matrix U contains three real parameters (Cabibbo-like mixing angles) and a phase factor $e^{i\delta}$ [see (12.116)]. The original parametrization was due to Kobayashi and Maskawa. Due to the phase δ, the matrix U is complex, unlike the 2×2 matrix of (12.110). That is, with the discovery of the b quark, complex elements U_{ij} enter the weak current. This has fundamental implications concerning CP invariance, which we discuss in the next section.

We illustrated how each element of the 2×2 Cabibbo matrix can be determined from experimental information on the corresponding quark flavor transition. The elements of the 3×3 matrix can also be studied in the same way. The present status of the experimental situation may be summarized as follows:

$$U = \begin{bmatrix} |U_{ud}| = 0.973 & |U_{us}| = 0.23 & |U_{ub}| \simeq 0 \\ |U_{cd}| \simeq 0.24 & |U_{cs}| \simeq 0.97 & |U_{cb}| \simeq 0.06 \\ |U_{td}| \simeq 0 & |U_{ts}| \simeq 0 & |U_{tb}| \simeq 1 \end{bmatrix}. \tag{12.120}$$

where $|U| \simeq 0$ means that the element is very small, but not yet determined. It is not surprising that some elements are not known, since there is no experimental information on the t quark. An incomplete sketch of how the results of (12.120) are obtained is shown in Table 12.1. The most striking feature of (12.120) is that the diagonal elements U_{ud}, U_{cs}, U_{tb} are clearly dominant. The large value of $|U_{cs}|$ simply reflects the experimental fact that charm particles preferentially decay into strange particles. The experimental observation that B mesons prefer to decay into charm particles implies $|U_{cb}| > |U_{ub}|$. Moreover (12.120) indicates that T mesons (when found) should preferentially decay into B mesons. All this has interesting consequences for the study of heavy quark states or for their detection in the case of T mesons. Spectacular experimental signatures result from the favored "cascade" decays, which may be characterized by the presence of multiple leptons (or strange particles) in the final state, for example:

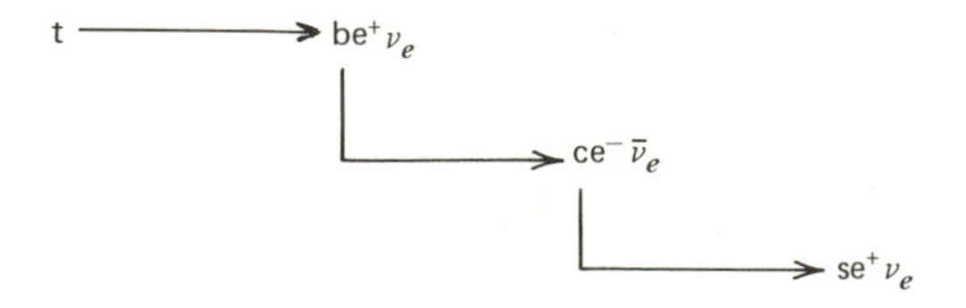

TABLE 12.1
A Summary of the Determination of the Elements $U_{qq'}$ of the Kobayashi–Maskawa Matrix

Element	Experimental Information
U_{ud}	β-Decay, generalize (12.107)
U_{us}	$K \to \pi e \nu$ and semileptonic hyperon decays, generalize discussion following (12.101)
U_{ub}	$b \to u e^- \bar{\nu}_e$, look for B meson decays with no K's in final state: gives $\lvert U_{ub}\rvert^2 < 0.02\lvert U_{cb}\rvert^2$.
U_{cd}	$\nu_\mu d \to \mu^- c$, charmed particle production by neutrinos
U_{cs}	$\nu_\mu s \to \mu^- c$ and $D^+ \to \overline{K}{}^0 e^+ \nu_e$, see (12.108); also constrained by the unitarity of U, which implies $\lvert U_{cd}\rvert^2 + \lvert U_{cs}\rvert^2 + \lvert U_{cb}\rvert^2 = 1$, together with the information on $\lvert U_{cd}\rvert$ and $\lvert U_{cb}\rvert$.
U_{cb}	The (long) lifetime of the B meson, $\tau_B \sim 10^{-12}$secs, and $\lvert U_{ub}\rvert$.
U_{tq}	From unitarity bounds

12.13 *CP* Invariance?

To investigate *CP* invariance, we first compare the amplitude for a weak process, say, the quark scattering process $ab \to cd$, with that for the antiparticle reaction $\bar{a}\bar{b} \to \bar{c}\bar{d}$. We take $ab \to cd$ to be the charged current interaction of Fig. 12.20a. The amplitude

$$\begin{aligned}\mathfrak{M} &\sim J^\mu_{ca} J^\dagger_{\mu bd} \\ &\sim \left(\bar{u}_c\gamma^\mu(1-\gamma^5)U_{ca}u_a\right)\left(\bar{u}_b\gamma_\mu(1-\gamma^5)U_{bd}u_d\right)^\dagger \\ &\sim U_{ca}U^*_{db}\left(\bar{u}_c\gamma^\mu(1-\gamma^5)u_a\right)\left(\bar{u}_d\gamma_\mu(1-\gamma^5)u_b\right),\end{aligned} \tag{12.121}$$

since $U^\dagger_{bd} = U^*_{db}$. $\mathfrak{M}$ describes either $ab \to cd$ or $\bar{c}\bar{d} \to \bar{a}\bar{b}$ (remembering the antiparticle description of Chapter 3).

On the other hand, the amplitude $\mathfrak{M}'$ for the antiparticle process $\bar{a}\bar{b} \to \bar{c}\bar{d}$ (or $cd \to ab$) is

$$\begin{aligned}\mathfrak{M}' &\sim (J^\mu_{ca})^\dagger J_{\mu bd} \\ &\sim U^*_{ca}U_{db}\left(\bar{u}_a\gamma^\mu(1-\gamma^5)u_c\right)\left(\bar{u}_b\gamma_\mu(1-\gamma^5)u_d\right);\end{aligned} \tag{12.122}$$

that is,

$$\mathfrak{M}' = \mathfrak{M}^\dagger.$$

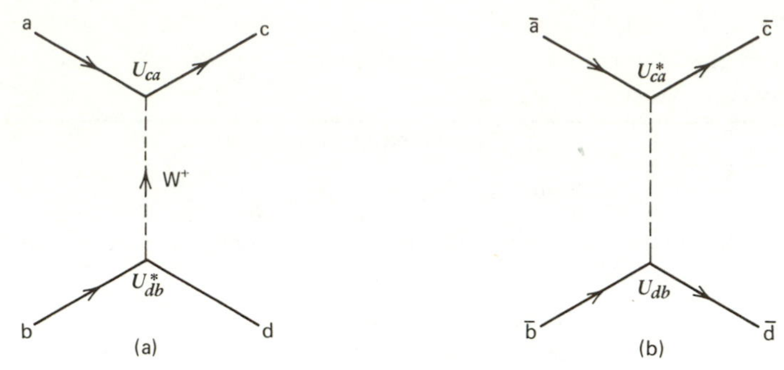

Fig. 12.20 The processes described by (a) the weak amplitude $\mathcal{M}(ab \to cd)$ and (b) its hermitian conjugate.

This should not be surprising. It is demanded by the hermicity of the Hamiltonian. By glancing back at (4.6) and (4.17), we see that $\mathcal{M}$ is essentially the interaction Hamiltonian V for the process. The total interaction Hamiltonian must contain $\mathcal{M} + \mathcal{M}^\dagger$, where $\mathcal{M}$ describes the $i \to f$ transition and $\mathcal{M}^\dagger$ describes the $f \to i$ transition in the notation of Chapter 4.

In Section 12.1, we have seen that weak interactions violate both P invariance and C invariance, but have indicated that invariance under the combined CP operation may hold. How do we verify that the theory is CP invariant? We calculate from $\mathcal{M}(ab \to cd)$ of (12.121) the amplitude $\mathcal{M}_{CP}$, describing the CP-transformed process, and see whether or not the Hamiltonian remains hermitian. If it does, that is, if

$$\mathcal{M}_{CP} = \mathcal{M}^\dagger,$$

then the theory is CP invariant. If it does not, then CP is violated.

$\mathcal{M}_{CP}$ is obtained by substituting the CP-transformed Dirac spinors in (12.121):

$$u_i \to P(u_i)_C, \qquad i = a,\ldots,d \tag{12.123}$$

where u_C are the charge-conjugate spinors of Section 5.4,

$$u_C = C\bar{u}^T. \tag{12.124}$$

Clearly, to form $\mathcal{M}_{CP}$, we need $\bar{u}_C$ and, also, to know how $\gamma^\mu(1-\gamma^5)$ transforms under C. In the standard representation of the γ-matrices, we have [see (5.39)]

$$\begin{aligned} \bar{u}_C &= -u^T C^{-1}, \\ C^{-1}\gamma^\mu C &= -(\gamma^\mu)^T, \\ C^{-1}\gamma^\mu\gamma^5 C &= +(\gamma^\mu\gamma^5)^T. \end{aligned} \tag{12.125}$$

EXERCISE 12.24 Verify (12.125) using (5.39).

With the replacements (12.123), the first charged current of (12.121) becomes

$$
\begin{aligned}
(J^{\mu}_{ca})_C &= U_{ca}(\bar{u}_c)_C \gamma^{\mu}(1-\gamma^5)(u_a)_C \\
&= -U_{ca}u_c^T C^{-1}\gamma^{\mu}(1-\gamma^5)C\bar{u}_a^T \\
&= U_{ca}u_c^T\left[\gamma^{\mu}(1+\gamma^5)\right]^T\bar{u}_a^T \\
&= (-)U_{ca}\bar{u}_a\gamma^{\mu}(1+\gamma^5)u_c.
\end{aligned}
\tag{12.126}
$$

The above procedure is exactly analogous to that used to obtain the charge-conjugate electromagnetic current, (5.40).

The parity operation $P = \gamma^0$, see (5.62), and so

$$P^{-1}\gamma^{\mu}(1+\gamma^5)P = \gamma^{\mu\dagger}(1-\gamma^5),$$

see (5.9)–(5.11). Thus,

$$(J^{\mu}_{ca})_{CP} = (-)U_{ca}\bar{u}_a\gamma^{\mu\dagger}(1-\gamma^5)u_c,$$

and hence

$$\mathfrak{M}_{CP} \sim U_{ca}U_{db}^*\left[\bar{u}_a\gamma^{\mu}(1-\gamma^5)u_c\right]\left[\bar{u}_b\gamma_{\mu}(1-\gamma^5)u_d\right]. \tag{12.127}$$

We can now compare $\mathfrak{M}_{CP}$ with $\mathfrak{M}^{\dagger}$ of (12.122). Provided the elements of the matrix U are real, we find

$$\mathfrak{M}_{CP} = \mathfrak{M}^{\dagger},$$

and the theory is *CP* invariant. At the four-quark (u, d, c, s) level, this is the case, as the 2×2 matrix U, (12.106), is indeed real. However, with the advent of the b (and t) quarks, the matrix U becomes the 3×3 Kobayashi–Maskawa (KM) matrix. It now contains a complex phase factor $e^{i\delta}$. Then, in general, we have

$$\mathfrak{M}_{CP} \neq \mathfrak{M}^{\dagger},$$

and the theory necessarily violates *CP* invariance.

In fact, a tiny *CP* violation had been established many years before the introduction of the KM matrix. The violation was discovered by observing the decays of neutral kaons. These particles offer a unique "window" through which to look for small *CP* violating effects. We discuss this next.

12.14 *CP* Violation: The Neutral Kaon System

The observations of neutral kaons have led to several fundamental discoveries in particle physics. K^0 and $\bar{K}^0$, with definite I_3 and Y, are the states produced by strong interactions (see Chapter 2). For example,

$$
\begin{aligned}
\pi^- p &\to K^0\Lambda, \\
\pi^+ p &\to \bar{K}^0 K^+ p.
\end{aligned}
$$

However, experimentally it is found that K^0 decay occurs with two different lifetimes:

$$\tau(K_S^0 \to 2\pi) = 0.9 \times 10^{-10}\ \text{sec}$$
$$\tau(K_L^0 \to 3\pi) = 0.5 \times 10^{-7}\ \text{sec}. \tag{12.128}$$

In other words, the K^0 produced by the strong interactions seems to be two different particles (K_S^0 and K_L^0) when we study its weak decays. The same dilemma appears for the $\overline{K}^0$, and it was therefore proposed that the K^0 and $\overline{K}^0$ are nothing but two different admixtures of the K_S^0 and K_L^0, the particles associated with the short- and long-lived 2π and 3π decay modes.

In the absence of orbital angular momentum, the 2π and 3π final states differ in parity, with $P = +1$ and -1, respectively. We mentioned that these observations were in fact instrumental in the discovery of parity violation in weak interactions in 1957. For some time, it was thought that weak interactions were at least invariant under the combined CP operation. We make the conventional choice of phase of $|K^0\rangle$ and $|\overline{K}^0\rangle$ such that

$$CP|K^0\rangle = |\overline{K}^0\rangle.$$

Since the final 2π and 3π states are eigenstates of CP with eigenvalues $+1$ and -1, respectively (see Exercise 12.26), it is tempting to identify the neutral kaon CP eigenstates with K_S^0 and K_L^0:

$$\begin{aligned} |K_S^0\rangle &= \sqrt{\tfrac{1}{2}}\left(|K^0\rangle + |\overline{K}^0\rangle\right) \qquad [CP = +1] \\ |K_L^0\rangle &= \sqrt{\tfrac{1}{2}}\left(|K^0\rangle - |\overline{K}^0\rangle\right) \qquad [CP = -1]. \end{aligned} \tag{12.129}$$

To a very good approximation, this is true. However, in 1964 it was demonstrated that $K_L^0 \to \pi^+\pi^-$ with a branching ratio of order 10^{-3}. Therefore, a small CP violating effect is indeed present. An excellent summary of the experiments and other K^0 phenomena is given, for example, by Perkins (1982).

EXERCISE 12.25 Show that $C = -1$ for a photon and hence that $C = +1$ for a π^0.

Hint Under $e \to -e$, the amplitude corresponding to Fig. 1.9a will change sign.

EXERCISE 12.26 Show that in the absence of angular momentum, a $\pi^+\pi^-$ or $\pi^0\pi^0$ state is an eigenstate of CP with eigenvalue $+1$. Further, show that by adding an S wave π^0, we obtain CP eigenstates $\pi^+\pi^-\pi^0$ or $3\pi^0$ with eigenvalue -1.

Hint A $\pi^+\pi^-$ state is totally symmetric under the interchange of the π-mesons, by Bose statistics. Interchange of the particles corresponds to the operation C followed by P.

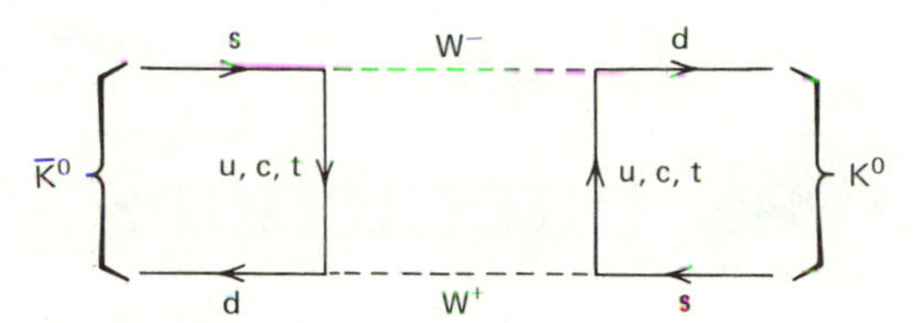

Fig. 12.21 Diagram responsible for $K^0 \leftrightarrow \overline{K}^0$ mixing.

The appearance of complex elements U_{ij} in the KM matrix may be related to the $\Delta S = 2$ transition mixing the $S = \pm 1$ K^0 and $\overline{K}^0$ states. The connection is not simple; the diagram of Fig. 12.21 is responsible for K^0–$\overline{K}^0$ mixing. With only u and c quarks exchanged (in the four-quark theory), the mixing would conserve *CP* and the resulting mass matrix would have eigenstates, (12.129), that differ by a small mass. In a six-quark theory, the values are slightly perturbed and $|K_S^0\rangle$ and $|K_L^0\rangle$ are no longer exactly *CP* eigenstates. Finally, it is widely believed that *CP* nonconservation in the early universe is the source of the apparent imbalance between matter and antimatter which we observe around us.

13

Electroweak Interactions

The picture of weak interactions we have discussed so far is only satisfactory at a superficial level. We have only calculated lowest-order graphs where the momentum exchanged between the weak currents satisfies $|q^2| \ll (100 \text{ GeV})^2$. The results do not depend on whether or not massive intermediate vector bosons $W^\pm$ exist. The existence of $W^\pm$ simply leads to a reinterpretation of the Fermi coupling G, see (12.15). However, calculations of anything other than lowest-order, low-energy amplitudes lead to very serious problems. For many years, the current–current interaction was regarded merely as a phenomenological prescription rather than a proper theory. With hindsight, we can see that to assume that massive weak bosons exist is indeed a step toward converting weak interaction phenomenology into a respectable (that is, renormalizable) theory. This seemed impossible until the discovery of "spontaneously broken non-Abelian gauge symmetries," an idea we introduce in the next chapter. Suffice it to say here that a necessary ingredient is that the weak currents form a symmetry group. We therefore continue our discussion of weak interactions with this objective in mind.

13.1 Weak Isospin and Hypercharge

Can the weak neutral current (J_μ^{NC} of Section 12.10), taken together with the charged currents (J_μ and $J_\mu^\dagger$), form a symmetry group of weak interactions? First, we recall the form of the charge currents,

$$\left.\begin{aligned} J_\mu \equiv J_\mu^+ &= \bar{u}_\nu \gamma_\mu \tfrac{1}{2}(1-\gamma^5) u_e \\ &\equiv \bar{\nu}\gamma_\mu \tfrac{1}{2}(1-\gamma^5) e = \bar{\nu}_L \gamma_\mu e_L \end{aligned}\right\} \qquad \left.\begin{aligned} J_\mu^\dagger \equiv J_\mu^- &= \bar{u}_e \gamma_\mu \tfrac{1}{2}(1-\gamma^5) u_\nu \\ &= \bar{e}_L \gamma_\mu \nu_L \end{aligned}\right\} \tag{13.1}$$

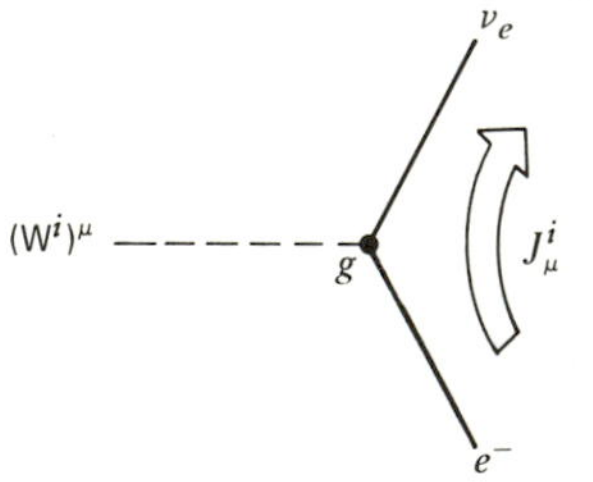

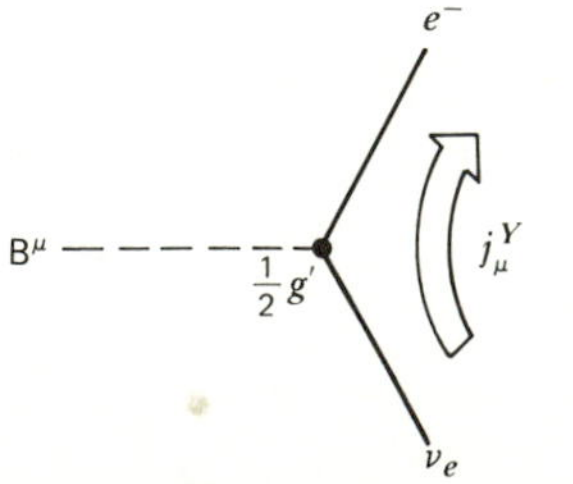

ν_e
$(W^i)^\mu$
g
J_μ^i
e^-
e^-
B^μ
$\frac{1}{2}g'$
j_μ^Y
ν_e

where the + and − superscripts are to indicate the charge-raising and charge-lowering character of the currents, respectively. The subscript L is used to denote left-handed spinors and records the V–A nature of the charged currents. Here, we have used the particle names to denote the Dirac spinors ($\bar{u}_\nu \equiv \bar{\nu}$, $u_e \equiv e$, etc.).

We can rewrite these two charged currents in a suggestive two-dimensional form. We introduce the doublet

$$\chi_L = \begin{pmatrix} \nu \\ e^- \end{pmatrix}_L \tag{13.2}$$

and the "step-up" and "step-down" operators $\tau_\pm = \frac{1}{2}(\tau_1 \pm i\tau_2)$:

$$\tau_+ = \begin{pmatrix} 0 & 1 \\ 0 & 0 \end{pmatrix}, \qquad \tau_- = \begin{pmatrix} 0 & 0 \\ 1 & 0 \end{pmatrix}, \tag{13.3}$$

where the τ's are the usual Pauli spin matrices. The charged currents (13.1) then become (with x dependence as in (6.6))

$$\begin{aligned} J_\mu^+(x) &= \bar{\chi}_L \gamma_\mu \tau_+ \chi_L, \\ J_\mu^-(x) &= \bar{\chi}_L \gamma_\mu \tau_- \chi_L. \end{aligned} \tag{13.4}$$

Anticipating a possible $SU(2)$ structure for the weak currents, we are led to introduce a neutral current of the form

$$J_\mu^3(x) = \bar{\chi}_L \gamma_\mu \tfrac{1}{2}\tau_3 \chi_L \tag{13.5}$$

$$= \tfrac{1}{2}\bar{\nu}_L \gamma_\mu \nu_L - \tfrac{1}{2}\bar{e}_L \gamma_\mu e_L \tag{13.6}$$

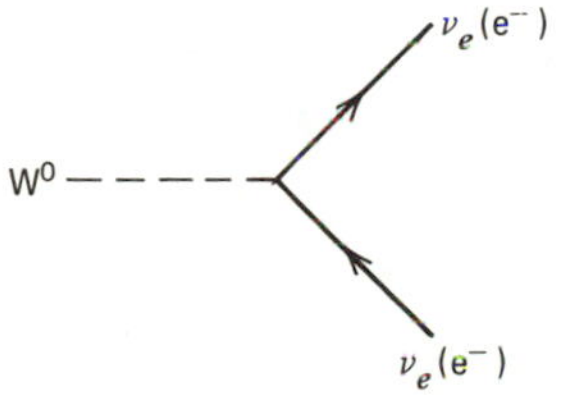

We have thus constructed an "isospin" triplet of weak currents,

$$J_\mu^i(x) = \bar{\chi}_L \gamma_\mu \tfrac{1}{2}\tau_i \chi_L, \qquad \text{with } i = 1, 2, 3, \tag{13.7}$$

whose corresponding charges

$$T^i = \int J_0^i(x)\, d^3x \tag{13.8}$$

generate an $SU(2)_L$ algebra

$$[T^i, T^j] = i\varepsilon_{ijk} T^k. \tag{13.9}$$

The subscript L on $SU(2)$ is to remind us that the weak isospin current couples only left-handed fermions.

Can the current $J_\mu^3(x)$ which we have just introduced be identified directly with the weak neutral current of Section 12.10? Unfortunately, we see immediately that this attractive idea does not work; the observed weak neutral current J_μ^{NC} has a right-handed component, see (12.97). However, the electromagnetic current is a neutral current with right- as well as left-handed components. For example, for the electron we have from (6.35)

$$j_\mu^{em}(x) = -\bar{e}\gamma_\mu e = -\bar{e}_R\gamma_\mu e_R - \bar{e}_L\gamma_\mu e_L. \tag{13.10}$$

Note, in passing, that we have omitted the coupling e in our definition of j_μ^{em}. This will simplify our discussion of electroweak interactions. Thus, the electromagnetic current j_μ of (6.5) is written

$$j_\mu \equiv e j_\mu^{em} = e\bar{\psi}\gamma_\mu Q\psi, \tag{13.11}$$

where Q is the charge operator, with eigenvalue $Q = -1$ for the electron. Technically speaking, we say that Q is the generator of a $U(1)_{em}$ symmetry group of electromagnetic interactions (see Section 14.2).

We include j_μ^{em} in an attempt to save the $SU(2)_L$ symmetry. Note that neither of the neutral currents J_μ^{NC} or j_μ^{em} respects the $SU(2)_L$ symmetry. However, the idea is to form two orthogonal combinations which do have definite transformation properties under $SU(2)_L$; one combination, J_μ^3, is to complete the weak isospin triplet J_μ^i, while the second, j_μ^Y, is unchanged by $SU(2)_L$ transformations (i.e., is a weak isospin singlet). j_μ^Y is called the weak hypercharge current and is given by

$$j_\mu^Y = \bar{\psi}\gamma_\mu Y\psi, \tag{13.12}$$

where the weak hypercharge Y is defined by

$$Q = T^3 + \frac{Y}{2}. \tag{13.13}$$

That is,

$$\boxed{j_\mu^{em} = J_\mu^3 + \tfrac{1}{2}j_\mu^Y.} \tag{13.14}$$

Just as Q generates the group $U(1)_{em}$, so the hypercharge operator Y generates a symmetry group $U(1)_Y$. Thus, we have incorporated the electromagnetic interaction, and as a result the symmetry group has been enlarged to $SU(2)_L \times U(1)_Y$. In a sense we have unified the electromagnetic with the weak interaction. However, rather than a single unified symmetry group, we have two groups each with an independent coupling strength. So, in addition to e, we will need another coupling to fully specify the electroweak interaction. Thus, from an aesthetic viewpoint the unification is perhaps not completely satisfying (see Section 15.7).

The proposed weak isospin and weak hypercharge scheme is mathematically an exact copy of the original Gell-Mann–Nishijima scheme for arranging strange particles in $SU(2)$ hadronic isospin multiplets (see Section 2.9). The names "weak isospin" and "weak hypercharge" are taken from this analogy.

The $SU(2)_L \times U(1)_Y$ proposal was first made by Glashow in 1961, long before the discovery of the weak neutral current, and, as we describe in Chapter 15, was

TABLE 13.1
Weak Isospin and Hypercharge Quantum Numbers of Leptons and Quarks

Lepton	T	T^3	Q	Y
ν_e	$\frac{1}{2}$	$\frac{1}{2}$	0	-1
e_L^-	$\frac{1}{2}$	$-\frac{1}{2}$	-1	-1
e_R^-	0	0	-1	-2

Quark	T	T^3	Q	Y
u_L	$\frac{1}{2}$	$\frac{1}{2}$	$\frac{2}{3}$	$\frac{1}{3}$
d_L	$\frac{1}{2}$	$-\frac{1}{2}$	$-\frac{1}{3}$	$\frac{1}{3}$
u_R	0	0	$\frac{2}{3}$	$\frac{4}{3}$
d_R	0	0	$-\frac{1}{3}$	$-\frac{2}{3}$

extended to accommodate massive vector bosons ($W^{\pm}, Z^0$) by Weinberg (1967) and Salam (1968). It is frequently referred to as the "standard model" for electroweak interactions.

Since we have a product of symmetry groups, the generator Y must commute with the generators T^i. As a consequence, all the members of an isospin multiplet must have the same value of the hypercharge. For example, for the electron multiplets, (13.14) becomes

$$\begin{aligned} j_\mu^Y &= 2\,j_\mu^{em} - 2J_\mu^3 \\ &= -2(\bar{e}_R\gamma_\mu e_R + \bar{e}_L\gamma_\mu e_L) - (\bar{\nu}_L\gamma_\mu\nu_L - \bar{e}_L\gamma_\mu e_L) \\ &= -2(\bar{e}_R\gamma_\mu e_R) - 1(\bar{\chi}_L\gamma_\mu\chi_L), \end{aligned} \tag{13.15}$$

where we have used (13.6) and (13.10). Thus, the isospin doublet $(\nu, e)_L$ has $Y = -1$, and the isospin singlet e_R has $Y = -2$. These quantum numbers are summarized in Table 13.1.

We can readily incorporate quarks into the scheme. The weak isospin current J_μ^i couples only to doublets of left-handed quarks $(u, d)_L$; we assign $T = \frac{1}{2}$ to q_L and $T = 0$ to q_R states.

EXERCISE 13.1 Verify the quark quantum numbers given in Table 13.1.

13.2 The Basic Electroweak Interaction

To complete the unification of the electromagnetic and weak interactions, we must modify the current–current form of the weak interactions given in (12.13) and (12.85). We assume that the current–current structure is an effective interaction which results from the exchange of massive vector bosons with only a small momentum transfer.

In Section 6.1, we developed QED from the basic interaction

$$-ie(j^{em})^\mu A_\mu. \tag{13.16}$$

Just as the electromagnetic current is coupled to the photon, we assume that the electroweak currents of Section 13.1 are coupled to vector bosons. The "standard" model consists of an isotriplet of vector fields W_μ^i coupled with strength g to the weak isospin current J_μ^i, together with a single vector field B_μ coupled to the weak hypercharge current j_μ^Y with strength conventionally taken to be $g'/2$. The basic electroweak interaction is therefore

$$\boxed{-i\,g(J^i)^\mu W_\mu^i - i\,\frac{g'}{2}(j^Y)^\mu B_\mu.} \tag{13.17}$$

The fields

$$W_\mu^\pm = \sqrt{\tfrac{1}{2}}\left(W_\mu^1 \mp iW_\mu^2\right) \tag{13.18}$$

describe massive charged bosons $W^\pm$, whereas W_μ^3 and B_μ are neutral fields. The basic interactions are pictured in Fig. 13.1.

The electromagnetic interaction (13.16) is embedded in (13.17). Indeed, when we generate the masses of the bosons by symmetry breaking (see Chapter 15), the two neutral fields W_μ^3 and B_μ must mix in such a way that the physical states (i.e., the mass eigenstates) are

$$A_\mu = B_\mu \cos\theta_W + W_\mu^3 \sin\theta_W \qquad \text{(massless)}, \tag{13.19}$$

$$Z_\mu = -B_\mu \sin\theta_W + W_\mu^3 \cos\theta_W \qquad \text{(massive)}, \tag{13.20}$$

where θ_W is called the Weinberg or weak mixing angle (although Glashow was the first to introduce the idea). We may therefore write the electroweak neutral current interaction

$$\begin{aligned} -igJ_\mu^3(W^3)^\mu - i\frac{g'}{2}j_\mu^Y B^\mu &= -i\left(g\sin\theta_W J_\mu^3 + g'\cos\theta_W \frac{j_\mu^Y}{2}\right)A^\mu \\ &\quad -i\left(g\cos\theta_W J_\mu^3 - g'\sin\theta_W \frac{j_\mu^Y}{2}\right)Z^\mu. \end{aligned} \tag{13.21}$$

The first term is the electromagnetic interaction, (13.16), and so the expression in

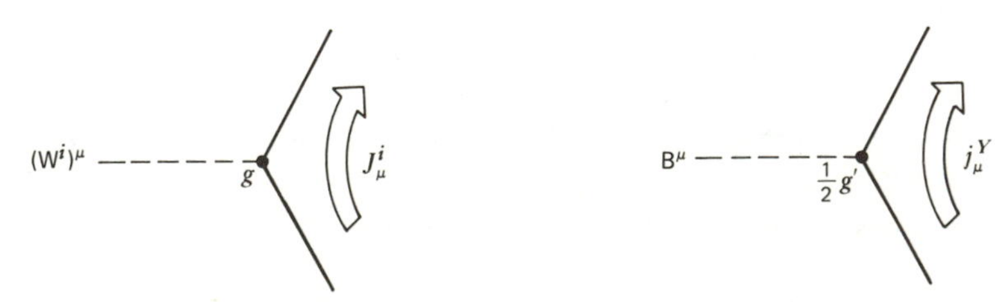

Fig. 13.1 The vector boson couplings to the weak isospin and hypercharge currents.

brackets must be

$$ej_\mu^{em} \equiv e\left(J_\mu^3 + \tfrac{1}{2}j_\mu^Y\right), \tag{13.22}$$

see (13.14). Therefore, we have

$$\boxed{g \sin\theta_W = g' \cos\theta_W = e.} \tag{13.23}$$

That is, the mixing angle in (13.19) and (13.20) is given by the ratio of the two independent group coupling constants, $\tan\theta_W = g'/g$.

Using (13.22) and (13.23), we may express the weak neutral current interaction of (13.21) in the form

$$-i\frac{g}{\cos\theta_W}\left(J_\mu^3 - \sin^2\theta_W j_\mu^{em}\right)Z^\mu \equiv -i\frac{g}{\cos\theta_W}J_\mu^{NC}Z^\mu. \tag{13.24}$$

It is this definition,

$$\boxed{J_\mu^{NC} \equiv J_\mu^3 - \sin^2\theta_W j_\mu^{em}}, \tag{13.25}$$

which relates the neutral current J^{NC} to the weak isospin current $\mathbf{J}$.

Very early in our discussion of the weak interaction, we speculated that perhaps it was not a new fundamental interaction but that it was a manifestation of the electromagnetic interaction. We suggested that this electroweak unification might be achieved by setting the coupling $g = e$. We now see that this simple unification is not true. The standard model relation,

$$g \sin\theta_W = g' \cos\theta_W = e,$$

is more complicated. The electromagnetic interaction (a $U(1)$ gauge symmetry with coupling e) "sits across" weak isospin (an $SU(2)_L$ symmetry with coupling g) and weak hypercharge (a $U(1)$ symmetry with coupling g'). We discuss these gauge symmetries in Chapters 14 and 15. Using (13.23), we see that the two couplings g and g' can be replaced by e and θ_W, where the parameter θ_W is to be determined by experiment.

To summarize, we have achieved our objective of expressing the observed neutral currents

$$\begin{aligned} j_\mu^{em} &= J_\mu^3 + \tfrac{1}{2}j_\mu^Y, \\ J_\mu^{NC} &= J_\mu^3 - \sin^2\theta_W j_\mu^{em} \end{aligned} \tag{13.26}$$

in terms of currents J_μ^3 and j_μ^Y belonging to symmetry groups $SU(2)_L$ and $U(1)_Y$, respectively. The right-handed component of J_μ^{NC} (the original problem) has been arranged to cancel with that in $\sin^2\theta_W j_\mu^{em}$ to leave a pure left-handed J_μ^3 of $SU(2)_L$, where $\sin^2\theta_W$ is to be determined by experiment. Of course, for the model to be successful, the same $\sin^2\theta_W$ must be found in all electroweak phenomena.

13.3 The Effective Current–Current Interaction

Before we continue, we should relate our basic interactions to the current–current structure of weak interactions that we used throughout Chapter 12. There, we saw that *charged current* phenomena could be explained by invariant amplitudes of the form

$$\mathcal{M}^{CC} = \frac{4G}{\sqrt{2}} J^{\mu} J_{\mu}^{\dagger}, \tag{13.27}$$

see (12.13), where in the new isospin notation [see (13.7)]

$$J_{\mu} \equiv J_{\mu}^{+} = \bar{\chi}_L \gamma_{\mu} \tau_{+} \chi_L = \tfrac{1}{2}\left(J_{\mu}^{1} + iJ_{\mu}^{2}\right). \tag{13.28}$$

For an interaction proceeding via the exchange of a massive charged boson, we can follow the QED procedure of Section 6.2 to calculate $\mathcal{M}$. The analogue of Fig. 6.2 is Fig. 13.2. First, we rewrite the basic charged current interaction of (13.17) in the form

$$-i\frac{g}{\sqrt{2}}\left(J^{\mu} W_{\mu}^{+} + J^{\mu\dagger} W_{\mu}^{-}\right), \tag{13.29}$$

where we have used the identity

$$\tfrac{1}{2}\left(\tau_1 W^1 + \tau_2 W^2\right) = \sqrt{\tfrac{1}{2}}\left(\tau_{+} W^{+} + \tau_{-} W^{-}\right) \tag{13.30}$$

with $W^{\pm}$ given by (13.18). Then, we copy the QED procedure leading to (6.8) and obtain

$$\mathcal{M}^{CC} \simeq \left(\frac{g}{\sqrt{2}} J_{\mu}\right)\left(\frac{1}{M_W^2}\right)\left(\frac{g}{\sqrt{2}} J^{\mu\dagger}\right), \tag{13.31}$$

where $1/M_W^2$ is the approximation to the W propagator at low q^2 (see Section 6.17). Comparison of (13.31) with (13.27) gives

$$\frac{G}{\sqrt{2}} = \frac{g^2}{8M_W^2}. \tag{13.32}$$

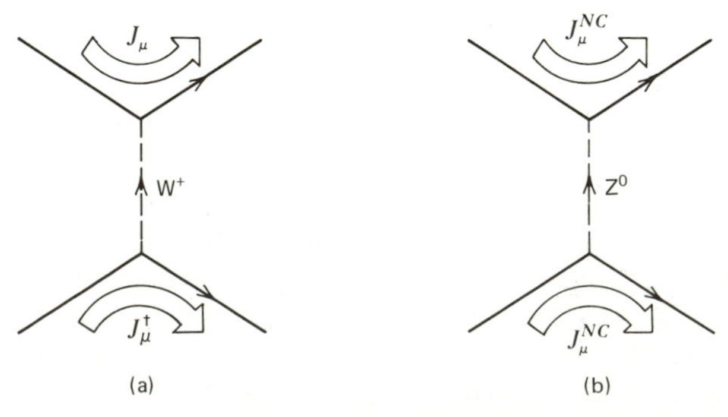

Fig. 13.2 Charged and neutral current weak interactions.

In an analogous way, we may express the amplitude for a *neutral current* process in terms of Z exchange, see Fig. 13.2b. Using (13.24), we have, for $|q^2| \ll M_Z^2$,

$$\mathfrak{M}^{NC} = \left(\frac{g}{\cos\theta_W} J_\mu^{NC}\right)\left(\frac{1}{M_W^2}\right)\left(\frac{g}{\cos\theta_W} J^{NC\mu}\right). \tag{13.33}$$

If we now compare (13.33) with the current–current form we used for the invariant amplitude, (12.85),

$$\mathfrak{M}^{NC} = \frac{4G}{\sqrt{2}} 2\rho J_\mu^{NC} J^{NC\mu}, \tag{13.34}$$

we can identify

$$\rho \frac{G}{\sqrt{2}} = \frac{g^2}{8M_Z^2 \cos^2\theta_W}. \tag{13.35}$$

From (13.32) and (13.35), we find that the parameter ρ, which specifies the relative strength of the neutral and charged current weak interactions, is given by

$$\boxed{\rho = \frac{M_W^2}{M_Z^2 \cos^2\theta_W}.} \tag{13.36}$$

We shall see that, experimentally, $\rho = 1$ to within a small error. This value is also predicted by the minimal model proposed by Weinberg and Salam (see Chapter 15).

13.4 Feynman Rules for Electroweak Interactions

To obtain the Feynman rules (for $-i\mathfrak{M}$) for electroweak interactions, we follow the procedure that we used for QED. In Section 6.1, we showed how the *electromagnetic interaction*

$$-ie(j^{em})^\mu A_\mu = -ie(\bar{\psi}\gamma^\mu Q\psi) A_\mu \tag{13.37}$$

led to the vertex factor

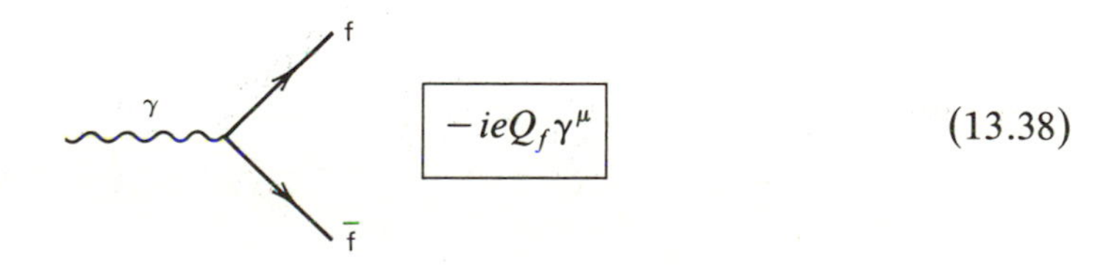

where Q_f is the charge of fermion f; $Q_f = -1$ for the electron. The outgoing $\bar{\text{f}}$ should be drawn as an ingoing f, and spinors attached to the fermion lines as in Section 6.17.

We copy this procedure for the *charged current interaction*, (13.29),

$$-i\frac{g}{\sqrt{2}}(\bar{\chi}_L\gamma^\mu\tau_+\chi_L)W^+_\mu = -i\frac{g}{\sqrt{2}}(\bar{\nu}_L\gamma^\mu e_L)W^+_\mu$$

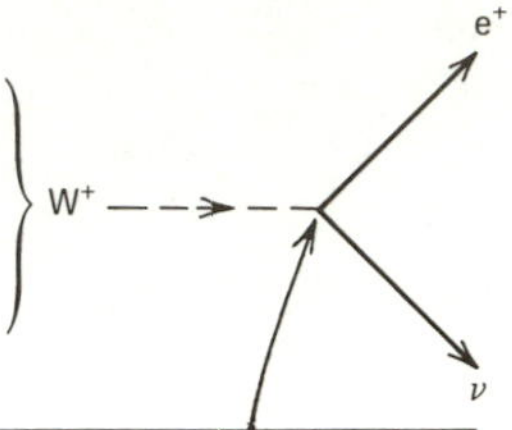

$$\boxed{-i\frac{g}{\sqrt{2}}\gamma^\mu\tfrac{1}{2}(1-\gamma^5)} \qquad (13.39)$$

$$-i\frac{g}{\sqrt{2}}(\bar{\chi}_L\gamma^\mu\tau_-\chi_L)W^-_\mu = -i\frac{g}{\sqrt{2}}(\bar{e}_L\gamma^\mu\nu_L)W^-_\mu$$

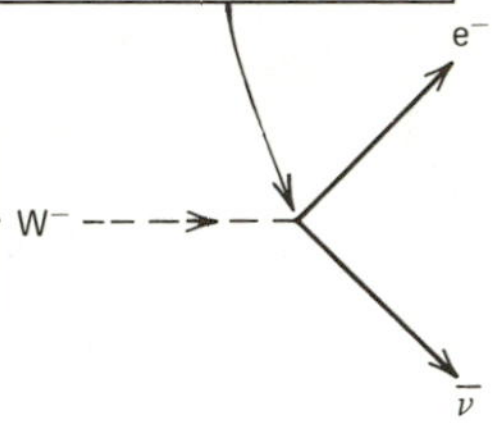

W⁺ e⁺ ν W⁻ e⁻ ν̄

For $\chi_L = (\nu_e, e^-)$, these interactions lead to the vertex factor shown; spinors are associated with the external fermion lines just as in QED. Clearly, the vertex factor will be the same for the $W^\pm$ couplings to the other fermion doublets (ν_μ, μ^-), (u, d'), and so on.

The *neutral current interaction* is given by (13.24),

$$-i\frac{g}{\cos\theta_W}\left(J^3_\mu - \sin^2\theta_W j^{em}_\mu\right)Z^\mu = -i\frac{g}{\cos\theta_W}\bar{\psi}_f\gamma^\mu\left[\tfrac{1}{2}(1-\gamma^5)T^3 - \sin^2\theta_W Q\right]\psi_f Z_\mu \qquad (13.40)$$

for the coupling $Z \to f\bar{f}$. It is customary to express the vertex factor in the general form

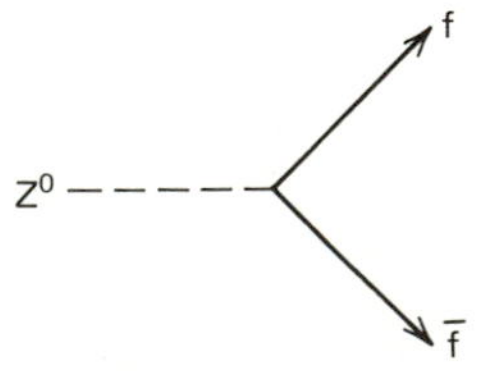

$$\Rightarrow \quad Z^0 \to f\bar{f} \qquad \boxed{-i\frac{g}{\cos\theta_W}\gamma^\mu\tfrac{1}{2}(c_V^f - c_A^f\gamma^5).} \qquad (13.41)$$

If we compare (13.40) with (13.41), we see that the vector and axial-vector couplings, c_V and c_A, are determined in the standard model (given the value of

TABLE 13.2
The Z → $f\bar{f}$ Vertex Factors, (13.41), in the Standard Model (with $\sin^2\theta_W = 0.234$)

f	Q_f	c_A^f	c_V^f
$\nu_e, \nu_\mu, \ldots$	0	$\frac{1}{2}$	$\frac{1}{2}$
$e^-, \mu^-, \ldots$	-1	$-\frac{1}{2}$	$-\frac{1}{2} + 2\sin^2\theta_W \simeq -0.03$
u, c, ...	$\frac{2}{3}$	$\frac{1}{2}$	$\frac{1}{2} - \frac{4}{3}\sin^2\theta_W \simeq 0.19$
d, s, ...	$-\frac{1}{3}$	$-\frac{1}{2}$	$-\frac{1}{2} + \frac{2}{3}\sin^2\theta_W \simeq -0.34$

$\sin^2\theta_W$). Their values are

$$c_V^f = T_f^3 - 2\sin^2\theta_W Q_f,$$
$$c_A^f = T_f^3, \tag{13.42}$$

where T_f^3 and Q_f are, respectively, the third component of the weak isospin and the charge of fermion f. The values of c_V and c_A are listed in Table 13.2.

The Feynman rules allow us to predict the decay properties of the $W^\pm$ and Z^0 bosons in the standard model, see Exercises 13.2–13.6 below.

EXERCISE 13.2 If the vertex factor for the decay of a vector boson X into two spin-$\frac{1}{2}$ fermions f_1 and $\bar{f}_2$ is

$$-ig_X\gamma^\mu \tfrac{1}{2}(c_V - c_A\gamma^5),$$

then show that

$$\Gamma(X \to f_1\bar{f}_2) = \frac{g_X^2}{48\pi}(c_V^2 + c_A^2)M_X, \tag{13.43}$$

where M_X is the mass of the boson and where we have neglected the masses of the fermions.

Hint Use (6.93) to show that after summing over the fermion and averaging over the boson spins,

$$\overline{|\mathfrak{M}|^2} = \tfrac{1}{4}g_X^2(c_V^2 + c_A^2)(-g_{\mu\nu})\,\mathrm{Tr}(\gamma^\mu \not{k}\gamma^\nu \not{k}') \tag{13.44}$$

where k, k' are the four-momenta of the fermions. Work in the boson rest frame. Use (4.37).

EXERCISE 13.3 Assuming the standard model coupling, show that

$$\Gamma(Z \to \nu_e\bar{\nu}_e) = \frac{g^2}{96\pi\cos^2\theta_W}M_Z, \tag{13.45}$$

see Exercise 13.2. Given that $\sin^2\theta_W = 0.25$ and $M_Z = 90$ GeV, predict the numerical value of the partial width.

EXERCISE 13.4 Calculate the partial widths of the three decay modes $Z \to e^+e^-, \bar{u}u, \bar{d}d$. Hence, predict the total width of the Z in the standard model, assuming $\sin^2\theta_W = \frac{1}{4}$ and $M_Z = 90$ GeV. Do not forget color.

EXERCISE 13.5 Repeat Exercise 13.3 for the $W^+ \to e^+\nu_e$ decay mode; take $M_W = 80$ GeV.

EXERCISE 13.6 Calculate the partial widths of the two decay modes $W^+ \to \bar{d}u, \bar{s}u$; use (12.102) and (12.103). Predict the total width of the W^+ in the standard model.

13.5 Neutrino–Electron Scattering

The $\nu_\mu e^-$ and $\bar{\nu}_\mu e^-$ elastic scattering processes can only proceed via a neutral current interaction, see Fig. 13.3. The current–current form of the invariant amplitude for the process $\nu_\mu e^- \to \nu_\mu e^-$ is analogous to (12.84) for $\nu q \to \nu q$ scattering:

$$\mathcal{M}^{NC}(\nu e \to \nu e) = \frac{G_N}{\sqrt{2}}\left(\bar{\nu}\gamma^\mu(1-\gamma^5)\nu\right)\left(\bar{e}\gamma_\mu\left(c_V^e - c_A^e\gamma^5\right)e\right), \qquad (13.46)$$

where $G_N = \rho G \simeq G$ [see (12.88)]. In fact, assuming electron–muon universality, we shall find that the four elastic scattering processes, $\nu_\mu e^-$, $\bar{\nu}_\mu e^-$, $\nu_e e^-$, and $\bar{\nu}_e e^-$, can all be explained in terms of the two parameters $c_V \equiv c_V^e$ and $c_A \equiv c_A^e$.

EXERCISE 13.7 If $\nu e \to \nu e$ scattering proceeds by Z exchange, show that (13.46) is obtained from the Feynman rules using (13.41) as the vertex factor. In particular, use the expression for the boson propagator (see Section 6.17) to verify that (13.46) is valid provided the four-momentum transfer q is such that $|q^2| \ll M_Z^2$.

Given that (13.46) is of identical form to that for $\nu q \to \nu q$ scattering, we may use the results of Section 12.10 to obtain the $\nu_\mu e^- \to \nu_\mu e^-$ cross section. We

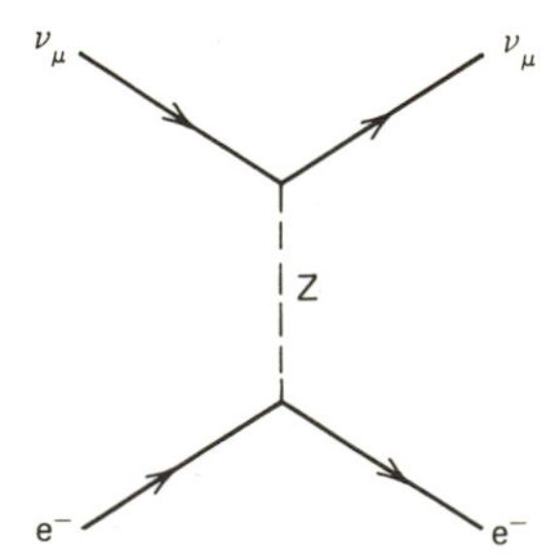

Fig. 13.3 The neutral current $\nu_\mu e^- \to \nu_\mu e^-$ interaction.

therefore have [see (12.90) and (12.91)]

$$\frac{d\sigma(\nu_\mu e)}{dy} = \frac{G_N^2 s}{4\pi}\left[(c_V + c_A)^2 + (c_V - c_A)^2(1-y)^2\right]. \tag{13.47}$$

Carrying out the y integration from 0 to 1 gives

$$\sigma(\nu_\mu e \to \nu_\mu e) = \frac{G_N^2 s}{3\pi}\left(c_V^2 + c_V c_A + c_A^2\right). \tag{13.48}$$

For $\bar{\nu}_\mu e^-$ elastic scattering, $c_A \to -c_A$ in (13.47), and so

$$\sigma(\bar{\nu}_\mu e \to \bar{\nu}_\mu e) = \frac{G_N^2 s}{3\pi}\left(c_V^2 - c_V c_A + c_A^2\right). \tag{13.49}$$

EXERCISE 13.8 Equation (13.47) is valid if $m^2/s \ll 1$. If the electron mass m is not ignored, show that the extra contribution to (13.47) is

$$-G^2 m^2 y\left(c_V^2 - c_A^2\right)/2\pi.$$

The process $\nu_e e^- \to \nu_e e^-$ offers the intriguing possibility of studying charged current and neutral current interference, see Fig. 13.4. The amplitude for diagram (a) is $\mathfrak{M}^{NC}$ of (13.46) with $\nu = \nu_e$. For diagram (b), we have

$$\mathfrak{M}^{CC} = -\frac{G}{\sqrt{2}}\left[\bar{e}\gamma^\mu(1-\gamma^5)\nu_e\right]\left[\bar{\nu}_e\gamma_\mu(1-\gamma^5)e\right], \tag{13.50}$$

where the minus sign relative to (13.46) arises from interchange of the outgoing leptons [see (6.9)]. We may use the Fierz reordering theorem, see, for example, Bailin (1982), to rewrite (13.50) as

$$\mathfrak{M}^{CC} = \frac{G}{\sqrt{2}}\left(\bar{\nu}_e\gamma^\mu(1-\gamma^5)\nu_e\right)\left(\bar{e}\gamma_\mu(1-\gamma^5)e\right). \tag{13.51}$$

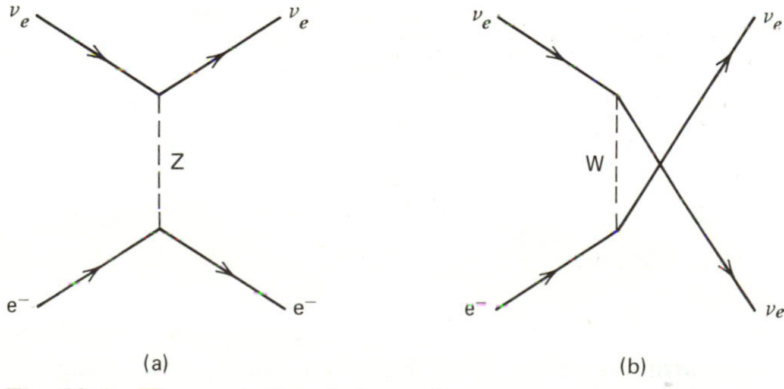

Fig. 13.4 The neutral and charged current $\nu_e e^- \to \nu_e e^-$ interaction.

EXERCISE 13.9 Show that (13.51) follows from (13.50). The invariance under reordering of the spinors is an important property of the $V-A$ interaction. The effect of reordering in the scalar product of bilinear covariants is, in general, much more involved. The answer is the Fierz theorem.

To obtain the amplitude $\mathfrak{M}(\nu_e e^- \to \nu_e e^-)$, we add the amplitudes ($\mathfrak{M}^{NC}$ and $\mathfrak{M}^{CC}$) for the two diagrams of Fig. 13.4. If we take $\rho = 1$, then $G_N = G$ [see (12.88)], and we find $\mathfrak{M} = \mathfrak{M}^{NC} + \mathfrak{M}^{CC}$ is given by (13.46) with

$$c_V \to c_V + 1, \qquad c_A \to c_A + 1. \tag{13.52}$$

Thus, the $\nu_e e^-$ and $\bar{\nu}_e e^-$ elastic scattering cross sections are in turn given by (13.48) and (13.49) with these replacements.

It is customary to present the results of a given neutrino–electron cross section measurement as an ellipse of possible values of c_V and c_A in the c_V, c_A plane. Recent results are shown in Fig. 13.5. The three "experimental" ellipses mutually intersect to give two possible solutions. The c_A dominant solution is

$$\begin{aligned} c_A^e &= -0.52 \pm 0.06, \\ c_V^e &= 0.06 \pm 0.08, \end{aligned} \tag{13.53}$$

in excellent agreement with the standard model and $\sin^2\theta_W \approx \frac{1}{4}$ (see Table 13.2).

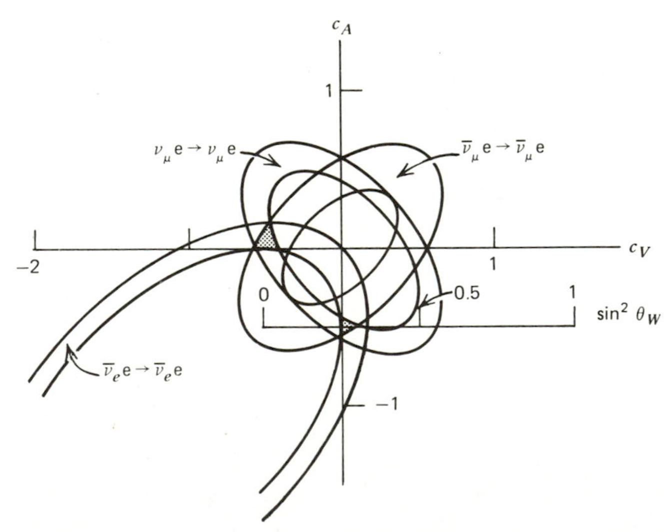

Fig. 13.5 Determination of the parameters c_V and c_A of (13.46) by neutrino–electron data. Figure is taken from Hung and Sakurai (1981). The absence of ν_e data is because reactor beams only have $\bar{\nu}_e$.

However, the best determinations of $\sin^2\theta_W$ come from the analyses of data for inclusive and exclusive neutrino–nucleon processes. A recent simultaneous analysis (Kim et al., 1981) of all available data gives

$$\sin^2\theta_W = 0.234 \pm 0.013, \quad \rho = 1.002 \pm 0.015. \tag{13.54}$$

13.6 Electroweak Interference in e$^+$e$^-$ Annihilation

Shortly after Bludman in 1958 first speculated about the existence of a weak neutral current, Zel'dovich gave a very simple argument to estimate the size of the asymmetry arising from the interference of the electromagnetic amplitude $\mathfrak{M}^{EM} \sim e^2/k^2$ with a small weak contribution. He predicted

$$\frac{|\mathfrak{M}^{EM}\mathfrak{M}^{NC}|}{|\mathfrak{M}^{EM}|^2} \approx \frac{G}{e^2/k^2} \approx \frac{10^{-4}k^2}{m_N^2}, \tag{13.55}$$

using $G \approx 10^{-5}/m_N^2$ and $e^2/4\pi = 1/137$.

The high-energy e$^+$e$^-$ colliding beam machines are an ideal testing ground for such interference effects. The e$^+$e$^-$ annihilations can occur through electromagnetic (γ) or weak neutral current (Z) interactions, see, for example, Fig. 13.6. With e$^\pm$ beam energies of 15 GeV we have $k^2 \simeq s = (30 \text{ GeV})^2$, and so (13.55) predicts about a 10% effect, which is readily observable.

To make a detailed prediction for the process $e^+e^- \to \mu^+\mu^-$, we assume that the neutral current interaction is mediated by a Z boson with couplings given by (13.41). Using the Feynman rules (Section 6.17), the amplitudes $\mathfrak{M}_\gamma$ and $\mathfrak{M}_Z$ corresponding to the diagrams of Fig. 13.6 are

$$\mathfrak{M}_\gamma = -\frac{e^2}{k^2}(\bar{\mu}\gamma^\nu\mu)(\bar{e}\gamma_\nu e), \tag{13.56}$$

$$\mathfrak{M}_Z = -\frac{g^2}{4\cos^2\theta_W}\left[\bar{\mu}\gamma^\nu(c_V^\mu - c_A^\mu\gamma^5)\mu\right]\left(\frac{g_{\nu\sigma} - k_\nu k_\sigma/M_Z^2}{k^2 - M_Z^2}\right)\left[\bar{e}\gamma^\sigma(c_V^e - c_A^e\gamma^5)e\right], \tag{13.57}$$

where k is the four-momentum of the virtual γ (or Z), $s \simeq k^2$. With electron–muon universality, the superscripts on $c_{V,A}$ are superfluous here, but we keep them so as to be able to translate the results directly to $e^+e^- \to q\bar{q}$ (where $c^q \neq c^\mu$). We

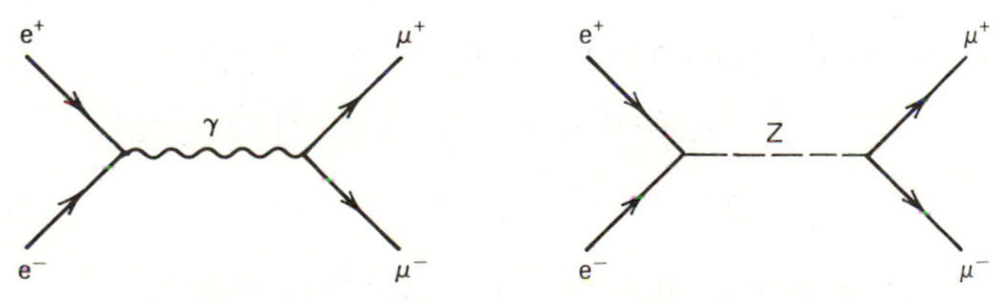

Fig. 13.6 Electromagnetic and weak contributions to $e^+e^- \to \mu^+\mu^-$.

ignore the lepton masses, so the Dirac equation for the incident positron reads $(\frac{1}{2}k_\sigma)\bar{e}\gamma^\sigma = 0$ and the numerator of the propagator simplifies to $g_{\nu\sigma}$. Thus, (13.57) becomes

$$\mathcal{M}_Z = \frac{-\sqrt{2}\,GM_Z^2}{s - M_Z^2}\left[c_R^\mu(\bar{\mu}_R\gamma^\nu\mu_R) + c_L^\mu(\bar{\mu}_L\gamma^\nu\mu_L)\right]\left[c_R^e(\bar{e}_R\gamma_\nu e_R) + c_L^e(\bar{e}_L\gamma_\nu e_L)\right], \tag{13.58}$$

using (13.32) and (13.36) with $\rho = 1$, and where

$$c_R \equiv c_V - c_A, \qquad c_L \equiv c_V + c_A. \tag{13.59}$$

That is, we have chosen to write

$$c_V - c_A\gamma^5 = (c_V - c_A)\tfrac{1}{2}(1 + \gamma^5) + (c_V + c_A)\tfrac{1}{2}(1 - \gamma^5).$$

The $\frac{1}{2}(1 \pm \gamma^5)$ are projection operators, see (5.78), which enable $\mathcal{M}_Z$ to be expressed explicitly in terms of right- and left-handed spinors. It is easier to calculate $|\mathcal{M}_\gamma + \mathcal{M}_Z|^2$ in this form. With definite electron and muon helicities, we can directly apply the results of the QED calculation of $e^-e^+ \to \mu^-\mu^+$ given in Sections 6.5 and 6.6. For instance,

$$\begin{aligned}\frac{d\sigma(e_L^- e_R^+ \to \mu_L^- \mu_R^+)}{d\Omega} &= \frac{\alpha^2}{4s}(1 + \cos\theta)^2|1 + rc_L^\mu c_L^e|^2,\\ \frac{d\sigma(e_L^- e_R^+ \to \mu_R^- \mu_L^+)}{d\Omega} &= \frac{\alpha^2}{4s}(1 - \cos\theta)^2|1 + rc_R^\mu c_L^e|^2\end{aligned} \tag{13.60}$$

[see (6.39) and (6.32)], where r is the ratio of the coefficients in front of the brackets in (13.58) and (13.56), that is,

$$r = \frac{\sqrt{2}\,GM_Z^2}{s - M_Z^2 + iM_Z\Gamma_Z}\left(\frac{s}{e^2}\right), \tag{13.61}$$

where we have included the finite resonance width Γ_Z, which is important for $s \sim M_Z^2$ [see $\chi(E)$ of (2.56) multiplied by $1/(E + M)$].

Expressions similar to (13.60) hold for the other two nonvanishing helicity configurations. To calculate the unpolarized $e^+e^- \to \mu^+\mu^-$ cross section, we average over the four allowed L, R helicity combinations. We find

$$\frac{d\sigma}{d\Omega} = \frac{\alpha^2}{4s}\left[A_0(1 + \cos^2\theta) + A_1\cos\theta\right], \tag{13.62}$$

where (assuming electron–muon universality $c_i^e = c_i^\mu \equiv c_i$)

$$\begin{aligned}A_0 &\equiv 1 + \tfrac{1}{2}\mathrm{Re}(r)(c_L + c_R)^2 + \tfrac{1}{4}|r|^2(c_L^2 + c_R^2)^2\\ &= 1 + 2\,\mathrm{Re}(r)c_V^2 + |r|^2(c_V^2 + c_A^2)^2\end{aligned} \tag{13.63}$$

$$\begin{aligned}A_1 &\equiv \mathrm{Re}(r)(c_L - c_R)^2 + \tfrac{1}{2}|r|^2(c_L^2 - c_R^2)^2\\ &= 4\,\mathrm{Re}(r)c_A^2 + 8|r|^2c_V^2c_A^2.\end{aligned} \tag{13.64}$$

The lowest-order QED result ($A_0 = 1$, $A_1 = 0$) gives a symmetric angular distribution. We now see that the weak interaction introduces a forward–backward asymmetry ($A_1 \neq 0$). Let us calculate the size of the integrated asymmetry defined by

$$A_{FB} \equiv \frac{F - B}{F + B} \qquad \text{with } F = \int_0^1 \frac{d\sigma}{d\Omega} d\Omega, \qquad B = \int_{-1}^0 \frac{d\sigma}{d\Omega} d\Omega. \tag{13.65}$$

Integrating (13.62), we obtain for $s \ll M_Z^2$ (i.e., $|r| \ll 1$)

$$A_{FB} = \frac{A_1}{(8A_0/3)} \simeq \frac{3}{2}\text{Re}(r)c_A^2 \simeq -\frac{3c_A^2}{\sqrt{2}}\left(\frac{Gs}{e^2}\right). \tag{13.66}$$

This is in agreement with the expectations of the order of magnitude estimate, Gs/e^2, of (13.55); an asymmetry which grows quadratically with the energy of the colliding e^+ and e^- beams (for $s \ll M_Z^2$).

We may use the standard model couplings ($c_A = -\frac{1}{2}$, $c_V = -\frac{1}{2} + \sin^2\theta_W \simeq 0$) to compare (13.62) with the experimental measurements of the high-energy $e^+e^- \to \mu^+\mu^-$ angular distribution, see Fig. 13.7. The agreement is good. Since $c_V \simeq 0$, these data do not, however, offer an accurate determination of $\sin^2\theta_W$.

Integrating (13.62) over $d\Omega$, we find

$$\sigma(e^+e^- \to \mu^+\mu^-) = \frac{\alpha^2}{4s}\left(2\pi\frac{8}{3}A_0\right) = \left(\frac{4\pi\alpha^2}{3s}\right)A_0 \equiv \sigma_0 A_0, \tag{13.67}$$

where σ_0 is the QED cross section of (6.32). That is,

$$\frac{\sigma(e^+e^- \to \mu^+\mu^-)}{\sigma_0} \equiv R_\mu = 1 + 2\,\text{Re}(r)c_V^2 + |r|^2(c_V^2 + c_A^2)^2. \tag{13.68}$$

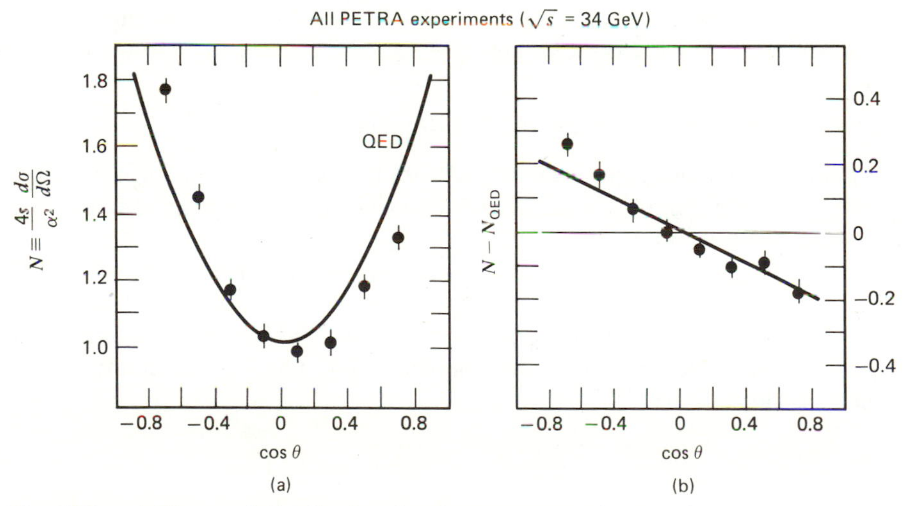

Fig. 13.7 (a) The cos θ distribution for the process $e^-e^+ \to \mu^-\mu^+$ does not follow the $1 + \cos^2\theta$ QED prediction. (b) The discrepancy is explained by the interference of the virtual Z and γ contributions. (Compilation by R. Marshall.)

EXERCISE 13.10 Using the couplings in the standard model, calculate R_μ at $s = M_Z^2$. Use $\sin^2\theta_W = \frac{1}{4}$, $M_Z = 90$ GeV, and $\Gamma_Z = 2.5$ GeV.

It is relevant to ask what electroweak effects occur in $e^+e^- \to \bar{q}q$. They are not the same as in $e^+e^- \to \mu^+\mu^-$, since $c_{V,A}^q \neq c_{V,A}^\mu$. However, the calculation proceeds exactly as above. For instance, the counterpart to (13.60) is

$$\frac{d\sigma(e_L^- e_R^+ \to q_L^- q_R^+)}{d\Omega} = 3\frac{\alpha^2}{4s}(1 + \cos\theta)^2 |Q_q + rc_L^q c_L^e|^2, \tag{13.69}$$

where Q_q is the charge of the quark and the factor 3 is for color. Following through the calculation, the analogous result to (13.68) is

$$\frac{\sigma(e^+e^- \to \bar{q}q)}{\sigma_0} \equiv R_q = 3\left[Q_q^2 + 2Q_q \mathrm{Re}(r) c_V^q c_V^e + |r|^2 \left(c_V^{q2} + c_A^{q2}\right)\left(c_V^{e2} + c_A^{e2}\right)\right]. \tag{13.70}$$

EXERCISE 13.11 Follow Exercise 13.10 and calculate R_u and R_d at $s = M_Z^2$. Hence, calculate $\sigma(e^+e^- \to \text{hadrons})$ at the Z resonance.

The numerical results of Exercises 13.10 and 13.11 give R's in the region 100–1000. This has crucial implications. Very large enhancements over σ_0 are therefore expected at beam energies $E \sim M_Z/2$, provided the neutral current interaction is mediated by a Z boson. This is a major motivation for the new 50 + 50 GeV e^+e^- collider being constructed at CERN, Geneva. Since $M_Z \approx 90$ GeV, the Z boson should be copiously produced at the new collider and its properties, and those of its decay products, studied in a clean environment, without the confusing background debris which accompanies a hadronic collision.

13.7 Other Observable Electroweak Interference Effects

Several other ingenious experiments have been performed to exhibit parity-violating effects arising from the interference between the electromagnetic and weak neutral current interactions.

One type of experiment involves the delicate measurement of *parity-violating effects in atomic transitions*. The idea is that Z^0 exchange between an electron and the nucleus leads to a modification of the Coulomb potential. At first sight, detection of the effect appears to be a hopeless task. Our rule of thumb, (13.55), predicts an effect of size

$$\frac{Gk^2}{e^2} \simeq \frac{10^{-4}k^2}{m_p^2} \simeq \frac{10^{-4}}{m_p^2 R^2} \simeq 10^{-15}, \tag{13.71}$$

where R is a typical atomic radius (~ 1 Å). However, a large enhancement can be achieved by studying highly forbidden electromagnetic transitions and by using atoms of high Z (in fact, the effect goes as Z^3). The general method is to look for optical rotation of transmitted LASER light induced by the parity-violating interference terms $c_V^e c_A^q$ and $c_A^e c_V^q$.

Another type of experiment measures the minute *parity-violating asymmetry in the inelastic scattering of longitudinally polarized electrons* (or muons) off nuclear targets. The asymmetry is defined by

$$A = \frac{\sigma_R - \sigma_L}{\sigma_R + \sigma_L} \tag{13.72}$$

where, for instance, σ_R is the cross section $d\Omega/dy$ for $e_R N \to e_R X$; e_R denotes a right-handed electron ($\lambda = +\frac{1}{2}$). In the laboratory frame, recall that y is the fractional energy loss of the electron, $y = (E - E')/E$. A nonzero asymmetry indicates the presence of a parity-violating effect. From (13.55), we anticipate an asymmetry from the interference between γ and Z^0 exchange of magnitude

$$A \sim \frac{Gk^2}{e^2} \sim \frac{10^{-4}}{m_p^2} k^2,$$

where k is the four-momentum carried by the exchanged boson.

For the deep inelastic scattering process $eN \to eX$, we can use the parton model to predict the asymmetry. Taking N to be an isoscalar target, we find (see Exercise 13.12)

$$A = \frac{6}{5}\left(\frac{\sqrt{2}\,Gk^2}{e^2}\right)\left(a_1 + a_2 \frac{1-(1-y)^2}{1+(1-y)^2}\right), \tag{13.73}$$

with

$$a_1 = c_A^e\left(2c_V^u - c_V^d\right)$$
$$a_2 = c_V^e\left(2c_A^u - c_A^d\right), \tag{13.74}$$

where the c's are the neutral-current couplings of (13.41) and Table 13.2. By measuring the asymmetry as a function of k^2 and y, the coefficients a_1 and a_2 can be determined. Experiments have been performed by scattering polarized electrons off deuterons and polarized $\mu^\pm$ off carbon. The results are in agreement with the standard model and with the values of $\sin^2\theta_W$ obtained by other data.

EXERCISE 13.12 Verify (13.73). Assume that $k^2 \ll M_Z^2$ and that the target contains equal numbers of u and d quarks (i.e., it is an isoscalar target), and neglect antiquarks. Show that, in the standard model,

$$a_1 = -\tfrac{3}{4}\left(1 - \tfrac{20}{9}\sin^2\theta_W\right),$$

$$a_2 = -\tfrac{3}{4}\left(1 - 4\sin^2\theta_W\right). \tag{13.75}$$

Hint As for $e^+e^- \to \mu^+\mu^-$, it is easiest to use definite helicity states. For example, in the electron–quark center-of-mass frame,

$$\frac{d\sigma(e_R u_L \to e_R u_L)}{d\Omega} = \frac{\alpha^2}{4s}(1 + \cos\theta)^2 |Q_u + r c_R^e c_L^u|^2, \qquad (13.76)$$

where $r = -\sqrt{2}\,Gk^2/e^2$, see (13.60) and (13.61). Rewrite (13.76) in terms of the parton model (invariant) variable y, using $1 - y = \frac{1}{2}(1 + \cos\theta)$, see (9.25).

14
Gauge Symmetries

14.1 The Lagrangian and Single-Particle Wave Equations

One of the most profound insights in theoretical physics is that interactions are dictated by symmetry principles. Einstein made great use of the predictive power of this idea. By considering invariance under general coordinate transformations (together with the equivalence principle), he was led to the general theory of relativity. The present belief is that all particle interactions may be dictated by so-called local gauge symmetries. We shall see that this is intimately connected with the idea that the conserved physical quantities (such as electric charge, color, etc.) are conserved in local regions of space, and not just globally.

The connection between symmetries and conservation laws is best discussed in the framework of Lagrangian field theory, which we have not introduced.* However, you are probably familiar with the fact that in classical mechanics the particle equations of motion can be obtained from Lagrange's equations [see, for example, Goldstein (1951)]

$$\frac{d}{dt}\left(\frac{\partial L}{\partial \dot{q}_i}\right) - \frac{\partial L}{\partial q_i} = 0, \tag{14.1}$$

where q_i are the generalized coordinates of the particles, t is the time variable, and $\dot{q}_i = dq_i/dt$. The Lagrangian is

$$L \equiv T - V, \tag{14.2}$$

where T and V are the kinetic and potential energies of the system, respectively. It is straightforward to extend the formalism from a discrete system, with coordinates $q_i(t)$, to a continuous system, that is, a system with continuously varying coordinates $\phi(\mathbf{x}, t)$. The Lagrangian

$$L(q_i, \dot{q}_i, t) \to \mathcal{L}\left(\phi, \frac{\partial \phi}{\partial x_\mu}, x_\mu\right), \tag{14.3}$$

*The formalism introduced in this book for computing particle interactions has been based entirely on the single-particle wave equations. Chapters 3 through 7 explain in detail how this can be done, despite the fact that we are routinely dealing with many-particle systems.

where the field ϕ itself is a function of the continuous parameters x_μ, and (14.1) becomes

$$\frac{\partial}{\partial x_\mu}\left(\frac{\partial \mathcal{L}}{\partial(\partial\phi/\partial x_\mu)}\right) - \frac{\partial \mathcal{L}}{\partial \phi} = 0, \tag{14.4}$$

the Euler–Lagrange equation. $\mathcal{L}$ is the Lagrangian density,

$$L = \int \mathcal{L}\, d^3x. \tag{14.5}$$

We shall follow common practice and call $\mathcal{L}$ itself the Lagrangian.

> ***EXERCISE 14.1*** If you are unfamiliar with this formalism, consult, for example, Goldstein (1977) or work through the example given in Sakurai (1967), page 3.

Instead of writing down a relativistic wave equation, we simply choose a Lagrangian $\mathcal{L}$. Provided our choice is a Lorentz scalar, the equation of motion resulting from (14.4) will be covariant. For example, substituting the Lagrangian

$$\mathcal{L} = \tfrac{1}{2}(\partial_\mu\phi)(\partial^\mu\phi) - \tfrac{1}{2}m^2\phi^2 \tag{14.6}$$

into (14.4) gives the Klein–Gordon equation

$$\partial_\mu\partial^\mu\phi + m^2\phi \equiv (\Box^2 + m^2)\phi = 0. \tag{14.7}$$

There is no mystery here. The choice of $\mathcal{L}$ was specifically designed to reproduce (14.7).

> ***EXERCISE 14.2*** Verify that the Dirac equation follows from
>
> $$\mathcal{L} = i\bar{\psi}\gamma_\mu\partial^\mu\psi - m\bar{\psi}\psi, \tag{14.8}$$
>
> where each of the four components of ψ and $\bar{\psi}$ is regarded as an independent field variable.

> ***EXERCISE 14.3*** Show that the substitution of the Lagrangian
>
> $$\mathcal{L} = -\tfrac{1}{4}F_{\mu\nu}F^{\mu\nu} - j^\mu A_\mu \tag{14.9}$$
>
> into the Euler–Lagrange equation for A_μ gives the Maxwell equations, (6.57),
>
> $$\partial_\mu F^{\mu\nu} = j^\nu, \tag{14.10}$$
>
> where $F^{\mu\nu} \equiv \partial^\mu A^\nu - \partial^\nu A^\mu$. Hence, show that the current is conserved, that is, $\partial_\nu j^\nu = 0$.

EXERCISE 14.4 With the addition of a term $\frac{1}{2}m^2 A_\mu A^\mu$, show that the Lagrangian of (14.9) leads to an equation of motion

$$(\Box^2 + m^2)A^\mu = j^\mu.$$

The new term is therefore a photon mass contribution. We shall see in Section 14.3 that it is forbidden by gauge invariance. The photon is massless.

What is the relation between the Lagrangian approach and the perturbative method based on Feynman rules obtained from single-particle wave equations? To each Lagrangian, there corresponds a set of Feynman rules; and so, once we identify these rules, the connection is made. We can then calculate physical quantities by just following the methods presented in Chapters 4 and 6.

The identification of the Feynman rules proceeds as follows:

1 We associate with the various terms in the Lagrangian a set of propagators and vertex factors.

2 The propagators are determined by the terms quadratic in the fields, that is, the terms in the Lagrangian containing ϕ^2, $\bar{\psi}\psi$, and so on, such as $\frac{1}{2}(\partial_\mu\phi)^2 - \frac{1}{2}m^2\phi^2$ and $\bar{\psi}(i\gamma_\mu\partial^\mu - m)\psi$. The propagators can then be identified from the Euler–Lagrange equations using the methods of Section 6.10.

3 The other terms in the Lagrangian are associated with interaction vertices. The Feynman vertex factor is just given by the coefficient of the corresponding term in $i\mathcal{L}$ containing the interacting fields. QED provides us with a simple illustration of this. As we shall see in Section 14.3, the second term in (14.9) describes the electron–photon interaction. The electron current is given by $j^\mu = -e\bar{\psi}\gamma^\mu\psi$, see (5.17); and so the interaction term in $i\mathcal{L}$ is

$$i\mathcal{L} = \cdots + ie\bar{\psi}\gamma^\mu\psi A_\mu$$

We recognize the coefficient $(ie\gamma^\mu)$ of the interacting fields $\bar{\psi}\psi A_\mu$ as the familiar vertex factor of QED, see Chapter 6.

In the orthodox approach to quantum field theory, we would now proceed to formally derive these assertions. In order to do this, the classical Lagrangian is quantized. Fields such as ψ and A_μ become operators describing the creation and annihilation of particles. Interactions are computed by evaluating a perturbation series in $i\mathcal{L}_{\text{int}}$, the interaction term(s) in $i\mathcal{L}$. The end result of this lengthy and formal approach can always be translated into a set of Feynman rules which are exactly those we just described. So, we might as well take these rules and proceed to investigate the physical implications of a given Lagrangian using the methods with which we are already familiar. The canonical formalism was formerly regarded as more rigorous and can be found in many books [for introductory discussions see, e.g., Mandl (1966), Muirhead (1965), and Sakurai (1967)]. We do not present it, as we will never use it. We hereby subscribe to the growing belief that "the diagrams contain more truth than the underlying formalism" ['t Hooft and Veltman (1973)].

14.2 Noether's Theorem: Symmetries and Conservation Laws

Invariance under translations, time displacements, rotations, and Lorentz transformations leads to the conservation of momentum, energy, and angular momentum. Rather than studying these conservation laws, we are interested here in "internal" symmetry transformations that do not mix fields with different space-time properties (that is, transformations that commute with the space-time components of the wavefunction).

For example, an electron is described by a complex field, and inspection of the Lagrangian (14.8) shows that it is invariant under the phase transformation

$$\psi(x) \rightarrow e^{i\alpha}\psi(x), \tag{14.11}$$

where α is a real constant. This can be easily checked by noting

$$\partial_\mu \psi \rightarrow e^{i\alpha}\partial_\mu \psi, \tag{14.12}$$

$$\bar{\psi} \rightarrow e^{-i\alpha}\bar{\psi}. \tag{14.13}$$

The family of phase transformations $U(\alpha) \equiv e^{i\alpha}$, where a single parameter α may run continuously over real numbers, forms a unitary Abelian group known as the $U(1)$ group. Abelian just records the property that the group multiplication is commutative:

$$U(\alpha_1)U(\alpha_2) = U(\alpha_2)U(\alpha_1). \tag{14.14}$$

You may think that the observation of $U(1)$ invariance of $\mathcal{L}$ is rather trivial and unimportant. This is not so. Through Noether's theorem, it implies the existence of a conserved current. To see this, it is sufficient to study the invariance of $\mathcal{L}$ under an infinitesimal $U(1)$ transformation,

$$\psi \rightarrow (1 + i\alpha)\psi. \tag{14.15}$$

Invariance requires the Lagrangian to be unchanged, that is,

$$\begin{aligned} 0 = \delta\mathcal{L} &= \frac{\partial\mathcal{L}}{\partial\psi}\delta\psi + \frac{\partial\mathcal{L}}{\partial(\partial_\mu\psi)}\delta(\partial_\mu\psi) + \delta\bar{\psi}\frac{\partial\mathcal{L}}{\partial\bar{\psi}} + \delta(\partial_\mu\bar{\psi})\frac{\partial\mathcal{L}}{\partial(\partial_\mu\bar{\psi})} \\ &= \frac{\partial\mathcal{L}}{\partial\psi}(i\alpha\psi) + \frac{\partial\mathcal{L}}{\partial(\partial_\mu\psi)}(i\alpha\partial_\mu\psi) + \cdots \\ &= i\alpha\left[\frac{\partial\mathcal{L}}{\partial\psi} - \partial_\mu\left(\frac{\partial\mathcal{L}}{\partial(\partial_\mu\psi)}\right)\right]\psi + i\alpha\partial_\mu\left(\frac{\partial\mathcal{L}}{\partial(\partial_\mu\psi)}\psi\right) + \cdots . \end{aligned} \tag{14.16}$$

The term in square brackets vanishes by virtue of the Euler–Lagrange equation, (14.4), for ψ (and similarly for $\bar{\psi}$), and so (14.16) reduces to the form of an equation for a conserved current:

$$\partial_\mu j^\mu = 0, \tag{14.17}$$

where

$$j^\mu = \frac{ie}{2}\left(\frac{\partial \mathcal{L}}{\partial(\partial_\mu \psi)}\psi - \bar{\psi}\frac{\partial \mathcal{L}}{\partial(\partial_\mu \bar{\psi})}\right) = -e\bar{\psi}\gamma^\mu\psi, \tag{14.18}$$

using (14.8). The proportionality factor is chosen so that j^μ matches up with the electromagnetic charge current density of an electron of charge $-e$, see (5.17). It follows from (14.17) that the charge

$$Q = \int d^3x\, j^0$$

must be a conserved quantity because of the $U(1)$ phase invariance.

EXERCISE 14.5 Show that $dQ/dt = 0$.

EXERCISE 14.6 Show that $U(1)$ phase invariance of the Lagrangian

$$\mathcal{L} = (\partial_\mu \phi)^*(\partial^\mu \phi) - m^2\phi^*\phi \tag{14.19}$$

of a complex scalar field implies the existence of a conserved current

$$j^\mu = -ie(\phi^*\partial^\mu\phi - \phi\partial^\mu\phi^*), \tag{14.20}$$

and compare with (3.25). Note that the Lagrangian, (14.19), for a complex field $\phi = (\phi_1 + i\phi_2)/\sqrt{2}$ is normalized to ensure that

$$\mathcal{L}(\phi) = \mathcal{L}(\phi_1) + \mathcal{L}(\phi_2),$$

where $\mathcal{L}(\phi_i)$, the Lagrangian for a real field ϕ_i, is given by (14.6).

From a physicist's point of view, the existence of a symmetry implies that some quantity is unmeasurable. For example, translation invariance means that we cannot determine an absolute position in space. Similarly, (14.11) implies that the phase α is unmeasurable, it has no physical meaning and can be chosen arbitrarily. α is a constant; therefore, once we fix it, the value is specified for all space and time. We speak of *global* "gauge" (a historical misnomer for "phase") invariance. This is surely not the most general invariance, for it would be more satisfactory if α could differ from space-time point to point, that is, $\alpha = \alpha(x)$.

14.3 $U(1)$ Local Gauge Invariance and QED

The message of the last paragraph is that we should generalize (14.11) to the transformation

$$\psi(x) \to e^{i\alpha(x)}\psi(x), \tag{14.21}$$

where $\alpha(x)$ now depends on space and time in a completely arbitrary way. This is known as local gauge invariance. However, this does not work. The Lagrangian, (14.8),

$$\mathcal{L} = i\bar{\psi}\gamma^\mu\partial_\mu\psi - m\bar{\psi}\psi,$$

is not invariant under such local phase transformations. From (14.21),

$$\bar{\psi} \to e^{-i\alpha(x)}\bar{\psi},$$

so the last term of $\mathcal{L}$ is invariant; however, the derivative of ψ does not follow (14.21). Rather,

$$\partial_\mu\psi \to e^{i\alpha(x)}\partial_\mu\psi + ie^{i\alpha(x)}\psi\,\partial_\mu\alpha \tag{14.22}$$

and the $\partial_\mu\alpha$ term breaks the invariance of $\mathcal{L}$.

If, on aesthetic grounds, we insist on imposing invariance of the Lagrangian under local gauge transformations, we must seek a modified derivative, D_μ, that transforms covariantly under phase transformations, that is, like ψ itself:

$$D_\mu\psi \to e^{i\alpha(x)}D_\mu\psi. \tag{14.23}$$

To form the "covariant derivative" D_μ, we must introduce a vector field A_μ with transformation properties such that the unwanted term in (14.22) is canceled. This can be accomplished by the construction

$$\boxed{D_\mu \equiv \partial_\mu - ieA_\mu,} \tag{14.24}$$

where A_μ transforms as

$$A_\mu \to A_\mu + \frac{1}{e}\partial_\mu\alpha. \tag{14.25}$$

It is easy to check that D_μ satisfies (14.23). Invariance of the Lagrangian (14.8) is then achieved by replacing ∂_μ by D_μ:

$$\begin{aligned}\mathcal{L} &= i\bar{\psi}\gamma^\mu D_\mu\psi - m\bar{\psi}\psi \\ &= \bar{\psi}(i\gamma^\mu\partial_\mu - m)\psi + e\bar{\psi}\gamma^\mu\psi A_\mu.\end{aligned} \tag{14.26}$$

Hence, by demanding local phase invariance, we are forced to introduce a vector field A_μ, called the *gauge field*, which couples to the Dirac particle (charge $-e$) in exactly the same way as the photon field; compare (14.24) with (6.1). Indeed, the new interaction term in (14.26) may be written $-j^\mu A_\mu$, where j^μ is the current density of (5.17).

If we are to regard this new field as the physical photon field, we must add to the Lagrangian a term corresponding to its kinetic energy, analogous to $\frac{1}{2}(\partial_\mu\phi)^2$ in (14.6). Since the kinetic term must be invariant under (14.25), it can only involve the gauge invariant field strength tensor

$$F_{\mu\nu} = \partial_\mu A_\nu - \partial_\nu A_\mu. \tag{14.27}$$

We are thus led to the Lagrangian of QED:

$$\boxed{\mathcal{L} = \bar{\psi}(i\gamma^\mu\partial_\mu - m)\psi + e\bar{\psi}\gamma^\mu A_\mu\psi - \tfrac{1}{4}F_{\mu\nu}F^{\mu\nu}.} \tag{14.28}$$

Note that the addition of a mass term $\frac{1}{2}m^2A_\mu A^\mu$ (see Exercise 14.4) is prohibited by gauge invariance. The gauge particle, the photon, must be *massless*.

It was clear from the outset that a new field would have to be introduced, since changing the phase locally will create phase differences which would be observable if not compensated in some way. The surprising result is that local gauge invariance can be restored, and restored by the photon field A_μ. We expect the gauge field to have infinite range (that is, the photon will be massless) since there is no limit to the distance over which the phases of the electron field might have to be reconciled.

In summary, we see that by imposing the "natural" requirement of local phase invariance on the free fermion Lagrangian, we are led to the interacting field theory of QED. Gauge invariance, which in your study of classical electrodynamics you may have conceived to be some formal curiosity of Maxwell theory, has become one of the most basic and essential ingredients.

EXERCISE 14.7 Read about the Bohm–Aharonov effect. Suggested references are the *Feynman Lectures on Physics*, Volume 2, or Wu, T. T. and Yang, C. N. (1975) *Phys. Rev.* **D12**, 3845.

14.4 Non-Abelian Gauge Invariance and QCD

In an analogous way, we can hope to infer the structure of quantum chromodynamics from local gauge invariance. QCD is based on the extension of the above idea, but with the $U(1)$ gauge group replaced by the $SU(3)$ group of phase transformations on the quark color fields. The free Lagrangian is

$$\mathcal{L}_0 = \bar{q}_j(i\gamma^\mu\partial_\mu - m)q_j, \tag{14.29}$$

where q_1, q_2, q_3 denote the three color fields. For simplicity, we show just one quark flavor.

We can explore the consequences of requiring $\mathcal{L}_0$ to be invariant under local phase transformations of the form

$$q(x) \to Uq(x) \equiv e^{i\alpha_a(x)T_a}q(x), \tag{14.30}$$

where U is an arbitrary 3×3 unitary matrix which we show parametrized by its general form. A summation over the repeated suffix a is implied. T_a with $a = 1,\ldots,8$ are a set of linearly independent traceless 3×3 matrices, and α_a are the group parameters. The matrices $\lambda_a/2$ of Section 2.8 are the conventional choice of the T_a matrices.

EXERCISE 14.8 Show that $\det U = e^{i\phi}$, where ϕ is real. We separate off such an overall phase by restricting the group transformations to those with $\det U = +1$, see Section 2.3. We called this the group of special unitary 3×3 matrices $SU(3)$. Show that the requirement $\det U = +1$ implies $\mathrm{Tr}(T_a) = 0$. Verify that $U^\dagger = U^{-1}$ requires

$$\alpha_a T_a = \alpha_a^* T_a^\dagger, \tag{14.31}$$

so that choosing T_a to be hermitian means the group parameters α_a are real.

The group is non-Abelian since not all the generators T_a commute with each other. It is easy to show that the commutator of any two is a linear combination of all the T's:

$$[T_a, T_b] = i f_{abc} T_c \tag{14.32}$$

where f_{abc} are real constants, called the structure constants of the group.

EXERCISE 14.9 Show that the structure constants f_{abc} are antisymmetric under interchange of any pair of indices.

To impose $SU(3)$ local gauge invariance on the Lagrangian (14.29), we follow the steps of Section 14.3. It is sufficient to consider infinitesimal phase transformations,

$$\begin{aligned} q(x) &\to [1 + i\alpha_a(x) T_a]\, q(x), \\ \partial_\mu q &\to (1 + i\alpha_a T_a)\,\partial_\mu q + i T_a q\, \partial_\mu \alpha_a. \end{aligned} \tag{14.33}$$

The last term spoils the invariance of $\mathcal{L}$. At first sight, it looks as if we can proceed exactly as for QED. That is, we introduce (eight) gauge fields G_μ^a, each transforming as [see (14.25)]

$$G_\mu^a \to G_\mu^a - \frac{1}{g}\partial_\mu \alpha_a, \tag{14.34}$$

and form a covariant derivative [see (14.24)]

$$D_\mu = \partial_\mu + i g T_a G_\mu^a. \tag{14.35}$$

We then make the replacement $\partial_\mu \to D_\mu$ in Lagrangian (14.29) and obtain

$$\mathcal{L} = \bar{q}\left(i\gamma^\mu \partial_\mu - m\right) q - g\left(\bar{q}\gamma^\mu T_a q\right) G_\mu^a. \tag{14.36}$$

This is the QCD analogue of (14.26). However, for a non-Abelian gauge transformation, this is not enough to produce a gauge-invariant Lagrangian. The problem is that

$$(\bar{q}\gamma^\mu T_a q) \rightarrow (\bar{q}\gamma^\mu T_a q) + i\alpha_b \bar{q}\gamma^\mu (T_a T_b - T_b T_a) q \rightarrow (\bar{q}\gamma^\mu T_a q) - f_{abc}\alpha_b(\bar{q}\gamma^\mu T_c q), \tag{14.37}$$

using (14.32). Taking note of this result, we see that we can achieve gauge invariance of $\mathcal{L}$ provided we rewrite (14.34) as

$$G_\mu^a \rightarrow G_\mu^a - \frac{1}{g}\partial_\mu \alpha_a - f_{abc}\alpha_b G_\mu^c. \tag{14.38}$$

Finally, we may add to $\mathcal{L}$ a gauge invariant kinetic energy term for each of the G_μ^a fields. The final gauge invariant QCD Lagrangian is then

$$\boxed{\mathcal{L} = \bar{q}(i\gamma^\mu \partial_\mu - m)q - g(\bar{q}\gamma^\mu T_a q)G_\mu^a - \tfrac{1}{4}G_{\mu\nu}^a G_a^{\mu\nu}.} \tag{14.39}$$

EXERCISE 14.10 Due to the additional term in (14.38), $G_{\mu\nu}^a$ has a more complicated form than its counterpart in QED, (14.27). In order that the kinetic energy be invariant under (14.38), show that

$$G_{\mu\nu}^a = \partial_\mu G_\nu^a - \partial_\nu G_\mu^a - g f_{abc} G_\mu^b G_\nu^c. \tag{14.40}$$

Equation (14.39) is the Lagrangian for interacting colored quarks q and vector gluons G_μ, with coupling specified by g, which follows simply from demanding that the Lagrangian be invariant under local color phase transformations to the quark fields. Since we can arbitrarily vary the phase of the three quark color fields, it is not surprising that eight vector gluon fields (G_μ^a with $a = 1, \ldots, 8$) are needed to compensate all possible phase changes. Just as for the photon, local gauge invariance requires the gluons to be *massless*.

The field strength tensor $G_{\mu\nu}^a$ has a remarkable new property on account of the extra term in (14.40). Imposing the gauge symmetry has required that the kinetic energy term in $\mathcal{L}$ is not purely kinetic but includes an induced self-interaction between the gauge bosons. This becomes clear if we rewrite Lagrangian (14.39) in the symbolic form

$$\mathcal{L} = \text{“}\bar{q}q\text{”} + \text{“}G^2\text{”} + g\text{“}\bar{q}qG\text{”} + g\text{“}G^3\text{”} + g^2\text{“}G^4\text{”}.$$

The first three terms have QED analogues. They describe the free propagation of quarks and gluons and the quark–gluon interaction. The remaining two terms show the presence of three and four gluon vertices in QCD and reflect the fact that gluons themselves carry color charge. They have no analogue in QED and arise on account of the non-Abelian character of the gauge group. We emphasize that gauge invariance uniquely determines the structure of these gluon self-coupling terms. There is only one coupling g.

EXERCISE 14.11 Using the prescription for obtaining the Feynman rules from the Lagrangian that we mentioned in Section 14.1, show that the vertex factors for the quark–gluon and triple gluon vertices of Fig. 14.1 are, respectively,

$$-ig\gamma_\mu (T_a)_{ij},$$

$$-gf_{abc}\left[g_{\mu\nu}(p_1 - p_2)_\lambda + g_{\nu\lambda}(p_2 - p_3)_\mu + g_{\lambda\mu}(p_3 - p_1)_\nu\right].$$

Theories with non-Abelian gauge invariance are frequently referred to as Yang–Mills theories, as these authors were the first to study the implications of a non-Abelian gauge group.

14.5 Massive Gauge Bosons?

We have seen that both photons and gluons are required to be massless, since the presence of mass terms for gauge fields destroys the gauge invariance of the Lagrangian. This raises a serious problem if we want to apply these ideas also to the weak interaction. How can we apply gauge symmetry to interactions which are mediated by gauge bosons ($W^{\pm}$, Z) with masses of the order of 100 GeV? You could argue that we demanded local gauge invariance on the basis of pure aesthetics; so why not introduce mass terms of the form $M^2 W_\mu W^\mu$ into the Lagrangian and just ignore their symmetry-breaking effect? If we go ahead and do this, we encounter (unrenormalizable) divergences which make the theory meaningless.

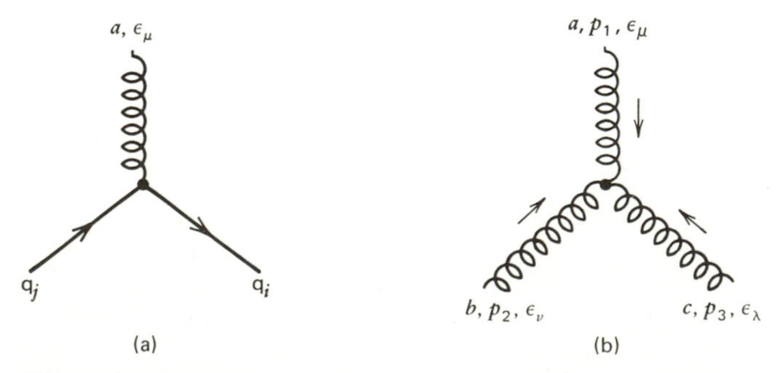

Fig. 14.1 (a) The quark–gluon vertex and (b) the triple-gluon vertex, showing the color, the four-momenta, and the polarization vectors of the gluons.

For example, let us calculate the diagram shown in Fig. 14.2. The momentum q circling around the loop may take any value. It must therefore be integrated over so that the amplitude is of the form

$$\int d^4q(\text{propagators})\ldots, \tag{14.41}$$

see Chapter 7. In QED, one can calculate the exchange of two photons between electrons (topologically the same as Fig. 14.2) and perform the integration (14.41). The answer is finite. The momentum in the loop can of course be arbitrarily large, but the $1/q^2$ behavior of the photon propagators makes the integral (14.41) well behaved. For massive gauge bosons, the result is different. The propagators (6.87) are of the form

$$i\frac{-g_{\mu\nu} + q_\mu q_\nu/M^2}{q^2 - M^2} \underset{q^2\to\infty}{\sim} i\frac{q_\mu q_\nu}{q^2M^2}, \tag{14.42}$$

and they no longer prevent (14.41) from diverging for large loop momenta. Our only hope is to introduce a cutoff in q^2. It represents a new parameter in the theory which might perhaps be determined by experiment. But this hope is shattered by inspecting diagrams containing more loops. New, ever more severe divergences appear in each order, and ultimately an infinite number of unknown parameters have to be introduced. Unlike the divergences discussed in Chapter 7, it is not possible to interpret these divergences as merely "renormalizing" the couplings in the Lagrangian. Such a theory is meaningless: no predictions are possible. It is said to be "unrenormalizable." Giving up local gauge invariance to introduce $W^\pm, Z$ masses "by hand" in this way is clearly no good. Is it possible then to introduce mass without breaking gauge invariance? Surprisingly, it can be done. The following sections describe this intriguing development.

14.6 Spontaneous Symmetry Breaking. "Hidden" Symmetry

The way to generate the mass of a particle by "spontaneous symmetry breaking," as opposed to putting it in by hand as we did in (14.6), can be seen in a very simple example. Consider a simple world consisting just of scalar particles described by the Lagrangian

$$\mathcal{L} \equiv T - V = \tfrac{1}{2}\left(\partial_\mu\phi\right)^2 - \left(\tfrac{1}{2}\mu^2\phi^2 + \tfrac{1}{4}\lambda\phi^4\right), \tag{14.43}$$

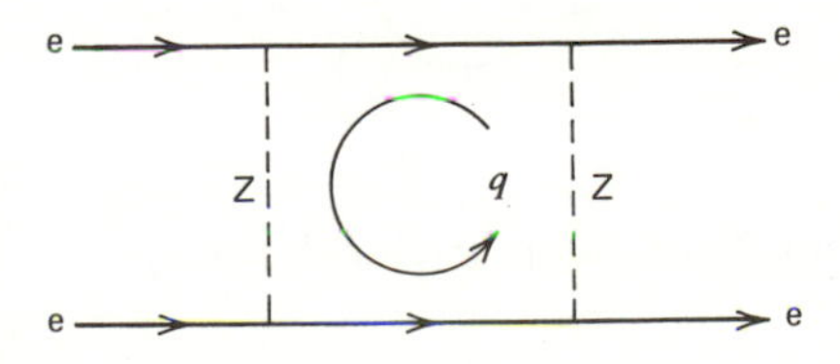

Fig. 14.2 A loop contribution to electron–electron scattering.

with $\lambda > 0$. Here, we have required $\mathcal{L}$ to be invariant under the symmetry operation which replaces ϕ by $-\phi$. It suffices to keep the first two allowed terms in the general expansion of V in powers of ϕ. The two possible forms of the potential are shown in Fig. 14.3. Case (a) for $\mu^2 > 0$ is already familiar; see (14.6). It simply describes a scalar field with mass μ. The ϕ^4 term shows that the four-particle vertex exists with coupling λ. We say that ϕ is a self-interacting field. The ground state (the vacuum) corresponds to $\phi = 0$. It obeys the reflection symmetry of the Lagrangian.

However, case (b) of Fig. 14.3, where $\mu^2 < 0$, is the possibility we really wish to explore. Now, the Lagrangian (14.43) has a mass term of the wrong sign for the field ϕ, since the relative sign of the ϕ^2 term and the kinetic energy T is positive. Unlike case (a), in case (b) the potential has two minima. These minima satisfy

$$\frac{\partial V}{\partial \phi} = \phi\left(\mu^2 + \lambda\phi^2\right) = 0$$

and are therefore at

$$\phi = \pm v \qquad \text{with} \quad v = \sqrt{-\mu^2/\lambda}\,. \tag{14.44}$$

The extremum $\phi = 0$ does not correspond to the energy minimum. Perturbative calculations should involve expansions around the classical minimum $\phi = v$ or $\phi = -v$. We therefore write

$$\phi(x) = v + \eta(x), \tag{14.45}$$

where $\eta(x)$ represents the quantum fluctuations about this minimum. We have chosen to translate the field to $\phi = +v$, but this does not imply any loss of generality since $\phi = -v$ can always be reached by reflection symmetry. (Nature has also to make such a choice.) Substituting (14.45) into Lagrangian (14.43), we

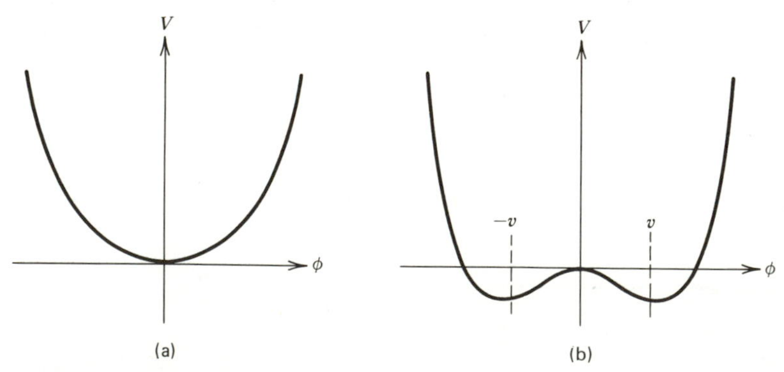

Fig. 14.3 The potential $V(\phi) = \frac{1}{2}\mu^2\phi^2 + \frac{1}{4}\lambda\phi^4$ for (a) $\mu^2 > 0$ and (b) $\mu^2 < 0$, and $\lambda > 0$.

obtain

$$\mathcal{L}' = \tfrac{1}{2}(\partial_\mu \eta)^2 - \lambda v^2 \eta^2 - \lambda v \eta^3 - \tfrac{1}{4}\lambda \eta^4 + \text{const.} \tag{14.46}$$

The field η has a mass term of the correct sign! Indeed, the relative sign of the η^2 term and the kinetic energy is negative. Identifying the first two terms of $\mathcal{L}'$ with (14.6) gives

$$m_\eta = \sqrt{2\lambda v^2} = \sqrt{-2\mu^2}\,. \tag{14.47}$$

The higher-order terms in η represent the interaction of the η field with itself.

There is of course a puzzle here. The Lagrangians, $\mathcal{L}$ of (14.43) and $\mathcal{L}'$ of (14.46), are completely equivalent. A transformation of the type (14.45) cannot change the physics. If we could solve the two Lagrangians exactly, they must yield identical physics. But in particle physics, we are not able to perform such a calculation. Instead, we do perturbation theory and calculate the fluctuations around the minimum energy. Using $\mathcal{L}$, we would discover that the perturbation series did not converge because we are trying to expand around the unstable point $\phi = 0$. The correct way to proceed is to use $\mathcal{L}'$ and expand in η around the stable vacuum $\phi = +v$. In perturbation theory, $\mathcal{L}'$ gives the correct picture of physics; $\mathcal{L}$ does not. The scalar particle (described by the in-principle-equivalent Lagrangians $\mathcal{L}$, $\mathcal{L}'$) therefore does have a mass!

We refer to the way this mass was "generated" (or, better, "revealed") as "spontaneous symmetry breaking." In the $\mathcal{L}'$ version of our scalar theory, the reflection symmetry of the Lagrangian has apparently been broken by our choice of the ground state $\phi = +v$ (rather than $\phi = -v$) around which to do our perturbation calculations.

Other physical systems are known in which the ground state does not possess the symmetry of the Lagrangian. An infinitely extended ferromagnet is described by a Lagrangian which is invariant under rotations in space. In the ground state, all the elementary spins are aligned in a particular direction, and the rotational symmetry is apparently broken. This direction is arbitrary, however; and by rotational symmetry, we can reach an infinite number of other ground states, each corresponding to a different alignment of the spins. In our scalar field example, only two ground states were possible. Other examples are superconductors, crystal lattices, and the buckling of a compressed needle! If we take a knitting needle and compress it with a force F along its axis (the z axis in Fig. 14.4), the obvious solution is that it stays in the configuration $x = y = 0$. However, if the force gets too large ($F > F_{\text{critical}}$), the needle will jump into the bent position shown in the figure. It does this because the energy in this state is lower than in the metastable state, where it stays aligned along the z axis. The cylindrical symmetry of the system around the z axis is apparently broken by the buckling of the needle. But the needle can buckle in any direction in the x–y plane, reaching a ground state with the same energy, so it is not possible to predict which way it will go.

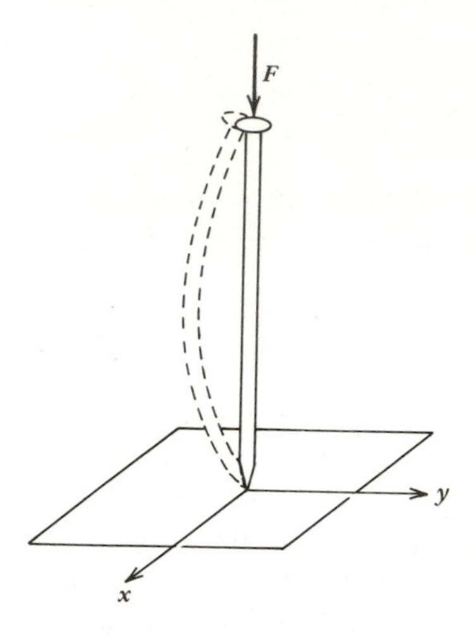

Fig. 14.4 An example of spontaneous symmetry breaking: the buckling of a knitting needle.

14.7 Spontaneous Breaking of a Global Gauge Symmetry

To approach our goal of generating a mass for the gauge bosons, we repeat the above procedure for a complex scalar field $\phi = (\phi_1 + i\phi_2)/\sqrt{2}$ described by Lagrangian

$$\mathcal{L} = (\partial_\mu \phi)^*(\partial^\mu \phi) - \mu^2 \phi^* \phi - \lambda(\phi^* \phi)^2, \tag{14.48}$$

which is invariant under $\phi \to e^{i\alpha}\phi$. That is, $\mathcal{L}$ possesses a $U(1)$ global gauge symmetry. As before, we consider the case when $\lambda > 0$ and $\mu^2 < 0$. We rewrite (14.48) in the form

$$\mathcal{L} = \tfrac{1}{2}(\partial_\mu \phi_1)^2 + \tfrac{1}{2}(\partial_\mu \phi_2)^2 - \tfrac{1}{2}\mu^2(\phi_1^2 + \phi_2^2) - \tfrac{1}{4}\lambda(\phi_1^2 + \phi_2^2)^2.$$

There is now a circle of minima of the potential $V(\phi)$ in the ϕ_1, ϕ_2 plane of radius v, such that

$$\phi_1^2 + \phi_2^2 = v^2 \qquad \text{with} \quad v^2 = -\frac{\mu^2}{\lambda}, \tag{14.49}$$

as shown in Fig. 14.5. Again, we translate the field ϕ to a minimum energy position, which without loss of generality we may take as the point $\phi_1 = v$, $\phi_2 = 0$. We expand $\mathcal{L}$ about the vacuum in terms of fields η, ξ by substituting

$$\phi(x) = \sqrt{\tfrac{1}{2}}\,[v + \eta(x) + i\xi(x)] \tag{14.50}$$

into (14.48) and obtain

$$\mathcal{L}' = \tfrac{1}{2}(\partial_\mu \xi)^2 + \tfrac{1}{2}(\partial_\mu \eta)^2 + \mu^2\eta^2 + \text{const.} + \text{cubic and quartic terms in } \eta, \xi. \tag{14.51}$$

The third term has the form of a mass term ($-\frac{1}{2}m_\eta^2\eta^2$) for the η-field. Thus, the

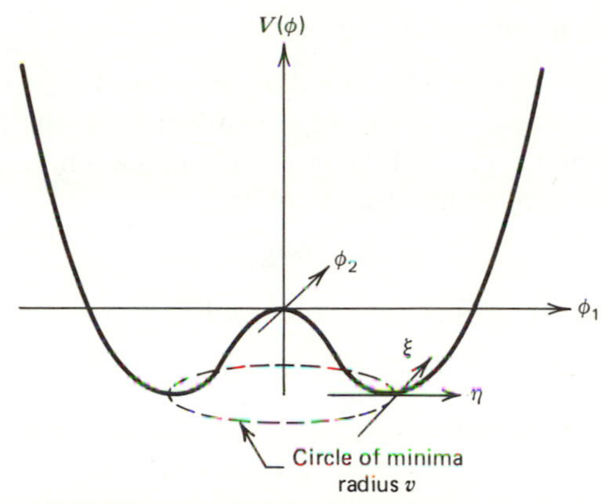

Fig. 14.5 The potential $V(\phi)$ for a complex scalar field for the case $\mu^2 < 0$ and $\lambda > 0$.

η-mass is $m_\eta = \sqrt{-2\mu^2}$, just as before, see (14.47). The first term in $\mathcal{L}'$ represents the kinetic energy of the ξ-field, but there is no corresponding mass term for ξ. That is, the theory also contains a massless scalar, which is known as a Goldstone boson. Thus, we have encountered a problem; in attempting to generate a massive gauge boson, we see that a spontaneously broken gauge theory appears to be plagued with its own massless scalar particle. Intuitively, it is easy to see the reason for its presence. The potential in the tangent (ξ) direction is flat, implying a massless mode; there is no resistance to excitations along the ξ-direction in Fig. 14.5.

Our Lagrangian is a simple example of the Goldstone theorem, which states that massless scalars occur whenever a continuous symmetry of a physical system is "spontaneously broken" (or, more accurately, is "not apparent in the ground state"). In the ferromagnet example, the analogue of our Goldstone boson is the long-range spin waves which are oscillations of the spin alignment.

EXERCISE 14.12 The Lagrangian for three interacting real fields ϕ_1, ϕ_2, ϕ_3 is

$$\mathcal{L} = \tfrac{1}{2}\left(\partial_\mu \phi_i\right)^2 - \tfrac{1}{2}\mu^2\phi_i^2 - \tfrac{1}{4}\lambda\left(\phi_i^2\right)^2$$

with $\mu^2 < 0$ and $\lambda > 0$, and where a summation of ϕ_i^2 over i is implied. Show that it describes a massive field of mass $\sqrt{-2\mu^2}$ and two massless Goldstone bosons.

Our hope of finding a gauge theory of weak interactions with massive gauge bosons looks forlorn. It appears that we shall also have unwanted (unobserved) massless scalar particles to worry about. Nevertheless, let us proceed from a global to a local gauge theory. A miracle is about to happen.

14.8 The Higgs Mechanism

The final step is to study spontaneous breaking of a local gauge symmetry. Here, we take the simplest example: a $U(1)$ gauge symmetry. In the following section, we extend the discussion to $SU(2)$. First, we must make our Lagrangian, (14.48), invariant under a $U(1)$ local gauge transformation,

$$\phi \to e^{i\alpha(x)}\phi. \tag{14.52}$$

As in Section 14.3, this requires ∂_μ to be replaced by the covariant derivative,

$$D_\mu = \partial_\mu - ieA_\mu,$$

where the gauge field transforms as

$$A_\mu \to A_\mu + \frac{1}{e}\partial_\mu \alpha. \tag{14.53}$$

The gauge invariant Lagrangian is thus

$$\mathcal{L} = (\partial^\mu + ieA^\mu)\phi^*(\partial_\mu - ieA_\mu)\phi - \mu^2\phi^*\phi - \lambda(\phi^*\phi)^2 - \tfrac{1}{4}F_{\mu\nu}F^{\mu\nu}. \tag{14.54}$$

If $\mu^2 > 0$, this is just the QED Lagrangian for a charged scalar particle of mass μ (apart from the ϕ^4 self-interaction term). We already obtained the analogous QED Lagrangian for a fermion field in (14.28). However, here we take $\mu^2 < 0$ since we want to generate masses by spontaneous symmetry breaking.

We repeat the by now familiar procedure of translating the field ϕ to a true ground state. On substituting (14.50), the Lagrangian (14.54) becomes

$$\mathcal{L}' = \tfrac{1}{2}(\partial_\mu \xi)^2 + \tfrac{1}{2}(\partial_\mu \eta)^2 - v^2\lambda\eta^2 + \tfrac{1}{2}e^2v^2A_\mu A^\mu$$

$$- evA_\mu \partial^\mu \xi - \tfrac{1}{4}F_{\mu\nu}F^{\mu\nu} + \text{interaction terms}. \tag{14.55}$$

The particle spectrum of $\mathcal{L}'$ appears to be a massless Goldstone boson ξ, a massive scalar η, and more crucially a massive vector A_μ for which we have sought so long. Indeed, from (14.55), we have

$$m_\xi = 0, \qquad m_\eta = \sqrt{2\lambda v^2}, \qquad m_A = ev$$

[see (14.6) and Exercise 14.4]. We have dynamically generated a mass for the gauge field, but we still have the problem of the occurrence of a massless Goldstone boson. But wait, the presence of a term off-diagonal in the fields, $A_\mu \partial^\mu \xi$, means we must take care in interpreting $\mathcal{L}'$. Indeed, the particle spectrum we assigned to $\mathcal{L}'$ cannot be correct. By giving mass to A_μ, we have clearly raised the polarization degrees of freedom from 2 to 3, because it can now have longitudinal polarization. But simply translating field variables, as in (14.50), does not create a new degree of freedom. We deduce that the fields in $\mathcal{L}'$ do not all correspond to distinct physical particles. This need not worry us; indeed, in QED we use a Lagrangian containing unphysical (longitudinal as well as scalar) massless photons. But which field in $\mathcal{L}'$ is unphysical? Can we find a particular gauge transformation which will eliminate a field from the Lagrangian? A clue is to note that

$$\begin{aligned}\phi &= \sqrt{\tfrac{1}{2}}\,(v + \eta + i\xi) \\ &\simeq \sqrt{\tfrac{1}{2}}\,(v + \eta)e^{i\xi/v}\end{aligned} \tag{14.56}$$

to lowest order in ξ. This suggests that we should substitute a different set of real fields h, θ, A_μ, where

$$\begin{aligned}\phi &\to \sqrt{\tfrac{1}{2}}\,(v + h(x))e^{i\theta(x)/v}, \\ A_\mu &\to A_\mu + \frac{1}{ev}\partial_\mu\theta\end{aligned} \tag{14.57}$$

into the original Lagrangian (14.54). This is a particular choice of gauge, with $\theta(x)$ chosen so that h is real. We therefore anticipate that the theory will be independent of θ. Indeed, we obtain

$$\begin{aligned}\mathcal{L}'' = {}& \tfrac{1}{2}(\partial_\mu h)^2 - \lambda v^2 h^2 + \tfrac{1}{2}e^2v^2A_\mu^2 - \lambda v h^3 - \tfrac{1}{4}\lambda h^4 \\ & + \tfrac{1}{2}e^2A_\mu^2h^2 + ve^2A_\mu^2h - \tfrac{1}{4}F_{\mu\nu}F^{\mu\nu}.\end{aligned} \tag{14.58}$$

The Goldstone boson actually does *not* appear in the theory. That is, the apparent extra degree of freedom is actually spurious, because it corresponds only to the freedom to make a gauge transformation. The Lagrangian describes just two interacting massive particles, a vector gauge boson A_μ and a massive scalar h, which is called a Higgs particle. The unwanted massless Goldstone boson has been turned into the badly needed longitudinal polarization of the massive gauge particle. This is called the "Higgs mechanism."

14.9 Spontaneous Breaking of a Local $SU(2)$ Gauge Symmetry

In the previous section, we studied the spontaneous breaking of a $U(1)$ gauge symmetry. It is necessary to repeat the procedure for an $SU(2)$ gauge symmetry. This final example will serve as a useful summary of the chapter. It encapsulates all the ideas we have introduced, as well as providing preparation for the discussion of the gauge symmetries of electroweak interactions to be given in Chapter 15.

We take a Lagrangian

$$\mathcal{L} = (\partial_\mu\phi)^\dagger(\partial^\mu\phi) - \mu^2\phi^\dagger\phi - \lambda(\phi^\dagger\phi)^2, \tag{14.59}$$

where ϕ is an $SU(2)$ doublet of complex scalar fields:

$$\phi = \begin{pmatrix}\phi_\alpha \\ \phi_\beta\end{pmatrix} = \sqrt{\tfrac{1}{2}}\begin{pmatrix}\phi_1 + i\phi_2 \\ \phi_3 + i\phi_4\end{pmatrix}. \tag{14.60}$$

$\mathcal{L}$ is manifestly invariant under global $SU(2)$ phase transformations

$$\phi \to \phi' = e^{i\alpha_a\tau_a/2}\phi \tag{14.61}$$

[see (14.30)]. To achieve local, that is, $\alpha_a(x)$, $SU(2)$ invariance of $\mathcal{L}$, we follow the steps of Section 14.4. We replace ∂_μ by the covariant derivative

$$D_\mu = \partial_\mu + ig\frac{\tau_a}{2}W_\mu^a, \tag{14.62}$$

and three gauge fields, $W_\mu^a(x)$ with $a = 1, 2, 3$, thereby join the cast. Under an infinitesimal gauge transformation

$$\phi(x) \to \phi'(x) = (1 + \boldsymbol{\alpha}(x)\cdot\boldsymbol{\tau}/2)\phi(x), \tag{14.63}$$

the three gauge fields transform as

$$\mathbf{W}_\mu \to \mathbf{W}_\mu - \frac{1}{g}\partial_\mu\boldsymbol{\alpha} - \boldsymbol{\alpha}\times\mathbf{W}_\mu \tag{14.64}$$

[see (14.38)]. The $\boldsymbol{\alpha}\times\mathbf{W}_\mu$ term occurs because $\mathbf{W}_\mu$ is an $SU(2)$ vector; it is "rotated" even if $\boldsymbol{\alpha}$ is independent of x. The gauge invariant Lagrangian is then

$$\mathcal{L} = \left(\partial_\mu\phi + ig\frac{1}{2}\boldsymbol{\tau}\cdot\mathbf{W}_\mu\phi\right)^\dagger\left(\partial^\mu\phi + ig\frac{1}{2}\boldsymbol{\tau}\cdot\mathbf{W}^\mu\phi\right) - V(\phi) - \frac{1}{4}\mathbf{W}_{\mu\nu}\cdot\mathbf{W}^{\mu\nu}, \tag{14.65}$$

with

$$V(\phi) = \mu^2\phi^\dagger\phi + \lambda(\phi^\dagger\phi)^2, \tag{14.66}$$

and where we have added the kinetic energy term of the gauge fields with

$$\mathbf{W}_{\mu\nu} = \partial_\mu\mathbf{W}_\nu - \partial_\nu\mathbf{W}_\mu - g\mathbf{W}_\mu\times\mathbf{W}_\nu. \tag{14.67}$$

The last term in (14.67), and in (14.64), arises from the non-Abelian character of the group; that is, it occurs because the τ's do not commute with each other. If $\mu^2 > 0$, the Lagrangian (14.65) describes a system of four scalar particles [ϕ_i of (14.60)], each of mass μ, interacting with three massless gauge bosons (W_μ^a).

We are interested in the case $\mu^2 < 0$ and $\lambda > 0$. The potential $V(\phi)$ of (14.66) then has its minimum at a finite value of $|\phi|$ where

$$\phi^\dagger\phi \equiv \frac{1}{2}(\phi_1^2 + \phi_2^2 + \phi_3^2 + \phi_4^2) = -\frac{\mu^2}{2\lambda}. \tag{14.68}$$

This manifold of points at which $V(\phi)$ is minimized is invariant under the $SU(2)$ transformations. We must expand $\phi(x)$ about a particular minimum. We can choose, say,

$$\phi_1 = \phi_2 = \phi_4 = 0, \qquad \phi_3^2 = -\frac{\mu^2}{\lambda} \equiv v^2. \tag{14.69}$$

The effect is equivalent to the spontaneous breaking of the $SU(2)$ symmetry. The symmetry, which, for example, was manifest in (14.68), has become hidden.

We now expand $\phi(x)$ about this particular vacuum:

$$\phi_0 \equiv \sqrt{\tfrac{1}{2}}\begin{pmatrix}0\\ v\end{pmatrix}. \tag{14.70}$$

The result is that, due to gauge invariance, we can simply substitute the expansion

$$\phi(x) = \sqrt{\tfrac{1}{2}}\begin{pmatrix}0\\ v + h(x)\end{pmatrix} \tag{14.71}$$

into the Lagrangian (14.65). That is, of the four scalar fields, the only one that remains is (the Higgs field) $h(x)$. At first sight, this may seem surprising. However, the reason is similar to that found for the spontaneously broken $U(1)$ gauge symmetry of the previous section. The argument is as follows. Proceeding in analogy to (14.57), we parametrize the fluctuations from the vacuum ϕ_0 in terms of four real fields θ_1, θ_2, θ_3, and h, using the form

$$\phi(x) = e^{i\boldsymbol{\tau}\cdot\boldsymbol{\theta}(x)/v}\begin{pmatrix} 0 \\ \dfrac{v+h(x)}{\sqrt{2}} \end{pmatrix}. \tag{14.72}$$

To verify that this is perfectly general, we examine small perturbations. We find

$$\begin{aligned}\phi(x) &\simeq \sqrt{\tfrac{1}{2}}\begin{pmatrix} 1+i\theta_3/v & i(\theta_1 - i\theta_2)/v \\ i(\theta_1+i\theta_2)/v & 1-i\theta_3/v \end{pmatrix}\begin{pmatrix} 0 \\ v+h(x)\end{pmatrix} \\ &\simeq \sqrt{\tfrac{1}{2}}\begin{pmatrix} \theta_2 + i\theta_1 \\ v+h-i\theta_3 \end{pmatrix}. \end{aligned} \tag{14.73}$$

We see that the four fields are indeed independent and fully parametrize the deviations from the vacuum ϕ_0. Now, the Lagrangian is locally $SU(2)$ invariant. Therefore, we can gauge the three (massless Goldstone boson) fields $\boldsymbol{\theta}(x)$ of (14.72). The Lagrangian will then contain no trace of the $\boldsymbol{\theta}(x)$. Hence, we arrive at the result of (14.71).

To determine the masses generated for the gauge bosons W_μ^a, it is sufficient to substitute ϕ_0 of (14.70) into the Lagrangian. The relevant term of (14.65) is

$$\begin{aligned}\left| ig\frac{1}{2}\boldsymbol{\tau}\cdot\mathbf{W}_\mu\phi\right|^2 &= \frac{g^2}{8}\left|\begin{pmatrix} W_\mu^3 & W_\mu^1 - iW_\mu^2 \\ W_\mu^1 + iW_\mu^2 & W_\mu^3 \end{pmatrix}\begin{pmatrix} 0 \\ v \end{pmatrix}\right|^2 \\ &= \frac{g^2v^2}{8}\left[\left(W_\mu^1\right)^2 + \left(W_\mu^2\right)^2 + \left(W_\mu^3\right)^2\right], \end{aligned} \tag{14.74}$$

where here $|\quad|^2$ has been used as shorthand for $(\quad)^\dagger(\quad)$. We compare these terms with a typical mass term of a boson, $\frac{1}{2}M^2B_\mu^2$, and find $M = \frac{1}{2}gv$. That is, the Lagrangian describes three massive gauge fields and one massive scalar h. In summary, the gauge fields have "eaten up" the Goldstone bosons and become massive. The scalar degrees of freedom become the longitudinal polarizations of the massive vector bosons. This is another example of the Higgs mechanism.

EXERCISE 14.13 Rather than (14.60), take instead ϕ to be an $SU(2)$ triplet of real scalar fields. For $\mu^2 < 0$ and $\lambda > 0$, show that in this case two gauge bosons acquire mass but that the third remains massless.

Hint Verify, and use, $(T_k)_{ij} = -i\varepsilon_{ijk}$ for the triplet representation of $SU(2)$.

The Higgs mechanism has enabled us to avoid massless particles. However, we need a second "miracle." Remember that the basic problem is not just to generate masses but to incorporate the mass of the weak bosons while still preserving the renormalizability of the theory. We noted in Section 14.5 that although *a priori* there is nothing to prevent us from brutally breaking the gauge symmetry by inserting explicit gauge mass terms into the Lagrangian, the resulting (unrenormalizable) theory loses all predictive power. In a spontaneously broken gauge theory, the symmetry is, in a sense, still present; it is merely "hidden" by our choice of ground state, and the theory can be shown to remain renormalizable. In a way, this is not so miraculous. It had already been conjectured to be true by Weinberg and Salam. However, the proof is far from simple. It was finally completed in 1971 by 't Hooft and goes beyond the scope of this book.

For the above reasons, it is widely believed that gauge principles may generate the structure of all particle interactions. The standard model for weak and electromagnetic interactions is constructed from a gauge theory with four gauge fields, the photon and the massive bosons, $W^{\pm}$ and Z^0. We generate the masses of the gauge fields (as well as the fermions) by spontaneous symmetry breaking, ensuring that one of them (the photon) remains massless. We also require that the theory must reproduce the low-energy (low q^2) phenomenology of Chapters 12 and 13. Such a theory will be renormalizable and will contain one (or possibly more) Higgs scalars but no Goldstone bosons. This exercise is the subject of the next chapter.

15

The Weinberg–Salam Model and Beyond

Our objective is to obtain a renormalized theory of electroweak interactions incorporating the massive gauge bosons ($W^{\pm}, Z^0$). This will be achieved by spontaneously breaking a local gauge symmetry. A pertinent example of this technique was described in Section 14.9. But what is the gauge symmetry of electroweak interactions? The data on weak and electromagnetic processes suggest that the interactions are invariant under weak isospin $SU(2)_L$ and weak hypercharge $U(1)_Y$ transformations (see Chapter 13). Our first task is therefore to cast the results of Chapter 13 into an $SU(2) \times U(1)$ invariant Lagrangian.

15.1 Electroweak Interactions Revisited

To begin, we recall QED. In Chapter 6, electromagnetic amplitudes ($-i\mathfrak{M}$) were calculated using an interaction

$$\boxed{-iej_{\mu}^{em}A^{\mu} = -ie\left(\bar{\psi}\gamma_{\mu}Q\psi\right)A^{\mu}} \qquad U(1)_{em}, \tag{15.1}$$

where Q is the charge operator (with eigenvalue -1 for the electron). Subsequently, we saw in Section 14.3 that precisely this interaction arose from demanding invariance of the Lagrangian for a free fermion,

$$\mathcal{L} = \bar{\psi}\left(i\gamma^{\mu}\partial_{\mu} - m\right)\psi,$$

under local gauge (or phase) transformations

$$\psi \to \psi' = e^{i\alpha(x)Q}\psi. \tag{15.2}$$

Indeed, by insisting on local gauge invariance, we were inevitably led to the Lagrangian of QED, (14.28):

$$\mathcal{L} = \underbrace{\bar{\psi}\left(i\gamma^{\mu}\partial_{\mu} - m\right)\psi}_{\substack{\text{Kinetic energy} \\ \text{and mass of } \psi}} - \underbrace{e\bar{\psi}\gamma^{\mu}Q\psi A_{\mu}}_{\text{Interaction}} - \underbrace{\tfrac{1}{4}F_{\mu\nu}F^{\mu\nu}}_{\substack{\text{Kinetic energy} \\ \text{of } A_{\mu}}}. \tag{15.3}$$

In Section 14.3, we considered only the electron field and we omitted the charge operator Q. However, to incorporate all the quark and lepton fields, charge conservation requires the presence of Q in (15.2), see (15.11). The electromagnetic coupling is therefore proportional to the charge of the appropriate field; $Q_e = -1$, $Q_u = +\frac{2}{3}$, and so on.

To include weak processes in the formalism, we must replace (15.1) by two basic interactions; first, an isotriplet of weak currents $\mathbf{J}_\mu$ coupled to three vector bosons $\mathbf{W}^\mu$,

$$\boxed{-ig\mathbf{J}_\mu \cdot \mathbf{W}^\mu = -ig\bar{\chi}_L\gamma_\mu \mathbf{T}\cdot\mathbf{W}^\mu\chi_L} \qquad SU(2)_L, \tag{15.4}$$

and second, a weak hypercharge current coupled to a fourth vector boson B^μ,

$$\boxed{-i\frac{g'}{2}j_\mu^Y B^\mu = -ig'\bar{\psi}\gamma_\mu\frac{Y}{2}\psi B^\mu} \qquad U(1)_Y, \tag{15.5}$$

see (13.17). The operators $\mathbf{T}$ and Y are the generators of the $SU(2)_L$ and $U(1)_Y$ groups of gauge transformations, respectively. Taken together, the $SU(2)\times U(1)$ transformations of the left- and right-hand components of ψ are

$$\begin{aligned} \chi_L &\to \chi_L' = e^{i\boldsymbol{\alpha}(x)\cdot\mathbf{T}+i\beta(x)Y}\chi_L, \\ \psi_R &\to \psi_R' = e^{i\beta(x)Y}\psi_R, \end{aligned} \tag{15.6}$$

where the left-handed fermions form isospin doublets χ_L and the right-handed fermions are isosinglets ψ_R. For example, for the electron and its neutrino, we have [see (13.15)]

$$\chi_L = \begin{pmatrix} \nu_e \\ e^- \end{pmatrix}_L \qquad \text{with } T = \tfrac{1}{2},\ Y = -1,$$

$$\psi_R = e_R^- \qquad \text{with } T = 0,\ Y = -2. \tag{15.7}$$

But for quarks,

$$\chi_L = \begin{pmatrix} u \\ d \end{pmatrix}_L, \qquad \psi_R = u_R, \text{ or } d_R.$$

Here, a right-handed up quark has been included since quarks, unlike neutrinos, have a finite mass and hence have both right- and left-handed components.

The electromagnetic interaction (15.1) is embedded in (15.4) and (15.5). Before we continue, it is appropriate to recall how this was achieved. The generators of the three groups satisfy

$$Q = T^3 + \frac{Y}{2},$$

so that

$$j_\mu^{em} = J_\mu^3 + \tfrac{1}{2}j_\mu^Y,$$

see (13.26). In other words, the electromagnetic current is a combination of the two neutral currents J_μ^3 and j_μ^Y occurring in (15.4) and (15.5). The two physical neutral gauge fields A_μ and Z_μ are thus orthogonal combinations of the gauge fields W_μ^3 and B_μ, with mixing angle θ_W, and the interaction in the neutral current sector can be rewritten in terms of these physical fields as in (13.21):

$$
\begin{aligned}
-igJ_\mu^3 W^{3\mu} - i\frac{g'}{2} j_\mu^Y B^\mu &= -i\left[g\sin\theta_W J_\mu^3 + g'\cos\theta_W \frac{j_\mu^Y}{2}\right]A^\mu \\
&\quad -i\left[g\cos\theta_W J_\mu^3 - g'\sin\theta_W \frac{j_\mu^Y}{2}\right]Z^\mu \\
&= -iej_\mu^{em}A^\mu - \frac{ie}{\sin\theta_W\cos\theta_W}\left[J_\mu^3 - \sin^2\theta_W j_\mu^{em}\right]Z^\mu .
\end{aligned}
\tag{15.8}
$$

The requirement that the electromagnetic interaction, (15.1), must appear on the right-hand side has specified the weak neutral current interaction and fixed the couplings g and g' in terms of e and θ_W, namely,

$$
e = g\sin\theta_W = g'\cos\theta_W . \tag{15.9}
$$

EXERCISE 15.1 As revision, derive (15.8) and (15.9) from the statements of the above paragraph.

Just as the QED Lagrangian, (15.3), resulted from imposing $U(1)_{em}$ local gauge invariance, so we are led to the electroweak Lagrangian by requiring an $SU(2) \times U(1)_Y$ invariant form. For example, for the electron–neutrino lepton pair, we have

$$
\begin{aligned}
\mathcal{L}_1 &= \bar{\chi}_L\gamma^\mu\left[i\partial_\mu - g\tfrac{1}{2}\boldsymbol{\tau}\cdot\mathbf{W}_\mu - g'(-\tfrac{1}{2})B_\mu\right]\chi_L \\
&\quad + \bar{e}_R\gamma^\mu\left[i\partial_\mu - g'(-1)B_\mu\right]e_R - \tfrac{1}{4}\mathbf{W}_{\mu\nu}\cdot\mathbf{W}^{\mu\nu} - \tfrac{1}{4}B_{\mu\nu}B^{\mu\nu},
\end{aligned}
\tag{15.10}
$$

where we have inserted the hypercharge values $Y_L = -1$, $Y_R = -2$ of (15.7). $\mathcal{L}_1$ embodies the weak isospin and hypercharge interactions of (15.4) and (15.5). The final two terms are the kinetic energy and self-coupling of the $\mathbf{W}_\mu$ fields, see (14.67), and the kinetic energy of the B_μ field, $B_{\mu\nu} \equiv \partial_\mu B_\nu - \partial_\nu B_\mu$.

Gauge invariance means that the Lagrangian transforms as a singlet under transformations of each gauge group. Take the $U(1)_{em}$ gauge group as an example. A term in the Lagrangian which is the product of fields $\phi_1, \phi_2 \ldots, \phi_n$ transforms as

$$
(\phi_1\phi_2\ldots\phi_n) \to e^{i\alpha(x)(Q_1+Q_2+\cdots+Q_n)}(\phi_1\phi_2\ldots\phi_n),
$$

see (15.2). Gauge invariance requires the Lagrangian to be neutral, that is, to transform as a $U(1)_{em}$ singlet; therefore,

$$Q_1 + Q_2 + \cdots + Q_n = 0. \tag{15.11}$$

This is charge conservation.

So far, so good. However, note that $\mathcal{L}_1$ describes massless gauge bosons and massless fermions. Mass terms such as $\frac{1}{2}M^2 B_\mu B^\mu$ and $-m\bar{\psi}\psi$ are not gauge invariant and so cannot be added. The requirement of a massless gauge boson is familiar (for example, the photon of Section 14.3). The electron mass term

$$\begin{aligned} -m_e \bar{e} e &= -m_e \bar{e}\left[\tfrac{1}{2}(1-\gamma^5) + \tfrac{1}{2}(1+\gamma^5)\right] e \\ &= -m_e(\bar{e}_R e_L + \bar{e}_L e_R). \end{aligned} \tag{15.12}$$

Since e_L is a member of an isospin doublet and e_R is a singlet, this term manifestly breaks gauge invariance.

To generate the particle masses in a gauge invariant way, we must use the Higgs mechanism. That is, we spontaneously break the gauge symmetry, which has the paramount virtue that the theory remains renormalizable.

15.2 Choice of the Higgs Field

We want to formulate the Higgs mechanism so that the $W^\pm$ and Z^0 become massive and the photon remains massless. To do this, we introduce four real scalar fields ϕ_i. We have to add to $\mathcal{L}_1$ an $SU(2) \times U(1)$ gauge invariant Lagrangian for the scalar fields

$$\mathcal{L}_2 = \left|\left(i\partial_\mu - g\mathbf{T}\cdot\mathbf{W}_\mu - g'\frac{Y}{2}B_\mu\right)\phi\right|^2 - V(\phi) \tag{15.13}$$

where $|\quad|^2 \equiv (\quad)^\dagger(\quad)$. The structure of $\mathcal{L}_2$ is explained in Section 14.9, see (14.65). To keep $\mathcal{L}_2$ gauge invariant, the ϕ_i must belong to $SU(2) \times U(1)$ multiplets. The most economical choice is to arrange four fields in an isospin doublet with weak hypercharge $Y = 1$:

$$\boxed{\phi = \begin{pmatrix} \phi^+ \\ \phi^0 \end{pmatrix}} \quad \text{with} \quad \begin{aligned} \phi^+ &\equiv (\phi_1 + i\phi_2)/\sqrt{2}, \\ \phi^0 &\equiv (\phi_3 + i\phi_4)/\sqrt{2}. \end{aligned} \tag{15.14}$$

This is in fact the choice originally made in 1967 by Weinberg. It completes the specification of the standard (or minimal) model of electroweak interactions. It is also called the "Weinberg–Salam model."

To generate gauge boson masses, we use the familiar Higgs potential $V(\phi)$ of (14.66) with $\mu^2 < 0$ and $\lambda > 0$ and choose a vacuum expectation value, ϕ_0, of $\phi(x)$. The appropriate choice is (14.70),

$$\phi_0 \equiv \sqrt{\tfrac{1}{2}}\begin{pmatrix} 0 \\ v \end{pmatrix}. \tag{15.15}$$

Why is an isospin doublet of complex scalar fields, with $Y = 1$ and vacuum expectation value of ϕ_0 of (15.15), suitable for the problem in hand? Any choice of ϕ_0 which breaks a symmetry operation will inevitably generate a mass for the corresponding gauge boson. However, if the vacuum ϕ_0 is still left invariant by some subgroup of gauge transformations, then the gauge bosons associated with this subgroup will remain massless. Now, the choice ϕ_0 with $T = \frac{1}{2}$, $T^3 = -\frac{1}{2}$, and $Y = 1$ breaks both $SU(2)$ and $U(1)_Y$ gauge symmetries. But since ϕ_0 is neutral, the $U(1)_{em}$ symmetry with generator

$$Q = T^3 + \frac{Y}{2}$$

remains unbroken. That is,

$$Q\phi_0 = 0, \tag{15.16}$$

so that

$$\phi_0 \to \phi_0' = e^{i\alpha(x)Q}\phi_0 = \phi_0$$

for any value of $\alpha(x)$. The vacuum is thus invariant under $U(1)_{em}$ transformations, and the photon remains massless. Out of the four $SU(2) \times U(1)_Y$ generators $\mathbf{T}$, Y, only the combination Q obeys relation (15.16). The other three break the symmetry and generate massive gauge bosons. To put it another way, due to the conservation of electric charge, we can only allow neutral scalars to acquire vacuum expectation values; hence, the choice of vacuum (15.15) for the charge assignments of (15.14).

The same Higgs isospin doublet is essential for the generation of fermion masses (see Section 15.4).

15.3 Masses of the Gauge Bosons

Just as in Section 14.9, the gauge boson masses are identified by substituting the vacuum expectation value ϕ_0 for $\phi(x)$ in the Lagrangian $\mathcal{L}_2$. The relevant term in (15.13) is

$$\begin{aligned}
&\left|\left(-ig\frac{\boldsymbol{\tau}}{2}\cdot\mathbf{W}_\mu - i\frac{g'}{2}B_\mu\right)\phi\right|^2 \\
&= \frac{1}{8}\left|\begin{pmatrix} gW_\mu^3 + g'B_\mu & g\left(W_\mu^1 - iW_\mu^2\right) \\ g\left(W_\mu^1 + iW_\mu^2\right) & -gW_\mu^3 + g'B_\mu \end{pmatrix}\begin{pmatrix} 0 \\ v \end{pmatrix}\right|^2 \\
&= \tfrac{1}{8}v^2g^2\left[\left(W_\mu^1\right)^2 + \left(W_\mu^2\right)^2\right] + \tfrac{1}{8}v^2\left(g'B_\mu - gW_\mu^3\right)\left(g'B^\mu - gW^{3\mu}\right) \\
&= \left(\tfrac{1}{2}vg\right)^2 W_\mu^+ W^{-\mu} + \tfrac{1}{8}v^2\left(W_\mu^3, B_\mu\right)\begin{pmatrix} g^2 & -gg' \\ -gg' & g'^2 \end{pmatrix}\begin{pmatrix} W^{3\mu} \\ B^\mu \end{pmatrix},
\end{aligned} \tag{15.17}$$

since $W^{\pm} = (W^1 \mp iW^2)/\sqrt{2}$. Comparing the first term with the mass term expected for a charged boson, $M_W^2 W^+ W^-$, we have

$$\boxed{M_W = \tfrac{1}{2} v g.} \tag{15.18}$$

The remaining term is off-diagonal in the W_μ^3 and B_μ basis:

$$\tfrac{1}{8}v^2\left[g^2\left(W_\mu^3\right)^2 - 2gg'W_\mu^3 B^\mu + g'^2 B_\mu^2\right] = \tfrac{1}{8}v^2\left[gW_\mu^3 - g'B_\mu\right]^2 + 0\left[g'W_\mu^3 + gB_\mu\right]^2. \tag{15.19}$$

One of the eigenvalues of the 2×2 matrix in (15.17) is zero, and we have included this term in (15.19) with a combination of fields that is orthogonal to the combination given in the first term. Now, the physical fields Z_μ and A_μ diagonalize the mass matrix so that (15.19) must be identified with

$$\tfrac{1}{2}M_Z^2 Z_\mu^2 + \tfrac{1}{2}M_A^2 A_\mu^2,$$

where the $\frac{1}{2}$ is appropriate for a mass term of a neutral vector boson (see Exercise 14.4). So, on normalizing the fields, we have

$$A_\mu = \frac{g'W_\mu^3 + gB_\mu}{\sqrt{g^2 + g'^2}} \qquad \text{with } M_A = 0, \tag{15.20}$$

$$Z_\mu = \frac{gW_\mu^3 - g'B_\mu}{\sqrt{g^2 + g'^2}} \qquad \text{with } \boxed{M_Z = \tfrac{1}{2} v \sqrt{g^2 + g'^2}\,.} \tag{15.21}$$

We can reexpress these results in the notation introduced in Chapter 13. From either (13.23) or (15.9),

$$\frac{g'}{g} = \tan\theta_W. \tag{15.22}$$

In terms of θ_W, (15.20) and (15.21) therefore become

$$\begin{aligned} A_\mu &= \cos\theta_W B_\mu + \sin\theta_W W_\mu^3, \\ Z_\mu &= -\sin\theta_W B_\mu + \cos\theta_W W_\mu^3; \end{aligned} \tag{15.23}$$

and from (15.18) and (15.21), we have

$$\boxed{\frac{M_W}{M_Z} = \cos\theta_W.} \tag{15.24}$$

The inequality $M_Z \neq M_W$ is due to the mixing between the W_μ^3 and B_μ fields. The mass eigenstates are then automatically a massless photon (A_μ) and a massive Z_μ

field with $M_Z > M_W$. In the limit $\theta_W = 0$, we see that $M_Z = M_W$. The model was constructed with the requirement that the photon be massless, and so the result $M_A = 0$ is just a consistency check on our calculation and not a prediction. However, the result (15.24) for M_W/M_Z is a prediction of the standard model (with its Higgs doublet). More complicated choices in the Higgs sector will lead to different mass relations (see Exercise 15.4).

The Weinberg–Salam model with a Higgs doublet therefore fixes the parameter ρ of (13.36) to be

$$\rho \equiv \frac{M_W^2}{M_Z^2 \cos^2 \theta_W} = 1.$$

Recall that ρ is the parameter which specifies the relative strength of the neutral and charged current weak interactions and that the data require $\rho = 1$ to within a small error, see (13.54).

EXERCISE 15.2 Show that in the Weinberg–Salam model

$$\frac{1}{2v^2} = \frac{g^2}{8M_W^2} = \frac{G}{\sqrt{2}} \tag{15.25}$$

and hence, using the empirical value of G of Chapter 12, verify that $v = 246$ GeV. Derive the mass relations

$$M_W = \frac{37.3}{\sin \theta_W} \text{ GeV}, \qquad M_Z = \frac{74.6}{\sin 2\theta_W} \text{ GeV} \tag{15.26}$$

and give the lower bounds for their masses. Predict M_W and M_Z using the experimental determination of $\sin^2 \theta_W$.

Very recently (1983) the W and Z bosons have been discovered at the CERN $\bar{p}p$ collider via the processes

$$\bar{p}p \to W^{\pm}X \to (e^{\pm}\nu)X$$
$$\bar{p}p \to ZX \to (e^{+}e^{-})X,$$

where X denotes all the other particles produced in the high-energy head-on collision. By studying the momentum distribution of the emitted decay electrons and positrons, the masses are measured to be

$$M_W = 81 \pm 2 \text{ GeV}$$
$$M_Z = 93 \pm 2 \text{ GeV},$$

which are in impressive agreement with the predictions of the standard electroweak model.

EXERCISE 15.3 Suppose that the Higgs scalar field $\phi(x)$ has weak isospin $T = 3$ and hypercharge $Y = -4$. If the neutral component ϕ^0 (with

$T^3 = 2$) develops a vacuum expectation value $v/\sqrt{2}$, show that

$$M_W^2 = \frac{g^2}{2}\phi^\dagger(T^+T^- + T^-T^+)\phi$$

$$= 4g^2v^2. \tag{15.27}$$

Further show that, as it happens, $M_W/M_Z = \cos\theta_W$, just as in the standard model. (However, we would still need a Higgs isodoublet to generate the fermion masses, see Section 15.4.)

EXERCISE 15.4 Suppose that there exist several representations ($i = 1,\ldots,N$) of Higgs scalars whose charge-zero members acquire vacuum expectation values v_i. Show that

$$\rho \equiv \left(\frac{M_W}{M_Z\cos\theta_W}\right)^2 = \frac{\sum v_i^2\left[T_i(T_i+1) - \frac{1}{4}Y_i^2\right]}{\sum \frac{1}{2}v_i^2 Y_i^2}, \tag{15.28}$$

where T_i and Y_i are, respectively, the weak isospin and hypercharge of representation i. Show that $\rho = 1$ if only Higgs doublets with $Y_i = +1$ exist.

Hint Use (15.27) and note that the neutral scalars have $T^3 = -Y/2$.

EXERCISE 15.5 The Lagrangian for the scalar field, (15.13), contains trilinear hW^+W^- and quadrilinear hhW^+W^- Higgs boson couplings. Use

$$\phi = \sqrt{\tfrac{1}{2}}\begin{pmatrix} 0 \\ v + h(x) \end{pmatrix} \tag{15.29}$$

[see (14.71)] to show that in the standard model the vertex factors are

$$igM_W \quad \text{and} \quad \tfrac{1}{4}ig^2, \tag{15.30}$$

respectively. Determine the hZZ and hhZZ vertex factors.

15.4 Masses of the Fermions

Recall that in the original Lagrangian, (15.10), a fermion mass term $-m\bar{\psi}\psi$ was excluded by gauge invariance. An attractive feature of the standard model is that the same Higgs doublet which generates $W^\pm$ and Z masses is also sufficient to give masses to the leptons and quarks. For example, to generate the electron mass, we include the following $SU(2)\times U(1)$ gauge invariant term in the

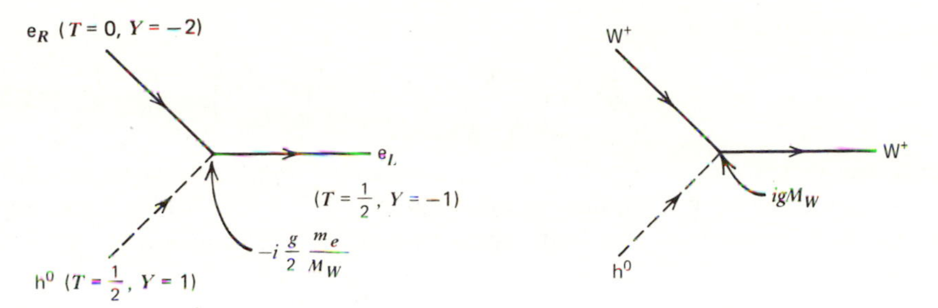

Fig. 15.1 The Higgs scalar h^0 coupling to the electron and W boson in the standard model.

Lagrangian:

$$\mathcal{L}_3 = -G_e\left[(\bar{\nu}_e, \bar{e})_L\begin{pmatrix}\phi^+\\ \phi^0\end{pmatrix}e_R + \bar{e}_R(\phi^-, \bar{\phi}^0)\begin{pmatrix}\nu_e\\ e\end{pmatrix}_L\right]. \tag{15.31}$$

The Higgs doublet has exactly the required $SU(2) \times U(1)$ quantum numbers to couple to $\bar{e}_L e_R$, see Fig. 15.1. We spontaneously break the symmetry and substitute

$$\phi = \sqrt{\tfrac{1}{2}}\begin{pmatrix}0\\ v + h(x)\end{pmatrix}$$

into (15.31). The neutral Higgs field $h(x)$ is the only remnant of the Higgs doublet, (15.14), after the spontaneous symmetry breaking has taken place. The other three fields can be gauged away, see (14.72). On substitution of ϕ, the Lagrangian becomes

$$\mathcal{L}_3 = -\frac{G_e}{\sqrt{2}}v(\bar{e}_L e_R + \bar{e}_R e_L) - \frac{G_e}{\sqrt{2}}(\bar{e}_L e_R + \bar{e}_R e_L)h.$$

We now choose G_e so that

$$m_e = \frac{G_e v}{\sqrt{2}} \tag{15.32}$$

and hence generate the required electron mass,

$$\mathcal{L}_3 = -m_e\bar{e}e - \frac{m_e}{v}\bar{e}eh, \tag{15.33}$$

using (15.12). Note however that, since G_e is arbitrary, the actual mass of the electron is not predicted. Besides the mass term, the Lagrangian contains an interaction term coupling the Higgs scalar to the electron. However, since $v = 246$ GeV, the coupling m_e/v is very small and so far has not produced a detectable effect in electroweak interactions. The he^+e^- vertex factor is shown in Fig. 15.1 together with (15.30) for the much stronger hW^+W^- coupling.

The quark masses are generated in the same way. The only novel feature is that to generate a mass for the upper member of a quark doublet, we must construct a new Higgs doublet from ϕ:

$$\phi_c = -i\tau_2\phi^* = \begin{pmatrix} -\bar{\phi}^0 \\ \phi^- \end{pmatrix} \xrightarrow[\text{breaking}]{} \sqrt{\tfrac{1}{2}}\begin{pmatrix} v+h \\ 0 \end{pmatrix}. \tag{15.34}$$

Due to the special properties of $SU(2)$, ϕ_c transforms identically to ϕ, see (2.41), (but has opposite weak hypercharge to ϕ, namely, $Y = -1$). It can therefore be used to construct a gauge invariant contribution to the Lagrangian,

$$\begin{aligned} \mathcal{L}_4 &= -G_d(\bar{u}, \bar{d}\,)_L \begin{pmatrix} \phi^+ \\ \phi^0 \end{pmatrix} d_R - G_u(\bar{u}, \bar{d}\,)_L \begin{pmatrix} -\bar{\phi}^0 \\ \phi^- \end{pmatrix} u_R + \text{hermitian conjugate} \\ &= -m_d\bar{d}d - m_u\bar{u}u - \frac{m_d}{v}\bar{d}dh - \frac{m_u}{v}\bar{u}uh. \end{aligned} \tag{15.35}$$

Here, we have just considered the $(u, d)_L$ quark doublet. However, weak interactions operate on $(u, d')_L, (c, s')_L, \dots$ doublets, where the primed states are linear combinations of the flavor eigenstates, see Sections 12.11 and 12.12. Using the notation of Section 12.12, the quark Lagrangian is therefore of the form

$$\mathcal{L}_4 = -G_d^{ij}(\bar{u}_i, \bar{d}_i')_L \begin{pmatrix} \phi^+ \\ \phi^0 \end{pmatrix} d_{jR} - G_u^{ij}(\bar{u}_i, \bar{d}_i')_L \begin{pmatrix} -\bar{\phi}^0 \\ \phi^- \end{pmatrix} u_{jR} + \text{hermitian conjugate}, \tag{15.36}$$

with $i, j = 1,\dots,N$, where N is the number of quark doublets. Proceeding as in Section 12.12, we can rewrite the quark Lagrangian in diagonal form:

$$\mathcal{L}_4 = -m_d^i\bar{d}_i d_i\left(1 + \frac{h}{v}\right) - m_u^i\bar{u}_i u_i\left(1 + \frac{h}{v}\right). \tag{15.37}$$

Again, the masses depend on the arbitrary couplings $G_{u,d}$ and cannot be predicted. This has the desirable consequence that the Higgs coupling is flavor conserving and will give no contribution to processes such as $K_L \to \mu^+\mu^-$.

Notwithstanding the attractive features above, it is fair to say that the Higgs sector is the least satisfactory and least well-determined aspect of electroweak gauge theory. The minimal choice of a single Higgs doublet is sufficient to generate the masses both of the gauge bosons and of the fermions, but the masses of the fermions are just parameters of the theory and are not predicted; their empirical values must be input. However, on the positive side, the Higgs coupling to the fermions is proportional to their masses, a prediction which could be tested when, and if, the Higgs particle is actually observed. A second deficiency is that the mass m_h of the neutral Higgs meson itself is not predicted either. Using the first two terms of the effective potential,

$$V(\phi) = \mu^2\phi^\dagger\phi + \lambda(\phi^\dagger\phi)^2 + \cdots, \tag{15.38}$$

we find

$$m_h^2 = 2v^2\lambda. \tag{15.39}$$

EXERCISE 15.6 Derive (15.39), see (14.58).

Since v is fixed, larger values of m_h correspond to large λ. For a meaningful perturbation expansion, it turns out that m_h must be smaller than a few hundred GeV. On the other hand, corrections to (15.38) from loop diagrams will wash out the nontrivial minimum at $v \neq 0$ if m_h is too small. Theoretical limits indicate that $m_h \gtrsim 10$ GeV. Experimental confirmation of the existence of this particle is of course eagerly awaited.

The Higgs is a difficult particle to discover precisely because of its characteristic property of coupling to fermions in proportion to their mass. The most readily experimentally accessible particles are light fermions (electrons and u, d quarks in protons and neutrons), and they couple to the Higgs particle only very weakly. The heavier fermions (τ, c, b, t) couple more readily to the Higgs but are difficult to produce themselves.

15.5 The Standard Model: The Final Lagrangian

To summarize the standard (Weinberg–Salam) model, we gather together all the ingredients of the Lagrangian. The complete Lagrangian is:

$$\mathcal{L} = -\tfrac{1}{4}\mathbf{W}_{\mu\nu}\cdot\mathbf{W}^{\mu\nu} - \tfrac{1}{4}B_{\mu\nu}B^{\mu\nu} \qquad \left\{\begin{array}{l}\mathrm{W}^{\pm}, \mathrm{Z}, \gamma \text{ kinetic} \\ \text{energies and} \\ \text{self-interactions}\end{array}\right.$$

$$+\bar{L}\gamma^{\mu}\left(i\partial_{\mu} - g\frac{1}{2}\boldsymbol{\tau}\cdot\mathbf{W}_{\mu} - g'\frac{Y}{2}B_{\mu}\right)L$$

$$+\bar{R}\gamma^{\mu}\left(i\partial_{\mu} - g'\frac{Y}{2}B_{\mu}\right)R \qquad \left\{\begin{array}{l}\text{lepton and quark} \\ \text{kinetic energies} \\ \text{and their} \\ \text{interactions with} \\ \mathrm{W}^{\pm}, \mathrm{Z}, \gamma\end{array}\right. \qquad (15.40)$$

$$+\left|\left(i\partial_{\mu} - g\frac{1}{2}\boldsymbol{\tau}\cdot\mathbf{W}_{\mu} - g'\frac{Y}{2}B_{\mu}\right)\phi\right|^{2} - V(\phi) \qquad \left\{\begin{array}{l}\mathrm{W}^{\pm}, \mathrm{Z}, \gamma, \text{ and Higgs} \\ \text{masses and} \\ \text{couplings}\end{array}\right.$$

$$-(G_1\bar{L}\phi R + G_2\bar{L}\phi_c R + \text{hermitian conjugate}). \qquad \left\{\begin{array}{l}\text{lepton and quark} \\ \text{masses and} \\ \text{coupling to Higgs}\end{array}\right.$$

L denotes a left-handed fermion (lepton or quark) doublet, and R denotes a right-handed fermion singlet.

15.6 Electroweak Theory is Renormalizable

Future experiments will decide whether the simplest Higgs model is correct. However, remember that the motivation for introducing the Higgs scalar was entirely theoretical, and there is at present no evidence that this particle actually exists. Its importance arises from the fact that it allows us to generate the masses of the weak bosons without spoiling the renormalizability of the electroweak gauge theory, see Section 14.5. The renormalizability of the theory is not trivial. It was eventually demonstrated by 't Hooft some four years after the model was proposed.

The structure of the lowest-order amplitudes for weak processes hints that we may have trouble. Consider the cross section of (12.60),

$$\sigma(\nu_e e \to \nu_e e) = \frac{G^2 s}{\pi}. \tag{15.41}$$

It becomes infinite as $s \to \infty$. A careful discussion, see, for example, Bjorken and Drell (1964) or Pilkuhn (1979), shows that this behavior cannot be reconciled with probability conservation. The introduction of a finite-mass W boson removes the divergence, and for large s it can be shown that

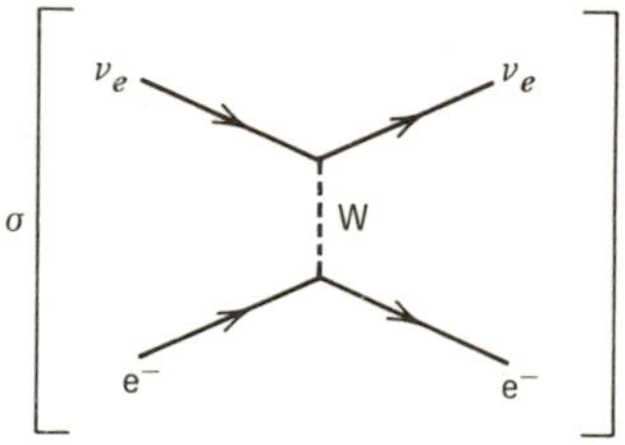

$$\sigma\left[\quad\right] = \frac{G^2 M_W^2}{\pi}. \tag{15.42}$$

However, the introduction of the W boson causes its own problems, for now we must consider

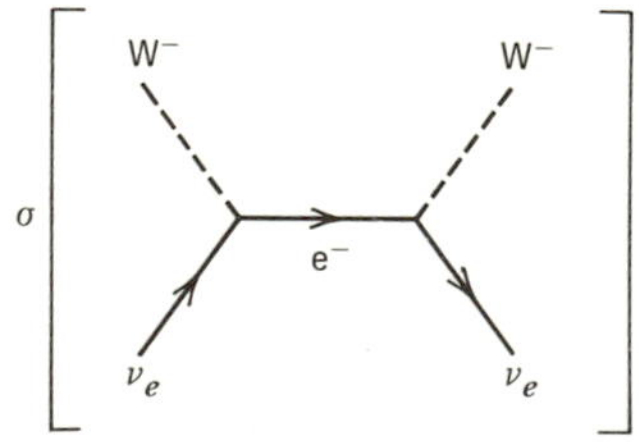

$$\sigma\left[\quad\right] = \frac{G^2 s}{3\pi}, \tag{15.43}$$

which similarly diverges at large s. It is here that we glimpse the beautiful structure of gauge theory. $\nu_e W^-$ scattering can proceed through a second diagram, Fig. 15.2, and in fact the standard model neutral current couplings are just such that this contribution cancels the divergence of the first diagram. We may be tempted to conclude that we are required to introduce the neutral current! This is

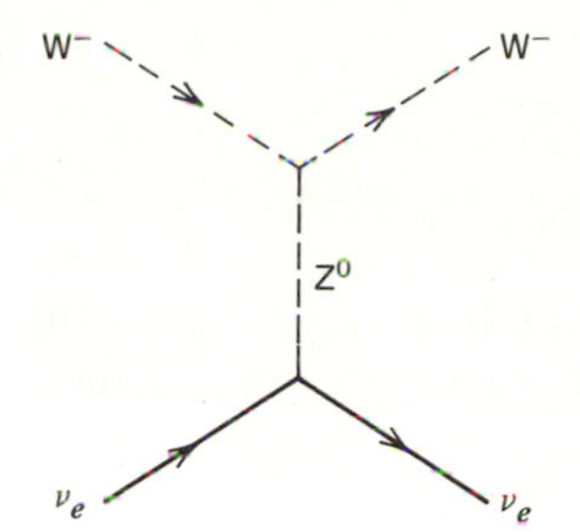

Fig. 15.2 The neutral current contribution to $\nu_e W^-$ scattering.

not quite true. We could in principle have introduced a new, heavy electron, thus providing us with a second diagram of the form (15.43) exchanging the new lepton. Then, a cancellation of the divergence between the two diagrams can be arranged. This is reminiscent of the introduction of a second u quark (the c quark) in the GIM mechanism. However, unlike GIM, the heavy-lepton option is not favored by experiment.

The process $e^-e^+ \to W^-W^+$ is another example where the self-coupling of gauge bosons ensures a finite answer. The individual diagrams of Fig. 15.3 diverge, but the sum is finite.

As a final example, we consider the scattering of charged W bosons,

Direct computation reveals that the individual diagrams diverge as s^2/M_W^4, but

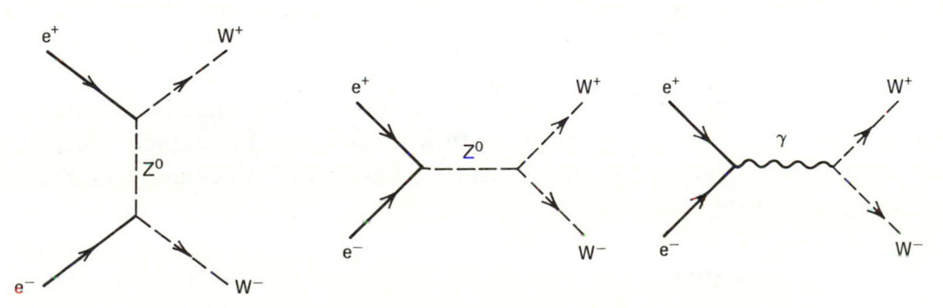

Fig. 15.3 Three contributions to the process $e^-e^+ \to W^-W^+$.

the divergence of the sum is more gentle:

$$-i\mathfrak{M}(\mathrm{W}^+\mathrm{W}^- \to \mathrm{W}^+\mathrm{W}^-) \sim \frac{s}{M_W^2} \qquad \text{as } s \to \infty .$$

Even after introducing similar diagrams with Z^0 exchange, the sum of all diagrams still diverges as s/M_W^2. Heavy leptons cannot help us, so the only solution is to introduce a scalar particle which cancels these residual divergences through diagrams of the type

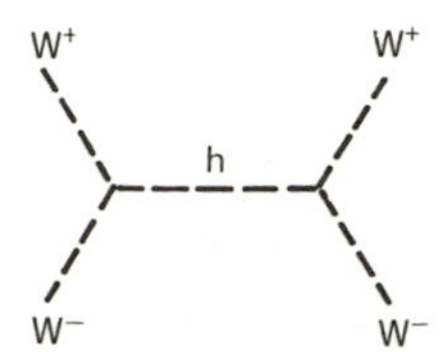

Here, h is just the Higgs particle. If we had not previously introduced it to generate the heavy boson masses, we would have been forced to invent it now to guarantee renormalizability. A detailed investigation of this point would reveal that the Higgs couplings are proportional to masses, a result we are also familiar with. Finally, we reiterate that Higgs particles have so far eluded experimental searches. This may be because we have not used the right experimental probes or because they are too heavy to be produced by existing accelerators. Alternatively, one might speculate that they do not exist as elementary fields, and the Higgs "particle" we have introduced actually corresponds to a more complex object, for example, a bound state of other particles which have yet to be incorporated in the theory.

The interested student is encouraged to perform this and previous calculations explicitly; for assistance, see, for example, Llewellyn Smith (1974), Abers and Lee (1973), and Bernstein (1974).

It is clear from the few examples mentioned that the various couplings must be closely related to ensure the "conspiratorial" cancellations of various divergences. Any electroweak theory must therefore meet the dual requirements that (1) it reproduces the phenomenology of Chapters 12 and 13, and (2) it has the correct relation between various couplings to ensure a desirable high-energy behavior. The Weinberg–Salam model (15.40) is the simplest model that achieves these ends.

In fact, 't Hooft has shown that for a theory to be renormalizable, it must be a Yang–Mills theory, that is, a theory with a local gauge invariance. Only if we have such a high degree of symmetry can we obtain the systematic cancellations of divergences order by order.

15.7 Grand Unification

The electroweak $SU(2) \times U(1)$ gauge theory is in impressive agreement with experiment. However, does it really unify the electromagnetic and weak interactions? The $SU(2) \times U(1)$ gauge group is a product of two disconnected sets of gauge transformations: the $SU(2)$ group with coupling strength g and the $U(1)$ group with strength g'. Therefore, these two couplings are not related by the theory; and, as we know from Chapter 13, their ratio

$$\frac{g'}{g} = \tan\theta_W \tag{15.44}$$

has to be measured experimentally.

Only if the $SU(2)$ and $U(1)$ gauge transformations are embedded into a larger set of transformations G can g and g' be related by the gauge theory. Symbolically, we write

$$G \supset SU(2) \times U(1), \tag{15.45}$$

and some of the new transformations in the group G will link the previously disconnected $SU(2)$ and $U(1)$ subsets of gauge transformations. So, g and g' are related by a number (actually a Clebsch–Gordan coefficient of G) whose value depends on the choice of "unifying" group G. In the quest for the "ultimate" theory, it seems natural to attempt to unify strong interactions with the electroweak $SU(2) \times U(1)$ interaction. That is, we seek a group G that also contains the color gauge group $SU(3)$ which successfully describes the strong interactions. Let us speculate that such a "grand unified group" G exists. Then (15.45) would be generalized to some grand unified group

$$G \supset SU(3) \times SU(2) \times U(1), \tag{15.46}$$

and gauge transformations in G also relate the electroweak couplings g, g' to the color coupling α_s. All the interactions would then be described by a grand unified gauge theory (GUT) with a single coupling g_G to which all couplings are related in a specific way once the gauge group G has been found.

This unification is pictured in Fig. 15.4, where we have denoted the couplings associated with the $SU(3)$, $SU(2)$, and $U(1)$ subgroups by $g_i(Q)$ with $i = 3, 2, 1$, respectively. The figure recalls the fact that the gauge couplings depend on the characteristic momentum Q (or distance $1/Q$) of the interactions. The couplings $g_2(Q)$ and $g_3(Q)$ of the non-Abelian groups are asymptotically free, whereas the Abelian coupling $g_1(Q)$ increases with increasing momentum Q similar to the conventional charge screening of electromagnetism. The figure suggests that for some large-momentum (or short-distance) scale $Q = M_X$, the three couplings merge into a single grand unified coupling g_G; that is, for $Q \geq M_X$,

$$g_i(Q) = g_G(Q), \tag{15.47}$$

and the group G describes a unified interaction with coupling $g_G(Q)$. When Q is decreased below M_X, the couplings $g_i(Q)$ separate and eventually give the phenomenological couplings g, g', and α_s, which describe the interactions observed in the present-day experiments for which $Q \simeq \mu \simeq 10$ GeV or thereabouts.

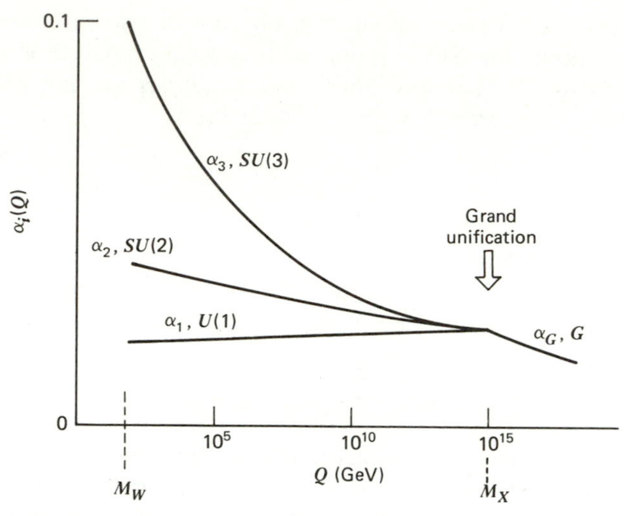

Fig. 15.4 The variation of $\alpha_i \equiv g_i^2/4\pi$ with Q, showing the speculative grand unification of strong $[SU(3)_{\text{color}}]$ and electroweak $[SU(2)_L \times U(1)_Y]$ interactions at very short distances $1/Q \approx 1/M_X$.

With the conventional choice of couplings, we have

$$\begin{aligned} \alpha_s(Q) &= \frac{g_3^2(Q)}{4\pi}, \\ g(Q) &= g_2(Q), \\ g'(Q) &= \frac{1}{C} g_1(Q), \end{aligned} \tag{15.48}$$

where the constant C is a Clebsch–Gordan coefficient of G. Equation (15.44) becomes, in terms of g_1 and g_2,

$$\frac{1}{C}\frac{g_1(Q)}{g_2(Q)} = \tan\theta_W(Q). \tag{15.49}$$

So, for $Q \gtrsim M_X$, where $g_1 = g_2$, C determines the Weinberg angle θ_W.

Assuming that there exists a group G, we can use the phenomenological values of the couplings at $Q \simeq \mu$ to estimate the unification mass M_X. This can be done because the Q dependence of the couplings $g_i(Q)$ is prescribed by the gauge theory, see Chapter 7. For instance, the Q dependence of α_s is given by (7.63). We use (15.48) to replace α_s by g_3, and after simple rearrangement, (7.63) becomes

$$\frac{1}{g_3^2(\mu)} = \frac{1}{g_3^2(Q)} + 2b_3 \log\frac{Q}{\mu}, \tag{15.50}$$

with

$$b_3 = \frac{1}{(4\pi)^2}\left(\frac{2}{3}n_f - 11\right), \tag{15.51}$$

where n_f is the number of quark flavors. For $Q = M_X$, we have $g_3 = g_G$; and so (15.50) gives

$$\boxed{\frac{1}{g_i^2(\mu)} = \frac{1}{g_G^2} + 2b_i \log\frac{M_X}{\mu}}, \tag{15.52}$$

with $i = 3$.

This equation applies equally well to the $SU(2)$ and $U(1)$ couplings, see (15.47). The different routes, pictured in Fig. 15.4, by which the three couplings $g_i(Q)$ reach g_G are due to the different b_i coefficients in (15.52). We have

$$b_1 = \frac{1}{(4\pi)^2}\left(\frac{4}{3}n_g\right), \tag{15.53a}$$

$$b_2 = \frac{1}{(4\pi)^2}\left(-\frac{22}{3}\right) + b_1, \tag{15.53b}$$

$$b_3 = \frac{1}{(4\pi)^2}(-11) + b_1, \tag{15.53c}$$

where n_g is the number of families (or generations) of fermions, that is, N of (12.118). Here, b_3 is just (15.51). In fact, for $SU(N)$,

$$b_N = \frac{1}{(4\pi)^2}\left[-\frac{11}{3}N + \frac{4}{3}n_g\right], \tag{15.54}$$

where the two terms correspond to the gauge boson and fermion loops, respectively, see (7.58). For further discussion and references, see Langacker (1981).

Given the Q dependence of the couplings, it is straightforward to estimate M_X. We eliminate n_g and g_G from the three equations (15.52) with $i = 1, 2, 3$. Using (15.53), we form the particular linear combination

$$\frac{C^2}{g_1^2} + \frac{1}{g_2^2} - \frac{(1 + C^2)}{g_3^2} = 2\left[C^2 b_1 + b_2 - (1 + C^2)b_3\right]\log\frac{M_X}{\mu}, \tag{15.55}$$

where $g_i^2 = g_i^2(\mu)$. The left-hand side has been chosen so that it can be expressed in terms of e^2 and g_3^2, or equivalently α and α_s. Indeed,

$$\frac{C^2}{g_1^2} + \frac{1}{g_2^2} = \frac{1}{g'^2} + \frac{1}{g^2} = \frac{1}{e^2}, \tag{15.56}$$

using (15.48) and (13.23). Inserting the explicit expressions for the b_i coefficients,

(15.53), into (15.55), we therefore have

$$\log \frac{M_X}{\mu} = \frac{3(4\pi)^2}{22(1+3C^2)}\left[\frac{1}{e^2} - (1+C^2)\frac{1}{g_3^2}\right]$$
$$= \frac{6\pi}{11(1+3C^2)}\left(\frac{1}{\alpha} - \frac{1+C^2}{\alpha_s}\right). \tag{15.57}$$

For $\mu \simeq 10$ GeV, we know that $\alpha \simeq \frac{1}{137}$ and $\alpha_s \simeq 0.1$, and we take $C^2 = \frac{5}{3}$ for reasons that will become apparent in the next section. So,

$$\boxed{M_X \simeq 5 \times 10^{14}\ \text{GeV}.} \tag{15.58}$$

Strictly speaking, we should use $\mu \sim M_W$, but the couplings are slowly varying and the order-of-magnitude estimates are not particularly sensitive to the "ordinary" mass chosen for μ, nor to the precise value of the Clebsch–Gordan coefficient C. The dependence of M_X on the value of α_s is shown in Table 15.1.

The unification scale M_X is very large. Can we still neglect the gravitational interaction? Recall that the gravitational force between two particles (Gm_1m_2/r^2) can be significant if the separation r becomes sufficiently small. Here, G is the gravitational constant. Gravitational effects become of order unity when the masses $(m_1 \simeq m_2 \simeq M)$ are such that the potential is comparable to the rest mass energy

$$\frac{GM^2/r}{Mc^2} \simeq 1.$$

For masses separated by a natural unit of length, $r = \hbar/Mc$ [see (1.3)], this happens when

$$Mc^2 = \left(\frac{\hbar c^5}{G}\right)^{1/2} = 1.22 \times 10^{19}\ \text{GeV}.$$

This value of M is known as the Planck mass, and it is the only dimensional quantity appearing in gravity. Comparing its value with (15.58), we see that it may still be a good approximation to neglect gravity, but it is intriguing that we are now contemplating mass scales so close to the Planck mass. For completeness,

TABLE 15.1
Values for M_X, $\sin^2\theta_W$, and the Proton Lifetime τ_p Evaluated from the "Order-of-Magnitude" Estimates (15.57), (15.60), and (15.66) Following Georgi, Quinn, and Weinberg

α_s	M_X (GeV)	$\sin^2\theta_W$	τ_p (years)
0.1	5×10^{14}	0.21	$\sim 10^{27}$
0.2	2×10^{16}	0.19	$\sim 10^{34}$

we have included $\hbar$ and c in the above discussion. From now on, we revert back to natural units with $\hbar = c = 1$.

The Weinberg angle θ_W is determined in a grand unified theory, and so we can compare the value with experiment. Indeed, from (15.49), we have

$$\sin^2 \theta_W = \frac{g_1^2(Q)}{g_1^2(Q) + C^2 g_2^2(Q)}. \tag{15.59}$$

Thus, if we take $C^2 = \frac{5}{3}$, we have $\sin^2 \theta_W = \frac{3}{8}$ at $Q = M_X$, where $g_1 = g_2$. However, at $Q \simeq \mu$, the value will be different because of the separation of g_1 and g_2 as Q decreases from M_X (see Fig. 15.4).

EXERCISE 15.7 Show that

$$\sin^2 \theta_W = \frac{1}{1 + 3C^2}\left(1 + 2C^2 \frac{\alpha}{\alpha_s}\right). \tag{15.60}$$

Hint By forming the combination $C^2(1/g_1^2 - 1/g_2^2)$, verify that

$$2C^2(b_1 - b_2)\log \frac{M_X}{\mu} = \frac{1}{e^2}\left(1 - (1 + C^2)\sin^2 \theta_W\right).$$

Then, use (15.54) and (15.53).

From (15.60), we can calculate $\sin^2 \theta_W$ at $Q \simeq 10$ GeV, using $C^2 = \frac{5}{3}$ as before. The results are entered in Table 15.1. We see that $\sin^2 \theta_W \simeq 0.2$. This is close to the experimental value discussed in Chapter 13.

15.8 Can the Proton Decay?

In Section 15.7, we speculated that a grand unified group G exists such that

$$G \supset SU(3) \times SU(2) \times U(1).$$

Georgi and Glashow have shown that the smallest such group of gauge transformations is the group $SU(5)$. Of course, different GUT's with larger groups than $SU(5)$ can be constructed.

Once a group is chosen for investigation, we have to assign the quarks and leptons to multiplets (i.e., irreducible representations) of the group. In the earlier chapters, we presented empirical evidence for distinct families (or generations) of fermions, (u, d; ν_e, e), (c, s; ν_μ, μ), ..., where in the first family, for instance, we have

$$\begin{aligned} &\left.\begin{pmatrix} \mathrm{u} \\ \mathrm{d} \end{pmatrix}_L, \mathrm{u}_R, \mathrm{d}_R \right\} \quad \text{each with three colors,} \\ &\begin{pmatrix} \nu_e \\ \mathrm{e}^- \end{pmatrix}_L, \mathrm{e}_R^-, \end{aligned} \tag{15.61}$$

together with their antiparticles. This grouping ensures that we can construct gauge theories free of anomalies, see Section 12.12. In a family, there are thus 15 left-handed states, for example, those in (15.61) together with $\bar{u}_L$, $\bar{d}_L$, and e^+_L. For the $SU(5)$ model, these can be accommodated in a fundamental $\bar{5}$- and a 10-representation (the 10 is the antisymmetric part of the product of two fundamental 5-representations, compare (2.58)). Explicitly, we have for the left-handed states,

$$\bar{5} = (1,2) + (\bar{3},1) = (\nu_e, e^-)_L + \bar{d}_L, \tag{15.62}$$

$$10 = (1,1) + (\bar{3},1) + (3,2) = e^+_L + \bar{u}_L + (u,d)_L, \tag{15.63}$$

where we have shown the $(SU(3)_{\text{color}}, SU(2)_L)$ decomposition of the multiplets.

What are the gauge bosons of $SU(5)$? An $SU(N)$ gauge theory has $N^2 - 1$ gauge bosons. For the $SU(5)$ model, these are

$$24 = \underset{\text{gluons}}{(8,1)} + \underbrace{(1,3) + (1,1)}_{W^{\pm}, Z, \gamma} + \underbrace{(3,2) + (\bar{3},2)}_{X, Y \text{ bosons}} \tag{15.64}$$

So, we have a new pair of superheavy gauge bosons, X and Y. They form a weak doublet and are colored. They mediate interactions which turn quarks into leptons:

$$(u,d)_L \to e^+_L + (\bar{Y}, \bar{X}), \tag{15.65}$$

or, in $SU(5)$ parlance,

$$(3,2) \to (1,1) \otimes (3,2).$$

Is the appearance of such transitions really surprising? First, recall that at energies above M_W, the distinction between weak and electromagnetic interactions disappears. Similarly, at the GUT-scale $M_{X,Y}$, which we identify with (15.58), the strong color force merges with the electroweak force, and the sharp separation of particles into colored quarks and colorless leptons, which interact only through the electroweak force, disappears. This leads to lepton/baryon number-violating interactions such as (15.65).

EXERCISE 15.8 What are the charges of the superheavy bosons X and Y?

EXERCISE 15.9 Comment on the behavior of $g_G(Q)$ for $Q > M_X$, see Fig. 15.4.

Can we ever hope to build accelerators to observe, via (15.65), the production of X particles weighing about 10^{-10} grams [see (15.58)]? This is most unlikely. However, low-Q effects associated with the weak gauge bosons ($W^{\pm}$, Z) were revealed long before accelerator technology achieved interactions with $Q \simeq M_W$. An example is μ-decay. The slow decay of the muon is a direct result of the large

TABLE 15.2
Low-Q^2 Phenomena Associated with the Scales $Q^2 = M_W^2$ and $Q^2 = M_X^2$

Muon Decay ($\mu^- \to e^- \bar{\nu}_e \nu_\mu$) at $Q^2 \ll M_W^2$	Proton Decay ($p \to \pi^0 e^+$) at $Q^2 \ll M_X^2$
$\frac{G}{\sqrt{2}} = \frac{g^2}{8M_W^2}$ (12.15)	$\frac{G_G}{\sqrt{2}} = \frac{g_G^2}{8M_X^2}$
$\Gamma(\mu \to e\bar{\nu}_e\nu_\mu) = \cdots G^2 m_\mu^5$ (12.42) $= \cdots \frac{m_\mu^5}{M_W^4}$	$\Gamma(p \to \pi e) = \cdots G_G^2 m_p^5 = \cdots \frac{m_p^5}{M_X^4}$

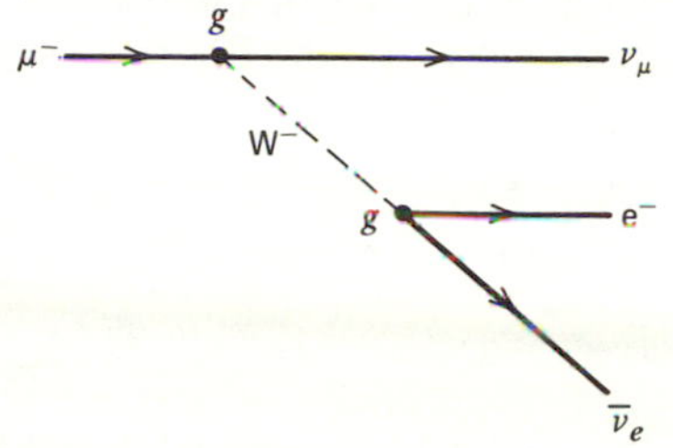

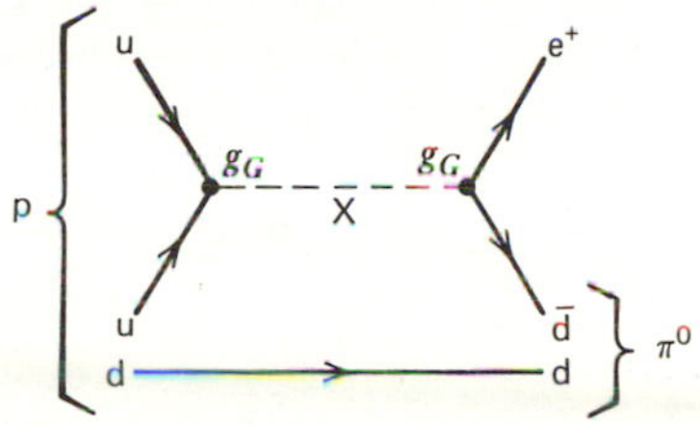

mass of the W. This argument, presented in detail in Chapter 12, is sketched again in Table 15.2. Using this analogy, we can infer low-Q manifestations of interactions with scale M_X. Such interactions will result in the (very) slow decay of the proton, see Table 15.2. The decay rate can be estimated in just the same way as the μ-decay rate. In the table, we find a proton lifetime of

$$\tau_p \simeq \frac{M_X^4}{m_p^5}, \tag{15.66}$$

and so its numerical value is very sensitive to the precise value of M_X (see Table 15.1). The estimated lifetime is in the vicinity of the present experimental limits, which are of order 10^{30} years. An accurate prediction of the lifetime requires calculations which are much more sophisticated than the order-of-magnitude estimates for M_X and τ_p presented above; see, for example, Langacker (1981). Such calculations give some hope that the $SU(5)$ GUT could be tested by experiments that detect the decay of the proton. In the meantime, the value of $\sin^2\theta_W \simeq 0.2$ computed from (15.60) with $C^2 = \frac{5}{3}$ (the $SU(5)$ value of the

Clebsch–Gordan coefficient relating g and g') is the main empirical evidence for this "great house of cards."

The GUT approach, however, offers a framework that permits the scientific pursuit of problems and goals that were until recently only idle dreams. Note for example the implications of $SU(5)$ for the electric charge of quarks and leptons. Since the photon is one of the gauge bosons of $SU(5)$, the charge operator Q is a generator of the group. But for a simple group, such as $SU(5)$, the trace of each generator vanishes for any representation. For example, for the $\overline{5}$ representation, (51.62), we have

$$\operatorname{Tr} Q = 3Q_{\bar{d}} + Q_{\nu} + Q_{e^-} = 0,$$

which implies that

$$Q_d = \tfrac{1}{3} Q_{e^-}.$$

This is an amazing result. Clearly, it implies that charges are quantized. A similar calculation for the 10 representation, (15.63), yields

$$Q_u = -2Q_d,$$

and the combined result resolves the mystery (Chapter 7) of why $Q_p = -Q_e$. Moreover, classifications like (15.62) and (15.63) also imply that the helicity structures of the weak interactions of quarks and leptons are very similar, an experimental fact that we repeatedly emphasized.

Clearly, proton decay would have profound implications, but one could argue that it would not be all that surprising. The conservation of charge is connected with the existence of a massless photon, but no massless particles are associated with the conservation of baryon number. Indeed, its nonconservation may already be apparent in the universe surrounding us. We discuss this next.

15.9 The Early Universe as a High-Energy Physics Experiment

Whereas the highest energies achieved with accelerators do not exceed 10^5 GeV, energies of 10^{19} GeV may have been commonplace at the very early times after the Big Bang. If we extrapolate our expanding universe backward in time toward the initial singularity in our past associated with the Big Bang, the matter and radiation get hotter, and eventually the average particle collision energies significantly exceed those which can be achieved in accelerator laboratories.

A speculative chronology of past events is shown in Table 15.3. A famous "tell-tale" event occurred when the universe was 10^6 years old. Then, energies had cooled to a fraction of an eV and photons were no longer sufficiently energetic to excite atoms. The photon radiation decoupled from matter and is still with us today in the form of the isotropic background radiation. Since decoupling, it has cooled from 10^3 K to 3 K through the expansion of the universe, see Table 15.3. The detection of this 3 K (microwave) background radiation in 1965 by Penzias and Wilson was crucial evidence for the expanding (or Big Bang) model of the universe.

TABLE 15.3
A Summary of Past Events in Standard Big Bang Cosmology

Past Events	Time	Temperature T (degrees Kelvin)	Energy kT (GeV)
Quantum gravity effects are large	10^{-45} sec	10^{32}	10^{19}
Matter–antimatter asymmetry through X–boson interactions	10^{-35} sec	10^{27}	10^{14}
Helium abundance established	10^{3} sec	10^{9}	10^{-4}
γ's decouple from matter, origin of photon background	10^{6} years	10^{3}	10^{-10}
NOW!	10^{10} years	3	10^{-12}

Making a bold extrapolation back past primordial "nucleosynthesis" at 10^3 sec, when such nuclei as helium and deuterium are believed to have been formed, all the way to 10^{-35} sec, we reach a phase of the universe where matter exists in the form of quarks and leptons in thermal equilibrium with the gauge bosons of (15.64). The thermal energies are of order of the GUT-scale M_X. But when the universe cools below this temperature, the superheavy X and $\overline{\text{X}}$ particles decay; and if C and CP are violated, their partial decay rates may differ,

$$\frac{\Gamma(\text{X} \to \text{q} + \text{q})}{\Gamma(\text{X} \to \text{all})} \neq \frac{\Gamma(\overline{\text{X}} \to \bar{\text{q}} + \bar{\text{q}})}{\Gamma(\overline{\text{X}} \to \text{all})}.$$

Thus, although we have equal densities of X and $\overline{\text{X}}$ before 10^{-35} sec, we could emerge from this equilibrium period with unequal numbers of q and $\bar{\text{q}}$ quarks. It is difficult to make quantitative predictions of the baryon asymmetry as the most important CP-violating processes involve the interaction and decay of the Higgs particles. Nevertheless, a GUT has the necessary ingredients so that we could emerge from the GUT era in Table 15.3 with a small net excess of matter over antimatter. The excess would survive subsequent matter–antimatter annihilation into photons, producing the observed ratio of baryons to photons:

$$\frac{N_B}{N_\gamma} \simeq \frac{N_B - N_{\bar{B}}}{N_B + N_{\bar{B}}} \simeq 10^{-9}.$$

The fact that our universe is made up of matter may therefore be accounted for by the same physical processes that are responsible for the eventual death of the proton. Though this explanation of the origin of matter in our present universe is still very speculative, it provides us with a good example of the inevitable symbiosis of astrophysics and the particle physics discussed in this book [see, e.g., Ellis in Mulvey (1981) or Ellis (1982)].

15.10 "Grander" Unification?

Although crucial tests remain to be done, it is widely believed that all matter is composed of spin-$\frac{1}{2}$ particles (quarks and leptons) whose interactions are a consequence of exact local gauge symmetries. The gauge bosons mediating these color and electroweak interactions have spin 1 (the photon, the gluons, and the weak bosons $W^{\pm}, Z$). Through spontaneous symmetry breaking, the weak bosons and the fermions acquire a mass, and the theory remains renormalizable. In this book, we have attempted to describe the details of this picture.

On the other hand, the concept of a grand unified theory is much more speculative. Even if you are willing to accept the underlying assumption that there is a "desert" with no new physics between $Q = M_W$ and $Q = M_X$, it brings many unanswered questions to the forefront. Why are the mass scales M_W and M_X separated by the enormous factor 10^{12}? What about gravity? Is there a principle which relates matter fields to gauge fields so they can be unified, too?

These questions take us well beyond the frontiers of our knowledge of the world of quarks and leptons. But we shall conclude with a brief mention of just two speculations regarding the possible structure of a more "ultimate" theory. One approach is to directly link matter and gauge fields. Such a symmetry is necessarily very different from any symmetry previously encountered since it links spin-$\frac{1}{2}$ Dirac matter fields with spin-1 boson gauge fields. A symmetry directly associating fields of integer and half-integer spin has been constructed mathematically and is called "supersymmetry" [see, e.g., Salam in Mulvey (1981)]. Such a symmetry can also encompass the spin-2 fields of Einstein gravity, the gravitons which mediate the interactions of masses.

This approach starts with GUT's and attempts to "tack on" gravity. A second idea follows the reverse approach. We could start with gravity, a force that is "well understood" classically as a result of the geometrical curvature of four-dimensional space-time, and ask the question, "What is the geometry associated with electromagnetism, color...?" Two answers are possible. One postulates that gravity is connected with the "gross" geometrical structure of space-time but that, at a much smaller scale, more complex topologies exist with which other interactions are associated. The alternative possibility is to introduce more space-time dimensions. This has been vigorously investigated since Kaluza and Klein suggested in the early twenties that electromagnetism may be a geometrical theory with a fifth dimension.

The most important achievement of the last two decades is not just to have established that our world is made of quarks, leptons, and gauge bosons, but to have brought us toward a new frontier where even more exciting questions can be raised. These speculations do inevitably include the possibility that quarks and leptons are themselves composite.

ANSWERS AND COMMENTS ON THE EXERCISES

CHAPTER 1

1.1 $\hbar c = 0.1973$ GeV F $= 1$ and 1 mb $= 0.1$ F^2.

1.3 Typical atomic energies and dimensions are, respectively, factors of α^2 and $1/\alpha$ different from natural units of energy and length, and $v/c \sim \alpha$ (with $\alpha \simeq \frac{1}{137}$). An illuminating discussion is given by Wichmann in *Quantum Physics* (Berkeley Physics Course Vol. 4, 1967).

CHAPTER 2

2.1 (i) States (2.1) are defined with respect to a given z axis. Now take an NN state $\uparrow\uparrow$ quantized along a different axis, say, z'. We can write this state as a linear combination of the original $S = 1$ states

$$|S = 1, M_S = 1\rangle' = \sum_{M_S} \alpha(M_S)|S = 1, M_S\rangle .$$

But the states $|M_S = 1\rangle'$ and $|M_S = \pm 1\rangle$ are manifestly symmetric and so $|S = 1, M_S = 0\rangle$ must also be symmetric. Orthogonality then demands $|S = 0, M_S = 0\rangle$ to be antisymmetric.

(ii) Applying the step-down operator to $|S = 1, M_S = 1\rangle = \uparrow\uparrow$, we obtain [see J_- of (2.18)]

$$\sqrt{2}\,|S = 1, M_S = 0\rangle = \uparrow\downarrow + \downarrow\uparrow .$$

2.2 $\psi = \psi_{\text{space}}\, \psi_{\text{spin}}\, \psi_{\text{isospin}}$, where for $L = 0$, ψ_{space} is symmetric under interchange of the two nucleons, whereas

$$\psi_{\text{spin}}(1,2) = (-1)^{S+1}\psi_{\text{spin}}(2,1)$$

$$\psi_{\text{isospin}}(1,2) = (-1)^{I+1}\psi_{\text{isospin}}(2,1),$$

see (2.1) and (2.2), respectively. For the overall ψ to be antisymmetric, $S + I$ must therefore be an odd integer.

2.3 pp $= |I = 1, I_3 = 1\rangle$ and np $= \sqrt{\frac{1}{2}}(|I = 1, I_3 = 0\rangle - |I = 0, I_3 = 0\rangle)$. Since the πd states are pure $I = 1$, the np $\to \pi^0$d reaction can only proceed with probability $(\sqrt{\frac{1}{2}})^2$ that of pp $\to \pi^+$d.

2.4 See, for example, Feynman (1961) or Rose (1957).

2.6 $j = \frac{1}{2}$ elements follow from (2.26). See Rose (1957).

2.7 Expand and use $(\sigma_2)^2 = 1$, $(\sigma_2)^3 = \sigma_2$, and so on.

2.8 The 3×3 matrices $\lambda_1, \lambda_2, \lambda_3$ are the 2×2 Pauli matrices supplemented by a third row and column constructed from zeros. Similarly, λ_4, λ_5 have zeros in the second row and column such that when these are removed, the remaining elements are those of σ_1, σ_2. The normalization of the λ_i is thus taken to be that of the σ_i,

$$\mathrm{Tr}(\lambda_i \lambda_j) = 2\delta_{ij}.$$

Examples of λ_i are

$$\lambda_4 = \begin{pmatrix} 0 & 0 & 1 \\ 0 & 0 & 0 \\ 1 & 0 & 0 \end{pmatrix}, \qquad \lambda_7 = \begin{pmatrix} 0 & 0 & 0 \\ 0 & 0 & -i \\ 0 & i & 0 \end{pmatrix}.$$

2.9 The main decay modes are $\phi \to K^+K^-$ and $\phi \to K^0\overline{K}^0$, which in the limit of exact symmetry would occur equally. However, the energy release is so small that the equality is broken by the mass difference of the K^+ and K^0. We have

$$\frac{\Gamma(\phi \to K^+K^-)}{\Gamma(\phi \to K^0\overline{K}^0)} \approx \left(\frac{p_{K^+}}{p_{K^0}}\right)^{2L+1}$$

where $p_i \equiv |\mathbf{p}_i|$ are the kaon momenta in the rest frame of the ϕ, and L is the orbital angular momentum of the $K\overline{K}$ system; ϕ and K have spin 1 and 0, respectively, and so $L = 1$. The observed branching ratios are

$$K^+K^- : K^0\overline{K}^0 : \pi^+\pi^-\pi^0 = 0.49 : 0.35 : 0.15.$$

The smallness of the 3π mode, despite being kinematically favored, supports the $\phi \simeq s\bar{s}$ identification. The $K^0\overline{K}^0$ mode is seen as $K_L^0 K_S^0$, since ϕ has $J^P = 1^-$ and $C = -1$, see eq. (12.129). The $\phi \to K\overline{K}$ decay rate can be estimated from the $SU(3)$-related $\rho \to \pi\pi$ decay. The kinematic suppression of $\phi \to K\overline{K}$ is apparent in the comparison of the observed widths $\Gamma(\phi) = 4$ MeV, $\Gamma(\rho) = 150$ MeV.

2.10 The production of the $(\pi\pi)$-system at small momentum transfer is dominated by π-exchange. The exchange (virtual) pion is almost real (i.e., on its mass shell), and so this experimentally accessible process is an excellent way of studying π"π" $\to \pi\pi$ scattering (see Fig. 4.8). Pions obey Bose statistics and a symmetric $\pi\pi$-wavefunction requires $(-1)^J(-1)^I = +1$.

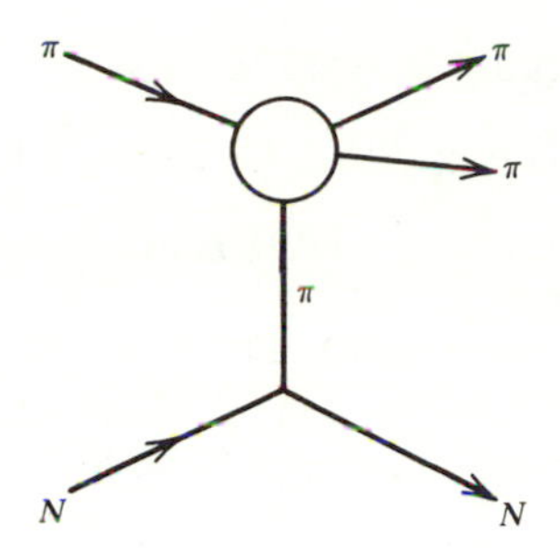

2.12 In addition to (2.63), we have

$$\Sigma(M_S) = \sqrt{\tfrac{1}{12}}\,((\mathrm{sd}+\mathrm{ds})\mathrm{u} + (\mathrm{su}+\mathrm{us})\mathrm{d} - 2(\mathrm{du}+\mathrm{ud})\mathrm{s})$$
$$\Sigma(M_A) = \tfrac{1}{2}((\mathrm{sd}-\mathrm{ds})\mathrm{u} + (\mathrm{su}-\mathrm{us})\mathrm{d})$$
$$\Lambda(M_S) = \tfrac{1}{2}((\mathrm{sd}+\mathrm{ds})\mathrm{u} - (\mathrm{su}+\mathrm{us})\mathrm{d})$$
$$\Lambda(M_A) = \sqrt{\tfrac{1}{12}}\,((\mathrm{sd}-\mathrm{ds})\mathrm{u} + (\mathrm{us}-\mathrm{su})\mathrm{d} - 2(\mathrm{du}-\mathrm{ud})\mathrm{s})$$
$$\Sigma(S) = \sqrt{\tfrac{1}{6}}\,((\mathrm{sd}+\mathrm{ds})\mathrm{u} + (\mathrm{su}+\mathrm{us})\mathrm{d} + (\mathrm{du}+\mathrm{ud})\mathrm{s})$$

where Σ and Λ denote states of $I = 1$ and $I = 0$, respectively.

2.14 $|\pi^+\rangle = \sqrt{\tfrac{1}{6}} \sum_{a=\mathrm{R,G,B}} (\bar{\mathrm{d}}_a \uparrow \mathrm{u}_a \downarrow - \bar{\mathrm{d}}_a \downarrow \mathrm{u}_a \uparrow)$

2.15 The members of a U-spin multiplet have the same charge. The charge operator Q commutes with the three generators of U spin.

$\Sigma^{\pm}$ and $\Sigma^*(1385)^+$ are members of $U = \frac{1}{2}$ multiplets, whereas $\Sigma^*(1385)^-$ is a member of a $U = \frac{3}{2}$ multiplet. Conservation of U spin therefore forbids $\Sigma^{*-} \to \Sigma^-\gamma$ if γ has $U = 0$.

2.16 The total angular momentum J and parity P are conserved in πN interactions and, since states with $L' = J \pm \frac{1}{2}$ have opposite parity, L' is also conserved. $P = \eta_\pi \eta_N (-1)^{L'}$, where the product of the intrinsic parities $\eta_\pi \eta_N = -1$. The quark model predicts that the first excited baryon level contains the following πN states: S_{11} (twice), D_{13} (twice), D_{15}, S_{31}, and D_{33}, where we have used the notation $L'_{2I,2J}$ where I is the isospin of the resonance. See also Close (1979).

2.17 $\mu_\Lambda = \mu_s$, $\mu_{\Sigma^+} = \frac{1}{3}(4\mu_u - \mu_s)$, $\mu_{\Xi^0} = \frac{1}{3}(4\mu_s - \mu_u)$, and so on. Using $\mu_p = 2.79$ nm, we have

$$\mu_d = \mu_s = -\tfrac{1}{2}\mu_u = -0.93 \text{ nm}.$$

We can obtain a better estimate of μ_s if we allow for quark mass differences (see Section 2.14)

$$\mu_s = \frac{m_d}{m_s}\mu_d.$$

2.20 $|\omega(M_J = 1)\rangle = \sqrt{\tfrac{1}{2}}(\bar{u}u + \bar{d}d)\uparrow\uparrow,$

$$|\pi^0\rangle = \sqrt{\tfrac{1}{2}}(\bar{u}u - \bar{d}d)\sqrt{\tfrac{1}{2}}(\uparrow\downarrow - \downarrow\uparrow),$$

$$(\mu\sigma_-)_i(q\uparrow)_i = \mu_q(q\downarrow)_i, \qquad (\mu\sigma_-)_i(q\downarrow)_i = 0, \qquad \text{and } \mu_{\bar{q}} = -\mu_q.$$

2.22 $D^0 \to K^-\pi^+$ due to $c \to s(\bar{d}u)$, $D^0 \to \pi^-\pi^+$ due to $c \to d(\bar{d}u)$, with, in each case, $\bar{u}$ as a spectator. In Section 12.11, we see that the $c \to d$ transmutation is "Cabibbo-suppressed" in comparison to $c \to s$.

2.23 The hadronic state must have $I = 0$ and $C = +1$. The dominant hadronic decay modes are observed to be $\psi' \to \psi\pi^+\pi^-$, $\psi\pi^0\pi^0$, $\psi\eta$ with branching ratios 33, 17, and 3%, respectively.

2.24 Radiative transitions occur between states of opposite C parity; the ($\Delta J =$ 1) transitions are E_1 or M_1 according to whether the relative parity of the levels is odd or even. The approximate equality of the $\psi' \to \chi\gamma$ branching ratios is due to the balancing compensation of the $2J + 1$ and phase space (k^3) factors which occur in the E_1 transition probability formula.

2.25 $\Gamma(q\bar{q} \to e^+e^-)$ is proportional to e_q^2; $|\rho\rangle$ and $|\omega\rangle$ are $(u\bar{u} \mp d\bar{d})/\sqrt{2}$, respectively. Since the $u\bar{u}$ and $d\bar{d}$ annihilations are indistinguishable, we must add amplitudes; and so

$$\Gamma_{\rho,\,\omega} \propto \tfrac{1}{2}\left(\tfrac{2}{3} \mp \left(-\tfrac{1}{3}\right)\right)^2.$$

2.26 See, for example, Close (1979).

2.27 See, for example, Close (1979).

2.28 The strong $\Sigma_c^{++} \to \Lambda_c^+\pi^+$ decay, followed by the weak decays $\Lambda_c^+ \to \Lambda\pi^+\pi^+\pi^-$ and $\Lambda \to p\pi^-$.

2.29 The eigenvalues of $\boldsymbol{\sigma}_1\cdot\boldsymbol{\sigma}_2$ are -3 and $+1$ for $S = 0$ and 1, respectively.

2.32 $$\sum_{i>j} \frac{\boldsymbol{\sigma}_i\cdot\boldsymbol{\sigma}_j}{m_i m_j} = \frac{4}{m_u^2}\left(\mathbf{S}_1\cdot\mathbf{S}_2 + \frac{m_u}{m_Q}(\mathbf{S}_1 + \mathbf{S}_2)\cdot\mathbf{S}_3\right).$$

CHAPTER 3

3.2 $$\sum_{\mu\nu} g_{\mu\nu}g^{\mu\nu} = \sum_{\mu}(g_{\mu\mu})^2 = 4.$$

3.3

$$\text{Center-of-Mass Frame: } p_1 = \left(\tfrac{1}{2}E_{\text{cm}}, \mathbf{p}\right), p_2 = \left(\tfrac{1}{2}E_{\text{cm}}, -\mathbf{p}\right)$$

$$\text{Laboratory Frame: } p_1 = (E_{\text{lab}}, \mathbf{p}_{\text{lab}}), p_2 = (M, \mathbf{0})$$

$$\therefore s = (E_{\text{lab}} + M)^2 - (E_{\text{lab}}^2 - M^2)$$

Advantages of Fixed Target: higher flux because of the density of the target;

choice of reaction πp, Kp, πd, and so on; need detectors over less solid angle about the interaction point and hence a less expensive experiment.

3.4

$$\sum_{\substack{n \neq i \\ m \neq n}} V_{fn} \frac{1}{E_m - E_n + i\varepsilon} V_{nm} \frac{1}{E_i - E_m + i\varepsilon} V_{mi}$$

3.5 (a) Let E and E' be the energies of the outgoing e^- and e^+, respectively; and so the matrix element contains

$$\int (e^{-iEt})^* e^{-i\omega t} e^{-i(-E')t}\, dt = 2\pi\, \delta(E + E' - \omega)$$

CHAPTER 4

4.1 To satisfy the periodic boundary conditions, the allowed values of p_x are such that $Lp_x = 2\pi n$, where n is an integer. Hence, the number of allowed states in the range p_x to $p_x + dp_x$ is $L\, dp_x/2\pi$.

4.2 From (4.30), we have

$$dQ = \frac{1}{4\pi^2} \frac{d^3p_C}{2E_C} \frac{d^3p_D}{2E_D} \delta^{(4)}(p_A + p_B - p_C - p_D)$$

$$= \frac{1}{4\pi^2} \frac{d^3p_C}{2E_C} \frac{1}{2E_D} \delta(E_A + E_B - E_C - E_D).$$

In the center-of-mass frame $(\sqrt{s} \equiv W = E_A + E_B)$,

$$dQ = \frac{1}{4\pi^2} \frac{p_f^2\, dp_f\, d\Omega}{4E_C E_D} \delta(W - E_C - E_D).$$

Using

$$W = E_C + E_D = \left(m_C^2 + p_f^2\right)^{1/2} + \left(m_D^2 + p_f^2\right)^{1/2},$$

we have

$$\frac{dW}{dp_f} = p_f\left(\frac{1}{E_C} + \frac{1}{E_D}\right).$$

Substituting for dp_f, we find

$$dQ = \frac{1}{4\pi^2} \frac{p_f}{4} \left(\frac{1}{E_C + E_D}\right) dW\, d\Omega\, \delta(W - E_C - E_D)$$

$$\boxed{dQ = \frac{1}{4\pi^2} \frac{p_f}{4\sqrt{s}}\, d\Omega.}$$

From (4.32),

$$F = 4(p_i E_B + p_i E_A) = 4p_i\sqrt{s}\,.$$

4.3 Neglecting masses, we may write in the center-of-mass frame

$$p_A = (p, \mathbf{p}), \qquad p_B = (p, -\mathbf{p}),$$
$$p_C = (p, \mathbf{p}'), \qquad p_D = (p, -\mathbf{p}'),$$

where $p = |\mathbf{p}| = |\mathbf{p}'|$. Substitute these values into (4.18), with $q = p_D - p_B$, and use (4.35).

4.5 Using (4.43) and $p_i^2 = m_i^2$,

$$s + t + u = \sum_i m_i^2 + 2p_A^2 + 2p_A \cdot (p_B - p_C - p_D) = \sum_i m_i^2.$$

4.6 If $e^-e^- \to e^-e^-$ is the s channel process $A + B \to C + D$, then

$$p_A = (E, \mathbf{k}_i), \qquad p_B = (E, -\mathbf{k}_i), \qquad p_C = (E, \mathbf{k}_f), \qquad p_D = (E, -\mathbf{k}_f)$$

where $E = (k^2 + m^2)^{1/2}$. So, for example,

$$t = (p_A - p_C)^2 = -(\mathbf{k}_i - \mathbf{k}_f)^2 = -2k^2(1 - \cos\theta)$$

since $\mathbf{k}_i \cdot \mathbf{k}_f = k^2 \cos\theta$. As $k^2 \geq 0$, we have $s \geq 4m^2$; and since $-1 \leq \cos\theta \leq 1$, we have $t \leq 0$ and $u \leq 0$.

4.7 If we keep $p_A, \ldots p_D$ defined as for the s channel $AB \to CD$ process, then for $A\overline{D} \to C\overline{B}$ in the center-of-mass frame,

$$p_A = (E, \mathbf{k}_i),\ -p_D = (E, -\mathbf{k}_i), p_C = (E, \mathbf{k}_f),\ -p_B = (E, -\mathbf{k}_f),$$

where $\mathbf{k}_i, -\mathbf{k}_i, \mathbf{k}_f, -\mathbf{k}_f$ are the three-momenta of A, $\overline{\mathrm{D}}$, C, and $\overline{\mathrm{B}}$, respectively. The required results now follow upon using (4.43).

4.8 See, for example, Martin and Spearman (1970), Chapter 4, Section 3.

4.9
$$-(p_A + p_C) \cdot (p_D + p_B) = -(2p_A + p_B - p_D) \cdot (p_D + p_B)$$
$$= -2p_A \cdot (p_D + p_B) = u - s.$$

CHAPTER 5

5.1 See, for example, Aitchison (1972), Chapter 8, Section 1.

5.2

$$0 = \gamma^\nu \partial_\nu (i\gamma^\mu \partial_\mu - m)\psi = i\tfrac{1}{2}(\gamma^\nu \gamma^\mu + \gamma^\mu \gamma^\nu)\partial_\nu \partial_\mu \psi - m\gamma^\nu \partial_\nu \psi$$
$$= ig^{\mu\nu}\partial_\nu \partial_\mu \psi + im^2\psi = i(\Box^2 + m^2)\psi.$$

5.3 The momentum $\mathbf{p}' = (p\sin\theta, 0, p\cos\theta)$ is obtained from $\mathbf{p} = (0, 0, p)$ by a rotation through θ around the y axis. The required helicity eigenspinor $u(\mathbf{p}')$ may therefore be obtained from $u^{(1)}(\mathbf{p})$ of (5.27) using (2.26):

$$u(\mathbf{p}') = N\left(\begin{array}{c|c} \cos\dfrac{\theta}{2} - i\sigma_2 \sin\dfrac{\theta}{2} & 0 \\ \hline 0 & \cos\dfrac{\theta}{2} - i\sigma_2\sin\dfrac{\theta}{2} \end{array}\right)\left(\begin{array}{c} 1 \\ 0 \\ \hline p/(E+m) \\ 0 \end{array}\right).$$

5.4 One way to proceed is to evaluate explicit components, for example,

$$[H, L_1] = [\boldsymbol{\alpha}\cdot\mathbf{P}, x_2 P_3 - x_3 P_2] = -i(\alpha_2 P_3 - \alpha_3 P_2) = -i(\boldsymbol{\alpha}\times\mathbf{P})_1.$$

5.5 Act on the first of eqs. (5.24) with the operator $(E + eA^0 + m)$,

$$(E + eA^0 + m)\boldsymbol{\sigma}\cdot(\mathbf{P} + e\mathbf{A})u_B = (E + eA^0 + m)(E + eA^0 - m)u_A$$
$$\simeq 2m(E_{NR} + eA^0)u_A.$$

If we were able to commute the two operators then, upon using the second of eqs. (5.24), the left-hand side would reduce to $(\mathbf{P} + e\mathbf{A})^2$. The lack of commutation gives rise to the extra term involving $\mathbf{B} = \nabla\times\mathbf{A}$.

5.6

$$\gamma^\mu = -(C\gamma^0)\gamma^{\mu *}(C\gamma^0)^{-1} = -C\gamma^0\gamma^{\mu *}\gamma^0 C^{-1} = -C\gamma^{\mu T}C^{-1}.$$

Thus, $\gamma^0 = -C\gamma^0 C^{-1}$, and so $(C\gamma^0) = (C\gamma^0)^T$ implies $C = -C^T$, and $(C\gamma^0)^2 = 1$ implies $C = -C^{-1}$. Also,

$$\bar{\psi}_C = \psi_C^\dagger\gamma^0 = (C\gamma^0\psi^*)^\dagger\gamma^0 = \psi^T C\gamma^0\gamma^0 = -\psi^T C^{-1}.$$

5.7

$$\bar{u}\gamma^0(\gamma^\mu p_\mu - m)u = 0$$
$$\underline{\bar{u}(\gamma^\mu p_\mu - m)\gamma^0 u = 0}$$

Adding: $2\bar{u}p_0 u - 2mu^\dagger u = 0$, since $\gamma^0\gamma^k = -\gamma^k\gamma^0$.

$$\bar{u}^{(r)}u^{(s)} = \frac{m}{E}u^{(r)\dagger}u^{(s)} = 2m\delta_{rs}.$$

5.9

$$u^{(s)} = N\begin{pmatrix} \chi^{(s)} \\ \dfrac{\boldsymbol{\sigma}\cdot\mathbf{p}}{E+m}\chi^{(s)} \end{pmatrix}, \qquad \bar{u}^{(s)} = N\left(\chi^{(s)\dagger}, \frac{-\boldsymbol{\sigma}\cdot\mathbf{p}}{E+m}\chi^{(s)\dagger}\right)$$

$$\sum_{s=1,2} u^{(s)}\bar{u}^{(s)} = N^2\begin{pmatrix} I & \dfrac{-\boldsymbol{\sigma}\cdot\mathbf{p}}{E+m} \\ \dfrac{\boldsymbol{\sigma}\cdot\mathbf{p}}{E+m} & -\left(\dfrac{\boldsymbol{\sigma}\cdot\mathbf{p}}{E+m}\right)^2 \end{pmatrix} = \begin{pmatrix} E+m & -\boldsymbol{\sigma}\cdot\mathbf{p} \\ \boldsymbol{\sigma}\cdot\mathbf{p} & m-E \end{pmatrix} = \not{p} + m,$$

since $\Sigma\chi^{(s)}\chi^{(s)\dagger} = I$ and $N^2 = E + m$.

5.10

$$\not{p}\not{p} = \gamma^\nu\gamma^\mu p_\nu p_\mu = -\gamma^\mu\gamma^\nu p_\nu p_\mu + 2g^{\mu\nu}p_\nu p_\mu = -\not{p}\not{p} + 2p^2.$$

5.11 The action of Λ_+ on an arbitrary spinor is

$$\Lambda_+\left(\sum_{r=1}^{4} a_r u^{(r)}\right) = \sum_{r=1}^{4} a_r\left(\sum_{s=1}^{2}\frac{u^{(s)}\bar{u}^{(s)}}{2m}\right)u^{(r)} = \sum_{r=1}^{2} a_r u^{(r)}.$$

$$\Lambda_+^2 = \frac{\not{p}\not{p} + 2m\not{p} + m^2}{4m^2} = \frac{p^2 + 2m\not{p} + m^2}{4m^2} = \frac{\not{p} + m}{2m} = \Lambda_+.$$

5.12 Particular examples of (5.59) (rotations and Lorentz boosts) are given by Sakurai (1967), but note the different metric and properties of γ-matrices.

5.14

$$\left[(\boldsymbol{\alpha}\cdot\mathbf{P} + \beta m), \gamma^5\right] = \left[\gamma^5, (\boldsymbol{\alpha}\cdot\mathbf{P} - \beta m)\right],$$

whereas

$$\left[(\boldsymbol{\alpha}\cdot\mathbf{P} + \beta m), \boldsymbol{\sigma}\cdot\hat{\mathbf{p}}\right] = \left[\boldsymbol{\sigma}\cdot\hat{\mathbf{p}}, (\boldsymbol{\alpha}\cdot\mathbf{P} + \beta m)\right].$$

5.15

$$\gamma^5\begin{pmatrix}\chi^{(s)} \\ \dfrac{\boldsymbol{\sigma}\cdot\mathbf{p}}{E+m}\chi^{(s)}\end{pmatrix} \simeq \begin{pmatrix}\boldsymbol{\sigma}\cdot\hat{\mathbf{p}}\chi^{(s)} \\ \chi^{(s)}\end{pmatrix} \simeq \boldsymbol{\sigma}\cdot\hat{\mathbf{p}}\begin{pmatrix}\chi^{(s)} \\ \dfrac{\boldsymbol{\sigma}\cdot\mathbf{p}}{E+m}\chi^{(s)}\end{pmatrix}$$

in the limit $E \gg m$.

CHAPTER 6

6.1 Start with $\bar{u}_f i\sigma^{\mu\nu}(p_f - p_i)_\nu u_i$ and rewrite the individual terms with the help of (5.9) so that you can make use of the Dirac equations

$$\gamma^\nu p_{i\nu}u_i = mu_i, \qquad \bar{u}_f\gamma^\nu p_{f\nu} = m\bar{u}_f.$$

6.3

$$\mathrm{Tr}(\not{a}\not{b}) = \tfrac{1}{2}\,\mathrm{Tr}(\not{a}\not{b} + \not{b}\not{a}) = \tfrac{1}{2}2g^{\mu\nu}a_\mu b_\nu\,\mathrm{Tr}(I) = 4a\cdot b.$$

$$\begin{aligned}\mathrm{Tr}(\not{a}_1\cdots\not{a}_n) &= \mathrm{Tr}(\widehat{\not{a}_1\ldots\not{a}_n\gamma^5}\gamma^5) \\ &= (-1)^n\,\mathrm{Tr}(\gamma^5\not{a}_1\ldots\not{a}_n\gamma^5) = (-1)^n\,\mathrm{Tr}(\not{a}_1\ldots\not{a}_n),\end{aligned}$$

and so, if n is odd, the trace vanishes.

6.4 Since the $\bar{\nu}_e$ is right-handed, the e^- must be left-handed (see Fig. 6.8). The e^+ from μ^+ decay is right-handed. See Chapter 12 for details.

6.5 From (2.21),

$$d^1_{00}(\theta) = \cos\theta \simeq \frac{t-u}{s}.$$

6.7

$$\int dp'_0 2p'_0 \theta(p'_0)\,\delta(p'^2 - M^2) = 1$$

where $p'^2 = p_0'^2 - \mathbf{p}'^2$. This result follows from the identity

$$\delta(p_0'^2 - a^2) = \frac{1}{2|a|}\left(\delta(p'_0 - a) + \delta(p'_0 + a)\right),$$

$$\delta\left((p+q)^2 - M^2\right) = \delta(2p\cdot q + q^2) = \frac{1}{2M}\delta\left(\nu + \frac{q^2}{2M}\right)$$

$$= \frac{1}{2M}\delta\left[E - E' - EE'(1-\cos\theta)/M\right].$$

6.9 Substitute (6.55) into (6.53). Show that the equations for $\nabla\cdot\mathbf{B}$ and $\nabla\times\mathbf{E}$ are automatically satisfied and that the remaining two equations can be arranged to give (6.54). $\partial_\mu\partial_\nu F^{\mu\nu} = 0$ follows on considering $\mu \leftrightarrow \nu$.

6.11 Show that under a rotation ϕ about the photon propagation direction (the z axis)

$$\boldsymbol{\varepsilon}_i \to \boldsymbol{\varepsilon}'_i = e^{-i\lambda_i\phi}\boldsymbol{\varepsilon}_i$$

where $\lambda_R = +1$ and $\lambda_L = -1$, see (2.12).

6.13

$$\left(Aq^2 g_{\mu\nu} + Bq_\mu q_\nu\right)\left(-g^{\nu\lambda}q^2 + q^\nu q^\lambda\right) = Aq^2\left(-\delta_\mu^\lambda q^2 + q_\mu q^\lambda\right)$$

which cannot be made equal to δ_μ^λ for any choice of the arbitrary functions A and B.

6.15 See Exercise 6.11; $\varepsilon^{(\lambda=0)}$ is chosen to satisfy (6.91) and to be suitably normalized.

6.16 Equation (6.93) can be checked explicitly component by component or, more elegantly, by writing the sum in its most general Lorentz form $Ag_{\mu\nu} + Bp_\mu p_\nu$. Then take the scalar product with p^μ to show $A = -BM^2$, and with $g^{\mu\nu}$ to show $A = -1$.

6.17 For verification of (6.101) itself, see, for example, Sakurai (1967), page 8, where it is shown that

$$\left(|\mathbf{q}|^2 + \mu^2\right)^{-1} \leftrightarrow e^{-\mu|\mathbf{x}|}/4\pi|\mathbf{x}|$$

are Fourier transforms.

6.18 To evaluate the $\mathfrak{M}_1$ contribution to $k_\mu T^{\mu\nu}$, note that

$$\left(\frac{(\not{p} + \not{k} + m)\not{k}}{(p+k)^2 - m^2}\right)u(p) = \left(\frac{-\not{k}\not{p} + 2k\cdot p + m\not{k}}{2k\cdot p}\right)u(p) = u(p),$$

since $\not{p}\not{k} + \not{k}\not{p} = 2k\cdot p$ and $\not{k}\not{k} = 0$ and $(\not{p} - m)u = 0$.

6.19 The variables (6.109) become

$$s = (k+p)^2 = 2k\cdot p - Q^2 = 2k'\cdot p', \text{ and so on.}$$

Repeat the derivation of (6.113) and show that $\overline{|\mathfrak{M}_1|^2}$, $\overline{|\mathfrak{M}_2|^2}$ are unchanged but that the interference contribution becomes $4e^4Q^2t/su$. Use (6.24).

6.20 At high energy, the dominant contribution to σ comes from

$$\overline{|\mathfrak{M}|^2} \simeq 2e^4\left(\frac{-s}{u - m^2}\right) \simeq 4e^4\left(\frac{2m^2}{s} + 1 + \cos\theta\right)^{-1}$$

and the $\cos\theta$ integration leads to the $\log(s/m^2)$ behavior.

6.23 See Aitchison and Hey (1982), Chapter 2, Section 10.

CHAPTER 8

8.1 Revision. For further discussion, see Section 7.1.

8.4 For a spherically symmetric potential, we can perform the angular integration in (8.3) and obtain

$$F = 2\pi\int\rho(r)\left(\frac{e^{iqr} - e^{-iqr}}{iqr}\right)r^2\,dr.$$

Substitution of $\rho = Ae^{-mr}$ and straightforward integration yield the result.

8.5 We might expect the most general form (for $x = 0$) to be

$$J^\mu = e\bar{u}(p')\big(\gamma^\mu K_1 + i\sigma^{\mu\nu}(p'-p)_\nu K_2 + i\sigma^{\mu\nu}(p'+p)_\nu K_3$$
$$+ (p'-p)^\mu K_4 + (p'+p)^\mu K_5\big)u(p)$$

where $K_i \equiv K_i(q^2)$. But using the Gordon decomposition, (6.7), we can reexpress the $(p'+p)^\mu$ terms as linear combinations of the γ^μ and $\sigma^{\mu\nu}(p'-p)_\nu$ terms. Thus, the most general form reduces to

$$J^\mu = e\bar{u}(p')\left(\gamma^\mu F_1 + \frac{i\kappa}{2M}F_2\sigma^{\mu\nu}q_\nu + q^\mu F_3\right)u(p)e^{iq\cdot x}.$$

Current conservation

$$q_\mu J^\mu = e\bar{u}(p')\left(\not{q}F_1 + \frac{i\kappa}{2M}F_2 q_\mu\sigma^{\mu\nu}q_\nu + q^2F_3\right)u(p) = 0$$

implies $F_3 = 0$, as the first term vanishes by virtue of the Dirac equation, and the second vanishes since $\sigma^{\mu\nu}$ is antisymmetric.

8.6 On squaring $p' = p + q$, we have

$$M^2 = M^2 + 2p \cdot q + q^2,$$

so $p \cdot q = -q^2/2$ is not an independent scalar variable.

8.7 Equation (8.18) follows from (8.13) on using the Gordon decomposition, (6.7). In the Breit frame $p^\mu = (E, \mathbf{p})$ and $p'^\mu = (E, -\mathbf{p})$. Therefore, (8.18) gives

$$\rho \equiv J^0 = e\bar{u}(p')\left(\gamma^0(F_1 + \kappa F_2) - \frac{E}{M}\kappa F_2\right)u(p).$$

But in this frame,

$$\bar{u}\gamma^0 u = 2M \qquad \text{and} \qquad \bar{u}u = 2E$$

with only states $\lambda = -\lambda'$ contributing. Thus,

$$\rho = 2Me\left[F_1 + \left(1 - \frac{E^2}{M^2}\right)\kappa F_2\right] = 2MeG_E$$

using (8.16) and noting $q^2 = 4\mathbf{p}^2$. Similarly,

$$\mathbf{J} = e\bar{u}(p')\boldsymbol{\gamma}u(p)G_M$$

in the Breit frame.

8.9 See Chapter 6; $L_{\mu\nu}$ is the product of two currents summed and averaged over spins. Use current conservation or check directly using the explicit form (6.25).

8.10 Using (8.24) and (8.26), we have

$$-W_1 q^\mu + \frac{W_2}{M^2}(p \cdot q)p^\mu + \frac{W_4}{M^2}q^2 q^\mu + \frac{W_5}{M^2}(q^2 p^\mu + (p \cdot q)q^\mu) = 0.$$

The coefficients of q^μ and p^μ must vanish separately.

8.11 $q^2 \leq 0$ and $\nu \geq 0$, so $x \geq 0$. $W^2 \geq M^2$, and so $x \leq 1$ from (8.29). y is an invariant variable, so evaluate in the target rest frame. We have $y = 1 - (E'/E)$, and hence $0 \leq y \leq 1$.

8.15 Substitute (8.27) into (8.52). For $\lambda = 0$, for example,

$$\sigma_L = \frac{4\pi^2\alpha}{K}\left(-\varepsilon_0^\mu \varepsilon_0^\nu g_{\mu\nu} W_1 + \varepsilon_0^\mu \varepsilon_0^\nu p_\mu p_\nu \frac{W_2}{M^2}\right),$$

since $\varepsilon \cdot q = 0$ (see Exercise 8.14). Evaluate in the laboratory frame using (8.50).

8.16 Solve (8.53) and (8.54) for $W_{1,2}$ and substitute into (8.34).

8.17 The poles at $q^2 = 0$ in (8.27) cannot be real but must be an artifact of the way we have written $W_{\mu\nu}$. To remove the leading pole, we must have

$$q^2 W_1 + \frac{(p \cdot q)^2}{M^2} W_2 \to 0 \qquad \text{as } q^2 \to 0$$

or

$$W_2 \to -\frac{q^2}{\nu^2} W_1 + O(q^4) \qquad \text{as } q^2 \to 0.$$

But W_1 can approach a constant as $q^2 \to 0$, and so $W_2 \to 0$. Hence, $\sigma_L \to 0$, as it must in the limit of real photons.

CHAPTER 9

9.3

$$\hat{s} = (k + xp)^2 \simeq x(2kp) \simeq xs,$$

$$\hat{t} = (k - k')^2 = t,$$

$$\hat{u} = (k' - xp)^2 \simeq xu.$$

Use (6.30) together with (4.35) and (4.45) to show that

$$\frac{d\sigma}{d\hat{t}} = \frac{2\pi\alpha^2 e_i^2}{\hat{t}^2}\left(\frac{\hat{s}^2 + \hat{u}^2}{\hat{t}^2}\right).$$

Equation (9.18) then follows upon using

$$-\frac{t}{s + u} = \frac{Q^2}{2M\nu} = x.$$

To derive (9.19), first verify that (8.31) may be written

$$(L^e)^{\mu\nu} W_{\mu\nu} = -2tW_1 - su\frac{W_2}{M^2}$$

$$= \frac{2}{M(s + u)}\left[(s + u)^2 xF_1 - usF_2\right].$$

Equation (9.19) follows on inserting this expression into (8.35) and using

$$d\Omega\, dE' = 2\pi\, d(\cos\theta)\, dE' = \frac{4\pi M^2}{su} dt\left(-\frac{1}{2M} du\right).$$

Here, we have used the laboratory kinematics of Section 6.8,

$$s = 2ME, \qquad u = -2ME', \qquad t = -Q^2 = -4EE'\sin^2\frac{\theta}{2}.$$

9.4 Equation (9.21) follows on using $Q^2 = 4EE'\sin^2\frac{\theta}{2}$ and $d\nu = -dE'$. Show that, in the laboratory frame,

$$\sin^2\frac{\theta}{2} = xy\frac{M}{2E'}, \qquad \cos^2\frac{\theta}{2} = \frac{E}{E'}\left(1 - y - \frac{Mxy}{2E}\right).$$

Insert these expressions into (8.34) and obtain

$$\frac{d\sigma}{dx\,dy} = \frac{8ME\pi\alpha^2}{Q^4}\left[xy^2F_1 + \left(1 - y - \frac{Mxy}{2E}\right)F_2\right].$$

Equation (9.23) then follows on verifying that

$$M\nu_{\max}x^2y^2 = \frac{Q^4}{4ME}$$

in the laboratory frame (where $\nu_{\max} = E$).

9.5 In the deep inelastic limit, the above equation for $d\sigma/dx\,dy$ becomes

$$\frac{d\sigma}{dx\,dy} = \frac{4\pi s\alpha^2}{Q^4}\left\{\frac{y^2}{2} + (1-y)\right\}F_2,$$

since $s = 2ME$ and $2xF_1 = F_2$.

9.7 In the ν, $Q^2 \to \infty$ limit, (8.54) and (8.53) imply

$$\frac{\sigma_L}{\sigma_T} \simeq \frac{\nu^2W_2 - Q^2W_1}{Q^2W_1} = \frac{(F_2/2x) - F_1}{F_1} \to 0.$$

9.9 This result is even more obvious using (9.32).

9.10 There is experimental evidence that $\nu W_2 \to$ constant as $x \to 0$. Then, (9.13) requires that $f_i(x) \sim 1/x$ as $x \to 0$. Let us therefore answer the question in reverse. From (8.53), (9.13), (9.14), and (8.48), we have, for fixed Q^2 and $\nu \to \infty$,

$$\sigma_T = \frac{4\pi^2\alpha}{MK}F_1 = \frac{4\pi^2\alpha}{M\nu}\frac{1}{2x}\sum_i e_i^2xf_i(x)$$

$$\sim \frac{4\pi^2\alpha}{Q^2}\sum_i e_i^2$$

if $f_i(x) \sim 1/x$ as $x \to 0$. That is, σ_T is independent of ν and approaches a constant at high energies and fixed Q^2.

9.11 The more spectators that share the initial momentum, the smaller is the chance of producing a parton with a large fraction of the momentum.

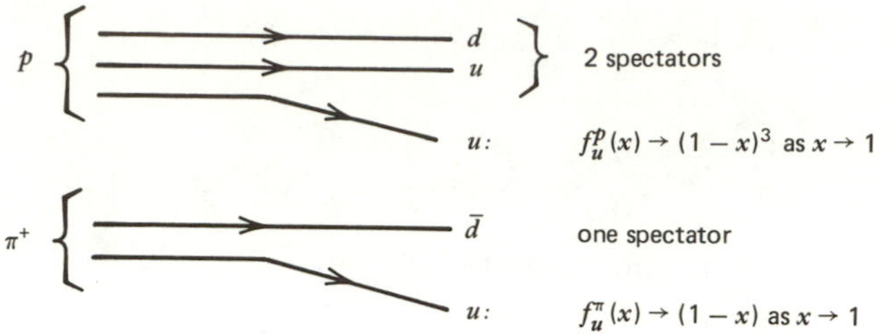

p — d, u — 2 spectators

u: $f_u^p(x) \to (1-x)^3$ as $x \to 1$

π^+ — $\bar{d}$ — one spectator

u: $f_u^\pi(x) \to (1-x)$ as $x \to 1$

CHAPTER 10

10.1 Equations (10.3) and (10.4) follow from (8.53) and (8.54) with $K \simeq \nu$.

10.2 For a γ^* with polarization vector ε_μ interacting with a quark of charge ee_i,

$$-i\mathfrak{M} = \bar{u}(p')(-iee_i\gamma^\mu)u(p)\varepsilon_\mu.$$

Thus, neglecting the mass of the quark, we have

$$\overline{|\mathfrak{M}|^2} = \tfrac{1}{4}2e_i^2e^2\,\mathrm{Tr}(\not{p}'\not{p}) = 2e_i^2e^2 p\cdot q,$$

using (6.93). From (10.13),

$$F\hat{\sigma}_T = \overline{|\mathfrak{M}|^2}2\pi\int d^3p'\,dp_0'\,\delta^4(p'-p-q)\,\theta(p_0')\,\delta(p'^2)$$

$$= 2\pi\overline{|\mathfrak{M}|^2}\delta\left[(p+q)^2\right] = 8\pi^2\alpha e_i^2\,\delta(1-z).$$

10.3

$$q = (q_0; 0, 0, k), \qquad q_2 = (k'; k'\sin\theta, 0, k'\cos\theta),$$

$$q_1 = (k; 0, 0, -k), \qquad g = (k'; -k'\sin\theta, 0, -k'\cos\theta).$$

Equations (10.20)–(10.23) follow from

$$\hat{s} = (q+q_1)^2 = (q_2+g)^2,$$

$$\hat{t} = (q-q_2)^2 = (g-q_1)^2, \qquad \hat{u} = (q_1-q_2)^2,$$

$$-\hat{t}-\hat{u} = Q^2 + 2k'q_0 + 2kk' = Q^2 + 4k'^2 = Q^2 + \hat{s},$$

where we have used $q_0 = 2k' - k$ (energy conservation). From (10.20)–(10.22), we have

$$\hat{s}\hat{t}\hat{u} = 4k'^2(2kk')^2\sin^2\theta = (4kk')^2 p_T^2$$

$$= (\hat{t}+\hat{u})^2 p_T^2 = (\hat{s}+Q^2)^2 p_T^2.$$

If $-\hat{t} \ll \hat{s}$, then $-\hat{u} = \hat{s} + Q^2$, and hence (10.25) follows. Finally,

$$dp_T^2 = k'^2\,d(\sin^2\theta) = 2k'^2\sin\theta\cos\theta\,d\theta$$

$$\simeq \frac{\hat{s}}{2}d(\cos\theta) \qquad \text{for } \cos\theta \simeq 1.$$

10.4 Adopting convention (8.48), the flux in the laboratory frame is

$$F = 4m_q K = 4m_q\left(q_0 - \frac{Q^2}{2m_q}\right) = 4m_q q_0 - 2Q^2,$$

where γ^* has $q = (q_0, 0, 0, k_L)$. But

$$\hat{s} = (m_q + q_0)^2 - k_L^2 = 2m_q q_0 - Q^2,$$

and so $F = 2\hat{s}$. Returning to the center-of-mass frame, (4.29) becomes

$$\frac{d\sigma}{d\Omega} = \frac{1}{64\pi^2 \hat{s}}\overline{|\mathfrak{M}|^2},$$

and (10.27) follows on using (10.26).

10.5

$$(p_T^2)_{\max} = k'^2 = \frac{\hat{s}}{4},$$

see (10.20). Substitute for $\hat{s}$ using (10.29).

10.6 The $\log Q^2$ term originates from the p_T^{-2} behavior of $d\hat{\sigma}/dp_T^2$, see (10.32) and (10.33). The p_T^{-2} results from the square of the propagator $t^{-2} \sim p_T^{-4}$ and a helicity suppression factor $(\sin\frac{1}{2}\theta)^2 \sim \theta^2 \sim p_T^2$. In the relativistic limit, the transition $q_L \to \text{gluon} + q_L$ is forbidden at $\theta = 0$ by angular momentum conservation as real spin-1 gluons flip q_L to q_R.

10.7

$$q(x, Q^2) \sim \int^{Q^2} dp_T^2 \frac{\alpha_s(p_T^2)}{p_T^2} \sim \int \frac{d\log p_T^2}{\log p_T^2} \sim \log(\log Q^2)$$

and so

$$\frac{d}{d\log Q^2} q(x, Q^2) \sim \frac{1}{\log Q^2} \sim \alpha_s(Q^2);$$

and, on following through the derivation, we recover (10.37) with $\alpha_s \to \alpha_s(Q^2)$.

10.8 The color factor $= (\frac{1}{2})^1(3 \times 3 - 1)/8 = \frac{1}{2}$. The first factor recovers the conventional definition of α_s, see (10.19); the denominator averages over the eight initial gluon colors; and the numerator sums over the final $q\bar{q}$ color states (excluding the colorless combination, as the initial γ^*g state is colored).

10.9 Follow the method used in Exercise 6.19. The terms in (10.38) correspond to $\overline{|\mathfrak{M}_1|^2}$, $\overline{|\mathfrak{M}_2|^2}$, and the interference contribution, respectively. Equation (10.41) follows upon replacing (10.17) with (10.38) and retracing the steps to (10.31).

10.10 Either repeat the calculation of Section 10.4 reordering q and g in the final state of $\gamma^* q \to gq$, or use (10.31) and (10.56). To calculate P_{gg}, we need to study the three-gluon vertex either directly or in, say, $\gamma^* g \to q\bar{q}g$ [see, e.g., Altarelli (1978, 1982)].

10.12 The second term in (10.42) does not contribute to the evolution of q_{NS}. On the contrary, the flavor singlet structure function contains quarks and gluons, and we have coupled evolution equations of the form

$$\frac{d}{d\log Q^2}\begin{bmatrix} q_S(x) \\ g(x) \end{bmatrix} = \frac{\alpha_s}{2\pi}\int_x^1 \frac{dy}{y}\begin{bmatrix} P_{qq} & 2n_f P_{qg} \\ P_{gq} & P_{gg} \end{bmatrix}\begin{bmatrix} q_S(y) \\ g(y) \end{bmatrix}.$$

10.13 and 10.14 See, for example, Altarelli (1978, 1982).

10.15 $P_{qq}(z)$ is related to the probability that a quark emits a quark with momentum fraction z and a gluon with momentum fraction $(1 - z)$; therefore, $P_{qq}(z) = P_{gq}(1 - z)$.

10.16 Multiply both sides of (10.37) by x^{n-1} and integrate over x:

$$\frac{d}{d\log Q^2}\int_0^1 x^{n-1} q(x)\,dx = \frac{\alpha_s}{2\pi}\int_0^1 y^{n-1} q(y)\,dy \int_0^1 z^{n-1} P_{qq}(z)\,dz$$

using $x = zy$. Denoting the $(n - 1)$th moment of $q(x)$ by M_n, this becomes

$$\frac{d}{d\log Q^2} M_n = \frac{A_n}{\log Q^2} M_n.$$

The solution is

$$M_n \sim (\log Q^2)^{A_n}.$$

CHAPTER 11

11.1

$$\langle z \rangle = \int_0^1 z D_q^h(z)\,dz = N\int_0^1 (1 - z)^n\,dz = \frac{N}{n + 1}.$$

$$n_h \propto N\int_{\frac{2m_h}{Q}}^1 \frac{dz}{z} + \cdots \propto N\log\left(\frac{Q}{2m_h}\right) + \cdots,$$

where ... represent terms which do not increase with Q.

11.2, 11.3, and 11.4 See Close (1979), Chapter 12, Section 2.

11.5 Subtract eqs. (11.18),

$$x_{\bar{q}}^2 + x_q^2 - x_g^2 = 2x_L x_q.$$

Square, and eliminate x_L^2 in favor of x_T^2,

$$4x_q^2 x_T^2 = 4x_q^2 x_{\bar{q}}^2 - \left(x_{\bar{q}}^2 + x_q^2 - x_g^2\right)^2$$

$$= \left((x_q + x_{\bar{q}})^2 - x_g^2\right)\left(x_g^2 - (x_q - x_{\bar{q}})^2\right)$$

$$= 16(1 - x_q)(1 - x_{\bar{q}})(1 - x_g).$$

Each of the last two steps uses the identity $a^2 - b^2 = (a + b)(a - b)$; also use (11.17). To obtain (11.20), start with $x_{\bar{q}} \sin\theta = x_T$, see Fig. 11.7. Square, and use (11.19) and (11.17),

$$\tfrac{1}{4}x_q^2 x_{\bar{q}}^2 \sin^2\theta = (1 - x_q)(1 - x_{\bar{q}})(x_q + x_{\bar{q}} - 1)$$

$$= -\left(1 - x_q - x_{\bar{q}} + \tfrac{1}{2}x_q x_{\bar{q}}\right)^2 + \tfrac{1}{4}x_q^2 x_{\bar{q}}^2.$$

Using $1 - \sin^2\theta = \cos^2\theta$, we have

$$1 - x_q - x_{\bar{q}} + \tfrac{1}{2}x_q x_{\bar{q}} = \tfrac{1}{2}x_q x_{\bar{q}} \cos\theta.$$

11.6 Use (11.30) and $\theta \simeq 2p_T/Q = x_T$.

11.7 See, for example, Cutler and Sivers (1978) *Phys. Rev.* **D17**, 196, where the diagrams are enumerated and evaluated in the appendix.

11.8 See, for example, Berman et al. (1971) *Phys. Rev.* **D4**, 3388, especially Appendix B.

11.9 When $Q^2/s \simeq 0$, $x \simeq y \simeq 0$ and sea quarks dominate, and there is no difference between $\pi^\pm$C. When $Q^2/s \simeq 1$, $x \simeq y \simeq 1$ and valence quarks dominate. Thus,

$$\frac{\sigma(\pi^+ \mathrm{C})}{\sigma(\pi^- \mathrm{C})} = \frac{e_d^2 f_{\bar{d}}^{\pi^+} f_d^C}{e_u^2 f_{\bar{u}}^{\pi^-} f_u^C} = \frac{e_d^2}{e_u^2} = \frac{1}{4},$$

since $f_{\bar{d}}^{\pi^+} = f_{\bar{u}}^{\pi^-}$ and the isoscalar C target has equal numbers of u and d quarks.

11.10 See, for example, Halzen and Scott (1978) *Phys. Rev.* **D18**, 3378.

CHAPTER 12

12.1 $\tau^- \to e^- \bar{\nu}_e \nu_\tau$ or $\mu^- \bar{\nu}_\mu \nu_\tau$ or ν_τ + hadrons. ν_τ is the tau neutrino (analogous to ν_e and ν_μ); the lepton numbers L_e, L_μ, and L_τ are separately conserved.

12.2 $\bar{u}_e \gamma^\mu \frac{1}{2}(1 - \gamma^5) u_\nu = \bar{u}_e \frac{1}{2}(1 + \gamma^5)\gamma^\mu u_\nu$ and use (6.36). See Exercise 5.15.

12.3 Use the Feynman rules of Chapter 6: u_e describes an ingoing e^- or an outgoing e^+.

12.4 The $I = 1$ states of ^{14}O and ^{14}N* are $|pp\rangle$ and $\sqrt{\frac{1}{2}}(|np\rangle + |pn\rangle)$, respectively. Each proton can decay, so the $^{14}\text{O} \to {}^{14}\text{N}^*$ amplitude contains the factor $2\sqrt{\frac{1}{2}}$.

12.5 For a detailed discussion of how to obtain G from data on lifetimes of β-emitters, see, for example, Källén (1964), Gasiorowicz (1967), or Commins (1973).

12.6 Using (12.15), we obtain $M_W^2 = \sqrt{2}\, e^2/8G \simeq (37.3 \text{ GeV})^2$. In the standard model with $\sin^2\theta_W = \frac{1}{4}$, we have $M_W = 2 \times 37.3 \text{ GeV} = 74.6 \text{ GeV}$.

12.7 Use the trace theorems of Section 6.4. The presence of γ^5 can be taken care of by using (6.23). An elegant derivation of (12.28) is given by Bjorken and Drell (1964), page 262. Equation (12.29) follows directly from (12.27) and (12.28).

12.8 $\int d^3k\, d\omega\, \theta(\omega)\, \delta(\omega^2 - |\mathbf{k}|^2) = \int d^3k/2\omega$, with $\omega^2 = |\mathbf{k}|^2$, see Solution 6.7.

12.9 Use (12.29).

12.10 $2k \cdot p' \simeq (k + p')^2 = (p - k')^2 = m^2 - 2p \cdot k'$.

12.11

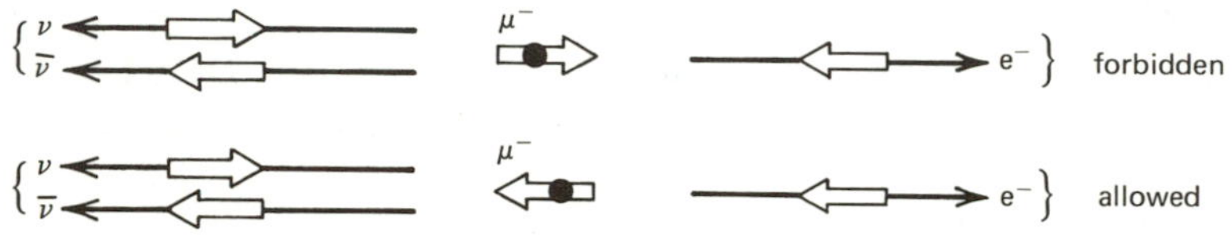

$\{\nu, \bar{\nu}\}$ μ^- $e^-\}$ forbidden

$\{\nu, \bar{\nu}\}$ μ^- $e^-\}$ allowed

12.12

$$\frac{\tau(\tau \to e\nu\bar{\nu})}{\tau(\mu \to e\nu\bar{\nu})} = \left(\frac{m_\mu}{m_\tau}\right)^5 \simeq 7 \times 10^{-7}$$

using (12.42). Thus,

$$\tau \text{ lifetime} = (2.2 \times 10^{-6})(7 \times 10^{-7})\tfrac{1}{5} = 3 \times 10^{-13} \text{ sec}$$

$$(\tau \to e) : (\tau \to \mu) : (\tau \to \text{hadrons}) = 1 : 1 : 3,$$

because of two lepton versus three ($u\bar{d}$) color decay modes.

12.13 The result is that $f_K \simeq f_\pi$.

12.14 For a point interaction, s is the only dimensional variable. As σ is proportional to G^2, the product $G^2 s$ is the only combination with the dimensions of a cross section. But a point interaction is an s-wave ($l = 0$) process for which $\sigma \leq \text{constant}/s$, and so this simple approach must fail when $s \sim \dfrac{1}{G} \sim 10^5 \text{ GeV}^2$.

12.15 Use $s \simeq 2m_e E_\nu$ and (12.60).

12.17 Note the similarity to the nuclear β-decay rate, Section 12.3.

12.18 To obtain the factor $\frac{5}{18}$, use (9.32) and (12.75), and neglect strange quarks:

$$\frac{F_2^{ep} + F_2^{en}}{F_2^{\nu p} + F_2^{\nu n}} = \frac{\frac{4}{9}(u + \bar{u}) + \frac{1}{9}(d + \bar{d}) + \frac{1}{9}(u + \bar{u}) + \frac{4}{9}(d + \bar{d})}{2(u + \bar{u} + d + \bar{d})} = \frac{5}{18}.$$

12.21 Draw quark diagrams and identify the Cabibbo-favored and Cabibbo-suppressed vertices. The amplitudes are in the ratio $\cos^2\theta_c : \sin\theta_c\cos\theta_c : \sin^2\theta_c$.

12.22 Use the $(\Delta m)^5$ rule of Exercise 12.17. To determine $\tau(D^0)$, estimate the leptonic branching ratio as was done for the τ-lepton in Exercise 12.12.

CHAPTER 13

13.2

$$\overline{|\mathcal{M}|^2} = \frac{g_X^2}{4}\left[\frac{1}{3}\sum_\lambda \varepsilon_\mu^{(\lambda)}\varepsilon_\nu^{(\lambda)*}\right]\left[(c_V^2 + c_A^2)T_1^{\mu\nu} - 2c_Vc_AT_2^{\mu\nu}\right]$$

where the first term in brackets is the average over the three helicity states of the X boson and the second brackets gives the sum over the fermion spin states performed as in (6.20), where T_1 and T_2 are the traces

$$T_1^{\mu\nu} = \mathrm{Tr}(\gamma^\mu \not{k}\gamma^\nu \not{k}'), \qquad T_2^{\mu\nu} = \mathrm{Tr}(\gamma^5\gamma^\mu \not{k}\gamma^\nu \not{k}').$$

We denote the X, $f_1, \bar{f}_2$ four-momenta by q, k, k', respectively. From (6.93) and (6.23), we see that the c_Vc_A term vanishes; the polarization sum is symmetric under $\mu \leftrightarrow \nu$, and T_2 is antisymmetric. In the X rest frame,

$$q = (M_X; 0,0,0), \qquad k = \tfrac{1}{2}M_X(1;0,0,1), \qquad k' = \tfrac{1}{2}M_X(1;0,0,-1);$$

and using (6.25), we find

$$-g_{\mu\nu}T_1^{\mu\nu} = 4M_X^2, \qquad q_\mu q_\nu T_1^{\mu\nu} = 0.$$

Finally, from (4.37), we have

$$\Gamma(X \to f_1\bar{f}_2) = \frac{1}{64\pi^2 M_X}\int \overline{|\mathcal{M}|^2}\,d\Omega = \frac{1}{16\pi M_X}\frac{g_X^2}{12}(c_V^2 + c_A^2)4M_X^2.$$

Note that if we were to consider the different polarization states, $\varepsilon^{(\lambda)}$, separately, then T_2 does not vanish. We find the widths

$$\Gamma^{(\pm)} \propto (c_V^2 + c_A^2)(1 + \cos^2\theta) \pm 2c_Vc_A 2\cos\theta$$

$$\Gamma^{(0)} \propto (c_V^2 + c_A^2)2\sin^2\theta$$

where $k = \frac{1}{2}M_X(1;\sin\theta, 0, \cos\theta)$. Check that these results embody angular momentum conservation for, say, $W^+ \to e^+\nu$.

13.3 Substitute $c_V = c_A = \frac{1}{2}$, $M_X = M_Z$, and $g_X = g/\cos\theta_W$ into (13.43). Calculate g from (12.15) with $G = 1.166 \times 10^{-5}\ \text{GeV}^{-2}$ and $M_W = M_Z \cos\theta_W$, and hence show that $\Gamma(Z \to \nu\bar{\nu}) = 159$ MeV.

13.4 Use Table 13.2 to calculate $c_V^2 + c_A^2$; for the $q\bar{q}$ modes, include a factor 3 for color. With three generations, $\Gamma(Z) = 2.5$ GeV (neglecting fermion masses).

13.5

$$\Gamma(W^+ \to e^+\nu) = \frac{g^2 M_W}{48\pi} = \frac{G}{\sqrt{2}}\frac{M_W^3}{6\pi} \simeq 224 \text{ MeV}.$$

13.7 In this limit, the propagator is $ig_{\mu\nu}/M_Z^2$; use (13.35).

13.9 For a derivation of the Fierz theorem, see, for example, Itzykson and Zuber (1980) or Bailin (1982).

13.10

$$R_\mu \simeq \tfrac{1}{8}\left(GM_Z^3/\Gamma_Z e^2\right)^2 = 173.$$

13.11 $R_u = 3(10/9)^2 R_\mu$ and $R_d = 3(13/9)^2 R_\mu$. With three generations,

$$R(\text{hadrons}) \simeq 3(R_u + R_d) \simeq 5000.$$

CHAPTER 14

14.2 Substitute (14.8) into (14.4) with $\phi = \psi, \bar{\psi}$, respectively. For $\phi = \bar{\psi}$, we first rewrite the Lagrangian density (14.8) as

$$\mathcal{L} = -i(\partial^\mu \bar{\psi})\gamma_\mu \psi - m\psi\bar{\psi}.$$

This leaves the action, $\int \mathcal{L}\, d^4x$, unchanged, as can be verified by partial integration. On substituting into (14.4), we obtain the Dirac equation for ψ, (5.7).

14.3 and 14.4

$$\begin{aligned}
\frac{\partial \mathcal{L}}{\partial(\partial_\mu A_\nu)} &= -\frac{1}{4}\frac{\partial}{\partial(\partial_\mu A_\nu)}\left((\partial_\alpha A_\beta - \partial_\beta A_\alpha)(\partial^\alpha A^\beta - \partial^\beta A^\alpha)\right)\\
&= -\frac{1}{2} g^{\alpha\alpha} g^{\beta\beta}\frac{\partial}{\partial(\partial_\mu A_\nu)}\left((\partial_\alpha A_\beta)^2 - (\partial_\beta A_\alpha)(\partial_\alpha A_\beta)\right)\\
&= -\partial^\mu A^\nu + \partial^\nu A^\mu = -F^{\mu\nu},\\
\frac{\partial \mathcal{L}}{\partial A_\nu} &= -j^\nu + m^2 A^\nu.
\end{aligned}$$

14.8 $U^\dagger U = I$ implies $(\det U)^*(\det U) = 1$. Also, if we write $U_{ij} = \delta_{ij} + \varepsilon_{ij}$, then $\det U = 1 + \text{Tr}(\varepsilon) + O(\varepsilon^2)$. Finally, for infinitesimal α_a, $U^\dagger = U^{-1}$ gives

$$1 - i\alpha_a^* T_a^\dagger = 1 - i\alpha_a T_a.$$

14.9 The conventional notation is $T_a \equiv \lambda_a/2$; compare (14.32) with (2.44). Just as for Pauli matrices, $\text{Tr}(\lambda_a\lambda_b) = 2\delta_{ab}$, which together with (2.44) gives

$$\text{Tr}(\lambda_c[\lambda_a, \lambda_b]) = 4if_{abc}.$$

It is now easy to show that f_{abc} is totally antisymmetric. Antisymmetry in a, b is obvious. For the pair b, c, we have

$$4if_{acb} = \text{Tr}(\lambda_b[\lambda_a, \lambda_c]) = -\text{Tr}(\lambda_c[\lambda_a, \lambda_b]) = -4if_{abc}$$

since $\text{Tr}(ABC) = \text{Tr}(CAB)$.

14.10 In QED, the field strength tensor, (14.27), can be introduced by

$$[D_\mu, D_\nu] = -ieF_{\mu\nu},$$

where D_μ is the covariant derivative (14.24). This construction is true for any gauge group [Itzykson and Zuber (1980)]. For QCD,

$$[D_\mu, D_\nu] = igT_a G^a_{\mu\nu}.$$

Substituting (14.38) into the left-hand side yields expression (14.40) for $G^a_{\mu\nu}$. Now under a gauge transformation,

$$\psi \to e^{i\alpha_a(x)T_a}\psi \equiv U\psi,$$

$D_\mu \to UD_\mu U^{-1}$, and so $G_{\mu\nu} \to UG_{\mu\nu}U^{-1}$. Thus, the gauge invariant quantity [see (14.39)] is

$$\text{Tr}(G_{\mu\nu}G^{\mu\nu}) \to \text{Tr}(UG_{\mu\nu}G^{\mu\nu}U^{-1}) = \text{Tr}(G_{\mu\nu}G^{\mu\nu}).$$

14.11 Substitute (14.40) into (14.39) and isolate the terms containing the three gluon fields. After appropriately relabeling the dummy indices in the different terms, we find

$$(i\mathcal{L})_{3g} = -\frac{g}{2} f_{abc}(g_{\mu\nu}p_{1\lambda} - g_{\lambda\mu}p_{1\nu})G_a^\mu G_b^\nu G_c^\lambda.$$

We have used $i\partial_\mu G_a^\mu = p_{1\mu}G_a^\mu$. We sum over all possible orderings of the gluons, bearing in mind that the three-gluon vertex must be completely symmetric. As f_{abc} is totally antisymmetric, we must make the factor in brackets totally antisymmetric.

14.13 The relevant term of the Lagrangian is [see (14.65) and (14.74)]

$$\tfrac{1}{2}|igT_k W_k \phi|^2 = \tfrac{1}{2}|ig(-i\varepsilon_{ijk})W_k\phi_j|^2$$
$$= \tfrac{1}{2}g^2v^2\varepsilon_{i3k}\varepsilon_{i3l}W_kW_l = \tfrac{1}{2}g^2v^2(W_1^2 + W_2^2),$$

using $\phi_j = v\delta_{j3}$. Two bosons acquire mass gv, and the third remains massless.

CHAPTER 15

15.2 Use (15.18) and the solution to Exercise 12.6.

15.3 The relevant term in the Lagrangian is

$$\left| \frac{g}{\sqrt{2}} \left(T^+ W_\mu^+ + T^- W_\mu^- \right) \phi_0 + \left(g T^3 W_\mu^3 + \frac{g'}{2} Y B_\mu \right) \phi_0 \right|^2 .$$

$$M_W^2 = \frac{g^2}{2} \phi_0^\dagger \left(T^+ T^- + T^- T^+ \right) \phi_0 = \frac{g^2 v^2}{2} \left[T(T+1) - (T^3)^2 \right] = 4 g^2 v^2 ,$$

since $T = 3$ and $T^3 = 2$. Also, following (15.19),

$$\frac{1}{2} M_Z^2 Z_\mu^2 = \frac{v^2}{2} 4 \left(g W_\mu^3 - g' B_\mu \right)^2 = \frac{2 g^2 v^2}{\cos^2 \theta_W} Z_\mu^2 .$$

15.8

$$Q_X = \tfrac{4}{3},\ Q_Y = \tfrac{1}{3}.$$

15.9 Investigate (15.54) for the full $SU(5)$ group and compare with (15.53).

Supplementary Reading

Chapter 1

Read the introductory chapters in Bransden, Evans, and Major (1973), Cheng and O'Neill (1979), Gasiorowicz (1967), Muirhead (1965), Perl (1974), and Perkins (1982). Also see articles by Adams, Peierls, and Wilkinson in Mulvey (1981), also Fabjan and Ludlam (1982), Harari (1978), and Kleinknecht (1982).

Chapter 2

Group Theory: Carruthers (1966), Georgi (1982), Hammermesh (1963), Lipkin (1966, 1973), Schiff (1955), and Wybourne (1974). *Quarks*: Close (1979), Feynman (1972), Leader and Predazzi (1982), Lipkin (1973), and Rosner (1974).

Chapter 3

Aitchison (1972), Aitchison and Hey (1982), Bjorken and Drell (1964), and Feynman (1961). *Perturbation Theory*: Merzbacher (1961) and Messiah (1962). *Relativistic Kinematics*: Byckling and Kajantie (1973) and Martin and Spearman (1970).

Chapter 4

Aitchison (1972), Aitchison and Hey (1982), and Feynman (1961). *Cross sections*: Martin and Spearman (1970), Pilkuhn (1967) and Scadron (1979).

Chapter 5

Aitchison (1972), Bethe and Jackiw (1968), Bjorken and Drell (1964), Gasiorowicz (1967), and Sakurai (1967).

Chapter 6

Aitchison (1972), Aitchison and Hey (1982), Bjorken and Drell (1964), Feynman (1961, 1962), Gasiorowicz (1967), Källén (1972), Sakurai (1967), and Scadron (1979).

Chapter 7

Bjorken and Drell (1964), Gastmans (1975), Gasiorowicz (1967), Feynman (1962), Itzykson and Zuber (1980), Jauch and Rohrlich (1976), Källén (1972), Lautrup (1975), Lee (1980), Lurie (1968), Mandl (1966), Ramond (1981), Sakurai (1967), and Scadron (1979). *Tests of QED*: Cheng and O'Neill (1979) and Perkins (1982). *QCD*: Aitchison and Hey (1982) and Field (1979).

Chapter 8

Close (1979), Feynman (1972), Gilman (1972), Llewellyn Smith (1972), Perl (1974), and West (1975). *Nonrelativistic Form Factors*: Bethe and Jackiw (1968).

Chapter 9

Close (1979), Feynman (1972), Leader and Predazzi (1982), Perl (1974), and Perkins (1982).

Chapters 10 and 11

Aitchison and Hey (1982), Altarelli (1978, 1982), Collins and Martin (1982), Ellis and Sachrajda (1979), Field (1979), Kogut and Susskind (1973), Mess and Wiik (1983), Pennington (1983), Reya (1981), Söding and Wolf (1981), Wiik and Wolf (1979), and Wilczek (1982).

Chapter 12

Bailin (1982), Bjorken and Drell (1964), Commins (1973), Feynman (1972), Gaillard and Maiani (1979), Gasiorowicz (1967), Harari (1978), Källén (1964), Maiani (1976), Marshak (1969), Muirhead (1965), Okun (1965), Perkins (1982), Pilkuhn (1979), and Steinberger (1976).

Chapter 13

Bailin (1982), Bilenky and Hosek (1982), Gaillard and Maiani (1979), Ellis et al. (1982), Harari (1978), Hung and Sakurai (1981), Maiani (1976), and Weinberg (1974).

Chapter 14

Abers and Lee (1973), Aitchison (1982), Aitchison and Hey (1982), Bailin (1982), Beg and Sirlin (1974), Bernstein (1974), Feynman (1977), Fritzsch and Minkowski (1981), Gaillard and Maiani (1979), Gastmans (1975), 't Hooft and Veltman

(1973), Iliopoulos (1976), Maiani (1976), Politzer (1975), Ramond (1981), Taylor (1976), and Weinberg (1974).

Chapter 15

Same as Chapters 13 and 14. Also Dolgov and Zeldovich (1981), Ellis (1982), Ellis and Salam in Mulvey (1981), Langacker (1981), Llewellyn Smith (1974), Steigman (1979), and Weinberg (1977).

References

Abers, E., and Lee, B. W. (1973) "Gauge Theories." *Phys. Rep.* **9C**, 1.

Aitchison, I. J. R. (1972) *Relativistic Quantum Mechanics*. Macmillan, London.

Aitchison, I. J. R. (1982) *An Informal Introduction to Gauge Field Theories*. Cambridge University Press, Cambridge, England.

Aitchison, I. J. R., and Hey, A. J. G. (1982) *Gauge Theories in Particle Physics*. Adam Hilger Ltd., Bristol.

Altarelli, G. (1978) In *New Phenomena in Lepton-Hadron Physics*. NATO Advanced Study Series, Series B, Vol. 49. Plenum Press, New York.

Altarelli, G. (1982) "Partons in Quantum Chromodynamics." *Phys. Rep.* **81C**, 1.

Bailin, D. (1982) *Weak Interactions*. Adam Hilger Ltd., Bristol.

Beg, M. A., and Sirlin, A. (1974) "Gauge Theories of Weak Interactions." *Ann. Rev. Nucl. Sci.* **24**, 379.

Bernstein, J. (1974) "Spontaneous Symmetry Breaking, Gauge Theories and All That." *Rev. Mod. Phys.* **46**, 7.

Bethe, H. A., and Jackiw, R. (1968) *Intermediate Quantum Mechanics*. Benjamin, Reading, Mass.

Bethe, H. A., and Salpeter, E. E. (1957) *Quantum Mechanics of One and Two Electron Atoms*. Academic Press, New York.

Bilenky, S. M., and Hosek, J. (1982) "Glashow–Weinberg–Salam Theory of Electroweak Interactions and the Neutral Currents." *Phys. Rep.* **90C**, 73.

Bjorken, J. D., and Drell, S. D. (1964) *Relativistic Quantum Mechanics*. McGraw-Hill, New York.

Bransden, B. H., Evans, D., and Major, J. V. (1973) *The Fundamental Particles*. Van Nostrand Reinhold, London.

Byckling, E., and Kajantie, K. (1973) *Particle Kinematics*. Wiley, New York.

Carruthers, P. (1966) *Introduction to Unitary Symmetries*, Wiley-Interscience, New York.

Cheng, D. C., and O'Neill, G. K. (1979) *Elementary Particle Physics*. Addison Wesley, Reading, Mass.

Close, F. E. (1979) *An Introduction to Quarks and Partons*. Academic Press, London.

Collins, P. D. B., and Martin, A. D. (1982) "Hadron Reaction Mechanisms," *Rep. Prog. Phys.* **45**, 335.

Commins, E. D. (1973) *Weak Interactions*. McGraw-Hill, New York.

Dokshitzer, Y. L., Dyakonov, D. I., and Trojan, S. I. (1980) "Hard Processes in Quantum Chromodynamics." *Phys. Rep.* **58C**, 269.

Dolgov, A. D., and Zeldovich, Ya. B. (1981) "Cosmology and Elementary Particles." *Rev. Mod. Phys.* **53**, 1.

Ellis, J. (1982) "Grand Unified Theories in Cosmology." *Phil. Trans. Roy. Soc. London*, **A307**, 21.

Ellis, J., Gaillard, M. K., Girardi, G., and Sorba, P. (1982) "Physics of Intermediate Vector Bosons." *Ann. Rev. Nucl. Particle Sci.* **32**, 443.

Ellis, J., and Sachrajda, C. T. (1979) In *Quarks and Leptons*. NATO Advanced Study Series, Series B, *Physics*, Vol. 61. Plenum Press, New York.

Fabjan, C. W., and Ludlam, T. (1982) "Calorimetry in High-Energy Physics." *Ann. Rev. Nucl. Particle Sci.* **32**, 335.

Feynman, R. P. (1961) *The Theory of Fundamental Processes*. Benjamin, New York.

Feynman, R. P. (1962) *Quantum Electrodynamics*. Benjamin, New York.

Feynman, R. P. (1963) *The Feynman Lectures on Physics*. Addison Wesley, Reading, Mass.

Feynman, R. P. (1972) *Photon–Hadron Interactions*. Benjamin, New York.

Feynman, R. P. (1977) In *Weak and Electromagnetic Interactions at High Energies*. Les Houches Session 29. North-Holland, Amsterdam.

Field, R. D. (1979) In *Quantum Flavordynamics, Quantum Chromodynamics and Unified Theories*. NATO Advanced Study Series, Series B, *Physics*, Vol. 54. Plenum Press, New York.

Fritzsch, H., and Minkowski, P. (1981) "Flavordynamics of Quarks and Leptons." *Phys. Rep.* **73C**, 67.

Gaillard, M. K., and Maiani, L. (1979) In *Quarks and Leptons*. NATO Advanced Study Series, Series B, *Physics*, Vol. 61. Plenum Press, New York.

Gasiorowicz, S. (1967) *Elementary Particle Physics*. Wiley, New York.

Gastmans, R. (1975) In *Weak and Electromagnetic Interactions at High Energies*. NATO Advanced Study Series, Series B, *Physics*, Vol. 13a, Plenum Press, New York.

Georgi, H. (1982) *Lie Algebras in Particle Physics*. Benjamin-Cummings, Reading, Mass.

Gilman, F. J. (1972) "Photoproduction and Electroproduction." *Phys. Rep.* **4C**, 95.

Goldstein, H. (1977) *Classical Mechanics*. Addison Wesley, Reading, Mass.

Halliday, D., and Resnick, R. (1970) *Fundamentals of Physics*. Wiley, New York.

Hammermesh, M. (1963) *Group Theory*. Addison Wesley, Reading, Mass.

Harari, H. (1978) "Quarks and Leptons." *Phys. Rep.* **42C**, 235.

Hung, P. Q., and Sakurai, J. J. (1981) "The Structure of Neutral Currents." *Ann. Rev. Nucl. Particle Sci.* **31**, 375.

Iliopoulos, J. (1977) "An Introduction to Gauge Theories." Proceedings of the 1977 CERN School of Physics, CERN Report 77-18. CERN, Geneva.

Itzykson, C., and Zuber, J. B. (1980) *Quantum Field Theory*. McGraw-Hill, New York.

Jauch, J. M., and Rohrlich, F. (1976) *Theory of Photons and Electrons*. Springer-Verlag, Berlin.

Källén, G. (1964) *Elementary Particle Physics*. Addison Wesley, Reading, Mass.

Källén, G. (1972) *Quantum Electrodynamics*. Addison Wesley, Reading, Mass.

Kim, J. E., Langacker, P., Levine, M., and Williams, H. H. (1981) "A Theoretical and Experimental Review of Neutral Currents." *Rev. Mod. Phys.* **53**, 211.

Kleinknecht, K. (1982) "Particle Detectors." *Phys. Rep.* **84C**, 85.

Kogut, J., and Susskind, L. (1973) "The Parton Picture of Elementary Particles." *Phys. Rep.* **8C**, 75.

Langacker, P. (1981) "Grand Unified Theories and Proton Decay." *Phys. Rep.* **72C**, 185.

Lautrup, B. (1975) In *Weak and Electromagnetic Interactions at High Energies*. NATO Advanced Study Series, Series B, *Physics*, Vol. 13a. Plenum Press, New York.

Leader, E., and Predazzi, E. (1982) *Gauge Theories and the New Physics*. Cambridge University Press, Cambridge, England.

Lee, T. D. (1980) *Particle Physics and Introduction to Field Theory*. Harwood Academic Publishers, Chur.

Lipkin, H. (1966) *Lie Groups for Pedestrians*. North-Holland, Amsterdam.

Lipkin, H. (1973) "Quark Models for Pedestrians." *Phys. Rep.* **18C**, 175.

Llewellyn Smith, C. H. (1972) "Neutrino Interactions at Accelerators." *Phys. Rep.* **3C**, 261.

Llewellyn Smith, C. H. (1974) In *Phenomenology of Particles at High Energy*. Academic Press, New York.

Lurie, D. (1968) *Particles and Fields*. Wiley-Interscience, New York.

Maiani, L. (1976) "An Elementary Introduction to Yang-Mills Theories and to their Applications to the Weak and Electromagnetic Interactions." Proceedings of the 1976 CERN School of Physics, CERN Report 76-20, CERN, Geneva.

Mandl, F. (1966) *Introduction to Quantum Field Theory*. Wiley-Interscience, New York.

Marshak, R. E., Riazuddin, and Ryan, C. P. (1969) *Theory of Weak Interactions in Particle Physics*. Wiley-Interscience, New York.

Martin, A. D., and Spearman, T. D. (1970) *Elementary Particle Theory*. North-Holland, Amsterdam.

Merzbacher, E. (1961) *Quantum Mechanics*. Wiley, New York.

Mess, K. H., and Wiik, B. H. (1983) In *Gauge Theories in High Energy Physics*. Les Houches Summer School Proc., 37.

Messiah, A. (1962) *Quantum Mechanics*. North-Holland, Amsterdam.

Muirhead, H. (1965) *The Physics of Elementary Particles*. Pergamon, London.

Mulvey, J. H. (1981) *The Nature of Matter*. Clarendon, Oxford.

Okun, L. B. (1965) *Weak Interactions of Elementary Particles*. Pergamon, London.

Omnès, R. (1970) *Introduction to Particle Physics*. Wiley, New York.

Pennington, M. R. (1983) "Cornerstones of QCD." *Rep. Prog. Phys.* **46**, 393.

Perkins, D. H. (1982) *Introduction to High Energy Physics*. Addison Wesley, Reading, Mass.

Perl, M. (1974) *High Energy Hadron Physics*. Wiley-Interscience, New York.

Pilkuhn, H. (1967) *The Interactions of Hadrons*. North-Holland, Amsterdam.

Pilkuhn, H. (1979) *Relativistic Particle Physics*. Springer-Verlag, New York.

Politzer, H. D. (1974) "Quantum Chromodynamics." *Phys. Rep.* **14C**, 129.

Ramond, P. (1981) *Field Theory, A Modern Primer*. Benjamin-Cummings, Reading, Mass.

Reya, E. (1981) "Perturbative Quantum Chromodynamics." *Phys. Rep.* **69C**, 195.

Rose, M. E. (1957) *Elementary Theory of Angular Momentum*. Wiley, New York.

Rosner, J. (1974) "Classification and Decays of Resonant Particles." *Phys. Rep.* **11C**, 193.

Sakurai, J. J. (1967) *Advanced Quantum Mechanics*. Addison Wesley, Reading, Mass.

Scadron, M. D. (1979) *Advanced Quantum Theory*. Springer-Verlag, New York.

Schiff, L. I. (1955) *Quantum Mechanics*. Third edition. McGraw-Hill, New York.

Söding, P., and Wolf, G. (1981) "Experimental Evidence on QCD." *Ann. Rev. Nucl. Particle Sci*. **31**, 231.

Steigman, G. (1979) "Cosmology Confronts Particle Physics." *Ann. Rev. Nucl. Sci*. **29**, 313.

Steinberger, J. (1976) "Neutrino Interactions." Proceedings of the 1976 CERN School of Physics, CERN Report 76-20, CERN, Geneva.

Taylor, J. C. (1976) *Gauge Theories of Weak Interactions*. Cambridge University Press, Cambridge, England.

't Hooft, G., and Veltman, M. (1973) "Diagrammar." CERN Report 73-9, CERN, Geneva.

Weinberg, S. (1974) "Recent Progress in the Gauge Theories of the Weak, Electromagnetic and Strong Interactions." *Rev. Mod. Phys*. **46**, 255.

Weinberg, S. (1977) *The First Three Minutes*. A. Deutsch and Fontana, London.

West, G. B. (1975) "Electron Scattering from Atoms, Nuclei, Nucleons." *Phys. Rep*. **18C**, 264.

Wiik, B. H., and Wolf, G. (1979) *Electron–Positron Interactions*. Springer Tracts in Modern Physics 86, Springer-Verlag, Berlin.

Wilczek, F. (1982) "Quantum Chromodynamics: The Modern Theory of the Strong Interaction." *Ann. Rev. Nucl. Particle Sci*. **32**, 177.

Wybourne, B. G. (1974) *Classical Groups for Physicists*. Wiley, New York.

Index

Page numbers in **boldface** type refer to principle information.

Klein Gordon eqn. 4.3 (4.4) p 85. Use hydrodynamically consistent A fields to produce mass spectrum this way. ~~write~~ 4.4 to solve for V, put into A, solve for m

Useful Formulae

$$\hbar = 6.58 \times 10^{-25} \text{ GeV sec} = 1$$
$$\hbar c = 0.197 \text{ GeV F} = 1$$
$$(1 \text{ GeV})^{-2} = 0.389 \text{ mb}$$
$$\alpha = \frac{e^2}{4\pi} = \frac{1}{137}$$

$$x^\mu = (t, \mathbf{x}), \qquad p^\mu = (E, \mathbf{p}) = i\left(\frac{\partial}{\partial t}, -\nabla\right) = i\partial^\mu$$

$$p \cdot x = Et - \mathbf{p}\cdot\mathbf{x}, \qquad p^2 \equiv p^\mu p_\mu = E^2 - \mathbf{p}^2 = m^2$$

$$(\square^2 + m^2)\phi = 0, \qquad (i\gamma^\mu \partial_\mu - m)\psi = 0.$$

In an electromagnetic field, $i\partial^\mu \to i\partial^\mu + eA^\mu$ (charge $-e$)

$$j^\mu = -ie(\phi^* \partial^\mu \phi - \phi \partial^\mu \phi^*), \qquad j^\mu = -e\bar{\psi}\gamma^\mu\psi$$

γ-Matrices

$\gamma^\mu\gamma^\nu + \gamma^\nu\gamma^\mu = 2g^{\mu\nu}, \qquad \gamma^{\mu\dagger} = \gamma^0\gamma^\mu\gamma^0.$

$\gamma^{0\dagger} = \gamma^0, \qquad \gamma^0\gamma^0 = I; \qquad \gamma^{k\dagger} = -\gamma^k, \qquad \gamma^k\gamma^k = -I, \qquad k = 1, 2, 3.$

$\gamma^5 = i\gamma^0\gamma^1\gamma^2\gamma^3, \qquad \gamma^\mu\gamma^5 + \gamma^5\gamma^\mu = 0, \qquad \gamma^{5\dagger} = \gamma^5.$

(Trace theorems on pages 123 and 261)

Standard representation:

$$\gamma^0 \equiv \beta = \begin{pmatrix} I & 0 \\ 0 & -I \end{pmatrix}, \qquad \boldsymbol{\gamma} \equiv \beta\boldsymbol{\alpha} = \begin{pmatrix} 0 & \boldsymbol{\sigma} \\ -\boldsymbol{\sigma} & 0 \end{pmatrix}, \qquad \gamma^5 = \begin{pmatrix} 0 & I \\ I & 0 \end{pmatrix}$$

$$\sigma_1 = \begin{pmatrix} 0 & 1 \\ 1 & 0 \end{pmatrix}, \qquad \sigma_2 = \begin{pmatrix} 0 & -i \\ i & 0 \end{pmatrix}, \qquad \sigma_3 = \begin{pmatrix} 1 & 0 \\ 0 & -1 \end{pmatrix}$$

Spinors

$$\begin{array}{l} (\not{p} - m)u = 0 \\ \bar{u}(\not{p} - m) = 0 \end{array} \qquad \left\{ \begin{array}{l} \bar{u} \equiv u^\dagger\gamma^0 \\ \not{p} \equiv \gamma_\mu p^\mu \end{array} \right.$$

$$u^{(r)\dagger}u^{(s)} = 2E\delta_{rs}, \qquad \bar{u}^{(r)}u^{(s)} = 2m\delta_{rs}, \qquad \sum_{s=1,2} u^{(s)}\bar{u}^{(s)} = \not{p} + m = 2m\Lambda_+$$

$\frac{1}{2}(1 - \gamma^5)u \equiv u_L, \qquad \frac{1}{2}(1 + \gamma^5)u \equiv u_R.$

If $m = 0$ or $E \gg m$, then u_L has helicity $\lambda = -\frac{1}{2}$, u_R has $\lambda = +\frac{1}{2}$.